Crossing the Straits: Prehistoric Obsidian Source Exploitation in the North Pacific Rim

Edited by

Yaroslav V. Kuzmin
Michael D. Glascock

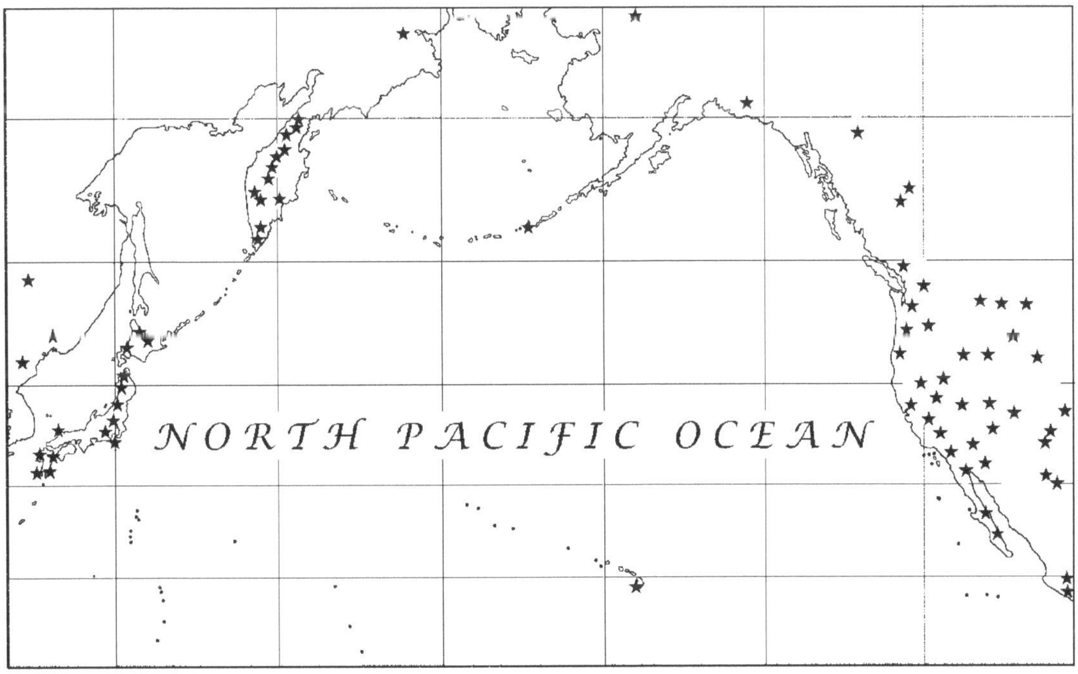

BAR International Series 2152
2010

Published in 2016 by
BAR Publishing, Oxford

BAR International Series 2152

Crossing the Straits: Prehistoric Obsidian Source Exploitation in the North Pacific Rim

ISBN 978 1 4073 0694 0

BAR Publishing is the trading name of British Archaeological Reports (Oxford) Ltd. British Archaeological Reports was first incorporated in 1974 to publish the BAR Series, International and British. In 1992 Hadrian Books Ltd became part of the BAR group. This volume was originally published by Archaeopress in conjunction with British Archaeological Reports (Oxford) Ltd / Hadrian Books Ltd, the Series principal publisher, in 2010. This present volume is published by BAR Publishing, 2016.

Printed in England

BAR
PUBLISHING

BAR titles are available from:

BAR Publishing
122 Banbury Rd, Oxford, OX2 7BP, UK
EMAIL info@barpublishing.com
PHONE +44 (0)1865 310431
FAX +44 (0)1865 316916
www.barpublishing.com

Table of Contents

List of Contributors

Jeffrey P. Blomster	Department of Anthropology, George Washington University, 2110 G St., N.W., Washington, D.C. 20052, USA (e-mail: blomster@gwu.edu)
Nam-Chul Cho	Department of Cultural Heritage, Conservation Science, Kongju National University, 182 Sinkwan-dong, Kongju, Chungnam 314–701, REPUBLIC OF KOREA (e-mail: nam1611@kongju.ac.kr)
J. Andrew Darling	Cultural Resource Management Program, Gila River Indian Community, Sacaton, AZ 85247, USA (e-mail: jadarlin@gilariver.com)
Carolyn D. Dillian	Center for Archaeology & Anthropology, Coastal Carolina University, P.O. Box 261954, Conway, SC 29528–6054, USA (e-mail: cdillian@coastal.edu)
Rodrigo Esparza López	Centro de Estudios Arqueológicos, El Colegio de Michoacán, Zamora, Michoacán, MEXICO (e-mail: wisarika@aol.com)
Mauricio Garduño Ambriz	Sección de Arqueologia, Centro INAH-Nayarit, Tepic, Nayarit, MEXICO (e-mail: mganyarit@hotmail.com)
Michael D. Glascock	230 K-9, Research Reactor, University of Missouri–Columbia, Columbia, MO 65211, USA (e-mail: glascockm@missouri.edu)
Andrei V. Grebennikov	Far Eastern Geological Institute, Far Eastern Branch of the Russian Academy of Sciences, 100-Letiya Vladivostoku Ave. 159, Vladivostok 690022, RUSSIA (e-mail: greandr@mail.ru)
Wataru Hirose	Geological Survey of Hokkaido, Kita 19, Nishi 12, Kita-ku, Sapporo 060–0819, JAPAN (e-mail: wathiro@gsh.pref.hokkaido.jp)
Richard E. Hughes	Geochemical Research Laboratory, 20 Portola Green Circle, Portola Valley, CA 94028, USA (e-mail: rehughes@silcon.com)
Masami Izuho	Faculty of Social Sciences & Humanities, Tokyo Metropolitan University, Minami-Osawa 1–1, Hachioji-shi, Tokyo 192–0397, JAPAN (e-mail: izuhom@tmu.ac.jp)
Susumu Kakubuchi	Department of Culture & Education, Saga University, 1 Honjo-machi, Saga 840–8502, JAPAN (e-mail: kakubuch@cc.saga-u.ac.jp)
Hyung-Tae Kang	Conservation Science Laboratory, National Museum of Korea, 168-6 Yongsan-dong, Yongsan-gu, Seoul 140–026, REPUBLIC OF KOREA (e-mail: akdntm@korea.kr)
Jong-Chan Kim	School of Physics, Seoul National University, Seoul 151–747, REPUBLIC OF KOREA (e-mail: jckim@phya.snu.ac.kr)
Yaroslav V. Kuzmin	Institute of Geology & Mineralogy, Siberian Branch of the Russian Academy of Sciences, Koptyug Ave. 3, Novosibirsk 630090, RUSSIA (e-mail: kuzmin@fulbrightmail.org)
Isao Morimoto	Okinawa Prefectural Archaeological Centre, 193-7 Uehara, Nishihara-cho, Nakagami-gun, Okinawa Prefecture 903–0125, JAPAN (e-mail: mormotoi@pref.okinawa.jp)
Joseph B. Mountjoy	Department of Anthropology, University of North Carolina–Greensboro, Greensboro, NC 27402, USA (e-mail: jbmountj@uncg.edu)
Hiroki Obata	Archaeological Operation Centre, Kumamoto University, 2-39-1 Kurokami-machi, Kumamoto 860–8555, JAPAN (e-mail: totori@kumamoto-u.ac.jp)
Michael A. Ohnersorgen	Department of Anthropology, University of Missouri–St. Louis, St. Louis, MO 63121, USA (e-mail: ohnersorgenm@umsl.edu)
S. Colby Phillips	Department of Anthropology, University of Washington, Seattle, WA 98195, USA (e-mail: colbyp@u.washington.edu)
Vladimir K. Popov	Far Eastern Geological Institute, Far Eastern Branch of the Russian Academy of Sciences, 100-Letiya Vladivostoku Ave. 159, Vladivostok 690022, RUSSIA (e-mail: vladpov@fegi.ru)
Andrei V. Ptashinsky	Vitus Bering's Kamchatka State University, Pogranichnaya St. 4, Petropavlovsk-Kamchatskiy 683032, RUSSIA (e-mail: ptashinski@mail.ru)
M. Steven Shackley	Geoarchaeological XRF Laboratory, Department of Anthropology, University of California–Berkeley, Berkeley, CA 94720, USA (e-mail: shackley@berkeley.edu)
Robert J. Speakman	Museum Conservation Institute, Smithsonian Institution, Suitland, MD 20746–2863, USA (e-mail: speakmanj@si.edu)

Takashi Tsutsumi Asama Jomon Museum, 1901-1 Maseguchi, Miyota, Nagano 389–0207, JAPAN (e-mail: jomon@mx2.avis.ne.jp)

Phil C. Weigand Centro de Estudios Arqueológicos, El Colegio de Michoacán, Zamora, Michoacán, MEXICO (e-mail: wisarika@aol.com)

List of Figures

List of Tables

Preface and Acknowledgements

Yaroslav V. Kuzmin

The original idea for developing this book as Proceedings from the Symposium on Obsidian Source Studies in Northeast Asia, held at the 70th Annual Meeting of the Society for American Archaeology (2005), was conceived sometime in the summer of 2004 when I was preparing to visit my partner and co-Principal Investigator in the CRDF Project (see Chapter 1), Dr Michael D. Glascock, head of the Archaeometry Group at the Research Reactor, University of Missouri (Columbia, MO, USA). I thought that we could finish the CRDF Project by gathering people from Russia, the USA, and adjacent countries (first of all, Japan and Korea), to give updated overviews on the subject. Soon the region under consideration expanded into the Western Hemisphere (Mexico and the US Southwest), and the geography became truly 'pan-North Pacific'. It fits nicely into the northern part of the 'Pacific Rim' where numerous obsidian sources are located; and many of them were exploited in prehistory (see volume's Logo on the cover page where each star corresponds to individual source of group of them).

The symposium was conducted successfully on 3 April 2005 at the Convention Centre, Salt Lake City (Utah, USA), with eight speakers and discussants attending. In total ten presentations were delivered. The proposition to contribute papers to an edited volume was announced at the meeting and was well-received. For some parts of Northeast Asia and North America, additional chapters were commissioned. It took us almost five years to prepare this collection, and now we have it in hand! Hopefully, this book will serve as a source of data and interpretation for the next decade or so, before new regional summaries will be created.

In this short Preface, I would like to point out some technical subjects. One of the most important and complicated issues was the style of references, written in non-Romance and non-Germanic languages, notably in Russian, Japanese, and Korean (and to some extent Chinese). If we were to give readers only translations of the original books and periodicals, this would be of little help in finding them in western libraries (for example, in the US Library of Congress, and most of the university libraries in North America and Western Europe), because to do it one must know the spelling of the original titles in Roman letters, and in many cases they are not available from references. Instead, only the translation of the original titles is often given. Therefore, it was decided that the original spelling of titles and sources (books, journals, etc.) would be provided in the Roman alphabet along with the translation of the articles' and monographs' titles only. This approach has already been used by some periodicals dealing with Oriental sources, including the *Journal of East Asian Archaeology*. Other examples are two edited volumes recently published, "Archaeology of the Russian Far East: Essays in Stone Age Prehistory" (Oxford, BAR Publishing, 2006) and "Origin and Spread of Microblade Technology in Northern Asia and North America" (Burnaby, B.C., Simon Fraser University, Archaeology Press, 2007). Although Romanisation of Russian and Japanese titles is difficult, at least it is consistent throughout this volume. To make the preparation of individual chapters easier, each of them has its own list of references. This is despite the fact that there is often overlap, especially for neighbouring regions like the Russian Far East, the Korean Peninsula, and to some extent Northeast China, the Kamchatka Peninsula, the Kurile Islands, and Hokkaido Island; and also in two chapters about the Mexican obsidian sources.

As for abbreviations used, the "BP" ("before present") is the most common one. It means the age in uncalibrated radiocarbon years, as received from the radiocarbon laboratories. Others are "a.s.l." ("above the sea level"); ^{14}C ("radiocarbon"); "part-per-million", or "per mill" ("ppm"); and different analytical methods: "X-ray Fluorescence" (XRF) and "Energy Dispersive X-ray Fluorescence" ("EDXRF"); "Proton-Induced X-ray Emission – Proton-Induced Gamma-ray Emission" ("PIXE-PIGME"); "Neutron Activation Analysis" ("NAA") and "Instrumental Neutron Activation Analysis" ("INAA"); "Electron Probe Microanalysis" ("EPMA"); and "Laser Ablation Inductively-Coupled-Plasma Mass Spectrometry" ("LA-ICP-MS"). Obvious abbreviations include "AD" ("*Anno Domini*", i.e. years after the birth of Jesus Christ); and "BC" ("before Christ"). Last two age determinations belong to the calendar time scale. Calibrated ^{14}C dates have as Zero Year AD 1950; therefore, calendar years are expressed in "cal BP" (years before AD 1950), "cal BC" (years before BC/AD boundary, 1 BC/AD 1); and "cal AD" (years after BC/AD boundary, 1 BC/AD 1). The Year 0 (Zero) is non-existent in the Gregorian calendar.

The editing of English for texts written originally in Japanese and Korean was a challenging task for both of the volume's editors and numerous reviewers (see below). Fortunately, several individuals were able to help us greatly improve the quality of the text. I still feel that in some parts it could be refined, but we all did our best. Despite this, the text is understandable and has no double meanings. What is more important is that each chapter is written by the original researcher or a group of researchers, who know the subject best. The spelling in the chapters follows British English, with reference to *The Concise Oxford Dictionary* (tenth edition, revised; J. Pearsall, editor; Oxford University Press, 2001). Synonyms are from *The Oxford American Writer's Thesaurus* (C. A. Lindberg, compiler; Oxford University Press, 2004). For creation of Index, *The Chicago Manual of Style* (fifteenth edition; University of Chicago Press, 2003) was used as a guideline. Geological terms in this volume follow the *Oxford Dictionary of Earth Sciences* (second edition; A. Allaby and M. Allaby, editors; Oxford University Press, 2003) and the *Dictionary of Geological Terms* (third edition; R. L. Bates and J. A. Jackson, editors; New York, Anchor Books, 1984). Geographic names and terms are from *The Times Atlas of the World* (seventh comprehensive edition; London, Times Books, 1989), the *Merriam Webster's Geographical Dictionary* (third edition; Springfield, MA, Merriam-Webster, Inc., 1997), and *Webster's Universal Encyclopedic Dictionary* (New York, Barnes and Noble Books, 2002).

Numerous colleagues took part in the process of reviewing and editing the chapters which constitute this volume. On behalf of both editors, I would like to thank them for all the help they provided. Included are the following scholars (in alphabetical order, without titles and academic degrees): Fumito Akai (Kagoshima City Board of Education, Kagoshima, Japan); Jeffrey P. Blomster (Department of Anthropology, George Washington University, Washington, D.C., USA); Matthew T. Boulanger (Research Reactor Center, University of Missouri–Columbia, Columbia, MO, USA); Geoffrey E. Braswell (Department of Anthropology, University of California–San Diego, San Diego, CA, USA); Jimmy D. Cassidy (Natural Resources and Environmental Affairs Division, Marine Corps Air Ground Combat Center, Twentynine Palms, CA, USA); Jeffrey R. Ferguson (Research Reactor Center, University of Missouri–Columbia, Columbia, MO, USA); Ted E. Goebel (Center for the Study of the First Americans, Texas A & M University, College Station, TX, USA); G. Tom Jones (Department of Anthropology, Hamilton College, Clinton, NY, USA); Charles T. Keally (Sophia University, Tokyo, Japan; retired); Susan G. Keates (London, UK); Candace Lindsey (Columbia, MO, USA); Fred W. Nelson (Department of Radiation Safety, Brigham Young University, Provo, UT, USA); Michael A. Ohnersorgen (Department of Anthropology, University of Missouri–St. Louis, St. Louis, MO, USA); Akira Ono (Centre for Obsidian and Lithic Studies, Meiji University, Tokyo, Japan); Richard J. Pearson (University of British Columbia, Vancouver, B.C., Canada; retired); Helen Pollard (Department of Anthropology, Michigan State University, East Lancing, MI, USA); Sergei A. Shcheka (Far Eastern Geological Institute, Far Eastern Branch of the Russian Academy of Sciences, Vladivostok, Russia; deceased); Robert J. Speakman (Museum Conservation Institute, Smithsonian Institution, Suitland, MD, USA); Anastasia Steffen (Valles Caldera National Preserve, Jemez Springs, NM, USA); Hiroto Takamiya (Sapporo University, Sapporo, Japan); Richard Veit (Department of History and Anthropology, Monmouth University, West Long Branch, NJ, USA); Carolyn White (Department of Anthropology, University of Nevada–Reno, Reno, NV, USA); Satoru Yamada (Kitami City Board of Education, Kitami, Japan); and Seonbok Yi (Seoul National University, Seoul, Republic of Korea).

Finally, I would like to express my sincere congratulations to all the 26 participants of this volume. Our joint efforts and patience bore fruit! *Finis coronat opus* [The end crowns the work].

Novosibirsk (Russia),

1 March 2010

Chapter 1
Introduction: Obsidian Sourcing in the North Pacific Rim Region and Beyond It

Yaroslav V. Kuzmin and Michael D. Glascock

'The stone which the builders refuses is become the head stone of the corner.'

The Book of Psalms, 118:22

'...we drove under a cliff of obsidian, which is black glass, some two hundred feet high; and the road at its foot was made of black glass that crackled. ...The glass cliff overlooks a lake...'

Rudyard Kipling (1900), From Sea to Sea and Other Sketches. Letters of Travel. London, Macmillan and Co., Ltd.

Obsidian was often a valuable commodity in human prehistory and early history; this is why many tales are devoted to this unusually sharp and shining rock. For example, the story (perhaps more like a fairy tale) of the discovery of Obsidian Cliff in Yellowstone National Park (Wyoming, USA) is that a hunter, Jim Bridger, was trying to shoot an elk which was standing behind a wall of transparent volcanic glass (i.e., obsidian). Of course the hunter could not kill the prey. It turned out that the elk was 20 miles away, and the layer of obsidian served as a magnifying glass! (Bright 1951, 5–6; see Skinner and Tremaine 1993). The Obsidian Cliff is a well-known place to acquire excellent quality volcanic glass. It was described by early travellers, such as Rudyard Kipling, in the North American 'Wild West' in the late nineteenth century (see Epigraph). In Egypt, in the tomb of Pharaoh Tutankhamen, dating back to about 1300 BC, several figurines of Egyptian gods and their eyes and royal insignia, are made of obsidian (Carter 2003, 178–229). These examples show clearly how important obsidian was in the early human societies.

The significance of obsidian can be expressed by using a biblical phrase about the stone which was first refused and later on became the cornerstone (see Epigraph). Literally, the refuse (obsidian flakes and other debris) of ancient humans now is the subject ("cornerstone") of study by archaeologists and geologists. This allows the establishment of several important features, especially in terms of human migrations and contacts. Unlike pure archaeological criteria (typological similarity of pottery, stone tools, and other items), obsidian source identification is the *direct* evidence of some kind of transport of commodities from one place to another. The famous British archaeologist, Howard Carter, said originally in the early twentieth century '...as self-respected historians, let us put aside the tempting "might have been" and "probablys" and come back to the cold hard facts of history' (Carter 2003, 3). This describes exactly the situation with primary and secondary verifications of these kinds of prehistoric phenomena. Finding the source for obsidian artefacts is always the primary indication of the transport and exchange of raw materials.

From the viewpoint of geology and petrology, "volcanic glass" is a general term for glassy compounds ('Natural glass produced by the cooling of molten lava, or some liquid fraction of it, too rapidly to permit cristallization. Examples are obsidian, pitchstone, tachylyte, and the glassy groundmass of many extrusive rocks'; Bates and Jackson 1984, 556). Obsidian ('A black or dark-colored volcanic glass, usually of rhyolite composition, characterized by conchoidal fracture. It has been used for making arrowheads, jewelry, and art objects'; Bates and Jackson 1984, 352), on the other hand is a pure (without even very small crystals) and waterless (with water content less than 1% weight) variety of volcanic glass, practically the only kind of it suitable for the manufacture of artefacts.

Obviously, not only obsidian is being used for the study of raw material movement. Flint (or chert) is another good example of a commodity which can shed a new light on issue of the degree of mobility of ancient humans (e.g., Slimak and Giraud 2007; see review: Shackley 2008, 197–198). In the North Pacific region, flint was also successfully used for provenance studies (e.g., Malyk-

Selivanova *et al.* 1998). However, obsidian has a higher potential for this kind of research because almost every single obsidian source has its own unique "geochemical portrait" (or "signature") (e.g., Glascock *et al.* 1998; also see Glascock *et al.*, this volume). This is the reason for concentrating on high quality volcanic glass. Its abundance in the North Pacific Rim also makes it possible to reveal patterns of human movements and migrations with a high degree of certainty. The well-known 'Pacific Rim of Fire', i.e. the area around the coast of the North Pacific, with high volcanic and seismic activity, was chosen as the focus region of obsidian provenance studies for this book. In fact, all of the Pacific Ocean belongs to the 'Rim of Fire' (e.g., Fairbridge 1975). In this volume the concentration will be on its northern part (see Logo on the front cover).

In the North Pacific region, obsidian was originally described in the middle to the end of eighteenth century AD. Stepan P. Krasheninnikov (1994) and Georg W. Steller (1999) (see also Waldman and Wexler 2004, 340, 558–559) described obsidian from the Bystraya [Khairyuzovka] River which runs from the Sredinny [Middle] Range of the Kamchatka Peninsula, a 'classical' region for volcanic glass studies (see Chapter 6 of this volume). Natural scientists were aware of obsidian in far eastern Russia since the second part of the eighteenth century. Well-known Russian (German-born) geologist of the day, Professor of the Imperial Moscow University, Johann Gotthelf Fischer von Waldheim (or just G. Fischer), described *marekanite*, a high quality volcanic glass from the Marekan River mouth on the northern coast of Sea of Okhotsk, west of the modern city of Magadan (Fischer 1818, 278–279), a perfect location in the Pacific Rim of Fire! Prior to the Second World War, when archaeological research in the Northeastern Siberia and Northwestern North America greatly accelerated, obsidian in this region was reported by Russian scholar Waldemar Jochelson (1928), within the framework of the famous Jesup North Pacific Expedition (1897–1902) (see Freed *et al.* 1988; Vakhtin 2001, 79–82), and by a few other researchers including N. N. Bilibin and S. I. Rudenko (see Glascock *et al.* 2006).

Monographs and edited volumes devoted to obsidian studies on an international scale are still comparatively rare. A book edited by Taylor (1976) represents the first collection of this kind. Leach and Davidson (1981) and Pollmann (1993) provided analysis of obsidian sources in the Pacific and Mediterranean regions. Very useful collections of both methodological and regional papers appeared in a volume edited by Shackley (1998). More examples of the use of obsidian as a commodity to study prehistory are in book by Cauvin *et al.* (1998) and a volume edited by Glascock (2002). Kuzmin and Popov (2000) summarised geological, geochemical, and archaeological data on obsidian for the Russian Far East and neighbouring regions. The 2004 Obsidian Summit meeting in Tokyo (Japan) unfortunately yielded only a volume of abstracts and short papers (Ambiru *et al.* 2004). One of the latest examples is an excellent monograph by Shackley (2005), completely dedicated to different aspects of obsidian sourcing and archaeology. There is also a forthcoming volume which includes obsidian-related studies (Shackley 2010). We hope that the current book will serve as source of primary information and some interpretations.

The beginning of the collaboration which finally resulted in the creation of this collection of articles goes back to 1991. Yaroslav V. Kuzmin contacted M. Steven Shackley about the offprints of obsidian sourcing data from the southwestern United States. Soon afterwards, Michael D. Glascock joined the informal team, and the first study of archaeological obsidian from the southern Russian Far East was conducted (see Glascock *et al.* 1996; Shackley *et al.* 1996). Fortunately, Vladimir K. Popov became part of the team in the mid-1990s with knowledge of 'geological' obsidian and other rocks in the region (see Shackley and Popov 1997); it appeared to be the proper interdisciplinary team with members from Russia, the USA, Japan, and the Republic of Korea (South Korea), including geologists, geochemists, archaeologists, analytical chemists, and geoarchaeologists. This is perhaps one of the most successful examples of ongoing international collaboration in the field of obsidian provenance in the North Pacific region.

The initial preparations for this volume were conducted within the framework of the joint Russian-US Project, "The Geochemistry of Volcanic Glasses and Sources of Obsidian of the Russian Far East" (No. RG1-2538-VL-03), supported by the US Civil Research and Development Foundation (CRDF) in 2003–5, with Yaroslav V. Kuzmin and Michael D. Glascock as co-Principal Investigators. Preliminary results of this CRDF Project's realisation were presented in a symposium "Crossing the Straits: Prehistoric Obsidian Exploitation in the Pacific Rim" on Sunday, 3 April 2005, within the Scientific Programme of the 70th Annual Meeting of the Society for American Archaeology, Salt Lake City (Utah, USA). It was organised and conducted by Yaroslav V. Kuzmin and Michael D. Glascock. Several papers delivered at this session constitute the core of this volume, with some additional chapters commissioned later. It took several years to evaluate, review, revise, and edit individual chapters which are now presented here for a wide scientific audience, especially for those scholars who are working in Northeast Asia and/or North and Central America.

In total, 26 individuals participated in the preparation and writing of chapters for this volume (see List of Contributors). This book is divided into two parts, based on the geography of the regions under investigation. The first part (chapters 2–8) contains data on Northeast Asia, including Japan, Korea, and the Russian Far East. Particular attention is given to the Japanese Islands as 'classical' territory for obsidian provenance research. The second part of this volume (chapters 9–12) deals with the American continent, including the Great Basin and Atlantic coast of North America and Mexico (Mesoamerica). The sequence of chapters is geographic, beginning with Northeast Asia and going clockwise to the Americas. Chapter 13 gives overview of the content of whole book.

In **Chapter 2**, Masami Izuho and Wataru Hirose present current knowledge of obsidian geology and geochemistry for one of the 'classical' regions in Northeast Asia, Hokkaido Island of Japan. They focus on the geology of the area with more than 20 obsidian sources. Several of them are important not only for Hokkaido but for neighbouring Northeast Asia (see also Chapter 8). Igneous rocks of Hokkaido are comprehensively studied due to the complete coverage of this large island by geological mapping (scale 1:50,000). However, little attention was given to volcanic glasses, especially in terms of their geochemistry. Only recently Instrumental Neutron Activation Analysis (hereafter – INAA) of obsidian was performed on major obsidian sources, Shirataki, Oketo, Tokachi-Mitsumata, and Akaigawa (Kuzmin and Glascock 2007; Kuzmin and Popov 2000; Kuzmin *et al.* 2002). An interesting feature on Hokkaido is the large number of secondary obsidian sources; their characterisation is an important task for the near future. Another key issue is the study of Tokoro Rover basin where the Oketo obsidian source is situated; here more than 20 archaeological sites have traces of obsidian exploitation. Finally, the earliest evidence of obsidian use by ancient humans on Hokkaido is dated to ca. 30,000 BP at the Wakabano Mori site (Obihiro-shi Kyoiku Iinkai 2004). This is among the first examples of obsidian use as a raw material in all of Northeast Asia (Jull *et al.* 2005). The presentation of Japanese geological data for obsidian sources on Hokkaido in this chapter, usually inaccessible to foreigners, is very welcomed. It is clear that Hokkaido is a vital part of a large-scale exchange network covered insular Northeast Asia. Peculiarities of the distribution of its obsidian are critical for reconstruction of human movements, migrations, and behaviour patterns in the Upper Palaeolithic, Neolithic, and Palaeometal epochs.

In **Chapter 3**, a summary of the current state of obsidian provenance research in prehistory of the Honshu Island of Japan is provided by Takashi Tsutsumi. It is written in quite a descriptive manner. This represents one of the best-studied regions in the entire world in terms of obsidian sources used by ancient people. A plethora of data is presented for the Upper Palaeolithic of Honshu, with meticulous characterisation of each important obsidian source and distribution of its material in different phases of the Upper Palaeolithic. For most of Northeast Asia and the Americas such study is practically impossible due to lack of primary data. The existence of long-distance obsidian exchange networks on Honshu Island in the Upper Palaeolithic is noteworthy (see also chapters 2 and 4–8 for other parts of Northeast Asia). Use of multiple obsidian sources at the same site is significant and shows the complexity of human behaviour (see also Chapter 6). Another important issue is the "phenomenon of Kozu-shima Island", i.e. the transport of large amounts of obsidian across the sea strait since ca. 30,000 BP, presumably with the help of watercraft. More attention should be given to this aspect of Honshu prehistory. Currently, it is hard to judge the extent and frequency of these journeys; but they definitely occurred. Although fewer obsidian source studies were done for the subsequent Jomon (pottery-bearing) phase of Japanese prehistory, a significant amount of data and its interpretation are provided. Perhaps, Honshu Island is a "model" of obsidian sourcing research as it is rich in information already accumulated and partly interpreted. Major parts of the original publications for Honshu are hardly known outside Japan, because they belong to the so-called "grey literature" written in Japanese (i.e., excavation reports); nevertheless, they are readily available in Japan. This review chapter with its comprehensive illustrations will hopefully serve as the basic publication for the region in the years to come.

In **Chapter 4**, Hiroki Obata and coauthors present the results of obsidian source determination for a remote part of the Japanese Archipelago, the Ryukyu Islands. This is a relatively less-known region of Northeast Asia for this kind of research, and introduction of data published previously almost exclusively in Japanese is an important new piece of primary information for the international scholarly community. The first comprehensive obsidian sourcing study for the Ryukyu prehistory resulted in revealing its major patterns. Obsidian was transported to the Ryukyus from sources on Kyushu Island for the first time across a wide water space in Late Jomon, ca. 4000 BP. For the islands neighbouring the Kyushu (Ōsumi group, including Tanega-shima and Yaku-shima islands), obsidian arrived even earlier, at ca. 8000–6000 BP. The long-distance transportation of obsidian from one of the main Japanese islands toward the south is therefore evident. This is an important case study of the prehistoric contacts and migrations, and one of the indications of extensive seafaring in the Late Neolithic of Northeast Asia.

In **Chapter 5**, two case studies on the Korean Peninsula, a region neglected by proper obsidian provenance research for a long time (see, for example, review: Kuzmin 2006, 67), are presented by Nam-Chul Cho and coauthors. The first case is based on the PhD dissertation of the leading coauthor (Cho 2005), concluding that obsidian from two well-known Upper Palaeolithic sites in the central part of the Korean Peninsula is from an unknown source and definitely not from the Paektusan Volcano on the North Korean/Chinese border, which is now known as a source for artefacts in this part of Northeast Asia (see Chapter 8). The second case study by Kim *et al.* (2007), conducted with the help of PIXE and NAA methods and briefly summarised here, resulted in a totally different identification. The majority of obsidian, including two sites analysed by Cho (2005), was obtained from the Paektusan source! It is obvious that these two studies contradict each other to some extent, demonstrating that obsidian provenance research in Korea is at the very beginning. Much more work is needed to determine the main geochemical parameters of Korean prehistoric obsidian. Scholars are now on the right track (e.g., Popov *et al.* 2005; Kim *et al.* 2007).

In **Chapter 6**, current progress with the identification of obsidian sources on the Kamchatka Peninsula in the northern Russian Far East is described by Andrei V. Grebennikov and coauthors. Kamchatka is a real bonanza for obsidian provenance studies, with dozens of primary sources. On the other hand, this is logistically an extremely difficult region, and this hampers the in-depth research of obsidian geochemistry and sourcing of prehistoric artefacts. Nevertheless, significant progress was achieved in the last four to five years. Seven sources of Kamchatkan archaeological obsidian were securely located. Several more sources which were used in prehistory were also identified, and their position is suggested on the basis of geological and petrological data. Additionally, the phenomenon of simultaneous use of several obsidian sources by the late Upper Palaeolithic and Neolithic populations was revealed. This is an important pattern in terms of the human strategy of raw material acquisition. Much more work needs to be done on Kamchatka to find the exact location for major obsidian sources used by ancient people, but a solid foundation created by the authors makes it possible in the near future.

In **Chapter 7**, S. Colby Phillips presents results on the determination of sources for obsidian from archaeological contexts in the Kurile Island of the Russian Far East. This is quite a new region for obsidian provenancing in the North Pacific Rim; this study is the most complete to date, compared to the first brief summary (Phillips and Speakman 2009). The importance of this still pilot research is that the 'traffic' of obsidian from both sides of the Kuriles, Hokkaido Island in the south and Kamchatka Peninsula in the north, to the island chain is determined with the help of modern analytical methods, based on source studies conducted recently in this part of Northeast Asia (Glascock *et al.* 2006; Hall and Kimura 2002; Kuzmin and Glascock 2007; Kuzmin *et al.* 2002; Sato *et al.* 2002; see also chapters 2 and 6 in this volume). The long-distance transport of obsidian to Kuriles from two regions with abundant sources, Kamchatka and Hokkaido, is now securely established. The distance between the place of origin and the utilisation sites varies from 150 to 1200km for the Hokkaido sources and from 600 to 1400km for the Kamchatkan ones. The age of sites investigated so far is about 2500–1000 BP. There is earlier evidence of obsidian use in the prehistory of Kurile Islands, since ca. 7000 BP (Yanshina *et al.* 2009); this indicates potential for future research.

In **Chapter 8**, Yaroslav V. Kuzmin provides an overview of the current state of obsidian provenance research in Northeast Asia, including the southern part of the Russian Far East (Primorye [Maritime] Province, the Amur River basin, and Sakhalin Island) and the Korean Peninsula, with some thoughts about neighbouring Northeast China (or Manchuria); and the two extremes of the of Japanese Archipelago, the northernmost part (Hokkaido Island) and the southernmost part (Ryukyu Islands). As for the territory of modern Russia, obsidian source studies were not common until the early 1990s, including regions with abundant obsidian resources like the Caucasus. One example of a scientific approach to obsidian from prehistoric sites in the Northern Caucasus was research by Nasedkin and Formozov (1965), somehow missing in the IAOS Obsidian Bibliography (Skinner and Tremaine 1993, 115–116). They used optical criteria such as the refractive index to separate obsidian from different sources. In the Russian Far East, the first attempts to establish sources of archaeological obsidian were undertaken in the late 1980s. However, they were based on very small amount of material and the limited number of chemical elements analysed; the same was true about the Korean Peninsula (see review: Kuzmin 2006, 67). Work began in the southern Russian Far East (Primorye and Amur River basin) in 1992 and on Sakhalin Island in the late 1990s. As for Korea, modern scientific investigations of obsidian sources were initiated only in the early to mid-2000s (Popov *et al.* 2005), and now are quickly developing (e.g., Kim *et al.* 2007; see also Chapter 5). The review of progress in obsidian sourcing for Northeast Asia achieved in the 2000s shows that a new level of research was recently reached, after extensive fieldwork and laboratory analyses. It is clear that more studies based on modern advanced methodological principles are necessary.

In **Chapter 9**, Carolyn D. Dillian presents an excellent example of the long-distance movement of obsidian, in this case for North America. She carefully evaluated the reliability of obsidian artefacts from the Atlantic coast of the USA in terms of their prehistoric transportation (and not modern discard), and concluded that the limited scale transport of obsidian from western North American sources (in the states of California, Utah, and Idaho) to the mid-Atlantic region took place in antiquity. The distances of this transportation are enormous, reaching several thousand kilometres. However, the author suggests that the movement was not very intentional and occurred in a rather 'stochastic' manner. Nevertheless, in this chapter the clear evidence of obsidian traffic on continent scale distances is well-illustrated. The investigation of circumstances for the discovery of obsidian artefacts in the eastern US states of Pennsylvania, New Jersey and New York in the late nineteenth and early twentieth centuries, based on archival data from different universities and societies, is noteworthy. It is interesting to mention that Frederic Ward Putnam of Harvard University who was corresponding with Charles Conrad Abbott, one of the pioneers of obsidian research on the US Atlantic coast, also communicated with Franz Boas, the head of Jesup North Pacific Expedition (Cole 2001, 65–66). The first scientific knowledge about obsidian from Northeastern Siberia came from activity of this research enterprise by Waldemar Jochelson.

In **Chapter 10**, Richard E. Hughes examines a relatively small part of the southern Great Basin of North America, located in southern Nevada (USA), known as Nellis Air Force Range. The Obsidian Butte source is located here. Its detailed study is an example of in-depth research of an area where several sub-sources are established based on obsidian geology and geochemistry. In addition to this, the author was able to identify the exact position of several previously unknown obsidian sources. This brings a better understanding of obsidian acquisition and use in the prehistory of the southern Great Basin, illustrated by the help of different chronological and typological kinds of points made of obsidian, although the data set size is still limited. This chapter is therefore an excellent case study of a specific area, similar to those in other parts of the world (e.g., Tykot 1998).

In **Chapter 11**, an important study of obsidian from Early Formative period (1500–850 BC) sites, located in the Oaxaca region of Mesoamerica, is presented by Jeffrey P. Blomster and Michael D. Glascock. Although the area surrounding the sites of Etlatongo, Yucuita, and Rancho Dolores Ortíz has abundant amounts of chert and other lithic materials, obsidian was imported from other distant regions in significant quantities. Previous obsidian studies in Oaxaca found that the Guadalupe Victoria source (near the Gulf of Mexico coast) was dominant in the Early Formative time. This larger study provides evidence that the two earliest sites of Rancho Dolores Ortíz and Yucuita had a Guadalupe Victoria focus while the later Early Formative site, Etlatongo, showed more obsidian sources but especially a shift to sources in central Mexico. Most importantly, the study demonstrates that interregional exchange between Oaxaca and other regions played a continuous role for obsidian and very likely for other goods much earlier than realised.

In **Chapter 12**, Michael D. Glascock and coauthors continue with the Mesoamerican theme by presenting results from a long-term geochemical study of more than 1000 samples from obsidian sources in the states of Jalisco, Nayarit, and Zacatecas in western Mexico. Although not one of the regions commonly studied by Mesoamerican archaeologists, western Mexico has some of the largest "mines" used to extract metal ores and obsidian in all of Mesoamerica. Use of the data presented in this chapter will enable archaeologists to make inferences concerning regional political dynamics and the extent of state control over mining activities.

In **Chapter 13**, M. Steven Shackley summarises the research presented in this volume and also highlights future perspectives in the field of obsidian source studies and their broader impact on archaeology and anthropology. Being one of the leaders in obsidian provenance studies in the southwestern United States and worldwide, M. S. Shackley recognises the importance of Northeast Asian data for the Old World archaeology. The overview of the current state-of-the-art with methods and equipment used in volcanic glass studies is very useful.

Perhaps, one of the most fascinating discoveries of the last two decades in the field of obsidian provenance is the establishment of long-distance obsidian transport and exchange in Northeast Asia, which was suggested in the 1970s but not properly justified. These "obsidian networks" covered not only neighbouring regions like Hokkaido and Sakhalin islands, but also extended from insular Northeast Asia to the Asian mainland as the distribution of obsidian from the Koshidake source (Kyushu Island) shows. This is the most reliable evidence of human contacts, migrations, and exchange of raw material over an extremely large space (order of thousand kilometres), including "mountains, rivers, and straits". The proof by scientific methods of extreme long-distance movement of obsidian in prehistory of Northeast Asia and the Americas is not exactly new, but it is explicitly demonstrated in this volume. People were truly "crossing mountains, rivers, and straits" to acquire valuable raw material. It seems that this may be one of the major contributions to the fields of archaeology and anthropology which scholars can derive from this book.

Unfortunately, two important regions of the North Pacific Rim in terms of obsidian sourcing, Alaska and the Northwest Coast of North America, are missing in this volume. However, on Alaska data obtained previously (Cook 1995) and new research on obsidian sources (see Speakman *et al.* 2007; Goebel *et al.* 2008; Reuther *et al.* 2008; Slobodina *et al.* 2009) creates solid background for in-depth studies; therefore, Alaska is catching up. For the Northwest Coast, an excellent review by Carlson (1994) still stands as the comprehensive summary of obsidian sources and their use in prehistory. As for the Chukotka region of Northeastern Siberia neighbouring Alaska, the obsidian provenance studies here have just begun; many sites with excellent quality volcanic glass (e.g., Dikov 1997, 2003; Kiryak 2005, 2010) will allow in the near future a connection between these two parts of the former Bering land bridge (another term is "Beringia") with the help of obsidian as evidence of contacts and migrations, as it has already been confirmed preliminarily (Cook 1995, 99).

The future of obsidian provenance studies in the North Pacific Rim seems to be more secure now compared with what it was in the 1970s – 1990s. Large data sets have been accumulated, and major patterns of obsidian transportation and exchange have been established. Further refinement of this large scale picture in the years to come will give us a better understanding of the prehistoric activity in terms of acquisition and use of raw material, human migrations and contacts, and ultimately some clues about the ways of the peopling of the New World.

References

Ambiru, M., K. Yajima, K. Sasaki, K. Shimada, and A. Yamashina (eds). 2004. *Obsidian and Its Use in Stone Age of East Asia (Obsidian Summit International Workshop Meiji University Session).* Tokyo, Meiji University Centre for Obsidian and Lithic Studies.

Bates, R. L., and J. A. Jackson (eds). 1984. *Dictionary of Geological Terms* (3rd edition). New York, Anchor Books.

Bright, V. 1951. Black Harris, Mountain Man, Teller of Tales. *Oregon Historical Quarterly* 52, 3–20.

Carlson, R. L. 1994. Trade and Exchange in Prehistoric British Columbia. In *Prehistoric Exchange Systems in North America*, edited by T. G. Baugh and J. E. Ericson, 307–361. New York, Plenum Press.

Carter, H. 2003. *The Tomb of Tutankhamen.* Washington, D.C., National Geographic Society.

Cauvin, M.-C., A. Tourgaud, B. Gratuze, N. Arnaud, P. Poupeau, J.-L. Poidevin, and C. Chataigner. 1998. *L'Obsidienne au Proche et Moyen Orient: Du Volcan à l'Outil* (B.A.R. International Series 738). Oxford, BAR Publishing.

Cho, N.-C. 2005. Classification of Obsidian Artifacts Found in Korean Peninsula Based on the Chemical Composition, Texture and Magnetic Property.

Unpublished PhD thesis, Kangwon National University, Ch'unch'on, Korea (in Korean with English summary).

Cole, D. 2001. "The Greatest Thing Undertaken by Any Museum"? Franz Boas, Morris Jesup, and the North Pacific Expedition. In *Gateways: Exploring the Legacy of the Jesup North Pacific Expedition, 1897–1902* (Contributions to Circumpolar Anthropology 1), edited by I. Krupnik and W. W. Fitzhugh, 29–70. Washington, D.C., Arctic Science Center, National Museum of Natural History.

Cook, J. P. 1995. Characterization and Distribution of Obsidian in Alaska. *Arctic Anthropology* 32(1), 92–100.

Dikov, N. N. 1997. *Asia at the Juncture with America in Antiquity (The Stone Age of the Chukchi Peninsula)*. Anchorage, AK, Shared Beringian Heritage Program.

Dikov, N. N. 2003. *Archaeological Sites of Kamchatka, Chukotka, and the Upper Kolyma*. Anchorage, AK, Shared Beringian Heritage Program.

Fairbridge, R. W. (ed.). 1975. *The Encyclopedia of World Regional Geology. Part 1. Western Hemisphere (including Antarctica and Australia)* (Encyclopedia of Earth Sciences. Vol. VIII). Stroudsburg, PA, Dowden, Hutchinson and Ross, Inc.

Fischer, G. 1818. *Oriknognoziya, ili Kratkoe Opisanie Vsekh Iskopaemykh Veshchestv, s Izyasneniyami Terminov. Chast 1* [The Orictognozy, or Short Description of All Fossil Compounds, with Explanation of Terms. Part 1]. Moscow: Tipografiya Imperatorskoi Mediko-Khirurgicheskoi Akademii.

Freed, S. A., R. S. Freed, and L. Williamson. 1988. Capitalist Philanthropy and Russian Revolutionaries: The Jesup North Pacific Expedition (1897–1902). *American Anthropologist* 90, 7–24.

Glascock, M. D. (ed.). 2002. *Geochemical Evidence for Long-Distance Exchange*. Westport, CT, Bergin and Garvey.

Glascock, M. D., G. E. Braswell, and R. H. Cobean. 1998. A Systematic Approach to Obsidian Source Characterization. In *Archaeological Obsidian Studies: Method and Theory*, edited by M. S. Shackley, 15–65. New York and London, Plenum Press.

Glascock. M. D., A. A. Krupianko, Y. V. Kuzmin, M. S. Shackley, and A. V. Tabarev. 1996. Geochemical Characterization of Obsidian Artefacts from Prehistoric Sites in the Russian Far East: An Initial Study. In *Arkheologiya Severnoi Patsifiki*, edited by A. L. Ivliev, N. N. Kradin, and I. S. Zhushchikhovskaya, 406–410. Vladivostok, Dalnauka Press.

Glascock, M. D., V. K. Popov, Y. V. Kuzmin, R. J. Speakman, A. V. Ptashinsky, and A. V. Grebennikov. 2006. Obsidian Sources and Prehistoric Obsidian Use on the Kamchatka Peninsula: Initial Results of Research. In *Archaeology in Northeast Asia: On the Pathway to Bering Strait* (University of Oregon Anthropological Papers 65), edited by D. E. Dumond and R. L. Bland, 73–88. Eugene, University of Oregon Press.

Goebel, T., R. J. Speakman, and J. D. Reuther. 2008. Obsidian from the Late-Pleistocene Walker Road Site, Central Alaska. *Current Research in the Pleistocene* 25, 88–90.

Hall, M. E., and H. Kimura. 2002. Quantitative EDXRF Studies of Obsidian Sources in Northern Hokkaido. *Journal of Archaeological Science* 29, 259–266.

Jochelson, W. 1928. *Archaeological Investigations in Kamchatka*. Washington, D.C., Carnegie Institution of Washington.

Jull, A. J. T., Y. V. Kuzmin, and G. S. Burr. 2005. Chronological Patterns of Obsidian Exploitation in Northeast Asian Prehistory. Paper presented at the 70th Annual Meeting of the Society for American Archaeology, Salt Lake City, UT.

Kim, J. C., D. K. Kim, M. Yoon, C. C. Yun, G. Park, H. J. Woo, M.-Y. Hong, and G. K. Lee. 2007. PIXE Provenancing of Obsidian Artefacts from Paleolithic Sites in Korea. *Bulletin of the Indo-Pacific Prehistory Association* 27, 122–128.

Kiryak (Dikova), M. A. 2005. *Kamenny Vek Chukotki (Novye Materialy)* [The Stone Age of Chukotka (New Materials)]. Magadan, KORDIS Press.

Kiryak (Dikova), M. A. 2010. *The Stone Age of Chukotka, Northeastern Siberia (New Materials)* (B.A.R. International Series 2099). Oxford, BAR Publishing.

Krasheninnikov, S. P. (1994) [1755]. *Opisanie Zemli Kamchatki* [The Description of the Land Kamchatka]. Moscow, Nauka Publishers.

Kuzmin, Y. V. 2006. Recent Studies of Obsidian Exchange Networks in Prehistoric Northeast Asia. In *Archaeology in Northeast Asia: On the Pathway to Bering Strait* (University of Oregon Anthropological Papers 65), edited by D. E. Dumond and R. L. Bland, 61–71. Eugene, University of Oregon.

Kuzmin, Y. V., and M. D. Glascock. 2007. Two Islands in the Ocean: Prehistoric Obsidian Exchange between Sakhalin and Hokkaido, Northeast Asia. *Journal of Island and Coastal Archaeology* 2, 99–120.

Kuzmin, Y. V., M. D. Glascock, and H. Sato. 2002. Sources of Archaeological Obsidian on Sakhalin Island (Russian Far East). *Journal of Archaeological Science* 29, 741–749.

Kuzmin, Y. V., and V. K. Popov (eds). 2000. *Vulkanicheskie Stekla Dalnego Vostoka Rossii: Geologicheskie i Arkheologicheskie Aspekty* [Volcanic Glasses of the Russian Far East: Geological and Archaeological Aspects]. Vladivostok, Dalnevostochny Geologichesky Institut Dalnevostochnogo Otdeleniya Rossiiskoi Akademii Nauk.

Leach, F., and J. Davidson (eds). 1981. *Archaeological Studies of Pacific Stone Resources* (B.A.R. International Series 104). Oxford, British Archaeological Reports.

Malyk-Selivanova, N., G. M. Ashley, R. Gal, M. D. Glascock, and H. Neff. 1998. Geological-Geochemical Approach to "Sourcing" of Prehistoric Chert Artifacts, Northwestern Alaska. *Geoarchaeology* 13, 673–708.

Nasedkin, V. V., and A. A. Formozov. 1965. Vulkanicheskoe Steklo iz Stoyanok Kamennogo Veka Krasnodarskogo Kraya i Checheno-Ingushetii [The Volcanic Glass from Stone Age Sites in the Krasnodar Province and the Republic of Checheno-Ingushetiya]. In *Arkheologiya i Estestvennye Nauki*, edited by B. A. Kolchin, 167–170. Moscow, Nauka Publishers.

Obihiro-shi Kyoiku Iinkai (eds). 2004. *Obihiro, Wakabano Mori Iseki* [Wakabano Mori Site, Obihiro]. Obihiro,

Japan, Obihiro-shi Kyoiku Iinkai.

Phillips, S. C., and R. J. Speakman. 2009. Initial Source Evaluation of Archaeological Obsidian from the Kuril Islands of the Russian Far East Using Portable XRF. *Journal of Archaeological Science* 36, 1256–1263.

Pollmann, H.-O. 1993. *Obsidian in Nordwestmediterranen Raum. Seine Verbreitutig und Nützung im Neolithikum und Äneolithikum* (BAR International Series 586). Oxford, BAR Publishing.

Popov, V. K., V. G. Sakhno, Y. V. Kuzmin, M. D. Glascock, and B.-K. Choi. 2005. Geochemistry of Volcanic Glasses of the Paektusan Volcano. *Doklady Earth Sciences* 403, 254–259.

Reuther, J., J. Rasic, N. Slobodina, and R. Speakman. 2008. Gaining Momentum – The Status of Obsidian Source Studies in Alaska and Their Importance from Developing a Better Understanding of the Regional Prehistory. In *Abstracts of the 73rd Annual Meeting of the Society for American Archaeology, Vancouver, B.C., Canada, March 26–30, 2008*, 465. Washington, D.C., Society for American Archaeology.

Sato, H., Y. V. Kuzmin, and M. D. Glascock. 2002. Sakhalin Tou Shutsudo no Senshi Jidai Kokuyosekisei Sekki no Gensanchi Bunseki to Kokuyoseki no Ryutsu [Source Analysis of Obsidian in Prehistoric Sakhalin and an Assessment of its Distribution in Northeast Asia]. *Hokkaido Kokogaku* 38, 1–13.

Shackley, M. S. (ed.). 1998. *Archaeological Obsidian Studies: Method and Theory*. New York and London, Plenum Press.

Shackley, M. S. 2005. *Obsidian: Geology and Archaeology in the North American Southwest*. Tucson, University of Arizona Press.

Shackley, M. S. 2008. Archaeological Petrology and the Archaeometry of Lithic Materials. *Archaeometry* 50, 194–215.

Shackley, M. S. (ed.). 2010. *X-Ray Fluorescence Spectrometry (XRF) in Geoarchaeology*. New York, Springer Publishing.

Shackley, M. S., M. D. Glascock, Y. V. Kuzmin, and A. V. Tabarev. 1996. Geochemical Characterization of Archaeological Obsidian from the Russian Far East: A Pilot Study. *International Association of Obsidian Studies Bulletin* 17, 16–19.

Shackley. M. S., and V. K. Popov. 1997. Obsidian Primorya: Pervye Rezultaty Arkheologo–Geologicheskoi Korrelyatsii [The Obsidian of Primorye: First Results of Archaeological–Geological Correlation]. *Vestnik Dalnevostochnogo Otdeleniya Rossiiskoi Akademii Nauk* 3(73), 77–85.

Skinner, C. E., and K. J. Tremaine. 1993. *Obsidian: An Interdisciplinary Bibliography* (International Association for Obsidian Studies Occasional Papers 1). San Jose, CA, San Jose State University.

Slimak, L., and Y. Giraud. 2007. Circulations sur Plusieurs Centaines de Kilomètres Durant le Paléolithique Moyen. Contribution à la Connaissance des Sociétés Néandertaliennes. *Comptes Rendus Palevol* 6, 359–368.

Slobodina, N., J. D., Reuther, J. Rasic, J. P. Cook, and R. J. Speakman. 2009. Obsidian Procurement and Use at the Dry Creek Site (HEA-005), Interior Alaska. *Current Research in the Pleistocene* 26, 115–117.

Speakman, R. J., C. E. Holmes, and M. D. Glascock. 2007. Source Determination of Obsidian Artifacts from Swan Point (XBD-156), Alaska. *Current Research in the Pleistocene* 24, 143–145.

Steller, G. W. (1999) [1774]. *Beschreibung von dem Lande Kamtschatka* [The Description of the Land Kamchatka]. Petropalvovsk-Kamchatskiy, Kamchatskiy Pechatny Dvor (in Russian).

Taylor, R. E. (ed.). 1976. *Advances in Obsidian Glass Studies: Archaeological and Geochemical Perspectives*. Park Ridge, NJ, Noyes Press.

Tykot, R. H. 1998. Mediterranean Islands and Multiple Flows: The Sources and Exploitation of Sardinian Obsidian. In *Archaeological Obsidian Studies: Method and Theory*, edited by M. S. Shackley, 67–82. New York and London, Plenum Press.

Vakhtin, N. 2001. Franz Boas and the Shaping of the Jesup Expedition Siberian Research, 1895–1900. In *Gateways: Exploring the Legacy of the Jesup North Pacific Expedition, 1897–1902* (Contributions to Circumpolar Anthropology 1), edited by I. Krupnik and W. W. Fitzhugh, 71–89. Washington, D.C., Arctic Science Center, National Museum of Natural History.

Waldman, C., and A. Wexler. 2004. *Encyclopedia of Exploration. Volume 1. The Explorers*. New York, Facts on File, Inc.

Yanshina, O. V., Y. V. Kuzmin, and G. S. Burr. 2009. Yankito, the Oldest Archaeological Site Cluster on the Kurile Islands (Russian Far East). *Current Research in the Pleistocene* 26, 30–33.

Chapter 2
A Review of Archaeological Obsidian Studies on Hokkaido Island (Japan)

Masami Izuho and Wataru Hirose

Abstract: *Twenty-one sources of geological obsidian are known on Hokkaido island. A majority of the sources, such as Shirataki, Oketo, Tokachi-Mitsumata, and Akaigawa, have been analysed using INAA; comparable data using EPMA and EDXRF techniques have also been generated in recent years. The use of obsidian for making stone tools on Hokkaido began in the early Upper Palaeolithic (around 30,000 BP), and by the late Upper Palaeolithic obsidian exchange networks existed throughout Hokkaido and Sakhalin Islands (with distances more than 300km). Discussions of archaeological obsidian studies on Hokkaido are, however, quite rare. Much of the current archaeological research in Northeast Asia is focused on an ecological approach in order to understand regional variability for the dispersal and adaptation processes of anatomically modern humans. This kind of study requires the reconstruction of long-distance networks of obsidian exchange that include investigations of igneous geology within the primary areas where obsidian occurs. The establishment of discrimination methods between obsidian sources based on differences among geochemical composition likewise is required.*

Keywords: *Obsidian, Geochemistry, Geology, Archaeology, Hokkaido Island, Japan*

Introduction

The existence of primary sources of high quality volcanic glass (obsidian), such as Shirataki and Oketo on Hokkaido Island (northern Japan) situated on the eastern margin of the Asian continent, is well-established. To date, at least 21 obsidian sources have been discovered on Hokkaido. A series of archaeological obsidian studies on Hokkaido, that originated from geological-archaeological collaborative research of the Shirataki archaeological site cluster (Shirataki Dantai Kenkyu Kai 1963), has demonstrated that the use of obsidian for making stone tools began during the early Upper Palaeolithic (around 30,000 BP), and the use of obsidian in this area has long history (Izuho and Sato 2007). Detailed discussions concerning lithic raw material procurement in the vicinity of the Shirataki primary obsidian sources (Kimura 1992); exchange networks of fine aphyric or cryptocrystalline raw materials, such as obsidian and siliceous hard shale (Kimura 1995, 1998); and the long-distance trade and exchange network that expanded from Hokkaido to Sakhalin Island (Kuzmin and Glascock 2007; Kuzmin *et al.* 2002; Sato *et al.* 2002), have been published in recent years.

The existence of extensive obsidian exchange networks in the Japan Sea Rim area—which were as long as 300km during the Upper Palaeolithic (Kuzmin and Glascock 2007; Kuzmin *et al.* 2002)—highlights the importance of a solid research framework and a standardisation of methods used for the systematic investigation of obsidian geochemistry beyond the political and regional boundaries. In anticipation of future research, we should now summarise the results generated for each region. This is particularly important for Japan where large amounts of contract archaeological excavations have been conducted but a few obsidian sourcing studies are available in English (e.g., Izuho and Sato 2007).

Current archaeological obsidian research has shown that several problems exist when geological source data and archaeological specimens are compared. Thus, it should be understood that studies of obsidian formation processes based on tectonics and geological descriptions are extremely important (e.g., Shackley 1998a, 2005). Without this basic information, we could not possibly achieve obsidian sources discrimination with a high degree of certainty. The ability to chemically discriminate among obsidian sources with a high probability is constrained by the analytical choice, the analyses of geologic sources, the selection of archaeological specimens, and the quantitative approaches to data interpretation.

This paper provides an outline of previous research that includes a discussion of several issues: (1) the tectonics of the Japan arc-trench system and the geological formation processes of obsidian; (2) the distribution of primary sources of obsidian and their geochemical characterisation; (3) distribution of secondary sources and archaeological inferences concerning raw material procurement; and (4) models of hunter-gatherer mobility and raw material trade and exchange networks. This is followed by a summary of archaeological obsidian studies in Hokkaido.

Tectonics of Arc-Trench System and the Obsidian Formation Process in Hokkaido

Hokkaido Island, located on western margin of the Pacific Ocean, is the northernmost land mass in the Japanese Archipelago. The geological environment of Hokkaido is considered as a plate subduction zone which explains the large amount of volcanism and the occurrence of obsidian sources.

Tectonic and Geological Settings of Hokkaido Island

Hokkaido is situated on the boundary of an oceanic plate

and two continental plates – the Pacific, the North American (Okhotsk), and the Eurasian (Amurian), respectively (Figure 2.1). The subduction of the Pacific Plate beneath the North American Plate occurs today at a rate of ca. 8.4–8.7cm/year in the Kuril and Japan trenches. This tectonic activity creates two subduction zones of the arc-trench system, called the Kuril Arc and the Northeast Japan Arc. These arcs are located on the continental earth crust, as well as the non-continental earth crust, where the mean thickness is about 30km and the maximal one is about 60km. The Pacific Plate has subducted obliquely westward against the Kuril Arc; the frontal portion of the Kuril Arc became a micro-plate that migrated westward and collided with the Northeast Japan Arc in central Hokkaido (Kimura 1986). However, the rate of oblique subduction lessened after the Pliocene (DeMets 1992; Hirose and Nakagawa 1999).

The arc-trench systems, which developed in the Kuril and the Northeast Japan arcs, were formed prior to the Cretaceous or early Palaeogene periods. The collision of the two arcs began sometime after the Neogene and was accompanied by the right-lateral strike-slip collision of the North American and the Eurasian plates (Kimura *et al.* 1983). Since the Cretaceous, the central axis of Hokkaido became the boundary area where two continental plates collided, and the thickest part of continental crust was

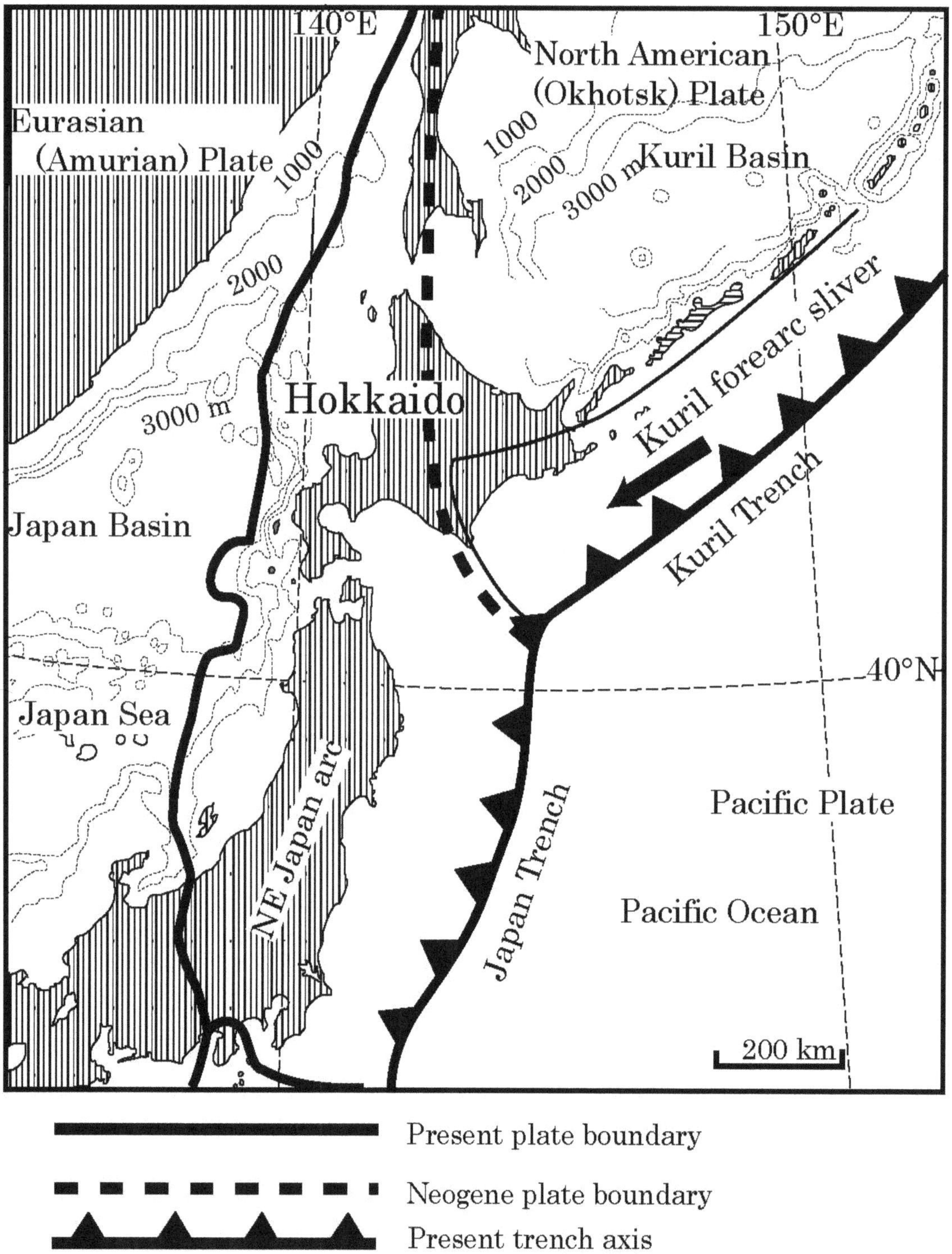

Figure 2.1. Geological setting of Hokkaido Island (after Hirose and Nakagawa 1999)

formed. In contrast, at the back-arc areas, the Kuril Basin and the Japan Sea (includes the Japan Basin and the Yamato Basin) were formed. Based on radiometric dates for basalt from the sea floor and the micropalaeontological ages for deposits of these basins, the Sea of Japan was formed during the late Palaeogene to middle Miocene (Kaneoka *et al.* 1992). As for the Kuril Basin (although there is no exact evidence for its geological age), its formation is estimated to have occurred sometime between the Cretaceous and the middle Miocene periods (Kimura and Tamaki 1985). These events established the basic geographical and geological setting on and around Hokkaido.

The westward migration of the Kuril Arc and its subsequent collision with the Northeast Japan Arc ceased sometime during the Miocene–Pliocene. These movements might be responsible for the change in stress field of the Kuril back-arc to intermediate. The boundary between the North American and Eurasian plates shifted westward to the eastern margin of Japan Sea during the Quaternary period. This event led to formation of an active crustal shortening zone from the western side of Hokkaido into the eastern margin of Japan Sea (Hirose and Nakagawa 1999).

The geology of Hokkaido consists of the arc-volcanic rocks formed between the Cretaceous and Palaeogene periods, and the fore-arc basin deposits and metamorphic rocks that originated from the accretionary complex zone. These basement rocks are overlain by sedimentary rocks and volcanic rocks that date from the Palaeogene to the Holocene. The Cretaceous, Palaeogene, and Neogene sedimentary rocks in central axis of Hokkaido are arranged in the north-south direction, and form a thrust and fold belt. In contrast to the axis zone, the geological structures and original topography of volcanic edifices are well-preserved in the back-arc side of Kuril and Northeast Japan arcs because deformations between the Neogene and Quaternary periods were relatively small when compared to the central axis part of Hokkaido.

An Outline of Volcanic Activity on Hokkaido Island

Volcanic activity on Hokkaido has occurred since the middle Miocene when the tectonic setting of arc-trench system first originated. The evolution of volcanic activity and eruption type (effusive or explosive, depending on rate of magma discharge) in this region corresponds to the changes in the tectonic setting. In this section, we summarise the evolution of volcanic activity on Hokkaido based on the research of Hirose and Nakagawa (1999). Roughly, volcanic activity on and around Hokkaido can be divided into three stages based on the type of volcanic activity, the chemical composition of magma, and the distribution of the eruptive centres. These stages include: (1) the early to middle Miocene (older than ca. 12,000,000 years ago); (2) from the late Miocene to the beginning of the Pleistocene (ca. 12,000,000–1,000,000 years ago); and (3) from the early Pleistocene to present day (less than ca. 1,000,000 years ago).

During the early–middle Miocene, large-scale volcanic activity extended throughout Hokkaido. It began after the alkaline volcanism and ended about 14,000,000–12,000,000 years ago. The bi-modal volcanism, which is dominated by basalt and rhyolite rocks and lack of andesite, is characteristic of this period. Ordinary arc volcanic activity did not exist because of influences from the large disturbance of the arc crust and mantle structure when formation of the back-arc basin took place. Obsidian formed during this period frequently occurs within large felsic pyroclastic sediments.

As for the late Miocene – early Pleistocene, the intensity of volcanism accompanied with the formation of the back-arc basin decreased around 12,000,000–9,000,000 years ago. In contrast, the arc type volcanoes became increasingly active in the Kuril and Northeast Japan arcs. Although bi-modal volcanic activity on Hokkaido continued in part, especially in the Kuril Arc (eastern Hokkaido), uni-modal (andesitic) volcanism gradually became dominant. In particular, this tendency in the evolution of volcanic activity is evident in the Kuril Arc, and it is explained by volcanic activity in the intermediate or tensional stress field of the earth crust caused by the oblique subduction. The rhyolitic lava domes and large-scale felsic pyroclastic flows were created across entire areas of volcanism. Most of the obsidian rock massifs on Hokkaido, including Shirataki, Oketo, and Engaru, were formed at the Kuril Arc during this period.

From the early Pleistocene to the present time, the stress field on Hokkaido shifts to more compressive. This change resulted in volcanism with uni-modal magmatic characteristics (e.g., the dominance of andesite). In contrast, bi-modal volcanism declined significantly. The large-scale felsic pyroclastic flows were still active in calderas, such as Shikotsu, Kutcharo, Kuttara, and Akan. However, sizable obsidian fields in this region are unknown. On the other hand, obsidian was formed during this period at Okushiri Island in western Hokkaido. This obsidian source is atypical in that it originates from a rhyolitic lava dome.

Distribution of Obsidian and Discrimination of Sources

Distribution of Obsidian

Geological survey for the purpose of creating geological maps (scale 1:50,000) was recently completed for all of Hokkaido Island, and most maps are now published (available at: http://www.gsh.pref.hokkaido.jp/zufuku.html). During the mapping process, the distribution and stratigraphy of volcanic rocks were also studied, and includes information on the 'larger' obsidian localities. Recent studies of the Neogene–Quaternary volcanism on Hokkaido provided numerous potassium-argon (K–Ar) dates and fission-track (FT) dates for the igneous rocks, including major obsidian sources such as Shirataki, Oketo, Akaigawa, and Engaru (Table 2.1). However, specific details about the geological feature of obsidians and their stratigraphy within the primary source area—which are required for the localisation of archaeological obsidian sources—are rare.

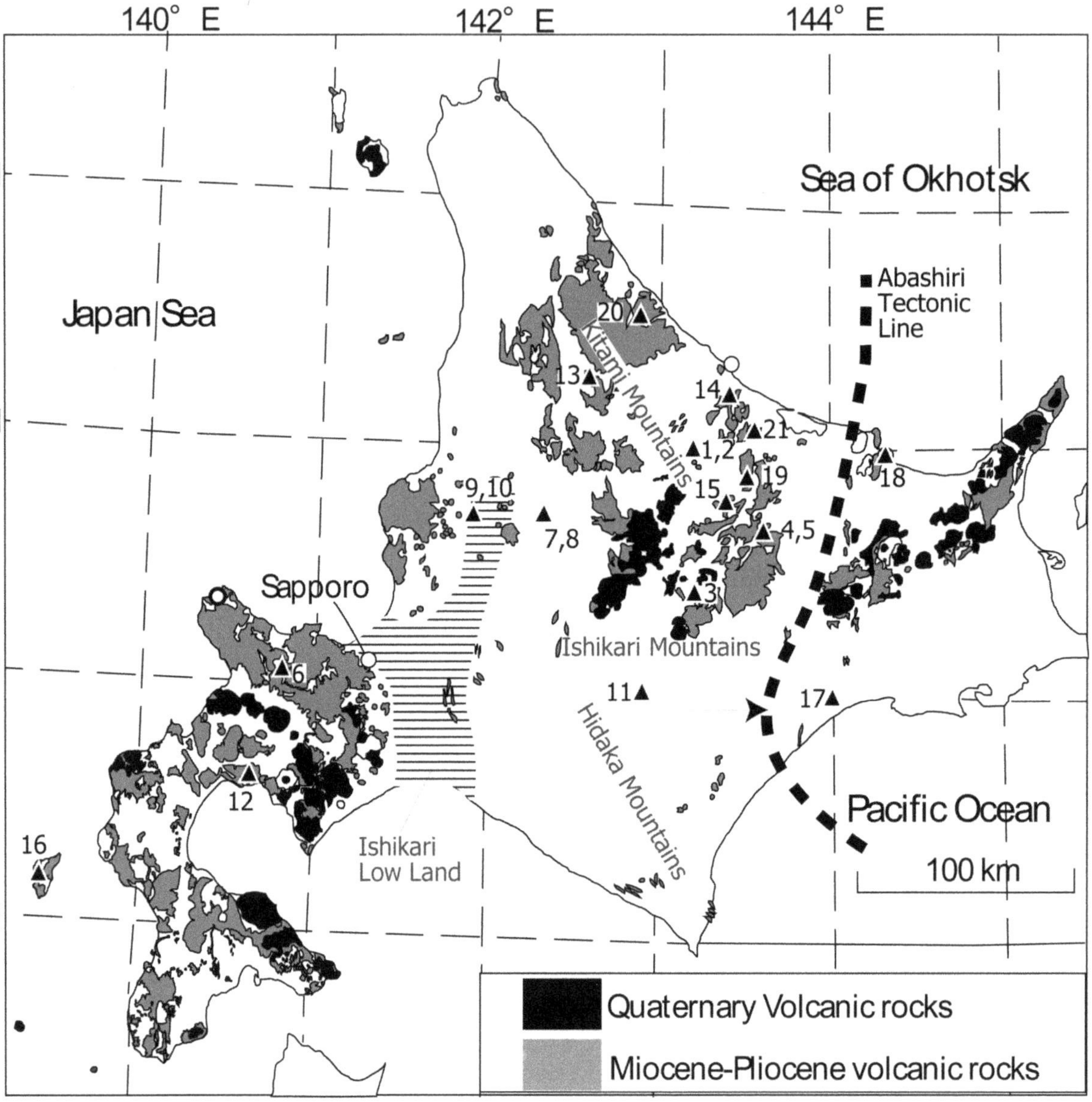

Figure 2.2: Distribution of the volcanic rocks and obsidian on Hokkaido (after Izuho and Sato 2007; Hirose 1999a, Hirose 1999b). The numbers correspond to the obsidian sources listed in Table 2.1. Obsidian source localities are shown as black triangles

Obsidian localities and the Tertiary–Quaternary volcanic rocks on Hokkaido are shown on Figure 2.2. The distribution of obsidian sources tends to occur in the back-arc side where volcanic front of the Pleistocene age located, rather than in the Shiretoko Volcanic Zone of eastern Hokkaido. However, obsidian sources are found throughout all Hokkaido. There is also a tendency for many obsidian sources, such as Shirataki, Oketo, Tokachi-Mitsumata, and Engaru, to be situated in the Kuril Arc area. Obsidian sources in the Northeast Japan Arc are scarce, and exceptions include sources of Akaigawa and Okushiri. Radiometric ages for major sources of obsidian such as Shirataki, Oketo, Tokachi-Mitsumata, and Akaigawa, indicate that most obsidian was formed ca. 4,000,000–1,000,000 years ago (Koshimizu 1981; Yokoyama *et al.* 2003). A few exceptions include Okushiri (late Miocene to Pleistocene); obsidian gravel from the Tokachi River (ca. 1,700,000 years ago) (Koshimizu 1981); and Engaru (late Miocene, ca. 4,000,000 years ago) (Yahata and Nishido 1995). Thus, the formation ages of obsidian sources on Hokkaido are clustered around the Pliocene and early Pleistocene periods (ca. 3,000,000–1,000,000 years ago), whereas volcanic activity on Hokkaido continues from the Neogene (middle Miocene) to the present time.

Acidic volcanic rocks represented by large felsic pyroclastic sediments are widely distributed from the volcanic front to back-arc zone of both the Kuril and Northeast Japan arcs. Although the uneven distribution of obsidian sources on Hokkaido is not well-explained, in particular for the Kuril Arc region, bi-modal volcanic activity zones and obsidian locations tend to be in agreement. Additional discussion concerning the formation mechanism of obsidian from experimental petrology is necessary, but it seems reasonable that the abrupt glass forming without gradual crystallisation may be possible because the temperature of the solidification of water saturated by silicate melt (magma) dramatically rises during the degassing of water (vapour) in the melt, or because of extremely rapid cooling of magma when acidic melt is erupted or welded. This model, however, does not explain why most obsidian sources are concentrated on the back-arc side of the Kuril Arc.

Currently, 21 obsidian sources are known for Hokkaido (Table 2.1; Figure 2.2) (Izuho and Sato 2007). They can be sub-divided into two types, primary and secondary (Glascock *et al.* 1998; Izuho 1997; Shackley 1998b, 2005). A primary source of obsidian is defined as the area where obsidian was formed and is exposed at the surface. It includes not only locales where outcrops are visible, but also the areas where the actual outcrop is not directly observed but instead represented by obsidian debris and gravel scatters. Large-scale rock collapses and transformations, such as landslides, frequently occur in the primary sources and its vicinities.

Eleven primary sources are known on Hokkaido: Shirataki-Akaishiyama, Shirataki-Tokachiishizawa, Tokachi-Mitsumata-Jyusan'nosawa, Oketo-Oketoyama, Oketo-Tokoroyama, Akaigawa-Dobokuzawa, Monbetsu-Kamimobetsu, Keshomappu-Michikozawa, Abashiri-Ponmoimisaki, Ikutahara-Nitappugawa, and Okushiri-Katsumayama. Among them, actual outcrops have been discovered at the Shirataki (Akaishiyama and Tokachiishizawa), Abashiri-Ponmoimisaki, Monbetsu-Kamimobetsu, Keshomappu-Michikozwa, and Okushiri-Katsumayama localities. At Shirataki, obsidian occurs in the large-scale felsic pyroclastic flows that erupted during the Pliocene. The Horoka-Yubetsu welded tuffs which include these pyroclastic flows are widely distributed throughout the Shirataki region (Matsui 1959). The Shirataki sources were actively used in prehistory, and many archaeological sites containing obsidian have been found in their vicinity (Kimura 1995). In contrast, the exploitation of Monbetsu-Kamimobetsu and Abashiri-Ponmoimisaki sources, where obsidian of poor quality occurs in small patches of rhyolite or basalt, has not been confirmed.

Outcrops have not yet been identified at five primary sources: Oketo (Oketoyama and Tokoroyama), Tokachimitsumata-Jyusan'nosawa, Akaigawa-Dobokuzawa, and Ikutahara-Nitappugawa. However, it is clear that these are also primary sources in the sense described above. These rocks are associated with rhyolitic domes, and there is large number of broken cobbles and flakes on the surface in the vicinity of these sources which suggest local exploitation of obsidian.

An investigation of the volcanic geology in and around the Akaigawa source was recently conducted (Yokoyama *et al.* 2003), and it may be possible that obsidian-bearing rocks were formed during the welding of erupted large-scale felsic pyroclastic flows when the Akaigawa Caldera was formed. We also can be certain that the primary source area of Akaigawa-Dobokuzawa was used in prehistory because a large number of artefacts and broken cobbles have been found in the vicinity along with debris and gravel.

In contrast, secondary sources occur when water streams erode material from the primary locales. This material is then transformed into rounded cobbles and pebbles and/ or transported by alluvial action away from the primary source. Secondary sources that have not been linked to specific primary sources are known in ten areas of Hokkaido: Chikabumidai, Ubundai, Hokuryu, Chippubetsu, Shikaribetsu, Toyoizumi, Nayoro, Kushiro, Oumu, and Engaru. Obsidian at these locales occurs in river beds and terrace deposits. Obsidian cobbles are known to have been exploited during the Jomon period (generally end of the early Holocene and the middle Holocene) at the Nayoro, Chikabumidai, and Toyoizumi localities (Tomoda 1996).

Geochemical Identifications

As in other areas of the world (e.g., Glascock *et al.* 1998; Shackley 1998c), geochemical characterisation of volcanic glass sources using a range of methods has been successful in discriminating among the various Hokkaido obsidian sources. These approaches include: Instrumental Neutron Activation Analysis (hereafter – INAA) (Kuzmin and Glascock 2007; Kuzmin and Popov 2000; Kuzmin *et al.* 2002), Energy Dispersive X-ray Fluorescence (hereafter – EDXRF) (Hall and Kimura 2002; Inoue 2003; Warashina 1999; Yoshitani *et al.* 2001), and Electron Probe Microanalysis (hereafter – EPMA) (Mukai 2005; Mukai and Wada 2001, 2003, 2004a, 2004b, Mukai *et al.* 2002, 2004a, 2004b). Notably, four major obsidian sources on Hokkaido (Shirataki, Oketo, Tokachi-Mitsumata, and Akaigawa) have been discriminated using data derived from 28 elements measured by INAA (Kuzmin and Glascock 2007; Kuzmin and Popov 2000; Kuzmin *et al.* 2002). We hope that INAA can be extended to other geologic sources and archaeological samples on Hokkaido. In contrast to these studies, most data concerning the geochemical composition of obsidian on Hokkaido has been independently accumulated from geological and archaeological perspectives and without close communication between the disciplines. We note two serious problems for these disparate studies: (1) obsidian studies undertaken for archaeological purposes tend to report only element ratios; and (2) geologists who are mostly interested in the study of the petrologic aspects for origin of magma have not paid sufficient attention to obsidian. In this section, we summarize previous obsidian geochemical studies on Hokkaido.

Warashina (2004) conducted analysis of obsidian using EDXRF. His identification approach was first to measure 12 trace elements (Al, Si, K, Ca, Ti, Mn, Fe, Rb, Sr, Y, Zr, and Nb) and then discriminate statistically the sources using Hotelling T^2 analysis based on ratios of elements (%/ppm) (Ca/K, Ti/K, Mn/Zr, Fe/Zr, Rb/Zr, Sr/Zr, Y/Zr, Nb/Zr, Al/K, and Si/K). T. Warashina generated large amounts of data after 30 years of collaboration with many archaeologists on Hokkaido. He identified a total 31 chemical groups of 'geological' obsidian representing 16 geological sources. These results have been published in more than 160 chapters in rescue excavation reports. However, the major sources of Akaigawa and Tokachi-Mitsumata which are different from the view of tectonic position and age were never identified as two separate sources. Akaigawa and Tokachi-Mitsumata are of primary importance for studies of prehistoric activities in terms of lithic raw material

Table 2.1. Obsidian sources on Hokkaido Island (after Izuho and Sato 2007)

#	*Source*	*Primary Locality*	*Secondary Distribution*	*Latitude, N*	*Longitude, E*	*Altitude, m a.s.l.*	*Geological age*	*References*[§]
1	Shirataki	Akaishiyama	Horokazawa R.* to Yubetsu R.	43° 56′	143° 08′	1172	2.90 ± 0.18 Ma** (FT***; Koshimizu 1981); 2.10 ± 0.15 Ma (FT; Koshimizu 1981)	Kawano (1950); Konoya *et al.* (1964); Kimura (1992, 2005)
2	Shirataki	Tokachi–Isizawa	Tokachi–Ishizawa R. to Yubetsu R.	43° 54′	143° 10′	872	2.90 ± 0.18 Ma (FT; Koshimizu 1981); 2.10 ± 0.15 Ma (FT; Koshimizu 1981)	Konoya *et al.* (1964); Kimura (1992, 2005)
3	Tokachi–Mitsumata	Jyusan'nosawa	Jyusan'nosawa R. to Tokachi R., Toshibetsu R. to Tokachi R.	43° 29′	143° 10′	725	1.70 ± 0.14 Ma (FT; Koshimizu 1981)	Kitazawa (1999)
4	Oketo	Oketoyama	Bochinosawa R. to Tokoro R., Kunneppu R. to Tokoro R.	43° 41'	143° 33′	550	Pliocene	Sawamura and Hata (1965); Izuho (1997); Mukai *et al.* (2002)
5	Oketo	Tokoroyama	Bochinosawa R. to Tokoro R., On'neanzu R. to Tokoro R.	43° 40′	143° 30′	582	3.90 ± 0.20 Ma (FT; Koshimizu 1981)	Izuho (1997); Mukai *et al.* (2002); Sawamura and Hata (1965)
6	Akaigawa	Dobokuzawa	Dobokuzawa R. to Yoichi R., Nakanosawa R. to Yoichi R.	43° 00′	140° 49′	550	2.10 ± 0.20 Ma (FT; Suzuki 1973); 2.06 ± 0.57 Ma (K–Ar****; Yokoyama *et al.* 2003)	Kimura (1978); Mukai *et al.* (2002); Yokoyama *et al.* (2003)
7	Chikabumidai	Unknown	Chikabumidai Terrace to Ishikari R.	43° 50′	142° 23′	145	–	Mukai *et al.* (2000)
8	Ubundai	Unknown	Ubundai Terrace to Ishikari R.	43° 46′	142° 17′	120	–	Mukai *et al.* (2000)
9	Hokuryu	Unknown	Hekisui Terrace to Uryu R.	43° 46′	141° 52′	51	–	Mukai *et al.* (2001)
10	Chippubetsu	Unknown	Nakayama Terrace to Chippubetsu Sakura R.	43°46′	141°58′	68	–	Mukai *et al.* (2001)
11	Shikaribetsu	Unknown	Shikaribetsu R.to Tokachi R.	43° 05′	142° 59′	180	Miocene? (Mukai *et al.* 2004b)	Mukai *et al.* (2004b)
12	Toyoizumi	Unknown	Toyoizumi R.	42° 36′	140° 39′	8	–	Mukai (2005)
13	Nayoro	Unknown	Chureppu R. to Teshio R., Tosei R. to Teshio R.	44° 15′	142° 34′	212	–	Nayoro–shi Kyoiku Iinkai (1988)
14	Monbetsu	Kamimobetsu	Kamimobetsu R. to Mobetsu R.	44° 10′	143° 22′	160	11.8 ± 0.60 Ma (K–Ar; Yahata and Nishido 1995)	Yahata *et al.* (1988); Mukai (2005)
15	Keshomappu	Michikozawa	Keshomappu R. to Muka R.	43° 45′	143° 18′	792	2.70 ± 0.20 Ma (FT; Koshimizu and Kim 1986)	Mukai (2005)
16	Okushiri	Katsumayama	Horonai R.	42° 11′	139° 27′	340	2.20 ± 0.06 Ma (FT; Ganzawa 1992); 0.71 ± 0.11 Ma (FT; NEDO 1994)	Hata *et al.* (1982); Mukai *et al.* (2004a)
17	Kushiro	Unknown	Koitoi R. to Akan R.	42° 59′	144° 09′	10	–	Okazaki (1966); Sawa (1978)
18	Abashiri	Ponmoimisaki	–	44° 01′	144° 16′	22	7.14 ± 0.37 Ma (K–Ar; Hirose and Nakagawa 1999)	Yahata (1999)
19	Ikutahara	Nitappugawa	Nitappu R.	43° 58′	143° 29′	293	Late Miocene	Mukai *et al.* (2004)
20	Oumu	Unknown	Otoineppu R.	44° 33′	142° 49′	144	Middle Miocene	Mukai and Wada (2003)
21	Engaru	Unknown	Sanabuchi R. to Yubetsu R.	44° 03′	143° 28′	109	Late Miocene	Tajika and Yahata (1991); Mukai and Wada (2003)

[§]Given in chronological order; *R. – River; **Ma – million years ago; ***FT – fission track dating method; ****K–Ar – potassium–argon dating method.

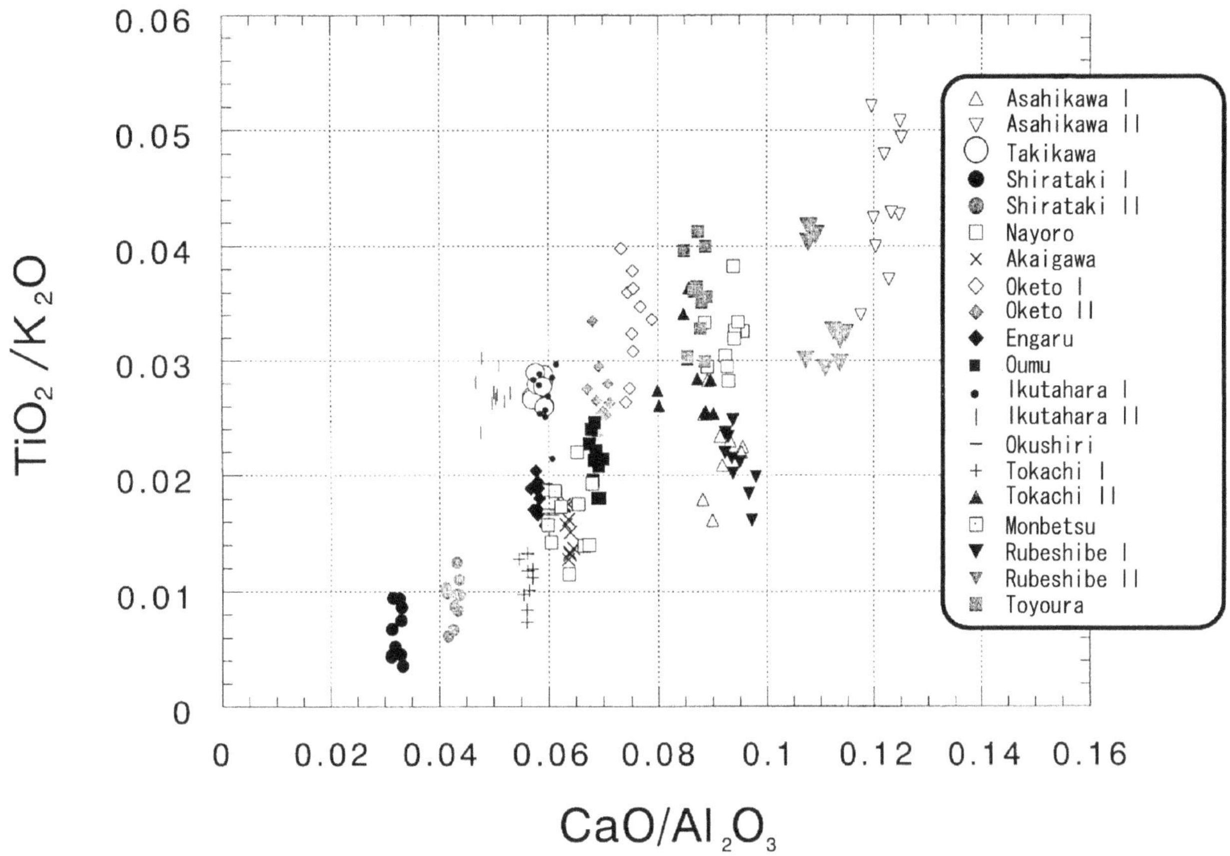

Figure 2.3. The CaO/Al_2O_3 — TiO_2/K_2O chemical composition diagram for obsidian glasses on Hokkaido (after Mukai 2005, modified)

procurement on and around Hokkaido Island. Moreover, T. Warashina failed to report original data for the measured elements, which has made it difficult to compare with data obtained by other researchers (Tsurumaru 2001).

Yoshitani *et al.* (2001) examined obsidian from the northern part of Tokachi Plain using wave length dispersive X-ray spectrometry (hereafter – WDX), EPMA, and EDXRF methods. Inoue (2003) used EDXRF for source identification of archaeological obsidian. Chemical composition for many obsidian sources on Hokkaido is expressed as the set of the oxidation density (weight %) between major elements and trace elements. Research by Yoshitani *et al.* (2001) and Inoue (2003) provides testable geochemical data; these results demonstrate that many obsidian sources can be discriminated on Hokkaido using the methods mentioned above.

In recent years, measurement techniques for chemical analysis at the micro-scale (less than ten microns) using EPMA have improved. Such analytical techniques have successfully been employed to study the petrology of volcanic rocks (Mukai 2005; Mukai and Wada 2001, 2003, 2004a, 2004b; Mukai *et al.* 2000, 2004). The results of these studies are reported in more than ten papers in Japanese, and they provide precise measurements of major elements and partly trace elements in volcanic glass including the sodium (Na) content. Although this analytical approach is well-established for tephra (fine volcanic glass particles), the application of this method to obsidian has occurred only recently. On Figure 2.3, for example, plots presented in Mukai (2005, Figure 1) are depicted. With the exception of a few overlapping chemical values, it is possible to discriminate the sources. These latest results have provided geochemists with the ability to assess accuracy and precision in petrology using analytical values derived from standard samples.

Obsidian analysis using INAA in order to discriminate among major obsidian sources on Hokkaido represents a significant breakthrough for archaeological obsidian studies (Kuzmin and Glascock 2007; Kuzmin and Popov 2000; Kuzmin *et al.* 2002). As it is shown on Figure 2.4, the determination of major obsidian sources on Hokkaido, including Shirataki, Oketo, Tokachi-Mitsumata, and Akaigawa, is possible. INAA technique provides measurements of many of the lanthanide and rare earth elements, and can be applied to the discussion of tectonics (life history of volcanoes and succession of magma composition) on Hokkaido. The discrimination between sources, which can be used to examine correlations among the Neogene volcanic rocks, represents a significant step forward in the systematical classification of archaeological obsidian studies on Hokkaido. INAA examination of archaeological obsidian is a promising endeavour for future research.

As mentioned above, various studies of obsidian including source discrimination by multiple analytical approaches have been conducted. Petrologic data have rapidly

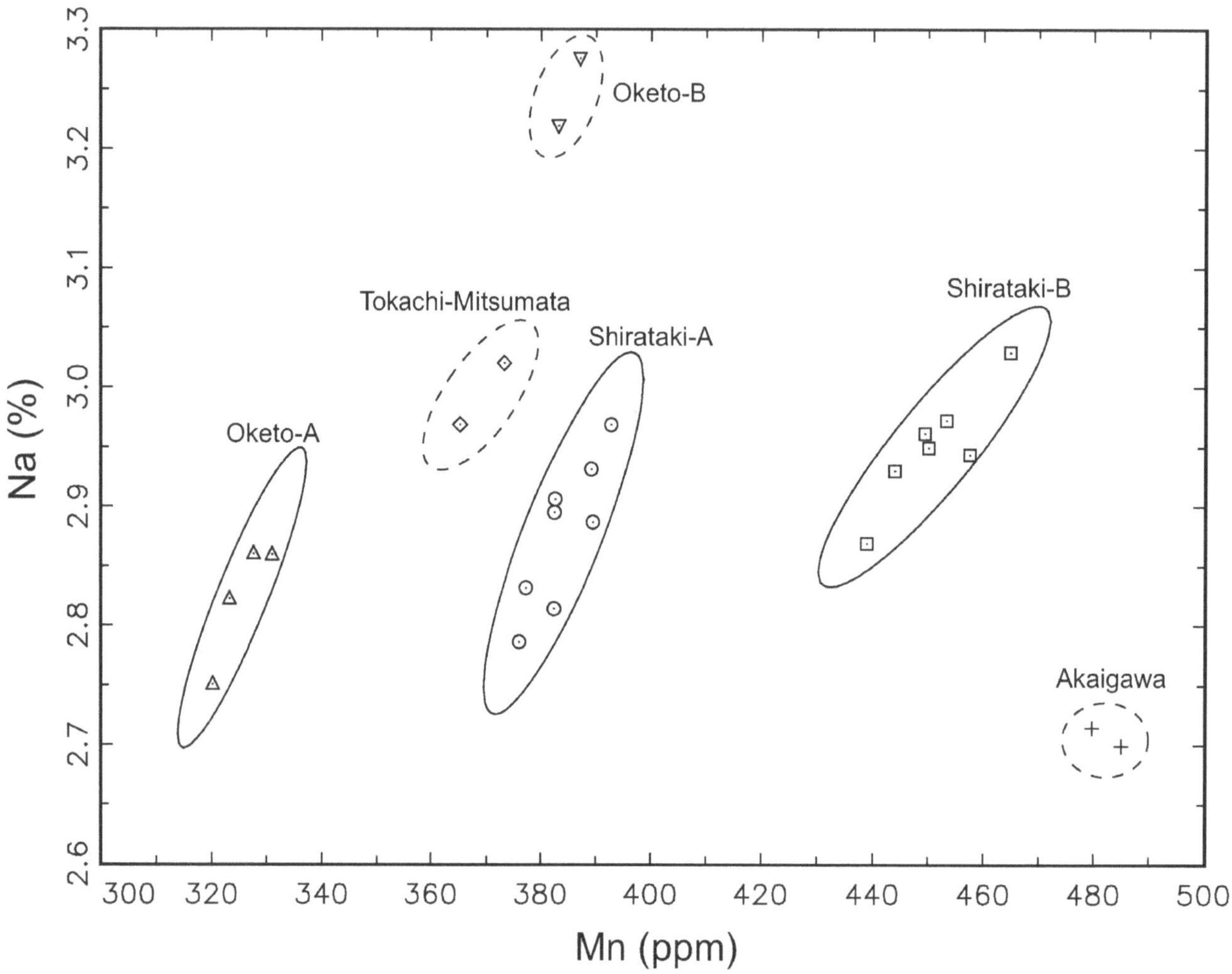

Figure 2.4. The Mn–Na diagram of the volcanic glass from major sources on Hokkaido (after Kuzmin and Glascock 2007, modified)

accumulated on Hokkaido during the past decade, and reliable archaeological obsidian source identification is now a reality. Conversely, fundamental petrologic data are not yet available. A complete description of the distribution and stratigraphy of obsidian is still needed because obsidian source samples described in several previous studies were collected without detailed description. With the exception of a few sources, most geologic samples are represented by small amount of obsidian from outcrops or river beds.

Secondary Distribution Areas and Lithic Raw Material Procurement Patterns

Secondary distribution of obsidian occurs when streams erode debris from primary distribution area(s). It is expected that the scale of secondary distribution varies depending on landforms, vegetation, and climatic conditions, such as temperature and precipitation. The form of obsidian gravel, expressed as the roundness and sphericity, varies due to transportation distance and steepness of river profile. The surface characteristics of obsidian at secondary distribution areas can result in obsidian becoming rounded fragments. These patterns provide the basis from which we can identify where lithic raw materials were procured by humans (Figure 2.5) (Izuho 1997).

Tectonic activity and earthquakes were common during the Pleistocene on Hokkaido. Earthquakes may have triggered some mountain collapses and landslides, and these geological phenomena provide conditions in which obsidian gravel was transported more easily. Moreover, severe freeze-thaw rupture of rocks and slope-wash processes affected Hokkaido which was positioned on the southern border of the continuous permafrost zone during the Last Glacial Maximum (hereafter – LGM). The transportation of obsidian by water was common given that mean annual precipitation ranges from 800 to 1600mm (data from 1971–2000; National Astronomical Observatory 2005).

Secondary obsidian sources on Hokkaido that have not been linked to a primary source are known at Chikabumidai, Ubundai, Hokuryu, Chippubetsu, Shikaribetsu, Toyoizumi, Nayoro, Kushiro, Oumu, and Engaru, where obsidian occurs in river beds and terrace deposits (Table 2.1). Obsidian cobbles are known to have been exploited at Nayoro, Chikabumidai, and Toyoizumi. Because of these conditions, the secondary distribution areas have changed according to environmental fluctuations. For example, there is a secondary distribution area which extends more than 100km along the Tokoro River, northeastern Hokkaido. At present, large-scale landslides and mountain collapses are distributed within the Oketo-Tokoroyama primary source

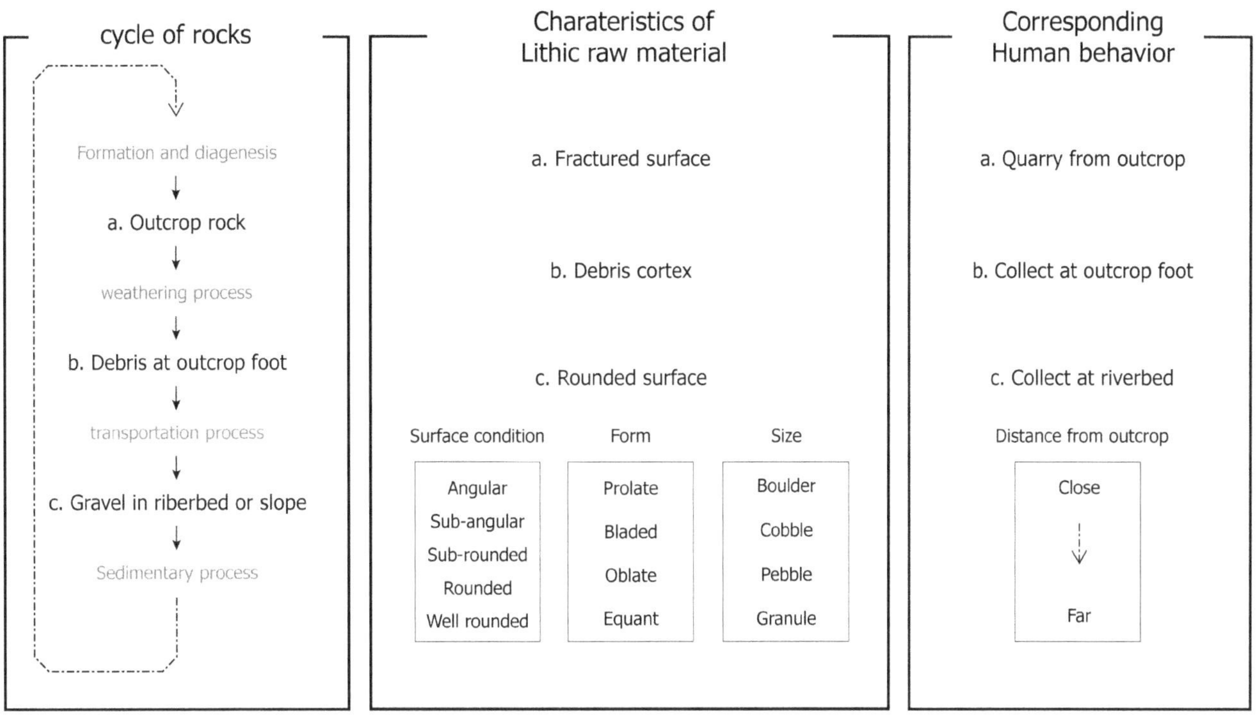

Figure 2.5. Model for lithic raw material procurement (after Izuho 1997)

area, and vast amount of obsidian debris and gravel are yielded from these deposits. Obsidian debris were also transported by water flow to the Bochinosawa River and the Onneanzu River where sub-angular to sub-rounded gravel dominates. Further downstream the Tokoro River, gravel dominates. At the LGM, sea level dropped to about 100m and the coastline was far offshore from its present position. It is expected that the course of Tokoro River extended to several tens of kilometres and profile of the river was steeper. These changes during the LGM resulted in an increased scale of secondary distribution area, and the sizes of obsidian nodules were larger than at present. However, surface conditions of obsidian gravels were not changed significantly. These distribution patterns influence the lithic raw material environment, and humans at that time developed technological strategies for the use of lithic raw materials of varying sizes.

Variation in the size of raw material is positively correlated to change in microblade assemblages in the Tokoro River basin (Izuho and Sato 2007; Nakazawa *et al.* 2005). In the upper course of Tokoro River near the primary sources of Oketo-Tokoroyama and Oketo-Oketoyama, obsidian is available as debris and gravel. By contrast, in the area of 100km along the middle and lower parts of the Tokoro River, obsidian occurs as secondary deposits dominated by sub-angular to well-rounded pieces (Figure 2.6). Although the cobbles were larger during the Pleistocene than they are today, the overall spatial pattern of angular versus rounded material remains stable.

Microblade assemblages with Hirosato-type cores which consist of a reduction sequence to produce large blades (up to 30cm long) are found in the upper reaches of the Tokoro River near the primary sources. Conversely, microblade assemblages with Oshorokko-type cores which produce blades of 10–20cm long are made mostly from middle-sized, well-rounded pieces. However, a few tools manufactured from large debris and angular pieces have also been found (Figure 2.7). The latter assemblages are known up to tens of kilometres from the secondary deposits along the middle stream of Tokoro River, i.e. the Kitagamidaichi (Kitami-shi Kyoiku Iinkai 1988) and Toyooka 7 (Bihoro-cho Kyoiku Iinkai 2002) sites. Although the quality of obsidian does not change, these examples illustrate that reduction sequences and tool types vary depending on the package size of raw material (Izuho and Sato 2007).

Models of Hunter-Gatherers Mobility, Lithic Raw Material Trade, Long-Distance Exchange Networks on and around Hokkaido

Procurement of obsidian as a raw material for manufacture of stone tools began during the early Upper Palaeolithic (ca. 30,000 BP). For example, a small flake industry based on the use of obsidian gravel (> 7cm in diameter) was identified at the Wakabano Mori site in eastern Hokkaido (Figure 2.8) (Izuho and Akai 2005; Obihiro-shi Kyoiku Iinkai 2004). Since the early Upper Palaeolithic, high-quality obsidian nodules have been used. This continuous procurement of obsidian seems to imply that its importance as lithic raw material was high throughout the prehistory of Hokkaido. Thus, it is expected that the role of obsidian provenance studies in archaeological research on Hokkaido is quite important.

Discussions of geochemistry for obsidian sources discrimination, as well as intersite and intrasite distribution patterns, became important for understanding the exchange

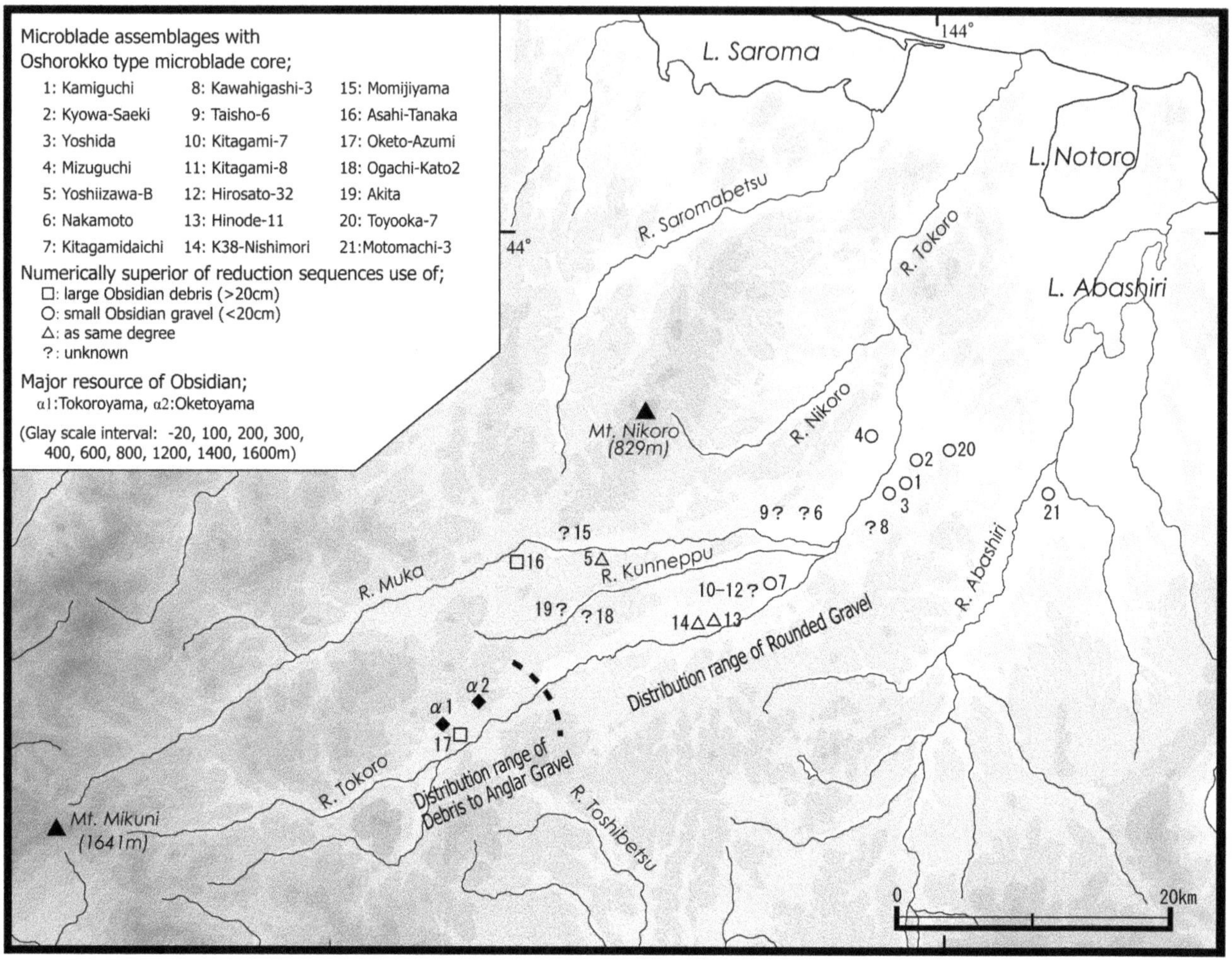

Figure 2.6. Site distribution and content of reduction sequence of the microblade assemblage with Oshorokko type microblade core in and around Tokoro River basin (after Izuho and Sato 2007, modified)

and interaction among Upper Palaeolithic groups on Hokkaido (e.g., Izuho and Sato 2007; Kimura 1992, 1995, 1998, 2002, 2005; Kuzmin and Glascock 2007; Kuzmin and Popov 2000; Kuzmin *et al.* 2002; Naoe and Nagasaki 2005; Sato 2005; Sato *et al.* 2002; Warashina 1999). Here we outline two main issues: extended exchange networks; and obsidian raw material procurement and processing (see, for example, Izuho and Sato 2007).

Extended Exchange Networks

Recent research using INAA has demonstrated that many artefacts on Sakhalin Island were made of Oketo and Shirataki obsidian. Use of this obsidian began during the Upper Palaeolithic and lasted throughout the Palaeometal period (Figure 2.9) (Kuzmin and Glascock 2007; Kuzmin and Popov 2000; Kuzmin *et al.* 2002; Sato 2004a). During the Upper Palaeolithic and Neolithic, the use of Shirataki sources was dominant (> 90% of total amount of obsidian), whereas the Oketo source was of less importance (< 10% of total obsidian). During the Palaeometal period, the Oketo source was dominant (38% of total obsidian), implying changes in exchange patterns (Kuzmin and Glascock 2007).

From wider perspective of an exchange models, Sakhalin and Hokkaido islands were integrated in large-scale trading networks. Similar features which began to exist during the Upper Palaeolithic period are also known on the Korean Peninsula and Kyushu Island in Japan, thus covering large distances up to 1000km from the sources (Kim *et al.* 2007; Kuzmin and Glascock 2007; Sato *et al.* 2002; Sato 2004b, 2005).

Obsidian Lithic Raw Material Procurement and Processing

In a series of studies, Kimura (1992, 1995, 1998, 2005) described changes in Upper Palaeolithic obsidian procurement patterns on Hokkaido inferred from the analyses of artefacts attributed to the Shirataki source. H. Kimura mentioned that before the appearance of microblades rounded gravel from secondary sources appears to have been procured for lithic raw material, and there is no clear evidence of raw material exploited from the primary sources. Exploitation of an outcrop near the top of Shirataki-Akaishiyama began during the period when microblade industries dominated and also when stemmed points first appeared in the archaeological record. Obsidian and/or artefacts made from this source were transported to Sakhalin Island. In addition there was interregional

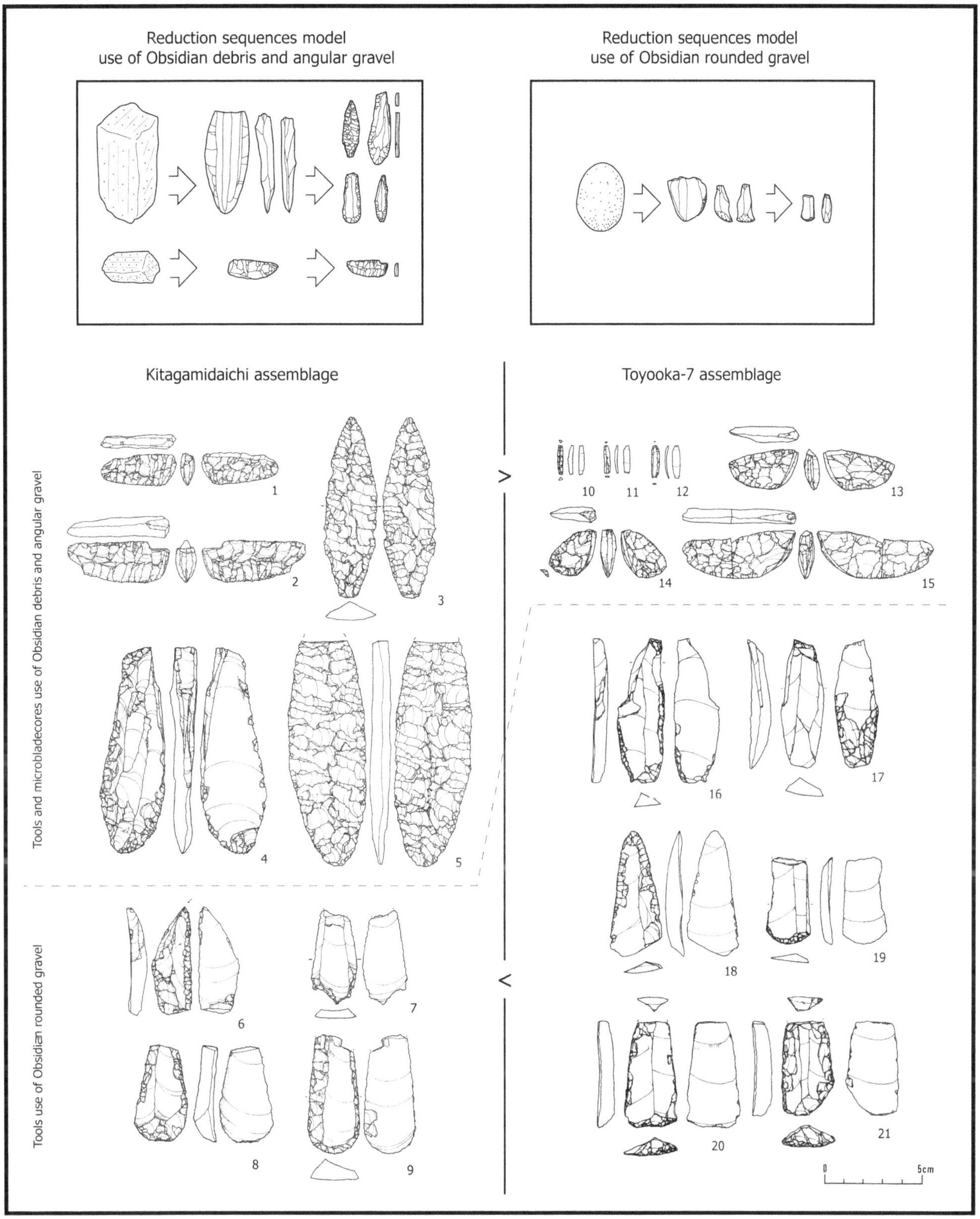

Figure 2.7. Blade reduction sequences in Oshorokko type microblade assemblages. 1, 2, 13–15 – Oshorokko type microblade cores; 3 – stemmed point; 4 – Hirosato type microblade core; 5 – bifacial leaf point; 6, 16–18 – burins; 7 – drill; 8–9, 19–21 – end scrapers; 10–12 – microblades; 18 – side scraper (after Izuho and Sato 2007, modified)

exchange of obsidian from Shirataki and fine-grained siliceous shale from southern Hokkaido between sites located more than 200km away from the Shirataki source. Further, during the period when Sakkotsu microblade cores were dominant, there seems to have been a division in behaviour that correlates with differences in altitude (Kimura 1995, 1998). Quarries were located at different elevations (higher than 800m above sea level, hereafter – a.s.l.), in intermediate areas (ca. 600m a.s.l.), and below 400m a.s.l. Different stages in the production processes, from raw material procurement to tool supply, occurred in each of these localities.

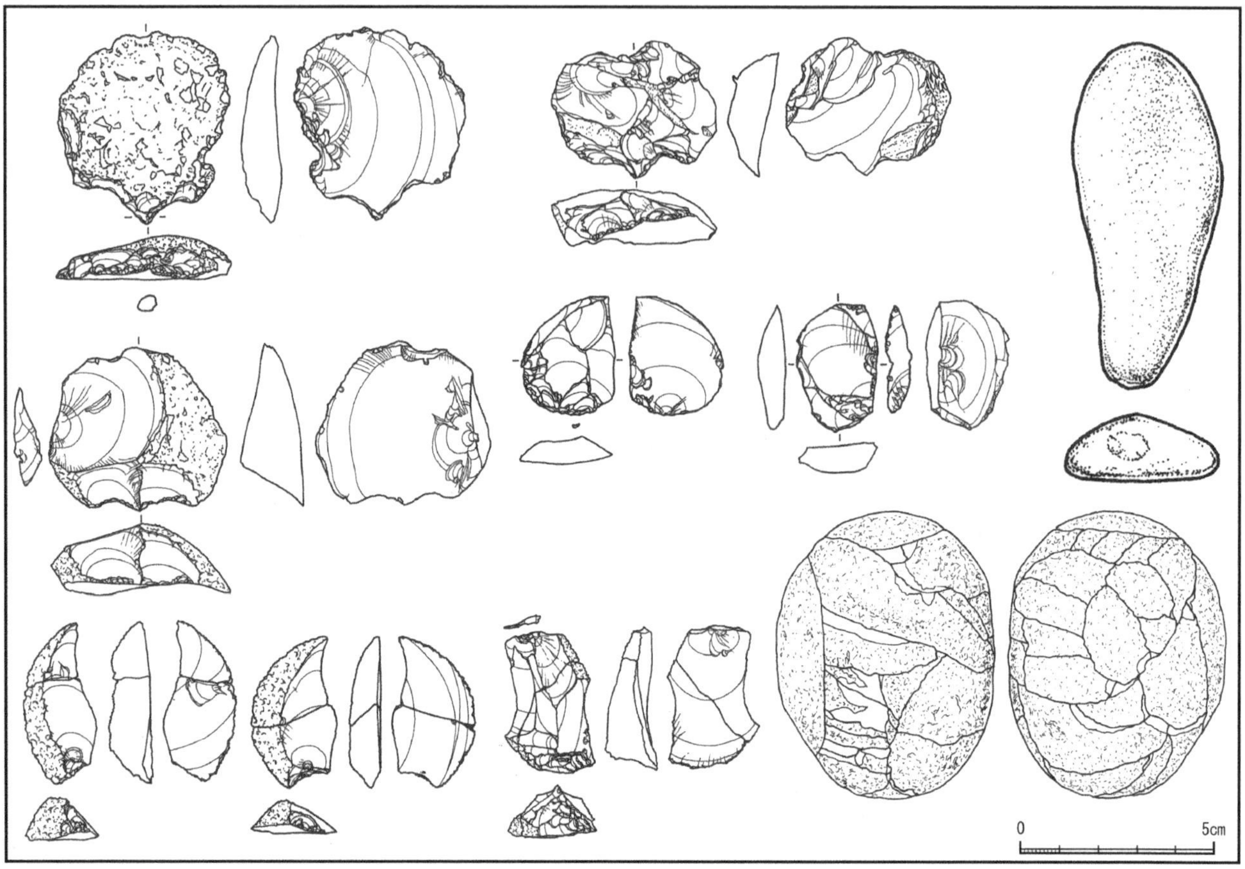

Figure 2.8. Chipped stone tools made using obsidian rounded gravel from the early Upper Palaeolithic site of Wakabano Mori (after Obihiro-shi Kyoiku Iinkai 2004, modified)

Future Directions of Archaeological Obsidian Studies on Hokkaido

As we presented in previous sections, systematic archaeological obsidian studies have been undertaken on Hokkaido during the past decade. Major results from these studies may be summarised as follows. There are 21 localities of 'geological' obsidian currently known on Hokkaido. The major sources, such as Shirataki, Oketo, Tokachi-Mitsumata, and Akaigawa, are characterised by geochemical data using INAA method (Kuzmin and Glascock 2007; Kuzmin *et al.* 2002); other sources have geochemical data derived from EPMA (Mukai 2005) and EDXRF methods. From an archaeological perspective, the long-distance movement of Oketo and Shirataki obsidian across Hokkaido and Sakhalin (distance more than 300km in straight line in the Upper Palaeolithic) has been demonstrated. In addition, several models regarding hunter-gatherer mobility patterns, lithic raw material exchange and processing, were proposed for Hokkaido.

Research Issues of Archaeological Obsidian Studies on Hokkaido

What is the ultimate objective of archaeological obsidian study on Hokkaido? Discussions regarding obsidian are still rare. Many of critical issues, on the other hand, are discussed in general archaeological literature and in sources related to the obsidian provenance studies in US Southwest (e.g., Glascock *et al.* 1998; Green 1998; Shackley 1998c, 2005). Although we do not consider them in detail here, many of these topics can be directly and/or indirectly applied to the obsidian studies on Hokkaido. As for Palaeolithic research on Hokkaido, we can highlight two main issues: (1) to demonstrate the dispersal and adaptation processes of anatomically modern humans in Northeast Asia and its regional variability; and (2) to provide a picture of the extended distribution networks in the Far East and to understand how such networks were formed, as well as the significance of obsidian exchange.

Models of behavioural strategy and exchange networks for Upper Palaeolithic hunter-gatherers on Hokkaido were recently put forward. The Upper Palaeolithic populations maintained well-adapted tool manufacturing system for various lithic raw materials that were induced by complicated tectonic activity on Hokkaido (Sato 2005). As it was previously stated, the use of obsidian as a high-quality raw material began during the early Upper Palaeolithic. Moreover, a variety of lithic reduction strategies and interassemblage variability were identified during the period of microblade assemblages (dated to ca. 20,000–12,000 BP), and this archaeological phenomenon implies indigenous regional pattern of Hokkaido (Nakazawa *et al.* 2005). We believe that the regional model of the indigenous pattern on Hokkaido might play important role for high resolution picture of adaptation process for anatomically

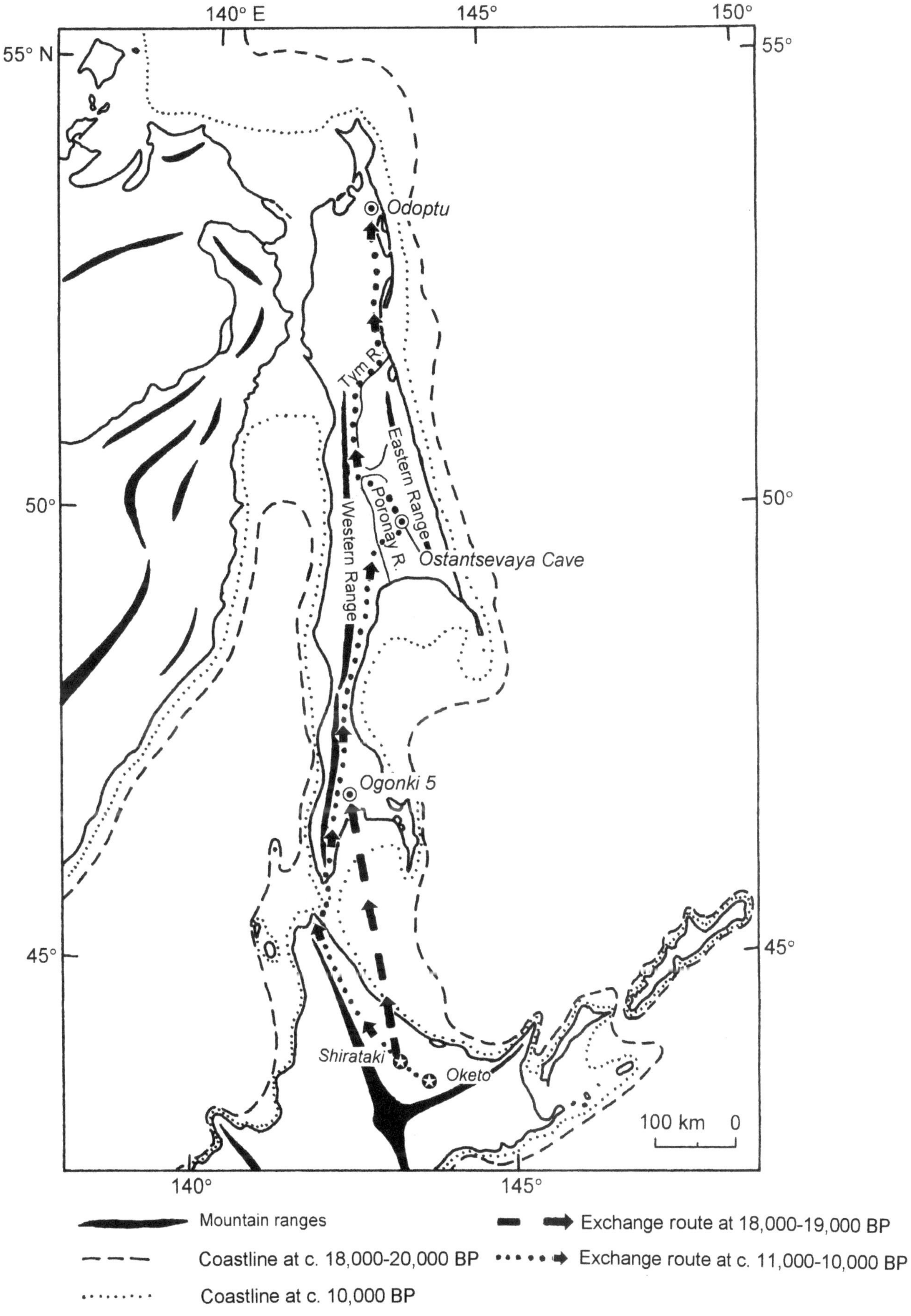

Figure 2.9. Probable obsidian exchange routes between Hokkaido and Sakhalin islands (after Kuzmin and Glascock 2007, modified)

modern humans and its regional variability in Northeast Asia represented by Goebel (1999, 2002).

On the other hand, the long-distance movement of Oketo and Shirataki obsidian across Hokkaido and Sakhalin (distance more than 300km), and obsidian exchange during the Upper Palaeolithic in the Japan Sea basin, including Hokkaido, Kyushu, and Korean Peninsula (Obata 2003), is so great that direct procurement by hunter-gatherer groups over a year, or even several years, is highly unlikely (Sato 2005; Sato *et al.* 2002). The processes by which such a distribution network was formed and the significance of exchange to the groups involved clearly require further investigation.

Objectives of Archaeological Obsidian Study on Hokkaido

Having briefly summarised two main issues of recent research, we can look to the future and highlight what will improve our understanding of obsidian procurement, stone tool technology, and exchange, in a systematic manner. Three fundamental research areas are presented: (1) in order to further plot the distribution of geological obsidian and provide the geological formation process, experimental petrology concerning geological formation of obsidian, as well as detailed geological description of obsidian rocks in primary distribution, is essential and should be positioned as urgent task; (2) in order to test the possibility of discrimination for all of primary obsidian sources on Hokkaido, accumulation of geochemical composition data by INAA method is necessary; and (3) to develop the non-destructive and low cost analytical routine with careful consideration on the geological formation process and variation of geochemical composition of obsidian on Hokkaido, more reliable study of source discrimination is needed.

So far, archaeological obsidian study on Hokkaido has a wide range of problems, such as lack of basic principles; disconnection between geologists and archaeologists; shortage of original data and analytical procedures; and irregularities in logic and interpretations. However, with solving these three subjects, we believe that archaeological obsidian investigations on Hokkaido could play important role for the prehistoric study in Northeast Asia.

Acknowledgements

We thank Dr Robin Torrence (Australian Museum, Sidney, Australia), Dr Steven M. Shackley (University of California, Berkeley, CA, USA), Dr Satoru Yamada (Kitami City Board of Education, Kitami, Japan), Dr Fumito Akai (Kagoshima City Board of Education, Kagoshima, Japan), Dr Hiroyuki Sato (University of Tokyo, Tokyo, Japan) and Dr Toshiaki Tsurumaru (Sapporo Gakuin University, Sapporo, Japan) for their help with both fieldwork and preparation of this paper. We specifically thank Dr Michael D. Glascock (University of Missouri, Columbia, MO, USA), Dr Yaroslav V. Kuzmin (Institute of Geology and Mineralogy, Siberian Branch of the Russian Academy of Sciences, Novosibirsk, Russia) for inviting us to Symposium "Crossing the Straits: Prehistoric Obsidian Source Exploitation in the Pacific Rim" held at the 70th Annual Meeting of the Society for American Archaeology (April 2005), and for help with editing of this paper along with Dr Robert J. Speakman (Smithsonian Institution, Washington, D.C., USA).

References

Bihoro-cho Kyoiku Iinkai. 2002. *Bihoro Cho Toyooka-7 Iseki* [Bihoro Toyooka-7 Site]. Bihoro, Japan, Bihoro-cho Kyoiku Iinkai.

DeMets, C. 1992. Oblique Convergence and Deformation along the Kuril and Japan Trenches. *Journal of Geophysical Research* 97, 17615–17625.

Ganzawa, Y. 1992. Neogene Stratigraphy and Paleogeography of the Oshima Peninsula, Southwest Hokkaido, Japan. *Memoirs of Geological Society of Japan* 37, 11–23.

Glascock, M. D., G. E. Braswell, and R. H. Cobean. 1998. A Systematic Approach to Obsidian Source Characterization. In *Archaeological Obsidian Studies: Method and Theory*, edited by M. S. Shackley, 15–65. New York and London, Plenum Press.

Goebel, T. 1999. Pleistocene Human Colonization of Siberia and Peopling of the Americas: An Ecological Approach. *Evolutionary Anthropology* 8, 208–227.

Goebel, T. 2002. The "Microblade Adaptation" and Recolonization of Siberia during the Late Upper Pleistocene. In *Thinking Small: Global Perspectives on Microlithization*, edited by R. G. Elston and S. L. Kuhn, 117–131. Arlington, VA, American Anthropological Association.

Green, R. C. 1998. A 1990s Perspective on Method and Theory in Archaeological Volcanic Glass Studies. In *Archaeological Obsidian Studies: Method and Theory*, edited by M. S. Shackley, 223–235. New York and London, Plenum Press.

Hall, M., and H. Kimura. 2002. Quantitative EDXRF Studies of Obsidian Sources in Northern Hokkaido. *Journal of Archaeological Science* 29, 259–266.

Hata, M., H. Segawa, and J. Yajima. 1982. *Okushiri* [Explanatory Text of the Geological Map of Japan (Scale 1:50,000) "Okushiri"]. Tokyo, Chitshitsu Chosa Jyo.

Hirose, W. 1999a. Hokkaido Chuobu Tobu niokeru Shin Daisanki Kazanganrui no Zengan Kagaku Sosei [Geochemistry of Neogene Volcanick Rocks in Central and Eastern Hokkaido]. *Chika Shigen Chosa Jo Hokoku* 70, 75–96.

Hirose, W. 1999b. Hokkaido Chuo kara Tobu niokeru Shin Daisanki Kazan Katsudo no Jiku Bunpu [Spatial and Temporal Distribution of Neogene Volcanism in Central and Eastern Hokkaido]. *Chika Shigen Chosa Jo Hokoku* 70, 97–111.

Hirose, W., and M. Nakagawa. 1999. Hokkaido Chuobu kara Tobu no Shin Daisanki Kazan Katsudo: Kazangakuteki Data karamita Toko Kazan Katsudo no Seritsu to Hensen [Neogene Volcanism in Central-Eastern Hokkaido: Beginning and Evolution of Arc Volcanism Inferred from Volcanological Parameters and Geochemistry]. *Chishitsugaku Zasshi* 105, 247–265.

Inoue, I. 2003. Shirataki Dai 30 Chiten Shutsudo Kokuyoseki no Kagaku Bunseki [Geochemical Analysis of Obsidian from Shirataki No. 30 Site]. In *Shirataki Dai 30 Chiten Iseki*, 245–258. Shirataki, Japan, Shirataki-mura Kyoiku Iinkai.

Izuho, M. 1997. Tokorogawa Ryuiki niokeru Sekkisekizai no Kiso Kenkyu [A Fundamental Study of the Lithic Raw Material in the Tokoro River Basin]. *Hokkaido Kyusekki Bunka Kenkyu* 2, 1–14.

Izuho, M., and F. Akai. 2005. Hokkaido no Kyusekki Hennen: Iseki Keisei Kateiron no Tekiyo [Geochronology of Palaeolithic Sites on Hokkaido]. *Kyusekki Kenkyu* 1, 39–55.

Izuho, M., and H. Sato. 2007. Archaeological Obsidian Study in Hokkaido (Japan): Retrospect and Prospects. *Bulletin of the Indo-Pacific Prehistory Association* 27, 114–121.

Kaneoka, I., Y. Takigami, N. Takaoka, S. Yamashita, and K. Tamaki. 1992. ^{40}Ar-^{39}Ar Analysis of Volcanic Rocks Recovered from the Japan Sea Floor: Constraints on the Age of Formation of the Japan Sea. In *Proceedings of the Ocean Drilling Program. Volumes 127/128, 2. Scientific Results, Part 2. Leg 127, Sites 794–797, Japan Sea; Leg 128, Sites 797, 798, 799, Japan Sea,* 819–836. College Station, TX, ODP Publication.

Kawano, Y. 1950. *Honpo San Harisitsu Ganseki no Kenkyu* [Natural Glasses in Japan] (Chishitsu Chosajyo Hokoku 134). Tokyo, Chishitsu Chosajo.

Kim, J. C., D. K. Kim, M. Youn, C. C. Yun, G. Park, H. J. Woo, M.-Y. Hong, and G. K. Lee. 2007. PIXE Provenancing of Obsidian Artefacts from Palaeolithic Sites in Korea. *Bulletin of the Indo-Pacific Prehistory Association* 27, 122–128.

Kimura, G. 1986. Oblique Subduction and Collision: Forearc Tectonics of the Kuril Arc. *Geology* 14, 404–407.

Kimura, G., S. Miyashita, and S. Miyasaka. 1983. Collision Tectonics on Hokkaido and Sakhalin. In *Accretion Tectonics in the Circum-Pacific Regions*, edited by M. Hashimoto and S. Ueda, 123–134. Tokyo, Terra Scientific Publishing.

Kimura, G., and K. Takami. 1985. Tectonic Framework of the Kuril Arc since Its Initiation. In *Advances in Earth and Planetary Sciences. Formation of Active Ocean Margins*, edited by N. Nasu, 641–676. Tokyo, Terra Scientific Publishing.

Kimura, H. 1978. Yoichigawa Akaigawa Ryuiki no Sendoki Sekkigun nitsuite [On the Pre-Ceramic Industry in the Yoichi River and Akaigawa River Basins]. *Hokkaido Kokogaku* 14, 23–48.

Kimura, H. 1992. *Reexamination of the Yubetsu Technique and Study of the Horokazawa Toma Lithic Culture.* Sapporo, Sapporo Daigaku Maizo Bunkazai Tenjishitsu.

Kimura, H. 1995. Kokuyouseki, Hito, Gijutsu [Obsidian, Humans, and Technology]. *Hokkaido Kokogaku* 31, 3–63.

Kimura, H. 1998. Obsidian, Humans, Technology. In *Paleoekologiya Pleistotseha i Kultury Kamennogo Veka Severnoi Azii i Sopredelnyikh Territorii. Tom 2*, edited by A. P. Derevianko, 302–314. Novosibirsk, Izdatelstvo Instituta Arkheologii i Etnografii Sibirskogo Otdeleniya Rossiiskoi Akademii Nauk.

Kimura, H. 2002. Hokkaido Chiiki ni okeru Kokuyoseki Kenkyuu no Tenbou; Tokuni Shirataki Kokuyouseki Gensanchi wo Chuusin ni [A Prospect of the Obsidian Study on Hokkaido District]. *Kokuyouseki Bunka Kenkyu* 1, 69–82.

Kimura, H. 2005. *Kita no Kokuyoseki no Michi: Shirataki Isekigun* [The Obsidian Road of the North: Shirataki Site Cluster]. Tokyo, Shinsensha Publishers.

Kitami-shi Kyoiku Iinkai. 1988. *Kitagami Daichi Iseki II* [Kitagami Daichi Site II]. Kitami, Japan, Kitami-shi Kyoiku Iinkai.

Kitazawa, M. 1999. Kamishihoro Cho 13 no Sawa Iseki no Bunpu Chosa. [General Survey at Sawa Site 13, Kamishihoro Town]. *Hokkaido Kyusekki Bunka Kenkyu* 4, 19–24.

Konoya, M., K. Hasegawa, and K. Matsui. 1964. *Shirataki* [Explanatory Text of the Geological Map of Japan (Scale 1:50,000) "Shirataki"]. Sapporo, Hokkaido Kaihatsu Cho.

Koshimizu, S. 1981. Ishikari Teichitai ni Shutsudo suru Kokuyosekihen no Gensanchi [Source Areas of Obsidian Found in the Prehistoric Sites in the Ishikari-Tomakomai Lowland Area]. *Chikyu Kagaku* 35, 267–273.

Koshimizu, S., and C. Kim. 1986. Hokkaido Chu kara Tobu Chiiki no Shinseikai no Fission-Track Nendai (Sono 1): Kamishiyubetsu, Kitamifuji Chiki [Fission-Track Dating of the Cenozoic Formations in Central-Eastern Hokkaido, Japan: Kamishiyuubetsu and Kitamifuji District]. *Chishitsugaku Zasshi* 92, 477–487.

Kuzmin, Y. V., and M. D. Glascock. 2007. Two Islands in the Ocean: Prehistoric Obsidian Exchange between Sakhalin and Hokkaido, Northeast Asia. *Journal of Island and Coastal Archaeology* 2, 99–120.

Kuzmin, Y. V., M. D. Glascock, and H. Sato. 2002. Sources of Archaeological Obsidian on Sakhalin Island (Russian Far East). *Journal of Archaeological Science* 29, 741–749.

Kuzmin, Y. V., and V. K. Popov (eds). 2000. *Vulkanicheskie Stekla Dalnego Vostoka Rossii: Geologicheskie i Arkheologicheskie Aspekty* [Volcanic Glasses of the Russian Far East: Geological and Archaeological Aspects]. Vladivostok, Dalnevostochny Geologichesky Institut Dalnevostochnogo Otdeleniya Rossiiskoi Akademii Nauk.

Matsui, K. 1959. Shirataki Mura no Keisan Hakudo Oyobi Kokuyoseki [Obsidian and Diatomaceous Earth in Shirataki Village]. *Chikashigen Chosajyo Hokoku* 21, 51–56.

Mukai, M. 2005. Monbetsu Chiiki, Rubeshibe Chiiki, Toyoura Chiiki kara Sanshutsu suru Kokuyoseki Glass no Kagaku Sosei [Chemical Compositions of the Obsidian Glasses from the Monbetsu, Rubeshibe, and the Toyoura Areas]. *Asahikawa Shi Hakubutsukan Kenkyu Hokoku* 11, 9–20.

Mukai, M., H. Hasegawa, and K. Wada. 2000. Asahikawa Shuhen Chiiki niokeru Kokuyoseki Glass no Kagaku Sosei: Kokuyoseki no Sanchi Tokutei heno Tekiyo [Chemical Compositions of the Obsidian Glasses around Asahikawa District]. *Asahikawa Shi Hakubutsukan Kenkyu Hokoku* 6, 51–64.

Mukai, M., R. Shibuya, and K. Wada. 2004. Ikutahara Chiiki kara Sanshutsu suru Kokuyoseki Glass no Kagaku Sosei [Chemical Compositions of the Obsidian Glasses from Ikutahara District]. *Asahikawa Shi Hakubutsukan Kenkyu Hokoku* 10, 35–40.

Mukai, M., and K. Wada. 2001. Asahikawa Seiho, Chippubetsu–Hokuryu Chiiki kara Sanshutsu suru Kokuyoseki Glass no Kagaku Sosei [Chemical Compositions of the Obsidian Glasses from Chippubetsu and Hokuryu District, Western Area of Asahikawa]. *Asahikawa Shi Hakubutsukan Kenkyu Hokoku* 7, 23–30.

Mukai, M., and K. Wada. 2003. Engaru Chiiki - Oumu Chiiki kara Sanshutsu suru Kokuyoseki Glass no Kagaku Sosei [Chemical Compositions of the Obsidian Glasses from Engaru and Oumu Districts]. *Asahikawa Shi Hakubutsukan Kenkyu Hokoku* 9, 19–26.

Mukai, M., and K. Wada. 2004a. Okushiri To kara Sanshutsu suru Kokuyoseki Glass no Kagaku Sosei [Chemical Compositions of the Obsidian Glasses from Okushiri Island]. *Asahikawa Shi Hakubutsukan Kenkyu Hokoku* 10, 41–46.

Mukai, M., and K. Wada. 2004b. Tokachi Chiho kara Sanshutsu suru Kokuyoseki Glass no Kagaku Sosei [Chemical Compositions of the Obsidian Glasses from Tokachi District]. *Asahikawa Shi Hakubutsukan Kenkyu Hokoku* 10, 47–56.

Mukai, M., K. Wada, and C. Okura. 2002. Oketo Chiiki – Akaigawa Chiiki kara Sanshutsu suru Kokuyoseki Glass no Kagaku Sosei [Chemical Compositions of the Obsidian Glasses from Oketo and Akaigawa Districts]. *Asahikawa Shi Hakubutsukan Kenkyu Hokoku* 8, 47–58.

Nakazawa, Y., M. Izuho, J. Takakura, and S. Yamada. 2005. Toward an Understanding of Technological Variability in Microblade Assemblages in Hokkaido, Japan. *Asian Perspectives* 44, 276–292.

Naoe, Y., and J. Nagasaki. 2005. Hokkaido Koki Kyusekki Jidai Zenhan Ki no Sekizai Shohi Senryaku; Shirataki Ia Gun to Shukubai Iseki Sankakuyama Chiten [Raw Material Consumption Strategy at Late Upper Palaeolithic on Hokkaido]. *Hokkaido Kyusekki Bunka Kenkyu* 10, 45–58.

National Astronomical Observatory (ed.). 2005. *Rika Nenpyo Heisei 18 Nen* [Chronological Scientific Tables for 2006]. Tokyo, Maruzen Publishers.

Nayoro-shi Kyoiku Iinkai (eds). 1988. *Nayoro Shi Nisshin 2 Iseki, Nisshin 31 Iseki* [Nisshin 2 and 31 Sites in Nayoro City]. Nayoro, Japan, Nayoro-shi Kyoiku Iinkai.

NEDO (New Energy and Industrial Technology Development Organization). 1994. *Report on the Data Processing for Exploration Survey for Geothermal Energy in the Okushiri Area during the Fiscal Year Heisei 4*. Tokyo, NEDO.

Obata, H. 2003. Kyoukutoni Okeru Kokuyouseki Shutsudoisekito Gensannchi Kenkyu [Study on the Prehistoric Obsidian Utilisation in Far East Asia – Review and Perspective]. *Sekki Gensanchi* 2, 67–88.

Obihiro-shi Kyoiku Iinkai (eds). 2004. *Obihiro, Wakabano Mori Iseki* [Wakabano Mori Site, Obihiro]. Obihiro, Japan, Obihiro-shi Kyoiku Iinkai.

Okazaki, Y. 1966. *Kushiro no Chishitsu* [Geology of Kushiro]. Kushiro, Japan, Kushiro-shi Kyoiku Iinkai.

Sato, H. 2004a. Russia Kyokuto Niokeru Senshi Jidai no Kokuyoseki no Riyo [Prehistoric Obsidian Exploitation in the Russian Far East]. *Kokuyouseki Bunka Kenkyu* 3, 45–55.

Sato, H. 2004b. Lithic Procurement and Reduction Strategy of Hirosato Industry in the Japan Sea Rim Area. *Seonsa Wa Kodae* 20, 205–221.

Sato, H. 2005. Hokkaido Kyusekki Bunka wo Fukan suru; Hokkaido to Sono Shuhen [Overlooking for Palaeolithic Culture on Hokkaido]. *Hokkaido Kyusekki Bunka Kenkyu* 10, 137–146.

Sato, H., Y. V. Kuzmin, and M. D Glascock. 2002. Sakhalin Tou Shutsudo no Senshi Jidai Kokuyosekisei Sekki no Gensanchi Bunseki to Kokuyoseki no Ryutsu [Source Analysis of Obsidian on Prehistoric Sakhalin and an Assessment of its Distribution in Northeast Asia]. *Hokkaido Kokogaku* 38, 1–13.

Sawa, S. 1978. *Kushiro Shi Higashi Kushiro Dai 4 Iseki Hakkutsu Hokoku* [The Excavation Report of Higashi Kushiro No. 4 Site in Kushiro City]. Kushiro, Japan, Kushiro-shi Kyoiku Iinkai.

Sawamura, K., and M. Hata. 1965. *Rubeshibe* [Explanatory Text of the Geological Map of Japan (Scale 1:50,000) "Rubeshibe"]. Sapporo, Hokkaido Kaihatsu Cho.

Shackley, M. S. 1998a. Current Issues and Future Directions in Archaeological Volcanic Glass Studies: An Introduction. In *Archaeological Obsidian Studies: Method and Theory*, edited by M. S. Shackley, 1–14. New York and London, Plenum Press.

Shackley, M. S. 1998b. Intrasource Chemical Variability and Secondary Depositional Processes: Lessons from the American Southwest. In *Archaeological Obsidian Studies: Method and Theory*, edited by M. S. Shackley, 83–102. New York and London, Plenum Press.

Shackley, M. S. (ed.) 1998c. *Archaeological Obsidian Studies: Method and Theory*. New York and London, Plenum Press.

Shackley, M. S. 2005. *Obsidian: Geology and Archaeology in the North American Southwest*. Tucson, University of Arizona Press.

Shirataki Dantai Kenkyu Kai (eds). 1963. *Shirataki Iseki no Kenkyu* [The Study of Shirataki Archaeological Sites]. Sapporo, Shirataki Dantai Kenkyu Kai.

Suzuki, M. 1973. Chronology of Prehistoric Human Activity in Kanto, Japan. Part 1. Framework for Reconstruction Prehistoric Human Activity in Obsidian. *Journal of the Faculty of Science, University of Tokyo. Section 5 (Anthropology)* 4(3), 241–318.

Tajika, J., and M. Yahata. 1991. *Engaru* [Explanatory Text of the Geological Map of Japan (Scale 1:50,000) "Engaru"]. Sapporo, Hokkaidoritsu Chikashigen Chosajo.

Tomoda, T. 1996. Kogata Genseki Sanshutsuchi niokeru Sekizai no Katsuyo ni Tsuite [Bipolar Flaking at a Source of Small Obsidian Pebbles]. *Hokkaido Kokogaku* 32, 63–74.

Tsurumaru, T. 2001. Hokkaido Kyusekki Kokogaku no Ronten: Kon-nichiteki Shoten to Tenbo [Contemporary Issues and Perspectives of Palaeolithic Archaeology on Hokkaido]. *Hokkaido Kokogaku* 37, 3–22.

Warashina, T. 1999. Sekki oyobi Tama Rui Genzairyo no

Sanchi Bunseki [Source Identification of Raw Material for Stone Tools and Beads]. In *Kokogaku to Nendai Sokuteigaku, Chikyu Kagaku*, edited by S. Matsu'ura, Y. Uesugi and T. Warashina, 259–293. Tokyo, Doseisha Publishers.

Warashina, T. 2004. Kyu Shirataki, Shimo Shirataki iseki Shutsudo no Kokuyousekisei-sekki no Genzai Sanchi Bunseki, Suiwasou Sokutei [Source Identification Analysis and Hydration Dating for Obsidian Lithic Artefacts Recovered from Kyu Shirataki and Shimo Shirataki Sites]. In *Shirataki Isekigun 5*, edited by Hokkaido Maizo Bunkazai Centre, 188–205. Sapporo, Hokkaido Maizo Bunkazai Centre.

Yahata, M. 1999. Abashiri – Notoro Misaki – Ponmoi Misaki [From Abashiri via Cape Notoro to Cape Ponmoi]. In *Doto no Shizen wo Aruku*, edited by Doto no Shizenshi Kenkyu Kai, 184–189. Sapporo, Hokkaido Daigaku.

Yahata, M., and H. Nishido. 1995. Chuo Hokkaido Hokutobu, Monbetsu–Engaru Chiki no Shin Daisanki Kazan Katsudo to Kozo Undo [The Neogene Volcanism and Tectonics in the Monbetsu–Engaru District in Northeastern Part of Central Hokkaido]. *Chishitsugaku Zasshi* 101, 685–704.

Yahata, M., J. Tajika, K. Kurosawa, and T. Matsunami. 1988. *Maruseppu Hokubu* [Explanatory Text of the Geological Map of Japan (Scale 1:50,000) "Maruseppu Hokubu"]. Sapporo, Hokkaidoritsu Chikashigen Chosajo.

Yokoyama, H., M. Yahata, S. Okamura, and H. Nishido. 2003. Volcano-Stratigraphy and Geological Development of Akaigawa Caldera in Southwest Hokkaido, Japan. *Japanese Journal of Mineralogical and Petrological Sciences* 32, 80–95.

Yoshitani, A., M. Kawabe, O. Suda, I. Mizufune, and K. Nagahata. 2001. Oribe 17 Iseki kara Shutsudo shita Kokuyogan no Gensanchi Nituite; Kokuyogan no Biryo Genso Sosei Karano [Approach on Resource of Obsidian from Oribe 17 Site]. *Kamishihoro Cho Higashi Taisetsu Hakubutukan Kenkyu Hokoku* 23, 1–19.

Chapter 3
Prehistoric Procurement of Obsidian from Sources on Honshu Island (Japan)

Takashi Tsutsumi

Abstract: *Based on the study of more than 5000 specimens, the exploitation of obsidian on Honshu Island during the Upper Palaeolithic and Jomon periods is discussed. X-ray Fluorescence analysis was used to identify the sources of volcanic glass in archaeological assemblages. The obsidian sources of Shinshu, Hakone, Amagi, Takaharayama, and Kozu-shima in central Honshu Island have been exploited since the beginning of the Upper Palaeolithic (ca. 35,000 BP). Obsidian transport from the Kozu-shima Island to the main Honshu Island indicates seafaring abilities. The obsidian supply zone in central Honshu during the Upper Palaeolithic was generally confined to the area within ca. 200km from the sources, suggesting a certain level of economic interaction. Obsidian transport on distances greater than 500km assumes the existence of large exchange networks. The change from collecting to quarrying during the transition from the Upper Palaeolithic to the Jomon is an important step forward in obsidian procurement. Many quarries in the Wada—Suwa area are dated to the Early Jomon and later. Mining was probably undertaken by local groups. Possibly, obsidian was stored at the main site and later distributed to satellite settlements. The range of obsidian distribution in the Jomon period was about 200km, similar to that during the Upper Palaeolithic.*

Keywords: *Obsidian, Sources, Utilisation, Upper Palaeolithic, Jomon, Honshu Island, Japan*

Introduction

More than 50 obsidian sources have been identified in the Japanese Archipelago, from Hokkaido Island in the north to Kyushu Island in the south, all located within the volcanic belt of the Pacific Rim (Figure 3.1). Shirataki on Hokkaido, and Himeshima and Koshidake on Kyushu are the primary sources which were exploited in prehistory. No obsidian sources are found on Shikoku Island. Honshu Island is located between Hokkaido and Kyushu, and its major sources of obsidian in the eastern part include (from north to south) Kizukuri, Fukaura, Oga, Kitakamigawa, Haguro, Takaharayama, Wada-toge, Suwa, Tateshina, Hakone, Amagi, and Kozu-shima [Kozu-jima] (Figure 3.1). The single obsidian source in the western part of Honshu is the Okino-shima. This study discriminates between two kinds of obsidian sources: (1) primary sources, i.e. outcrops in the form of cliffs or other exposures which consist of pure volcanic glass; and (2) secondary sources which are redeposited pieces of obsidian (e.g., in river gravel).

The study of sources for archaeological obsidian in Japan based on chemical composition began with Suzuki's (1973) pioneering work, followed by Higashimura and Warashina (1988), Osawa and Ninomiya (1991), and others. Today, this is an important research subject in Japanese prehistory. This author has completed a ten-year study working with analytical chemist Prof. Akihiko Mochizuki to research Upper Palaeolithic obsidian exploitation in the Chubu and Kanto regions of Honshu Island using the non-destructive X-ray Fluorescence (hereafter – XRF) method. Over 5000 Upper Palaeolithic specimens were analysed. This chapter discusses the issues of procurement and supply of obsidian on Honshu during the Upper Palaeolithic and Jomon periods with references to data published by others.

The main topics of archaeological obsidian studies in Japan today include: (1) the distribution of obsidian artefacts; (2) procurement methods; (3) the nature of archaeological sites located at the sources; (4) supply systems and supply zones [the latter term means the areas where obsidian taken from a particular source have been identified – *Editors*]; (5) temporal changes in obsidian exploitation; and (6) prehistoric social systems and economies of communities which exploited obsidian.

The key tool types of the Japanese Upper Palaeolithic are trapezoids, backed points, points, and microblades. The correlation of stone industries to typological phases is as follows: (1) Phase I: Trapezoid industry, including edge-polished axes associated with circular settlements; (2) Phase II: Backed point industry and the establishment of blade technology; (3) Phase III: Appearance of regionalism represented by regional backed point industries; (4) Phase IV: Size reduction of backed points in the Kanto and Chubu regions and development of the point industry; and (5) Phase V: Microblade industry in the entire Japanese Islands.

Obsidian Procurement in the Upper Palaeolithic of Honshu Island

Trends in Obsidian Exploitation

Obsidian was the preferred raw material for producing tools such as trapezoids, backed points, points, and microblades. It was also used to manufacture scrapers. In this section, the current state of the sources, their exploitation, and the supply area are discussed. Different researchers use different names for the same obsidian source. Therefore, references to published data require some caution.

The Kizukuri, Fukaura, and Oga sources are located in Aomori and Akita prefectures in the northernmost part of Honshu (Figures 3.1–3.2). However, as the area also

Figure 3.1. Major obsidian sources in Japan

yields good quality siliceous shale, the degree of obsidian exploitation was not high. Unfortunately, the amount of data for these sources is sparse compared to the Chubu and Kanto regions of central Honshu because very few source studies have been conducted.

The Kizukuri source is situated on the Sea of Japan coast in Aomori Prefecture. It is also called Dekijima (Sato and Tsutsumi 2007, 74). The obsidian outcrop has not been found yet; only pebbles can be observed on the beach. It is not known if the source was exploited by humans who lived in the vicinity, as no chemical analysis was done for the local sites. Figure 3.2 shows obsidian transport to the distant sites from this source. Three pieces from the Ohdaira B site in the Nojiri-ko [Lake Nojiri] cluster (Nagano Prefecture) 500km away have been identified as coming from this source (Mochizuki 2002a); but their exact age is unknown.

The Fukaura source is also located on the Sea of Japan's coast in Aomori Prefecture. The exact position of the obsidian-containing outcrop is unknown; and only pebbles were found on the coast. No research in terms of sourcing

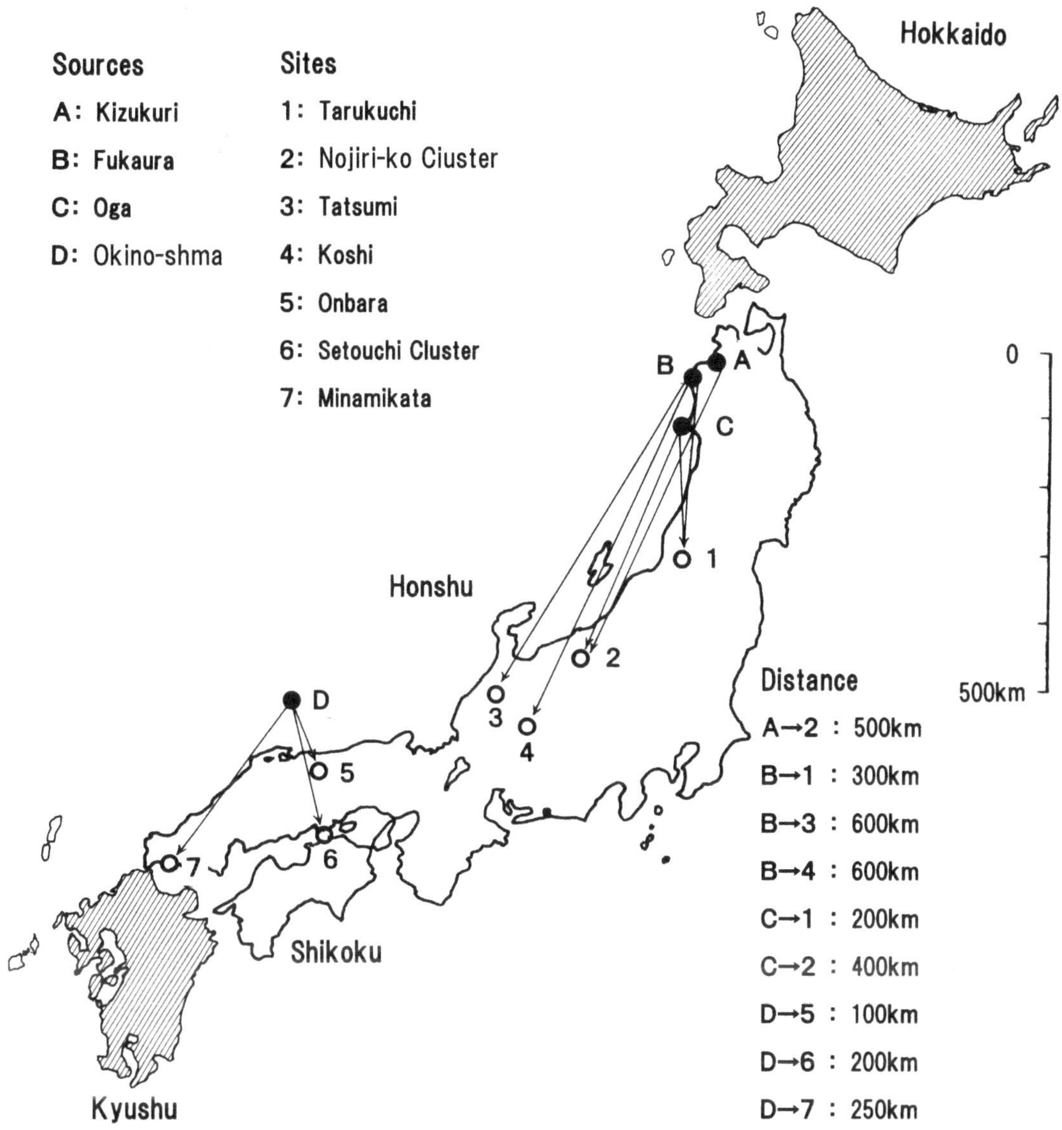

Figure 3.2. Long-distance obsidian transport in the Upper Palaeolithic of Honshu Island

has been done for the local archaeological sites. As for the distant sites, obsidian was used to make Phase IV points at the Tatsumi site in Toyama Prefecture, 600km away (Figure 3.2; Higashimura and Warashina 1988). This source was also used during Phase IV at the Tarukuchi site in Niigata Prefecture (Warashina and Higashimura 1996) (Figure 3.3, left). An artefact from the Hinatabayashi A site in the Nojiri-ko cluster, which is 460km away, has been also identified as being from this source (Mochizuki 2002a) (Figure 3.2). However, the exact age of this site is not clear.

The Oga source is situated along the Sea of Japan's coast in the Akita Prefecture (Figure 3.2). No quarry site has been found; only pebbles have been discovered on the beach. The use of Oga obsidian at the Kazanashidai 2 site and some other sites within a 50km range from the source has been confirmed by chemical analysis (Higashimura and Warashina 1988). As for more distant sites, Oga was the source for 17 artefacts belonging to the microblade industry (Phase V; see Figure 3.3, right) at the Tarukuchi site (Warashina and Higashimura 1996). In addition, three obsidian artefacts from the Hinatabayashi A site, which is 400km away, are from this source. These specimens can be attributed to Phase I, indicating obsidian exploitation since the beginning of the Upper Palaeolithic.

Details about the exploitation of obsidian from the Takaharayama source group were revealed by recent surveys (Tamura *et al.* 2006; Yaita City Board of Education 2006). The primary outcrops are located in the mountains at an elevation of 1300–1400m above sea level (hereafter – a.s.l.). According to Mochizuki (2002b), the obsidian from these sources is sub-divided into the Amayuzawa and the Nanahirozawa groups in terms of chemical composition; and the former one was exploited for stone tool production. About six Upper Palaeolithic sites were identified in the Takaharayama area in 2005 (Tamura *et al.* 2006). Trapezoids of Phase I, *kakusui-jo sekki* (unifacially pointed tools with a triangular or trapezoidal cross-section) of Phase III, and microblade cores of Phase V are among the lithic artefacts reported from these sites. Their presence indicates that obsidian tools were manufactured throughout

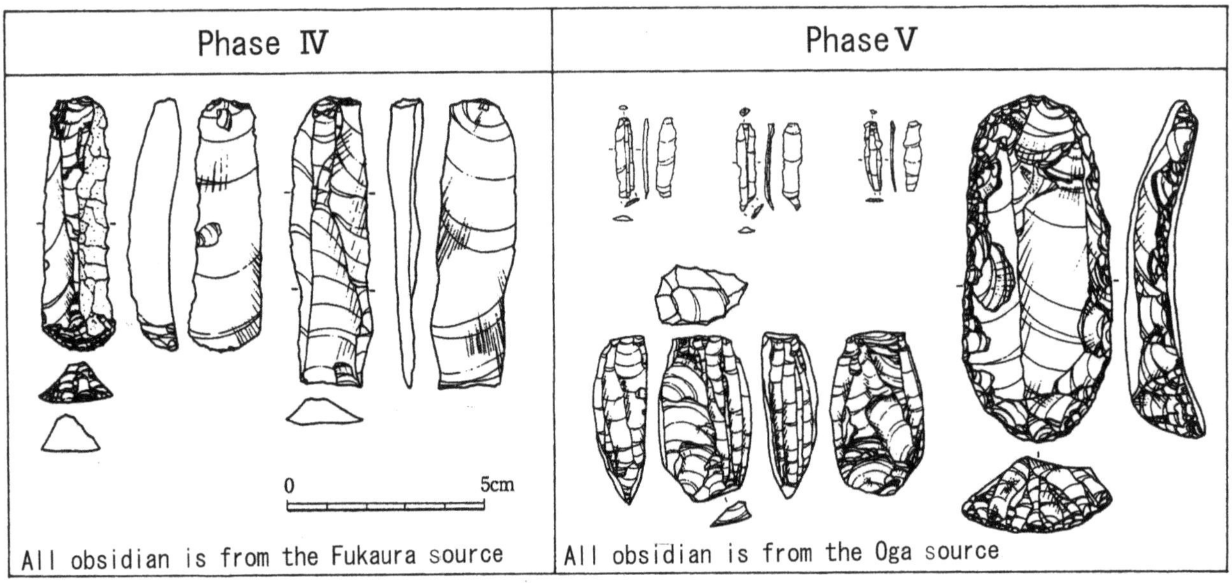

Figure 3.3. Artefacts made of the Fukaura and Oga obsidian

the Upper Palaeolithic. Exploitation of the Takaharayama source was confirmed by geochemical analysis of artefacts from phases I and IV at the Kambayashi site about 70km away (Figure 3.4, No. 3).

On Figure 3.4, the distribution of archaeological sites with Takaharayama obsidian artefacts is shown (Serizawa 2006). They are located primarily south of the source and within a 150km radius in the Kanto Plain; the Shimousa, Omiya, and Musashino uplands hold the majority of sites. Sites are distributed from the Takaharayama source down to the Kanto Plain along rivers which represent a lithic raw material transport route called the Shimotsuke–Hokusou corridor (Tamura *et al.* 2006). However, the number of Takaharayama obsidian artefacts drops off sharply in the Sagamino Upland south of the Musashino Upland. For example, in cultural horizon 9 of the Kashiwagaya-nagaosa site in the Sagamino Upland (Figure 3.4, No. 4) only eight pieces from Takaharayama were detected among 1120 obsidian specimens. To the north and west of this source, there are very few sites with Takaharayama obsidian (Figure 3.4). In the north, the Ohu Mountain range may have blocked the transport; and in the west there was not much demand for raw material because of the presence of the local Wada-toge and Suwa obsidian sources as well as the siliceous shale sources in the northern part of the region.

The Wada-toge – Suwa source group is located in the central highlands of Honshu (Nagano Prefecture). It includes the Kirigamine (elevation of 1500m a.s.l.), Hoshigatou (1500m), Wada-toge [Wada Pass] (1500m), Hoshikuso-toge (1500m), Omegura (1200m), Fuyo-lite (1200m), and Furu-toge (1500m) locales; all are within a 10km radius of the Wada Pass. They are also known as the Shinshu source group which consists of the Wada-toge – Suwa (hereafter Wada – Suwa) and Tateshina source groups, situated only 20km apart. On the basis of chemical composition, Mochizuki (2002b) sub-divided the obsidian from the Wada – Suwa source group into three types: 1) Wada-toge D; 2) Wada-toge O; and 3) Suwa. They are in turn further divided into 11 groups. The volcanic glass in this region was created by activities of the Kirigamine volcanoes, and primary obsidian outcrops are visible. A large quantity of debris is also found. The Wada – Suwa group was the most heavily used obsidian source on the central Honshu Island. Exploitation of the Wada – Suwa sources by prehistoric people is already known at the Hinatabayashi B site in the Nojiri-ko cluster (^{14}C-dated to ca. 27,900–31,400 BP; Figure 3.5, A, No. 2) and at other sites like Doteue (Figure 3.5, A, No. 4), all dated to the early Upper Palaeolithic (Phase I). The sources were continuously utilised afterwards, throughout phases II to V.

On Figure 3.5 (A), the distribution of obsidian from the Wada – Suwa source group in the Upper Palaeolithic is shown. The supply zone includes the Nobeyama Highland (Figure 3.5, A, letter "F") within a 50km radius; the Nojiri-ko cluster (Figure 3.5, A, letter "H") within a 100km radius; and the Kanto Plain. The latter includes the Sagamino Upland (Figure 3.5, A, letter "A"), Musashino Upland (Figure 3.5, A, letter "B"), Shimousa Upland (Figure 3.5, A, letter "D"), and the Hakone-Ashitaka flanks (Figure 3.5, A, letter "E"), located between 100 and 200km from sites. This demonstrates that the supply zone falls within a 200km radius. As for longer-distance transport, the Nijozan cluster in Nara Prefecture located about 300km from the source yielded three specimens of Wada – Suwa obsidian (Higashimura and Warashina 1988). Also, artefacts belonging to phases II and V at the Tarukuchi site 250km away are made of obsidian from these sources (Warashina and Higashimura 1996).

The Tateshina source group is located 20km from the Wada – Suwa source group, also in Nagano Prefecture. The primary sources are situated at an elevation of 1500–2000m a.s.l., including Tsumetayama (1800m), Mugikusa-toge

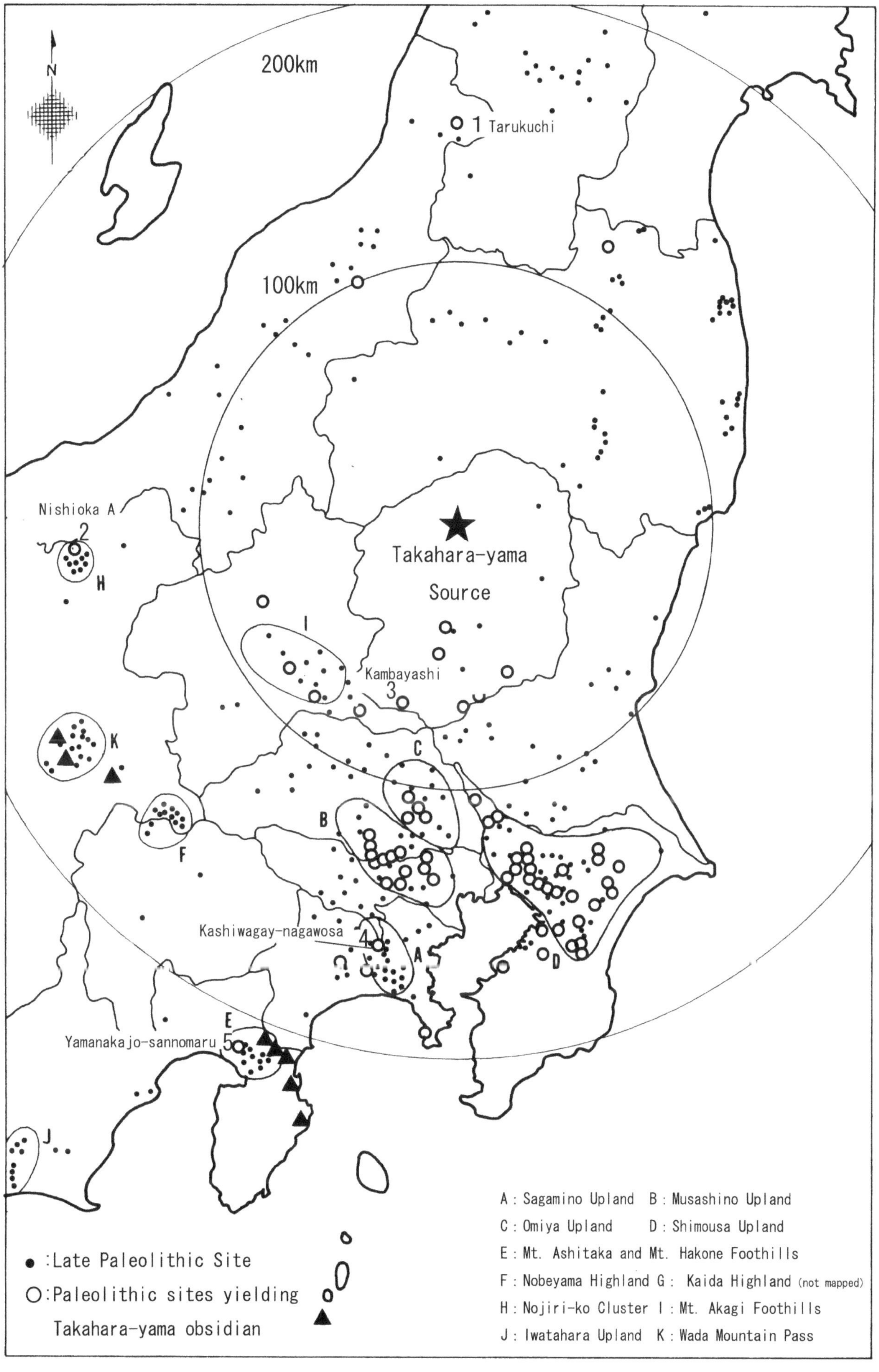

Figure 3.4. Distribution of Upper Palaeolithic sites with obsidian from the Takaharayama source

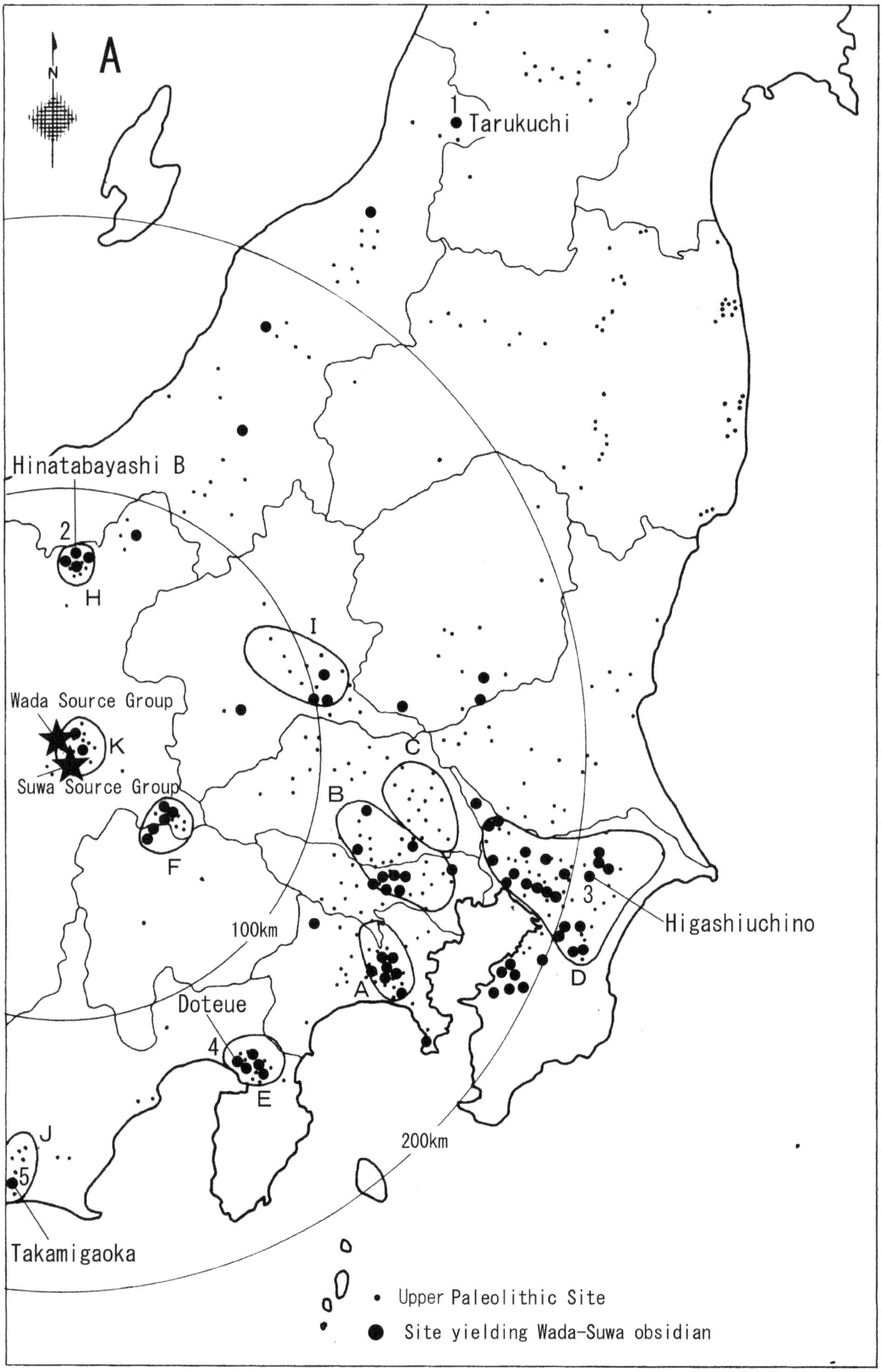

Figure 3.5, A. Distribution of Upper Palaeolithic sites with obsidian on central Honshu Island. Sites with Wada – Suwa obsidian

(2000m), and Futagoike (1500m). As for archaeological sites near the sources, the Ikenodaira cluster is listed as a workshop of Phase IV. The earliest exploitation of these sources is confirmed in Phase I (early Upper Palaeolithic) at the Hinatabayashi B site. They were continuously exploited from the middle to the end of the Upper Palaeolithic (phases II through V). The supply zone includes the Nobeyama Highland (50km radius; Figure 3.5, B, letter "F"); the Nojiri-ko cluster (100km distance; Figure 3.5, B, letter "H"); the Kanto Plain including the Sagamino (Figure 3.5, B, letter "A"), Musashino (Figure 3.5, B, letter "B"), and Shimousa (Figure 3.5, B, letter "D") uplands (distance between 100 and 200km); and the Hakone-Ashitaka flanks (Figure 3.5, B, letter "E"). All sites are positioned within a 200km radius, and the supply zone almost coincides with that of the neighbouring Wada – Suwa source group. Therefore, obsidian from these sources may have been transported by the same supply system. However, the inferior quality of this obsidian (mainly due to the presence of inclusions) made exploitation of Tateshina sources less intense compared to the Wada – Suwa ones.

The Hakone source group and Amagi source are located at the base of the Izu Peninsula (Shizuoka Prefecture); it consists of five individual sources: Hatajuku, Kajiya, Ashinoyu, Kuroiwa Basin, and Kamitaga (Mochizuki 2002b). Obsidian from the Hatajuku source (elevation of 500m a.s.l.) was primarily exploited. Because of their inferior quality (due to an abundance of inclusions compared to obsidian from the Wada – Suwa and the Kozu-shima groups), the Hakone and Amagi obsidian materials were not intensively used for making elaborate bifacial points or microblades. Also, the main supply zone is limited to the 100km radius, such as the Hakone-Ashitaka flanks (Figure 3.5, C, letter "E"), the Sagamino Upland (Figure 3.5, C, letter "A"), and the Musashino Upland (Figure 3.5, C, letter "B"). Occurrence of obsidian beyond this range is rare. The Hakone source group and the Amagi source are located within 30km distance; and their supply zones largely overlap. Therefore, the obsidian may have been transported by the same supply system.

The Kozu-shima source group is located about 50km off the coast of the Izu Peninsula (Figure 3.1). Even with a sea level drop of about 140m during the Last Glacial Maximum, this source was still 30km away from Honshu Island. This suggests that watercraft was used to reach this source (Suzuki 1973). The Kozu-shima group includes the Nukazaki source on the main island of Kozu-shima and the Ombase-jima [Ombase Islands] source located nearby. Obsidian is found here in outcrops and on the beaches, and is sub-divided into two large groups based on chemical composition (Mochizuki 2002b). To date, only the Ombase-jima obsidian has been identified among the archaeological materials on the Honshu mainland. No Upper Palaeolithic sites have been discovered in the Kozu-shima area. They may have been submerged during the post-glacial sea level rise.

Obsidian from the Kozu-shima source group is found in Phase I assemblages, demonstrating that the sources were already known to the inhabitants of Honshu in the early Upper Palaeolithic. However, obsidian from this source was exploited on a minor scale only in the subsequent phases II–IV. By contrast, it was used in large quantities in Phase V (late Upper Palaeolithic). The supply zone for Kozu-shima obsidian includes several regions along the eastern shore of Honshu Island: the Hakone-Ashitaka flanks (Figure 3.5, D, letter "E"); and the Sagamino (Figure 3.5, D, letter "A"), Musashino (Figure 3.5, D, letter "B"), and Shimousa (Figure 3.5, D, letter "D") uplands. The distances are within a 100–200km range. This overlaps with the supply zone of Hakone–Amagi obsidian (Figure 3.5, C). It was suggested that a human group based in the Hakone–Amagi area was also involved in the procurement and transport of the Kozu-shima obsidian situated offshore. The overlapping distribution of obsidian from these two source groups may support this hypothesis. The obsidian from Kozu-shima constitutes 30% of the volcanic glass at the Yadegawa site in the Nobeyama Highland in Phase V (Figure 3.5, D, letter "F"). The source is 200km from the site. This is an example of inland exploitation of this source group (Tsutsumi 2006).

The Okino-shima [Oki Islands] sources are situated in the Sea of Japan, Shimane Prefecture (Figure 3.1). The sources (including Kumi) are sub-divided into three groups based on chemical composition (Mochizuki 2002b). To date, only a few Upper Palaeolithic sites have been found on Okino-shima, but no sites specialising in obsidian exploitation have been discovered. The distance between Okino-shima and Honshu is about 60km; but the sea depth is less than 100m. Okino-shima and Honshu were presumably connected during the Last Glacial Maximum when the sea level dropped more than 100m and obsidian was transported across the land bridge. Source analysis indicates that obsidian was brought to a number of different locations. These include the Onbara site in Okayama Prefecture 100km away; the Oura, Hanamiyama, Usajima, and Yoshima site clusters in the Seto Naikai [Inland Sea] area 200km away; and as far as the Mianamikata site in Yamaguchi Prefecture 250km away (Figure 3.2; Higashimura and Warashina 1988). However, the extent of the supply zone and time periods of exploitation of this source are not well-known and further source analysis is necessary.

Formation of Site Clusters at Obsidian Sources

Source site clusters are strategic locales for the raw material transport networks formed around obsidian sources. At the Wada – Suwa group, obsidian in the form of chunks is found in outcrops at the Wada-toge source. At other sources, smaller volcanic glass debris can be seen on ground today. In this region, there is a dense cluster of Upper Palaeolithic sites in the Omegura Valley with an elevation of about 1200m a.s.l. This is called the Omegura cluster (Tsutsumi *et al.* 1993).

In this area a lot of obsidian debris including angular gravel up to the size of a human fist has been found. The Omegura cluster is believed to have been formed on the terrace of the Omegura Valley which was an optimal place for collecting obsidian from secondary deposits. Raw material pieces

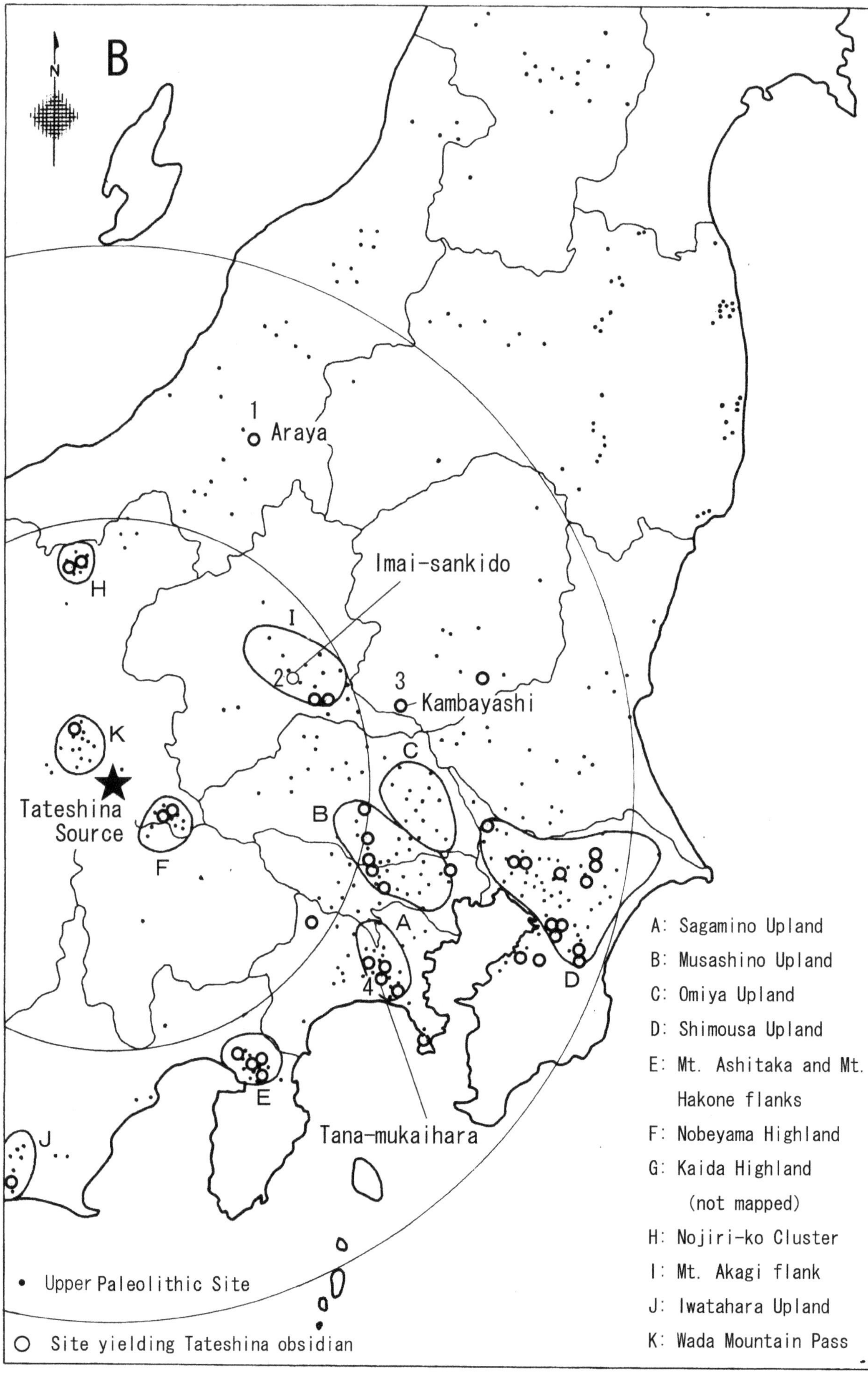

Figure 3.5, B. Distribution of Upper Palaeolithic sites with obsidian on central Honshu Island. Sites with Tateshina obsidian

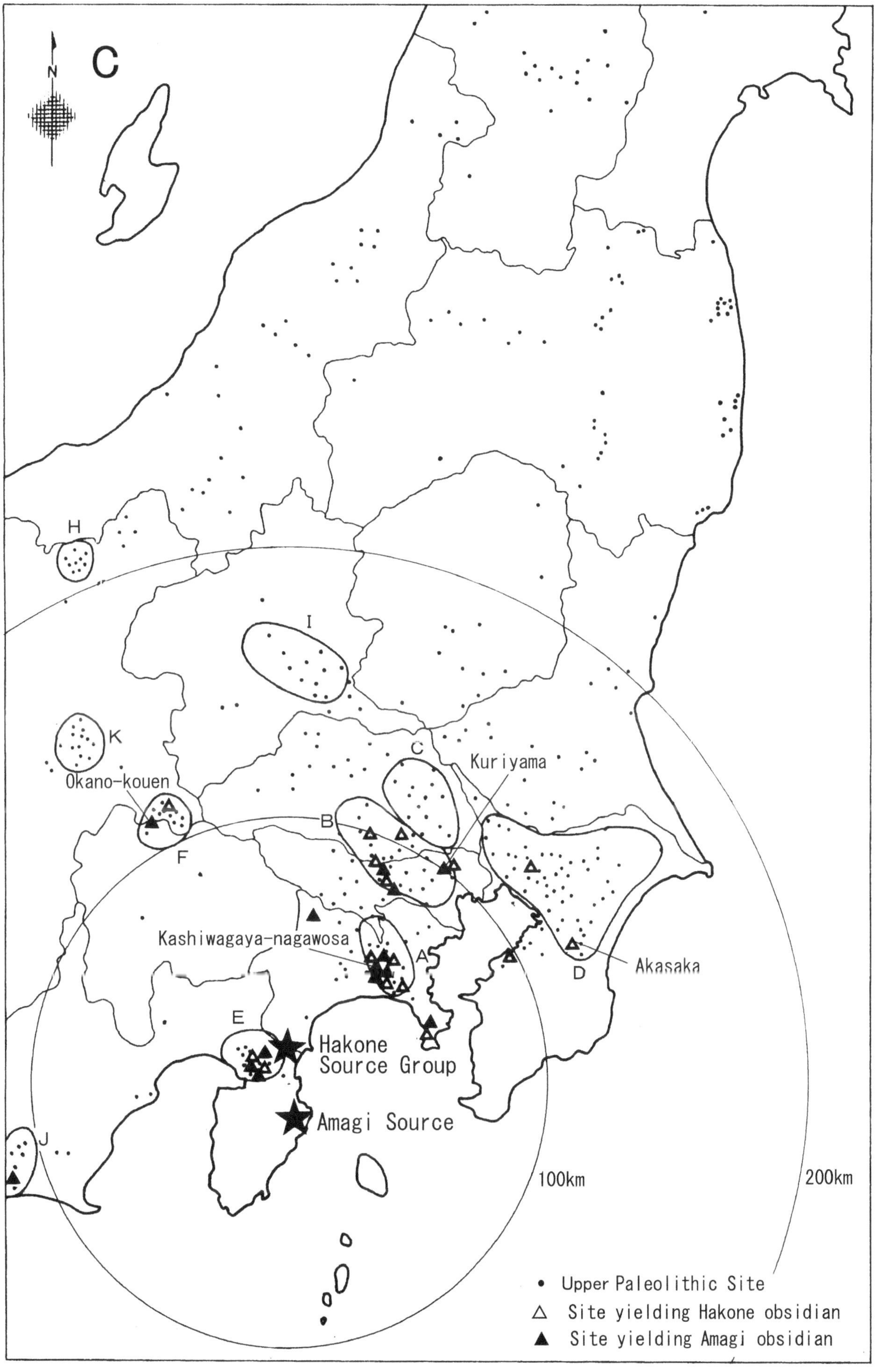

Figure 3.5, C. Distribution of Upper Palaeolithic sites with obsidian on central Honshu Island. Sites with Hakone and Amagi obsidians

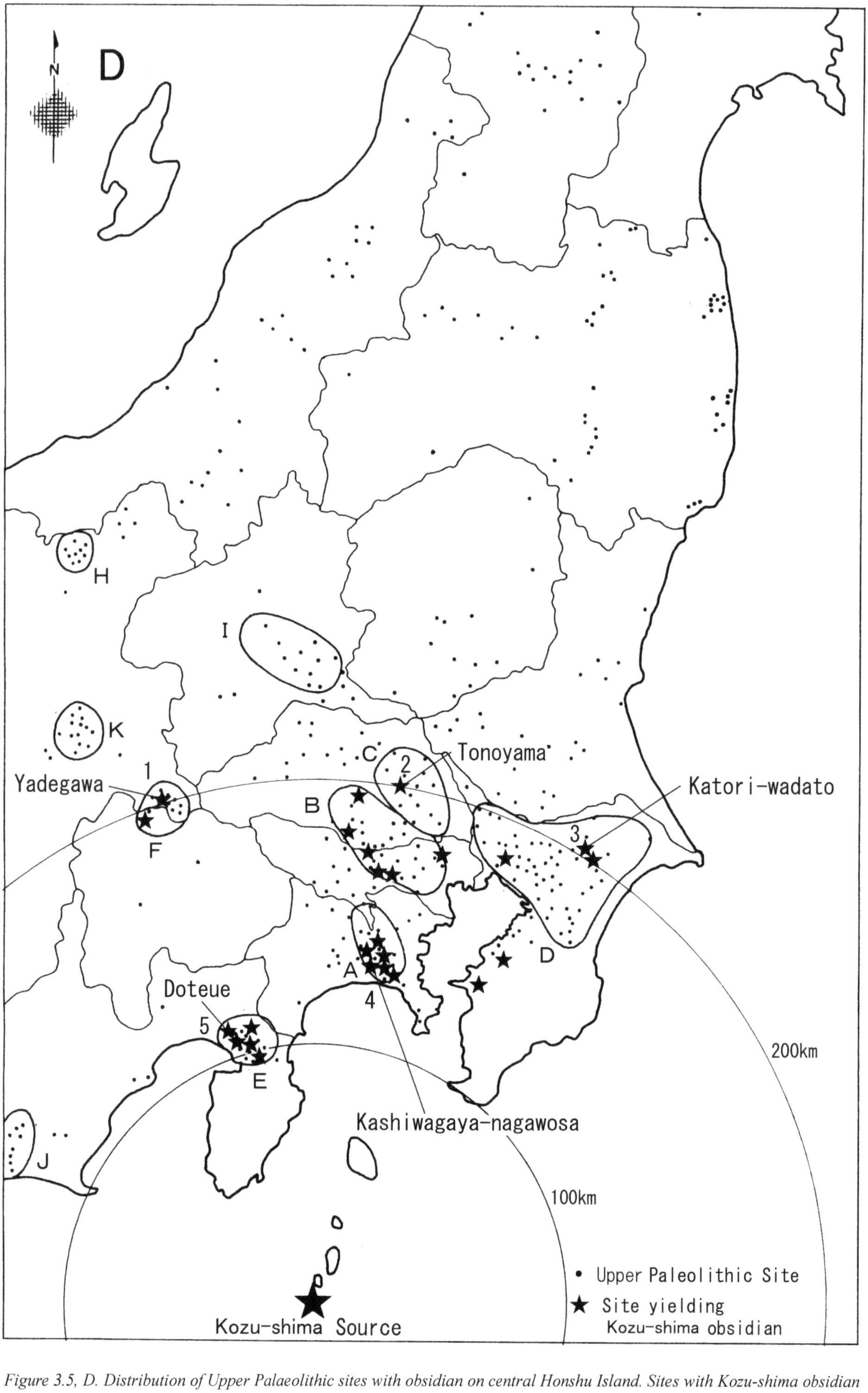

Figure 3.5, D. Distribution of Upper Palaeolithic sites with obsidian on central Honshu Island. Sites with Kozu-shima obsidian

and cores left at the Omegura cluster tend to retain cortical surfaces. Rather than knapping obsidian from quarries, people probably preferred to pick up gravel from the river bed which had already been naturally broken. After rolling in the river channel, the fragile portions of gravel containing many inclusions were removed and pieces of good quality survived. As it will be mentioned later, there are many Jomon period quarry pits in the vicinity of the Wada-toge source but none are known to be the Upper Palaeolithic one. Perhaps the demand for obsidian in the Upper Palaeolithic was not high. Collection of gravel from secondary deposits was the optimal procurement strategy in terms of time and energy (Tsutsumi 2002).

About 20 sites have been identified in the Omegura cluster to date, including sites A through O and I through IV (Figure 3.6). There are some distinctive features for source sites. First, there is a high occurrence of cores and original pebble/rubble in assemblages regardless of the typological phase. This pattern is observed at the Omegura B, C, I, and J sites. The Omegura J site has yielded 1001 cores from a total of 5095 lithic artefacts found, or one-fifth of the total (for example, see Figure 3.6), which is unusual. In addition, at the source site there are many incomplete artefacts as well as pieces which retain a large amount of cortex. The second important feature of source sites is the extremely large number of lithics per square unit. For example, the Omegura B site yielded a total of 17,149 artefacts from an area of 320m^2 which amounts to 53 items per 1m^2. By contrast, Kamiwada-jouyama II C, a residential site with Wada – Suwa obsidian located more than 100km from the sources, yielded 472 artefacts from 96m^2 (about 5 pieces per 1m^2). Therefore, the artefact density at Omegura B is ten times higher. The third feature is the spatial distribution of artefacts. The sub-cluster I in lithic cluster B at the Omegura J site includes 121 cores, 76 core blanks, 93 raw material pieces, and 17 hammerstones. These cores and related materials appear to have been deliberately accumulated there. Not only cores but caches of particular tool types tend to be found at source sites, suggesting some kind of caching strategy for planned export.

There are different types of artefact transportation from sources to residential sites in terms of what was brought: (1) bulk raw material; 2) cores; 3) blanks such as flakes; 4) incomplete (unfinished) artefacts; and (5) completed pieces. The presence of roughly shaped large bifaces is common at the Omegura B site (Figure 3.6). These incomplete bifacial points indicate transport from sources to residential sites in unfinished form. Many sites found at the sources belong to Phase IV. The manufacture of bifacial points requires extensive retouch. Therefore, it is believed that at source sites knappers pre-shaped points to reduce possible breakage during manufacture at residential sites. This strategy also minimised the transportation load by removing the waste portions of raw material. This lithic raw material procurement strategy is defined as *on-source production* type. There is a significant size reduction of bifacial points between the Omegura B site and Kannoki, a residential site with Wada – Suwa obsidian located 80km away. At the latter, the size of points was further reduced, probably by repeated cycles of use and rejuvenation (Figure 3.7).

On the other hand, there is a marked contrast with the *on-residence production* type of raw material procurement strategy seen in microblade industries. For example, at the Yokota site situated about 100km away from the sources, small pieces of obsidian were brought from the Wada – Suwa sources (Mochizuki 1997a); and the preparation of microblade cores and detachment of microblades was carried out at the site (Figure 3.8). Unlike bifacial points, the manufacture of microblades does not require elaborate retouch, thus it involves a low risk of failure. Even with breakage, the edges were still functional. Therefore, microblades did not need to be produced at the source sites. In fact, among 28,157 pieces from ten sites of the Omegura cluster only 13 microblades and three microblade cores have been counted. This phenomenon indicates that microblade manufacture is not an on-source production strategy. Even if raw material procurement groups camped at sources, their traces are difficult to identify due to the absence of diagnostic artefacts. Therefore, the apparent absence of microblade sites does not necessarily mean the true absence of on-source sites in the microblade phase.

Tendencies in Obsidian Exploitation at Residential Sites

Eighty-five sites in Japan correspond to the early Upper Palaeolithic (Phase I; ca. 30,000–35,000 BP) (Hashimoto 2006) while there are 1792 sites of the Phase V (end of the Upper Palaeolithic) (Sato and Tsutsumi 2007; Tsutsumi 2004, 2007) (Figure 3.9). There are no precise statistics for the entire Upper Palaeolithic in the Japanese Islands; but the total number of sites is estimated to be no fewer than 10,000. Numerous large-scale rescue excavations have been conducted in the densely populated regions of Japan because of construction work, with precise site documentation such as three-dimensional coordinates for every artefact.

Scientific analysis of Upper Palaeolithic obsidian artefacts in Japan has been carried out since the 1970s. In the 1990s, attempts at *exhaustive analysis* by the XRF method were made. This kind of analysis allows processing of large quantity of specimens in a short period utilising the data on the precise position of artefacts (Mochizuki *et al.* 1994). As a result, it became possible to discuss more accurately obsidian source composition in archaeological assemblages. Also, a method of identifying rare obsidian artefacts from distant sources in collections where these minor sources could easily be missed by 'regular' sampling was established. An exhaustive analysis approach of obsidian exploitation was followed by studying the microblade assemblages (Mochizuki and Tsutsumi 1997). A framework of Upper Palaeolithic chronology is presented to provide the basis for discussion (Figure 3.10). The Japanese Upper Palaeolithic was sub-divided into first and second halves, and further divided the former into two phases (I and II) and the latter into three phases (III, IV, and V). The Aira-Tanzawa (AT) tephra layer (dated to ca. 24,500 BP) marks the boundary between the first and second halves.

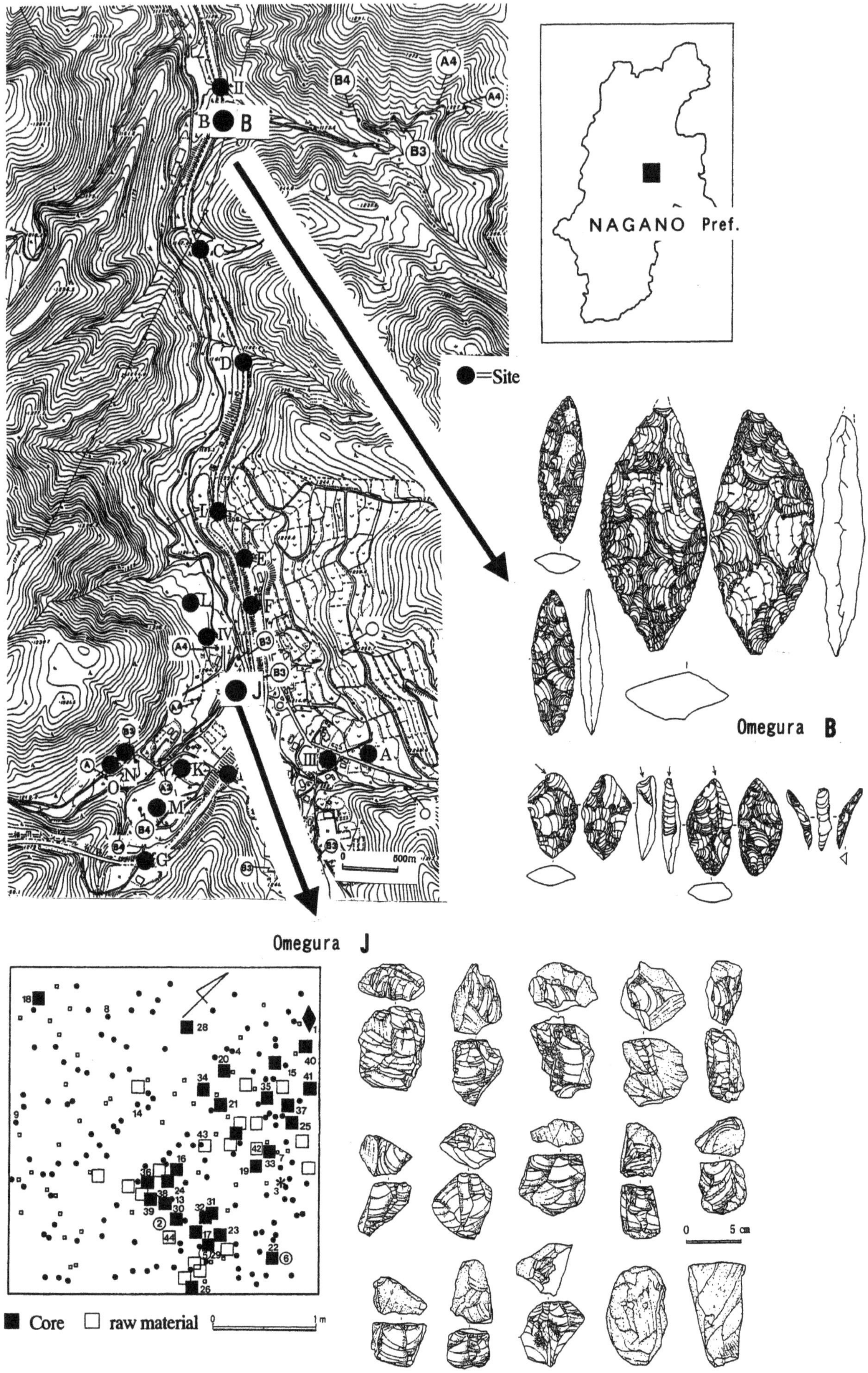

Figure 3.6. Distribution of Upper Palaeolithic sites at the Omegura source (Wada – Suwa source group)

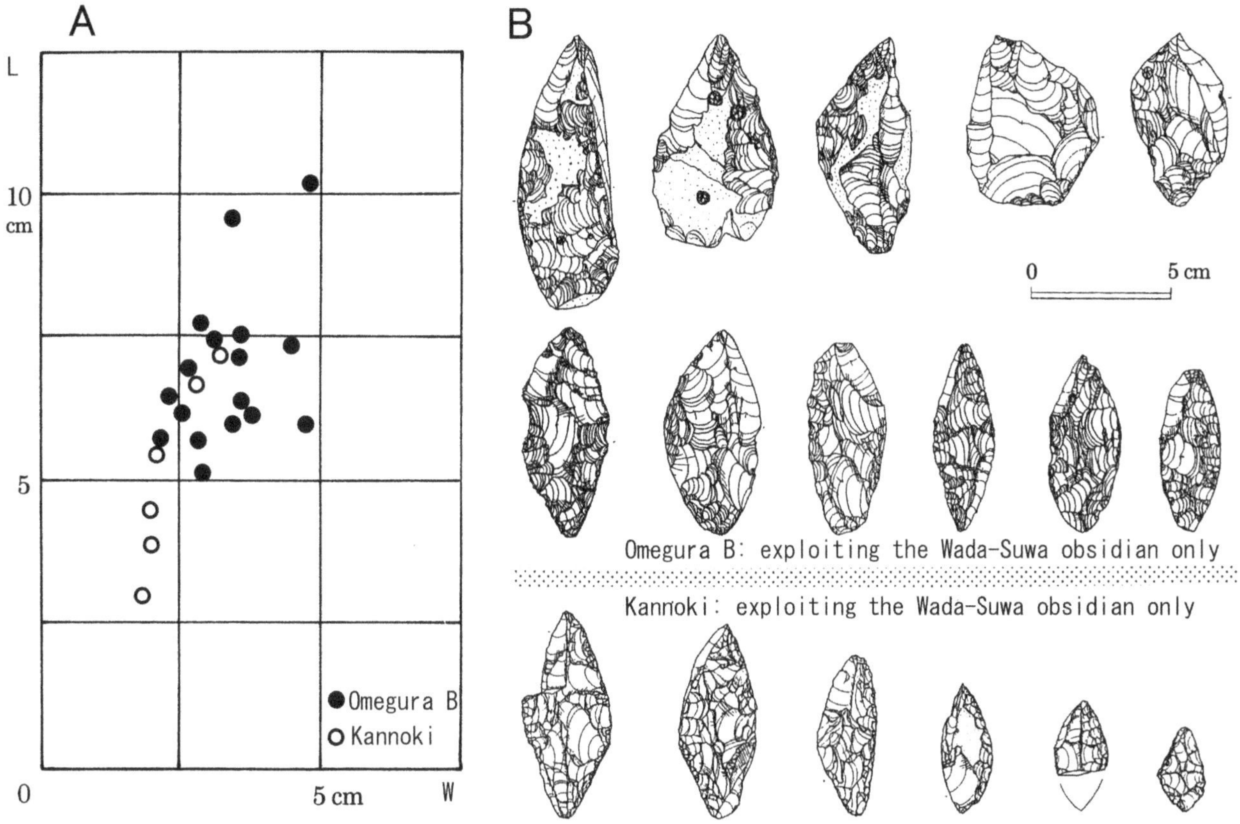

Figure 3.7. Variation in point size between source site (Omegura B) and residential site (Kannoki)). A – Chart of artefact width (W) versus length (L); B – artefacts

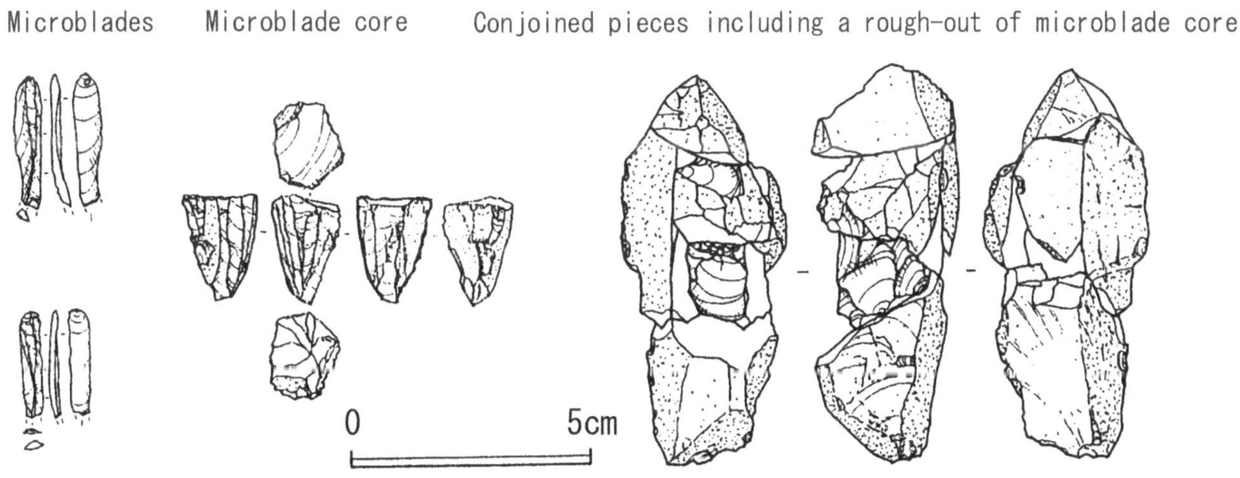

Figure 3.8. Microblade assemblage from the Yokota site

In the following paragraphs, obsidian exploitation in each phase of the Upper Palaeolithic is examined by comparing data from the Sagamino Upland with data from other parts of the Kanto and Chubu regions. The excellent stratigraphic profiles of the Sagamino Upland provide the basis for the chronological framework for the lithic assemblages; and obsidian sourcing is relatively well-developed here. The Sagamino Upland data are based on source analysis by Mochizuki and Tsutsumi (1997) and on research by Suwama (2006). Figure 3.11 shows the types of raw materials used at the sites, and Figure 3.12 indicates the amount of obsidian from different sources.

The investigation of the Phase I sites in the Sagamino Upland is still limited, and short of data (Figure 3.12, Nos. 1–3). However, in terms of the ratio between volcanic glass and other rock types, the degree of obsidian exploitation was relatively small except for the Yoshioka Locality D B4 on the Sagamino Upland (Figure 3.11, Nos. 1–5). The analysis of the Doteue site assemblage on the Ashitaka flanks of Mount Fuji (Figure 3.5, A, No. 4) shows that 81 pieces are from the Wada – Suwa sources, 13 from Tateshina, 923 from Hakone, 390 from Amagi, and 495 from Kozu-shima (the total number of specimens is 1902). This shows that human groups who occupied the site used obsidian

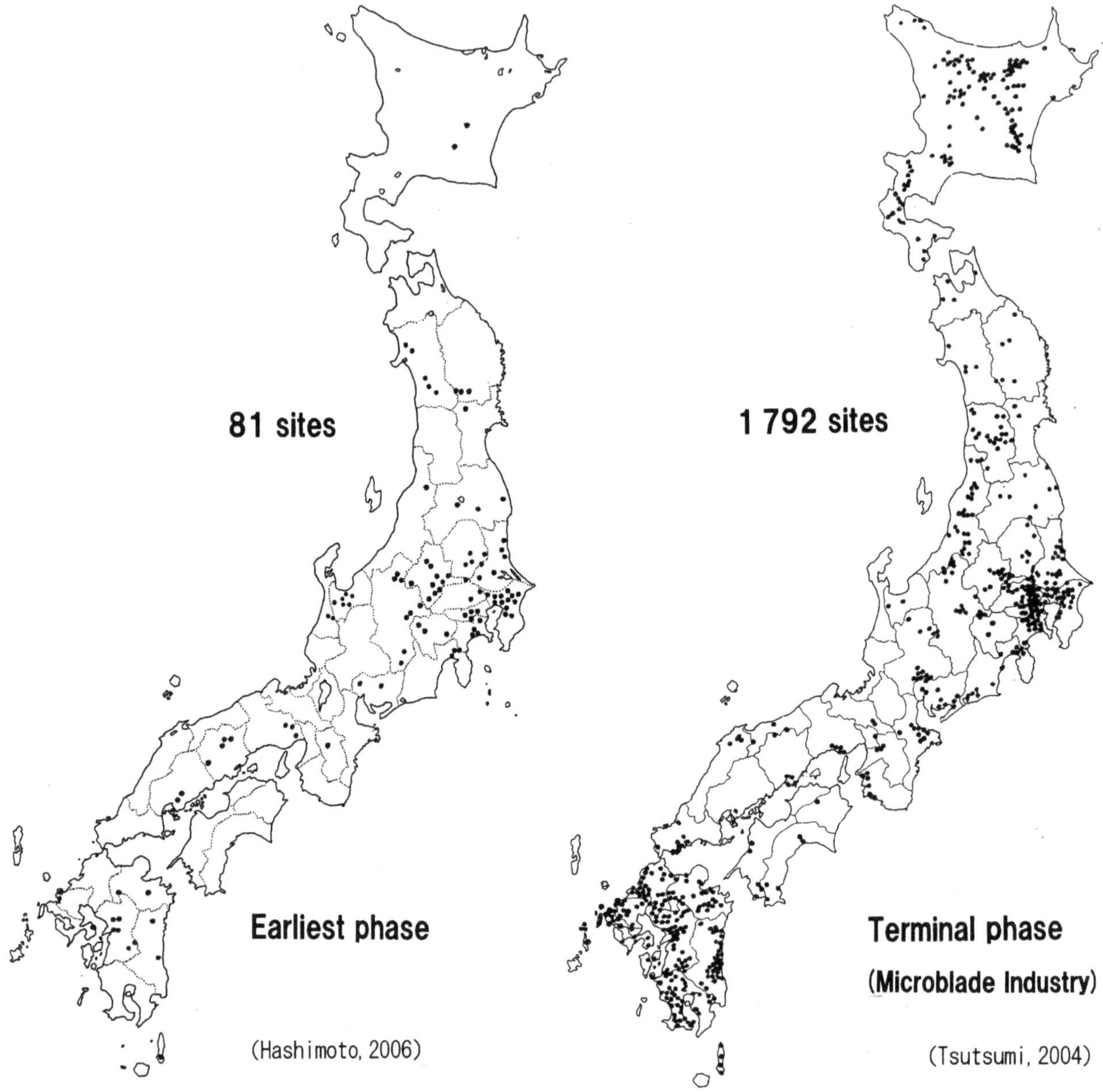

Figure 3.9. Distribution of early and late Upper Palaeolithic sites on the Japanese Archipelago (after Hashimoto 2006; Sato and Tsutsumi 2007; Tsutsumi 2004, 2007)

from diverse sources. It is believed that such a pattern of exploitation demonstrates the involvement of several groups based at each source rather than procurement by a single band visiting all of the sources (Mochizuki *et al.* 1994). Among the 3540 obsidian pieces from the Hinatabayashi B site in the Nojiri-ko cluster (Phase I; Figure 3.5, A, No. 2), the material from the Wada – Suwa sources constitutes the major part (3422 artefacts, or 96.7%), and the Tateshina obsidian is very rare (37 items, or about 1%); 45 artefacts were not analysed (Tani 2000).

Obsidian exploitation in the Sagamino Upland skyrocketed from Phase I to Phase II (Figure 3.11, Nos. 1–6), and other types of raw material became less popular. At the Terao site (Figure 3.12, No. 7), obsidian from the Shinshu source group represents the largest share with a small quantity of Amagi obsidian. By contrast, at the Kamitsuchidana site the Amagi obsidian is the most abundant, with a few pieces of Shinshu origin. In other regions of Honshu Island this kind of data is limited. In the second half of Phase II, the Shinshu obsidian is said to have been actively imported to the southern Kanto Plain (Kanayama 1989). However, there is great intersite variability in the Sagamino Upland; and more systematic analysis should be done.

Similar to Phase II, obsidian exploitation in the Sagamino Upland was active during Phase III. The other common rock used was volcanic tuff. The Kashiwagaya-nagaosa site has provided a substantial dataset (Mochizuki 1997b). This site has had a great deal of attention because five cultural horizons (5m thick) of Phase III were identified, with an excavation area of about 5000m^2 (Tsutsumi 1997a). A total of 1701 obsidian specimens were recovered, and 1636 pieces (96.1% of total) were analysed (Table 3.1). Over 98% of the artefact raw materials from each of the five cultural horizons were obtained from the Hakone or Amagi

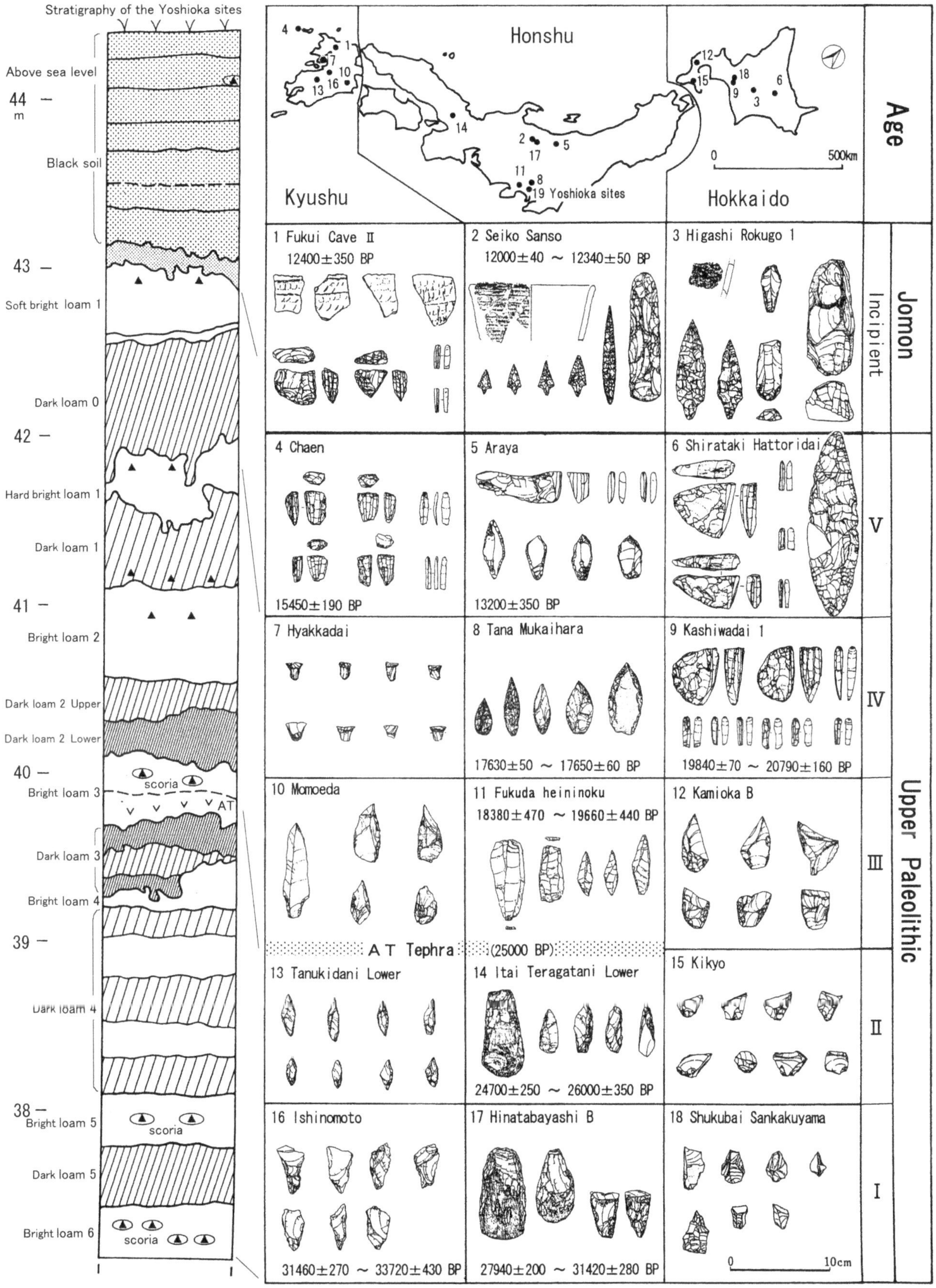

Figure 3.10. The chronology and stratigraphy of the Upper Palaeolithic complexes in the Sagamino Upland and its comparison with neighbouring regions

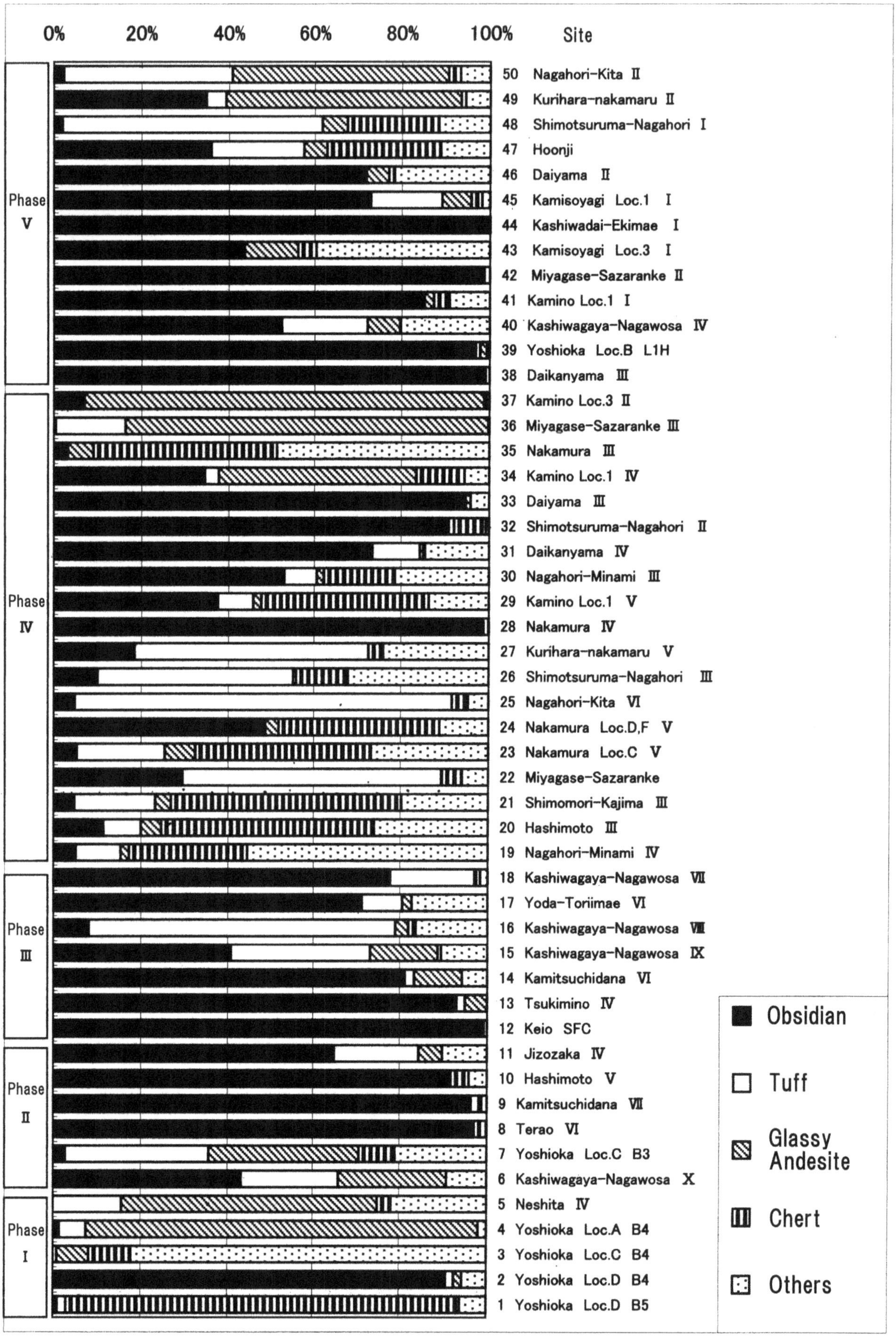

Figure 3.11. Lithic raw material composition at Upper Palaeolithic sites in the Sagamino Upland (after Suwama 2006)

Table 3.1. Obsidian source composition at the Kashiwagaya-nagaosa site, middle Upper Palaeolithic (after Mochizuki and Tsutsumi 1997)

Cultural horizon	*Total number of specimens*	*Amount of obsidian by source*						*Number of specimens identified*	*Percentage of specimens identified*
		Hakone	*Amagi*	*Wada – Suwa*	*Tateshina*	*Kozu-shima*	*Takaharayama*		
VI	139	131	2	1				134	96.4
VII	182	3	175					178	96.2
VIII	91	15	69					84	92.3
IX	1164	616	489	4	1	2	8	1120	96.2
X	125	62	58					120	96.0
Total	1701	827	793	5	1	2	8	1636	96.1

areas located within a 70km radius. By contrast, there are few obsidian artefacts from the Shinshu (Wada – Suwa and Tateshina) and Kozu-shima sources which are located over 100km away. However, there is no clear indication regarding which of these two sources, Hakone or Amagi, was more important as the ratio varies by cultural horizon. The absence or very low occurrence of Shinshu obsidian at the Kashiwagaya-nagaosa site reflects more of a general trend for the Sagamino Upland in this phase.

As for other regions of Honshu Island, among the 100 specimens from cultural horizon 1 at the Shimo-yanagisawa site in the Musashino Upland, only seven Shinshu obsidian artefacts have been identified; the remaining 93 pieces are from the Hakone source (Mochizuki 2002b). Among the 43 specimens from cultural horizon 4 at the Musashi-kokubunji-kanren site, 24 are from the Hakone sources and 19 from Amagi. No other sources have been identified (Warashina 2004). The prevalence of the Hakone and Amagi obsidians appears to be a common phenomenon in the Musashino and Sagamino uplands in Phase III.

Obsidian use was not active in the Sagamino Upland in the first half of Phase IV (Figure 3.11, Nos. 19–27). This was a period when tuff, chert, and other non-obsidian rocks were exploited in large quantities. Obsidian exploitation increased in the second half of Phase IV (Figure 3.11, Nos. 28–33); but it declined at the end of Phase IV. Although the import of Shinshu obsidian ceased during Phase III, it increased again in Phase IV. However, there is almost no obsidian from the Kozu-shima source in Phase IV, indicating that the supply remained inactive compared to the previous phase.

At the Tsukimino 2 site in the first half of Phase IV, 301 artefacts of Shinshu obsidian, 204 of Hakone, and four of Amagi obsidian were identified, with no obsidian from either the Kozu-shima or Takaharayama sources (Mochizuki *et al.* 2003). At the Tana-mukaihara site belonging to the second half of Phase IV (Figure 3.5, B, No. 4), an Upper Palaeolithic dwelling has been discovered which is rare in Japan. Using source analysis of 1400 obsidian pieces from this dwelling, it was determined that 1096 items are of Shinshu origin, 110 are from the Hakone source, 158 pieces are from the Amagi source, and 36 artefacts are from Takaharayama (Mochizuki 2004). Among the Shinshu sources, the amount of Tateshina obsidian exceeds that from Wada – Suwa by almost an order of magnitude (910 items of the former versus 186 of the latter). There are also several artefacts made from Hakone and Amagi obsidian and a small quantity of Takaharayama obsidian, with a total lack of volcanic glass from the Kozu-shima source. Tools made of obsidian include 181 small points, 15 backed points, and 34 scrapers. Source analysis revealed that the obsidian used during Phase IV at the Kamiwada-jo-yama site – which is practically contemporaneous to the Tana-mukaihara site – is also almost exclusively composed of Shinshu obsidian (Mochizuiki and Tsutsumi 1997).

In other regions of central Honshu, 129 of 135 specimens (except for six items identified as not analysable) from cultural horizon 3 at the Shimo-yanagisawa site in the Musashino Upland are made of Shinshu obsidian (Mochizuki 2000a). At the Yokota site in the Musashino Upland, 57 out of 72 small obsidian points are made of Shinshu obsidian; 49 points are from the Wada – Suwa sources; and eight artefacts are from Tateshina (Mochizuki 1997a).

The results of a sourcing study for 199 points from the Boso Peninsula, mainly from the Shimousa Upland, show that 105 items are from the Shinshu obsidian source, 88 from the Takaharayama, two from the Hakone, and three from the Kozu-shima source. Although the Shinshu obsidian artefacts are the most common (52%), the Takaharayama source specimens total 44 %. These were the major sources in this region (Ninomiya and Shimadate 2001). In cultural horizon 2 at the Kambayashi site (Phase IV) (Figure 3.4, No. 3) situated 70km from the Takaharayama locality, 21 out of 22 pieces are from this source and a single artefact is from the Wada – Suwa source over 100km away. All of the four obsidian points in the assemblage are of Takaharayama origin (Sano City Board of Education 2004).

As for obsidian exploitation in the microblade industry (Phase V) which manifests at the end of the Upper Palaeolithic, the results of the analysis of 4040 specimens from the Chubu and Kanto regions are presented in Table

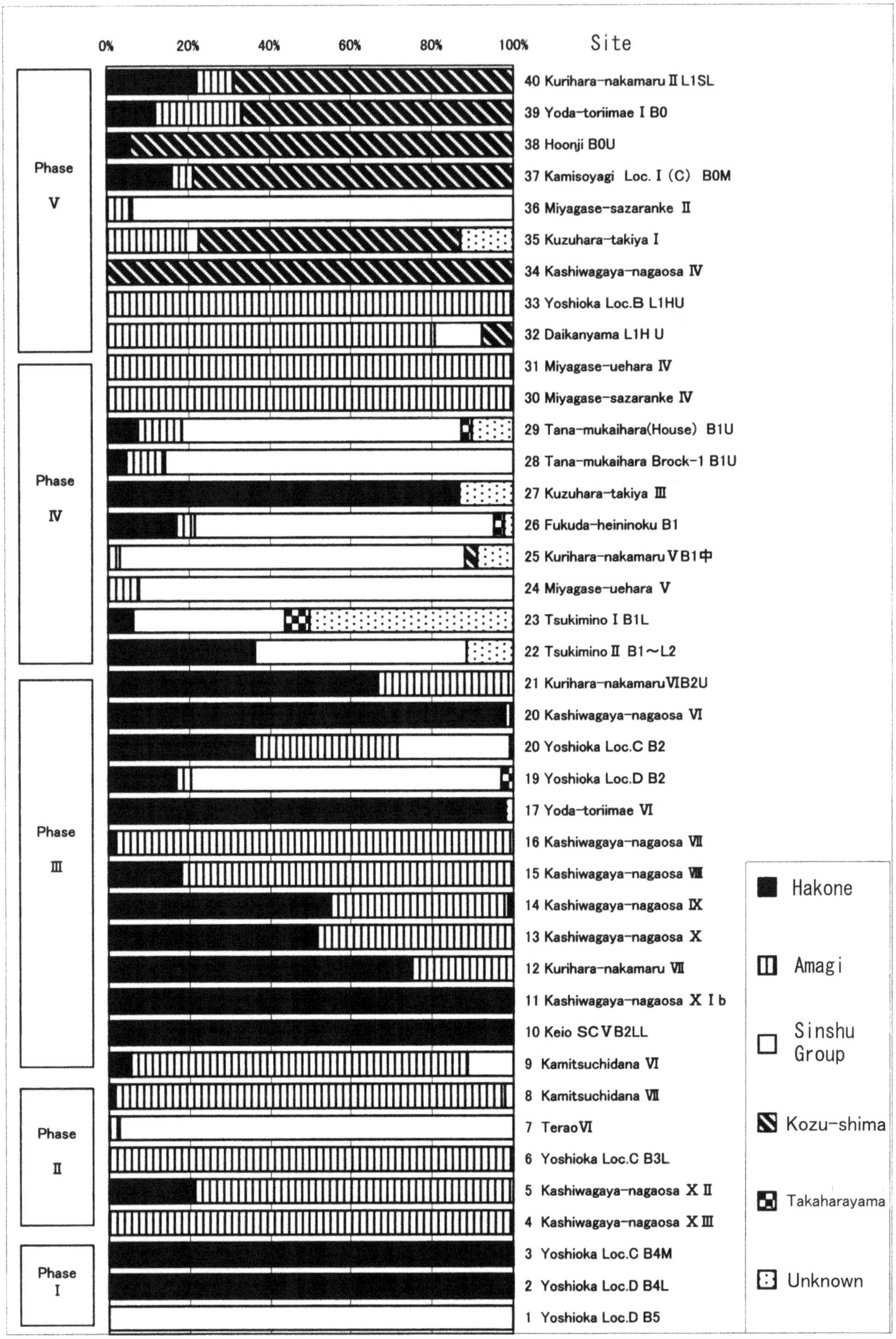

Figure 3.12. Obsidian sources for Upper Palaeolithic complexes in the Sagamino Upland (after Suwama 2006)

Table 3.2. The microblade assemblages in Chubu and Kanto regions and their obsidian sources (after Mochizuki 1998; Mochizuki and Tsutsumi 1997)

No.	Region	Microblade assemblage	Shinshu sources		Hakone sources	Amagi sources	Kozu-shima sources	NK source	XO source	Oga source	Total
			Wada – Suwa	Tateshina	Hatajuku	Kashiwa-toge	Onbase-jima				
1	Sagamino Upland	Kamisoyagi Loc. 1			52	18	260				330
2		Kamisoyagi Loc. 3-C	138								138
3		Kamisoyagi Loc. 3-E	2								2
4		Kamisoyagi Loc. 4					2				2
5		Fukuda-Fudanotsuji					1				1
6		Nagahori-Minami II	2								2
7		Daiyama II	338	29		380	1				748
8		Kashiwagaya-nagaosa IV					349				349
9		Kamiwada-Joyama I	130								130
10		Soyagi-Nakamura II	20	58	1		1				80
11		Kashiwadai-Ekimae I				7	273				280
12		Kashiwadai-Ekimae II					97				97
13		Hoonji		1	12		185				198
14	Musashino Upland	Yokota	138	2							140
15		Tamaranzaka	154				1				155
16	Hakone and Ashitaka mountain flanks	Nakamiyo III	1	1	1	2	139				144
17		Yamanakajo-Sannomaru	25	4	2	5	36				72
18		Uehara I	14	6		4	616				640
19	Nobeyama Highland	Yadegawa	112	133			157	48	1		451
20	Kaida Highland	Koshi	80							1	81
Total			1154	234	68	416	2118	48	1	1	4040

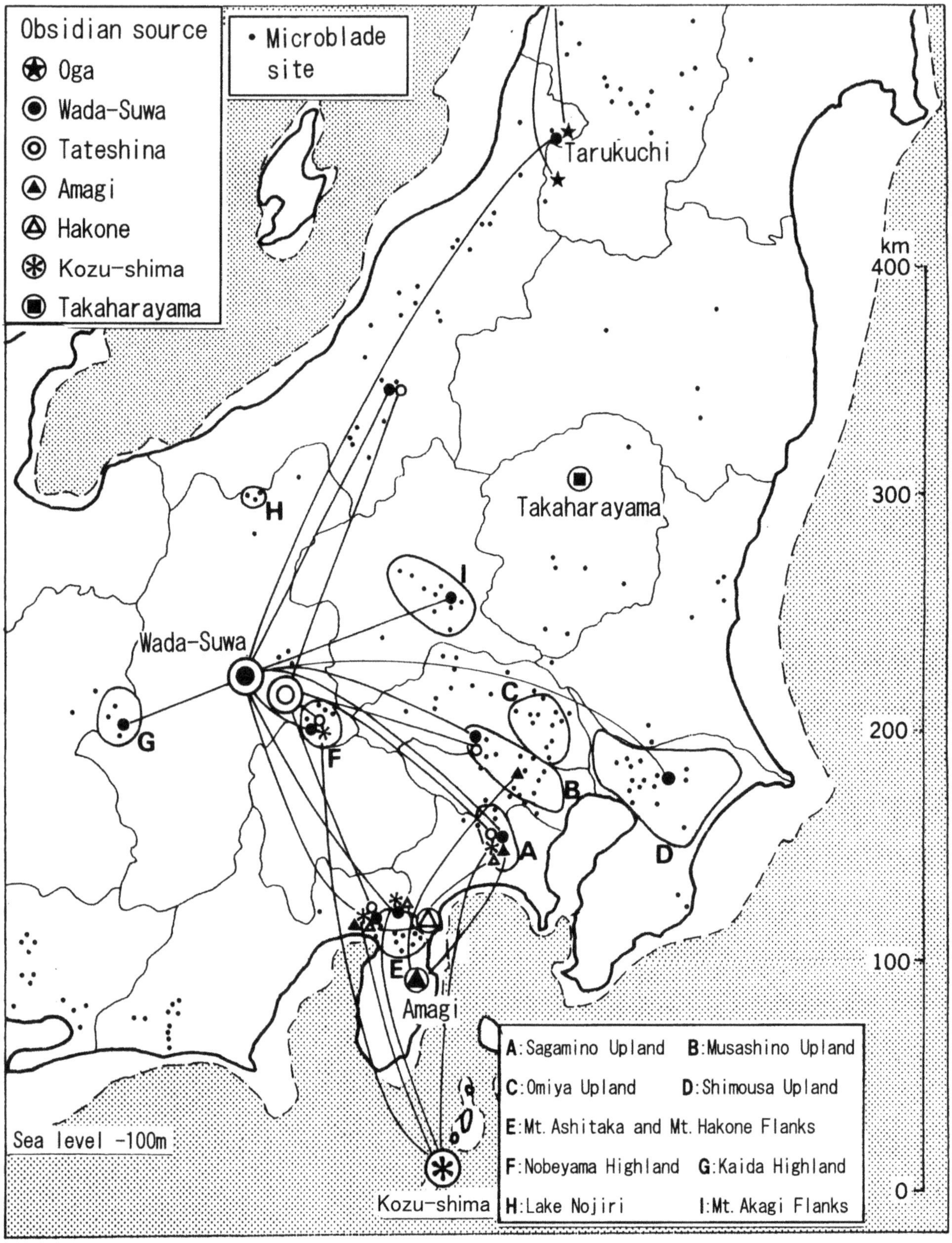

Figure 3.13. Obsidian sources and residential sites with obsidian artefacts of the microblade industries (after Sato and Tsutsumi 2007, modified)

3.2 (see also Figure 3.13). At the Sagamino Upland, obsidian use in this period skyrocketed in comparison to other raw materials (Figure 3.11, Nos. 38–46). In the second half of Phase V, however, obsidian exploitation tended to decline (Figure 3.11, Nos. 47–50). The most distinctive phenomenon is the import of Kozu-shima obsidian in large quantities compared to phases I through IV (Figure 3.12, Nos. 34–40). Source analyses have been conducted on materials from 13 sites (Mochizuki 1998; Mochizuki and Tsutsumi 1997). Five patterns have been recognised in terms of source composition: (1) Shinshu obsidian only; (2) Amagi and Hakone obsidians; (3) Kozu-shima obsidian only; (4) Shinshu, Amagi, and Hakone obsidians; and (5) Kozu-shima, Amagi, and Hakone obsidians. Note that there is no example of the joint presence of the Kozu-shima and the Shinshu obsidians, and this is a distinctive feature in obsidian transport. Rather than correlating it to differences in age or phase, it is interpreted as an effect of seasonality. That is, rocks from mountains (Shinshu region) and from the sea (Kozu-shima source) had different supply seasons in the annual cycle. The Shinshu obsidian in the highlands became inaccessible in the cold times of the year and thus was "closed for winter".

With regard to the Musashino Upland, 295 specimens belonging to Phase V from two sites (Yokota and Tamaranzaka) have been analysed (Mochizuki 1998, 2000b). Obsidian from the Wada – Suwa source group located about 100km away constitutes the majority at both sites. On the Hakone-Ashitaka flanks, 856 specimens from three sites (Nakamiyo III, Yamanakajo-Sannomaru, and Uehara I) have been studied (Mochizuki 1998). Most of the obsidian is from the Kozu-shima source, with some items from the Wada – Suwa source group. However, exploitation of obsidian from the neighbouring Amagi and Hakone source groups was not intensive. For the Nobeyama Highland, 451 microblade cores from the Yadegawa site (Table 3.2; Figure 3.13, F) have been analysed (Tsutsumi 2006). One hundred and twelve pieces are from the Wada – Suwa source group located 40km away, and 133 items are from the Yatsugatake source group situated at a distance of 20km. In addition, 157 obsidian artefacts are from the Kozu-shima source on a small island 200km away, and 49 items are from the NH (an unknown source group originally sampled from the Nakappara sites in the Nagano Prefecture) and XO (an unknown source group originally sampled from the Okanokoen site in the Yamanashi Prefecture) sources.

Acquisition of Obsidian Resources

This study has chronologically examined the topics, including obsidian source exploitation on Honshu Island in the Upper Palaeolithic, the formation of source sites, and the patterns of obsidian use at residential sites. This section will summarise the available data and discuss existing problems. The obsidian sources of Takaharayama, Shinshu, Hakone, Amagi, and Kozu-shima were already used at the beginning of the Upper Palaeolithic (Phase I). However, when considering the Sagamino Upland and other regions, long-distance obsidian transport in this phase was not common compared to subsequent phases. Residents largely depended on local raw materials other than obsidian. Nevertheless, the transport of obsidian from the Kozu-shima source situated in the open sea suggests that early Upper Palaeolithic people had acquired seafaring skills.

The exploitation of several obsidian sources is observed in Phase I. In the middle Upper Palaeolithic (Phase II), only Hakone and Amagi obsidians were exploited in the Sagamino and Musashino uplands, and the transport of Shinshu obsidian was interrupted. There are two hypotheses to explain why this happened. One of these suggests that due to the cold climate the sustainable procurement of the Shinshu obsidian located in the highlands became difficult (Sato 1996). This is dated to a time interval before the Last Glacial Maximum, but the quantity of endscrapers used for hide processing (a sign of cold weather) increased; this indicates cold climate (Tsutusmi 2000). The second hypothesis is that the obsidian supply ceased because of changes in the social relationship among groups on the supply routes between the Shinshu sources and the residential sites. In either case, it is certain that during Phase III access to obsidian sources for groups which occupied the Sagamino Upland became limited. In addition, the supply of Kozu-shima obsidian across the open sea also stopped.

The first half of Phase IV showed a decrease in obsidian exploitation in the Sagamino Upland. It is intriguing why such a shift in obsidian use took place. In Phase IV, obsidian from Shinshu sources is the dominant raw material and shows a sharp contrast with Phase III in which Shinshu obsidian was absent, with Hakone and Amagi obsidians dominating. Many bifacial points were produced in Phase IV; and these points were made of Shinshu obsidian in the Sagamino and Musashino uplands. On the other hand, a certain number of points made of Takaharayama obsidian have been found in the Shimousa Upland in addition to Shinshu obsidian. The supply of Kozu-shima obsidian did not continue; and the exploitation of Hakone and Amagi obsidians was uncommon.

At the end of the Upper Palaeolithic (Phase V) when microblade industries spread all over Honshu Island, Kozu-shima obsidian, that had otherwise been rarely exploited before (except in Phase I), was used in large quantities (Tsutsumi 2007). This demonstrates that the procurement of obsidian by seafaring was undertaken on a large scale. This must have been a high risk activity with the potential loss of lives. Although, according to Weissner (1982) risk reduction is one of the fundamentals in resource procurement strategies.

As demonstrated in the above example of the Omegura case, there were clusters of source sites which served as logistic bases for obsidian source exploitation. Also, demonstrated is the supply zone of each obsidian source. Major sources like Shinshu were exploited in high frequency within a 200km radius; whereas the supply zones for minor sources such as Hakone and Amagi were within a 100km radius. The question raised here is the relationship between the obsidian supply zone and the actual territories of Palaeolithic populations. In the Upper Palaeolithic, territories are believed to have been much more extended because of higher mobility of groups in comparison to those of sedentary people in the Jomon period. In some cases, a lithic raw material supply zone and a group's territory may have coincided. In any case, how were obsidian resources that were located in a few limited places used by Upper Palaeolithic people?

For the procurement of lithic raw material like obsidian, the following models have been put forward: (a) the direct procurement strategy in which each group including those who lived in distant areas dispatched a procurement band (Ono 1975); (b) the exchange strategy in which groups acquired raw materials by trade and exchange (Harunari 1976); and (c) embedded procurement in which lithic raw material acquisition was part of the regular substantial activities (Binford 1979; Tamura 1992; see Tsutsumi 2007). The most important problem is what kind of socio-economic system existed in this period. Whether each group needed directly procured lithic raw materials or whether a social network had been established to obtain distant resources by exchange. At this point, there is no conclusive answer. It may not necessarily be true that only one of these two strategies existed, but each of them may have been employed depending on the rock type. The actual situation must have been complex.

Regarding the 400–500km long transport range of the Kizukuri, Fukaura, and Oga obsidians in the northernmost part of Honshu Island to the Nojiri-ko cluster in central Honshu (Figure 3.2), direct procurement appears unlikely considering the substantial distance. Other human groups also must have been involved. This indicates existence of social networks which covered a large area.

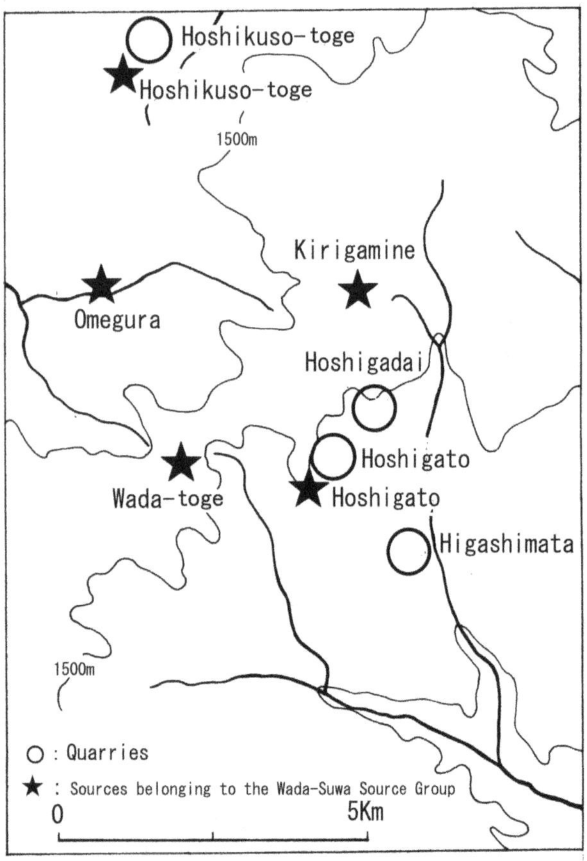

Figure 3.14. Location of obsidian quarry clusters from the Jomon period on central Honshu Island

Obsidian Exploitation in the Jomon Period

In the Jomon epoch of Japanese prehistory that is almost a synonym of the term "Neolithic", the invention of the bow caused a change in hunting strategies while the emergence of pottery enhanced the diet. During the later phases of Jomon, primitive cultivation may have been practiced in some areas. In the Jomon, people experienced a transition from mobile to sedentary life with the development of complex social organisation. Some scholars even suggest the existence of a social hierarchy (e.g., Watanabe 1990). In accordance with these innovations, the mode of obsidian exploitation changed dramatically. Mining pits were dug at obsidian sources to obtain large amounts of good quality raw material. It must be said that studies of Jomon obsidian exploitation are not particularly advanced compared to Upper Palaeolithic. This is because Jomon studies have focused mostly on pottery, settlement patterns, and faunal and botanical remains for which Jomon sites have provided a wealth of data. Less attention has been given to the study of lithic raw material use (e.g., Yamamoto 1990). Therefore, the discussion here is limited to obsidian quarries that have recently been discovered, and to the exploitation of the Shinshu and Kozu-shima obsidians at residential sites.

Jomon Obsidian Quarries

At the obsidian sources near the Wada Pass in Nagano Prefecture, there are four Jomon obsidian quarry clusters—Hoshidagai, Higashimata, Hoshigato, and Hoshikuso-toge—located within a 10km range (Figure 3.14); and three of them were excavated. The Higashimata quarry cluster is situated on a narrow ridge with an elevation of 1150m a.s.l., southwest of the Wada Pass. Eighteen crater-like depressions have been found on the surface. They were confirmed by excavations in 2000 to be obsidian quarry pits 3 to 4m in diameter (Miyasaka and Tanaka 2001). The quarries can be dated to the end of Early Jomon because potsherds of this phase were found around the pits. However, it is not certain if the quarrying was limited to this phase. Most of the obsidian debris found in the non-cultural layers around the pits is less than 30g in weight, and people appear to have targeted small size pieces for making little tools such as arrowheads. Jomon people collected fragments of obsidian mainly from the surface. One of the Final Jomon quarry pits was exploited because it was an obsidian vein. A test calculation suggested that quarrying of 1m^3 of vein yielded more than 500kg of obsidian, thus demonstrating its high productivity. No quarrying tools have been discovered. A large quantity of artefacts including hammerstones and obsidian flakes and cores was found around the pit indicating that people had knapped obsidian on site. Potsherds belonging to the end of the Early Jomon as well as to the Late and Final

Jomon were found around the pits indicating the periods of quarrying (Miyasaka and Tanaka 2001).

The Hoshigato quarry cluster is located on the slope of Mount Hoshigato at an elevation of about 1500m, south of the Wada Pass. About ten crater-like depressions were found, identified by excavations as obsidian quarry pits. General surveys subsequently revealed a dense cluster of depressions in an area of about 40,000m^2; the precise number cannot be counted.

The Hoshikuso-toge quarry cluster is positioned 5km to the east of the Wada Pass, on the slope of Mount Mushikura at an elevation of about 1500–1550m. About 195 crater-like depressions were found on the surface, and are thought to be obsidian quarry pits. Excavations uncovered numerous quarry pits about 3m in diameter and 3m deep, below Quarry 1, initially identified as a crater-like feature 10m in diameter. All these quarries contain obsidian in a white clay layer (Nagano City Board of Education 1999); and they can be associated with the Late Jomon because potsherds of Kasori B type (dated to ca. 3500 BP) were discovered here. Also, many potsherds of the second half of Incipient Jomon have been found on the neighbouring terrace suggesting that quarrying dates back to this phase. The Hoshikuso-toge cluster was designated as a National Historic Site because it is valuable evidence of Japanese heritage in lithic raw material procurement. The Centre for Obsidian and Lithic Studies of Meiji University (Tokyo) and the Obsidian Museum of Nagano City were built here to provide a base for obsidian research in Japan.

Obsidian Supply at Residential Sites and its Temporal Changes

What was the path for transportation of obsidian collected at the Wada – Suwa sources? This author spent five years excavating a group of sites which appear to be on the obsidian supply route. This is called the Shiono-nishi cluster, and is situated on the southern flanks of Mount Asama 40km from the Wada – Suwa sources. It is located directly on the way connecting the obsidian sources of highlands in central Honshu and the Kanto Plain (Tsutsumi 1997b).

Four major lithic raw materials – obsidian, glassy andesite (sanukite), chert, and hard shale – used for the manufacture of flake tools such as the arrowheads, drills, and tanged stone knives, are found at dwellings of the early Early Jomon (the Shimidou site), the middle Early Jomon (the Tsukata site), and the middle Middle Jomon (the Kawarada site) in the Shiono-nishi cluster (Figure 3.15). All these types of raw material have sources within a 40km radius (Figure 3.16). As for the frequency of rock types, hard shale constitutes over 50% of the total weight of flake tools, while obsidian accounts for less than 10%. However, during the middle Early Jomon and later the ratio of hard shale dropped to 10–20% whereas obsidian increased to 30–50%.

Caches of obsidian raw material have been found sporadically on sites from the second half of the Early Jomon along the transport route of the Wada – Suwa obsidian, including the Shiono-nishi cluster. At the Shimebikibara site in Gunma Prefecture located at the entrance to the Kanto Plain down from the Mount Asama flanks, seven chunks of obsidian have been found in a cache. Another hoard of seven obsidian chunks was found at the Nakano-matsubara site. It is suggested that in this period the nature of obsidian distribution changed from exchange to trade, and middlemen were involved to promote the trade (Daikuhara 2002).

On the other hand, a large quantity of obsidian was found at the Shakado site in the Kofu Basin, along the route south of the Wada – Suwa sources. Ikeya's (2006) comparison of obsidian quantities in pit-dwellings demonstrated that the total weight of obsidian reached a peak in the middle Early Jomon and then drastically reduced to one-fifth of its maximum in the Middle Jomon. It is interesting to observe that at the same time hunting weaponry such as arrowheads also tended to decrease in the Middle Jomon. In the Middle Jomon in general, an extraordinarily large quantity of lithic digging tools ("chipped stone axes") is found at residential sites. This is interpreted as a sign of transition to the specialised subsistence activity of collecting tubers such as taro. Ikeya (2006) argues that during the Middle Jomon a shift in the basic subsistence activity from hunting to the gathering of plant food such as tubers took place. This reduced the demand for small flake tools such as arrowheads and resulted in the decline of obsidian exploitation. During the Late Jomon, a large quantity of Shunshu obsidian was imported to residential sites again (Ikeya 2006). This phenomenon correlates with the intensive quarrying activities taking place at that time.

At the Tsukagoshi-mukouyama site in Chichibu region (Saitama Prefecture) located on the trade route of Shinshu obsidian, a ceremonial deposit of obsidian was found. Artefacts, including ten pieces of polished axes and obsidian chunks, were placed in a pot dated to the second half of the Middle Jomon. The pot was found at the indoor hearth. It appears to represent some kind of ritual activity, but may also demonstrate the prestigious nature of obsidian.

How does the distribution of the Kozu-shima obsidian located in the open sea differ from that of the Shinshu obsidian from the mountains? Let us review the results of Ikeya's (2005) research covering usage of the Kozu-shima source. During the Initial Jomon, Kozu-shima obsidian was exploited predominantly in the coastal region of central Honshu, but in the succeeding Early Jomon a decrease in the obsidian supply from this source occurred. Ikeya (2005) infers that the Akahoya tephra that fell over parts of Honshu at the end of the Initial Jomon caused a shift in the ecological system. This may have changed the behavioural pattern of human populations and reorganised the distribution network of Kozu-shima and Shinshu obsidians.

During the middle Early Jomon, the use of Shinshu obsidian resumed. This was a definite turning point in raw material distribution. A large quantity of Kozu-shima obsidian was

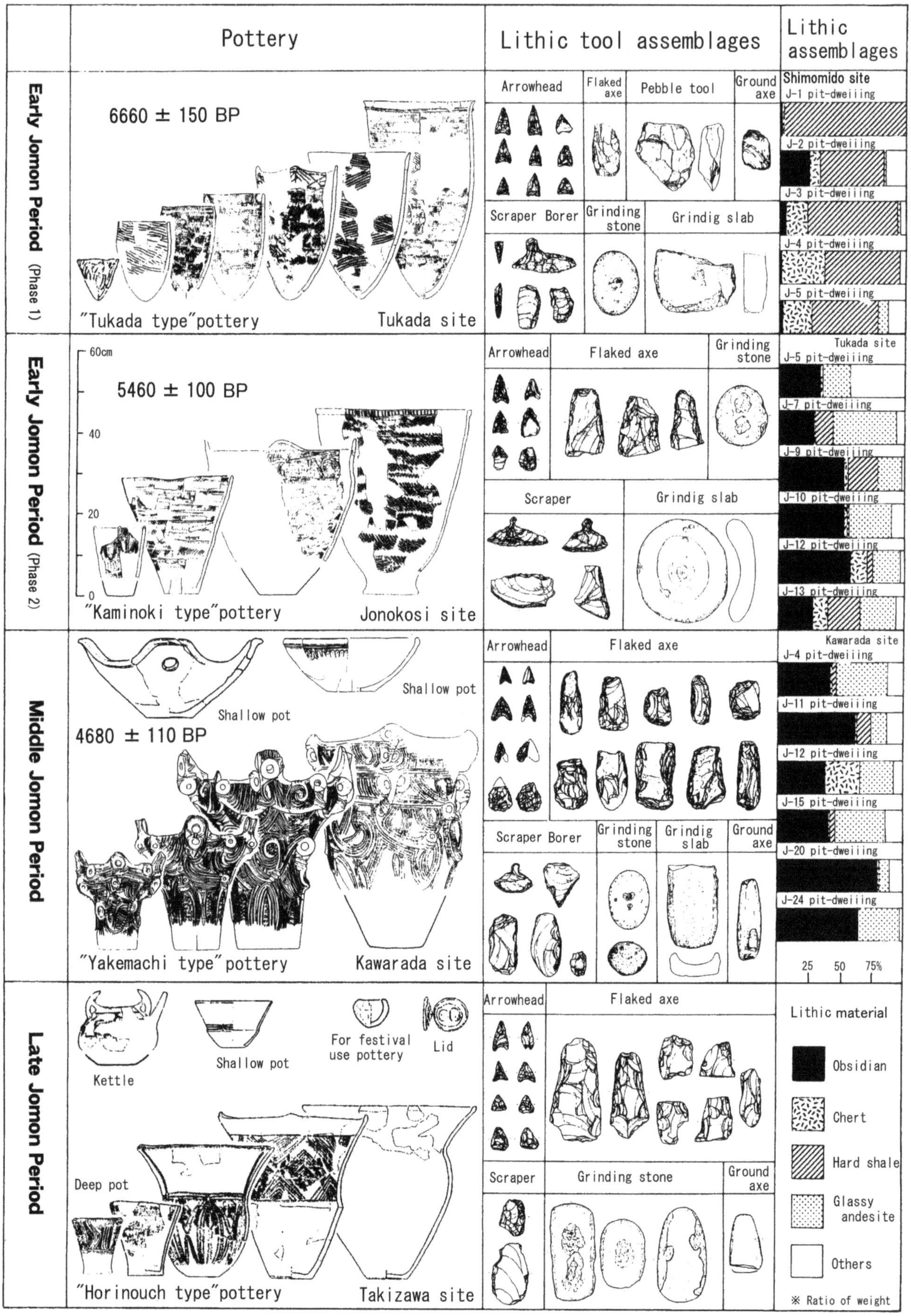

Figure 3.15. Pottery, stone artefacts, and composition of raw materials at the Shiono-nishi cluster during the Jomon period

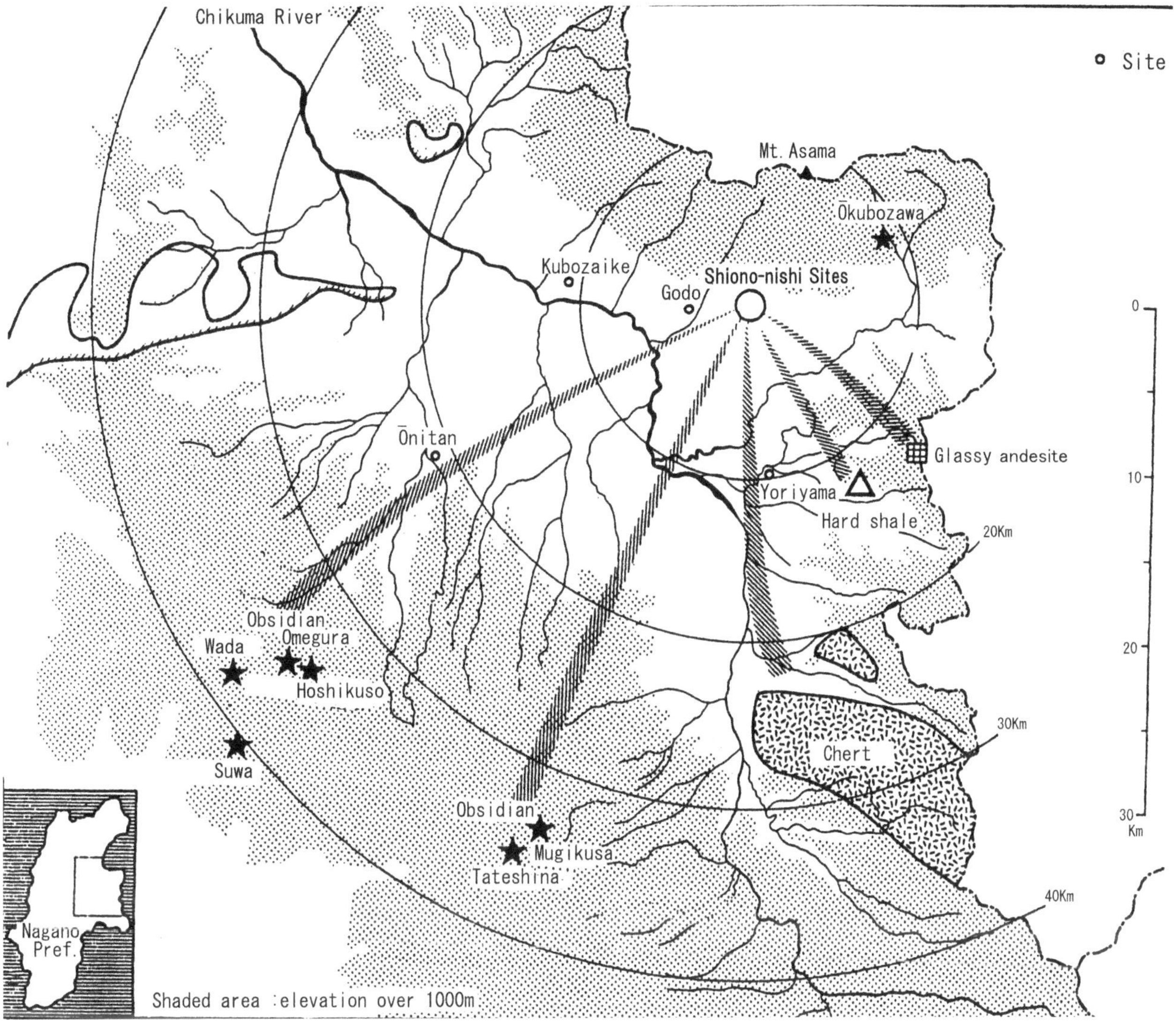

Figure 3.16. Obsidian sources used by inhabitants of the Shiono-nishi sites during the Jomon period

transported to the main island of Honshu and brought to various places (Figure 3.17). However, at the end of the Middle Jomon and in the Late Jomon the quantity of Shinshu obsidian (Wada – Suwa and Tateshina sources) increased in relation to the Kozu-shima one (Figure 3.18) In addition, Amagi obsidian was exploited in the area neighbouring the source.

Several Jomon sites have been discovered on Kozu-shima Island; but no structures directly related to obsidian procurement have been found yet. On the other hand, on Honshu, the Mitaka-danma site in the Izu Peninsula, 60km away from Kozu-shima, was identified as an obsidian "landing place". Ten pit-dwellings of the Middle Jomon were discovered, and a total of 254kg of obsidian was found in the excavation of a 740m^2 area (Kawazu Town Board of Education 1980). It was suggested that obsidian transported from Kozu-shima Island by boats was brought ashore and stored at the Mitaka-danma site as a base camp, and then distributed in the southern Kanto region (Ikeya 2005).

Problems in Jomon Obsidian Procurement

Previously, the obsidian quarries around the Wada-toge source in central Honshu and its supply system and the distribution of Kozu-shima obsidian have been described. Finally, some problems in obsidian procurement and supply research are presented here. The first problem is how far back the obsidian quarries can be dated. At the Hoshikuso-toge cluster, Incipient Jomon potsherds were discovered, but there is no firm evidence of quarrying at that time. Full-scale mining is believed to have originated in the Early Jomon as confirmed at the Hoshigato and Higashimata clusters. The second problem is how intensive was the quarrying at each mine during the obvious decline in obsidian exploitation at the Chubu and Kanto residential sites during the Middle Jomon. Active mining in the succeeding Late Jomon has been confirmed at the Hoshikuso-toge cluster.

The third problem is the ownership of quarries and lithic resources. Did the people who were engaged in mining monopolise the obsidian resources? There is a hypothesis that ownership of resources was not developed in the Jomon and all groups had access to obsidian sources, travelling from residential sites to quarries when raw materials were running low. However, there are no artefacts found at mines which indicate visits by distant groups. Considering the sedentarisation as well as the resultant fixation of territories

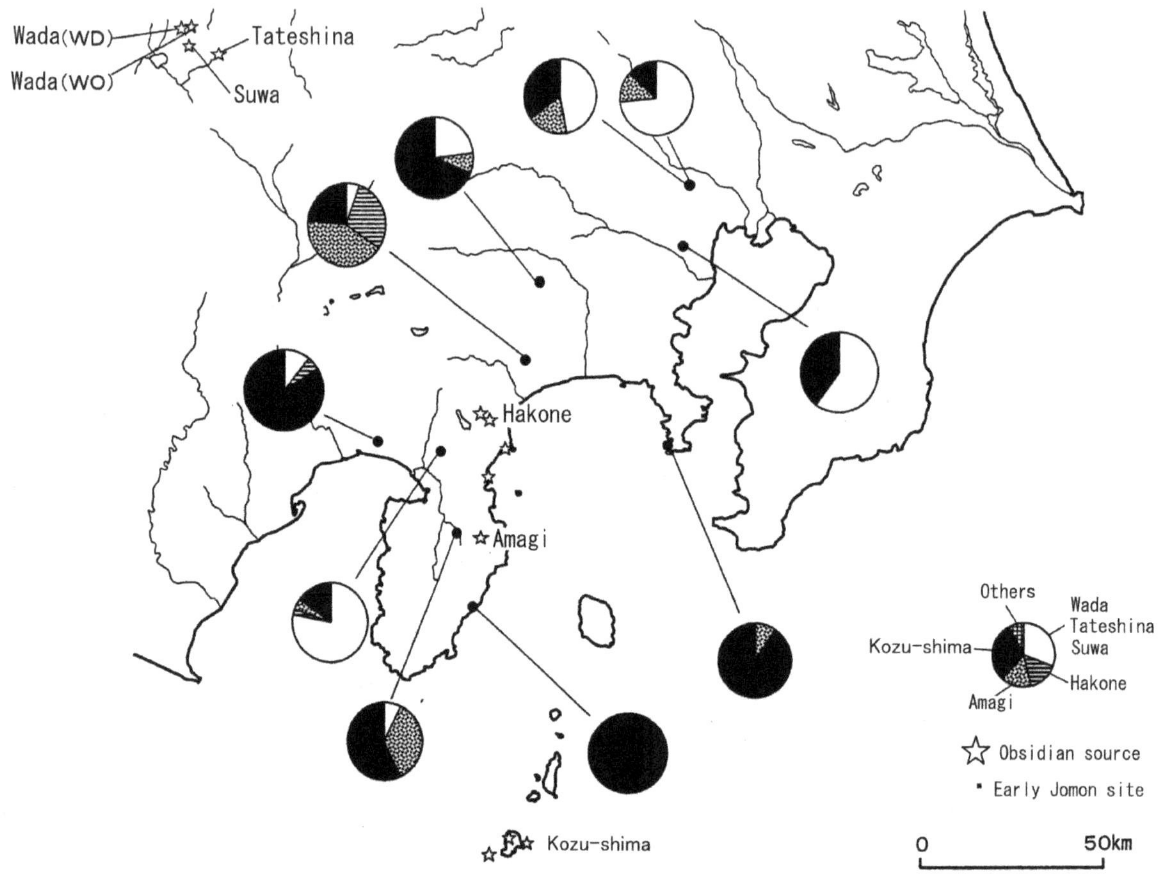

Figure 3.17. Ratios of different obsidian sources exploited at Early Jomon sites in the southern Kanto Region (after Ikeya 2005)

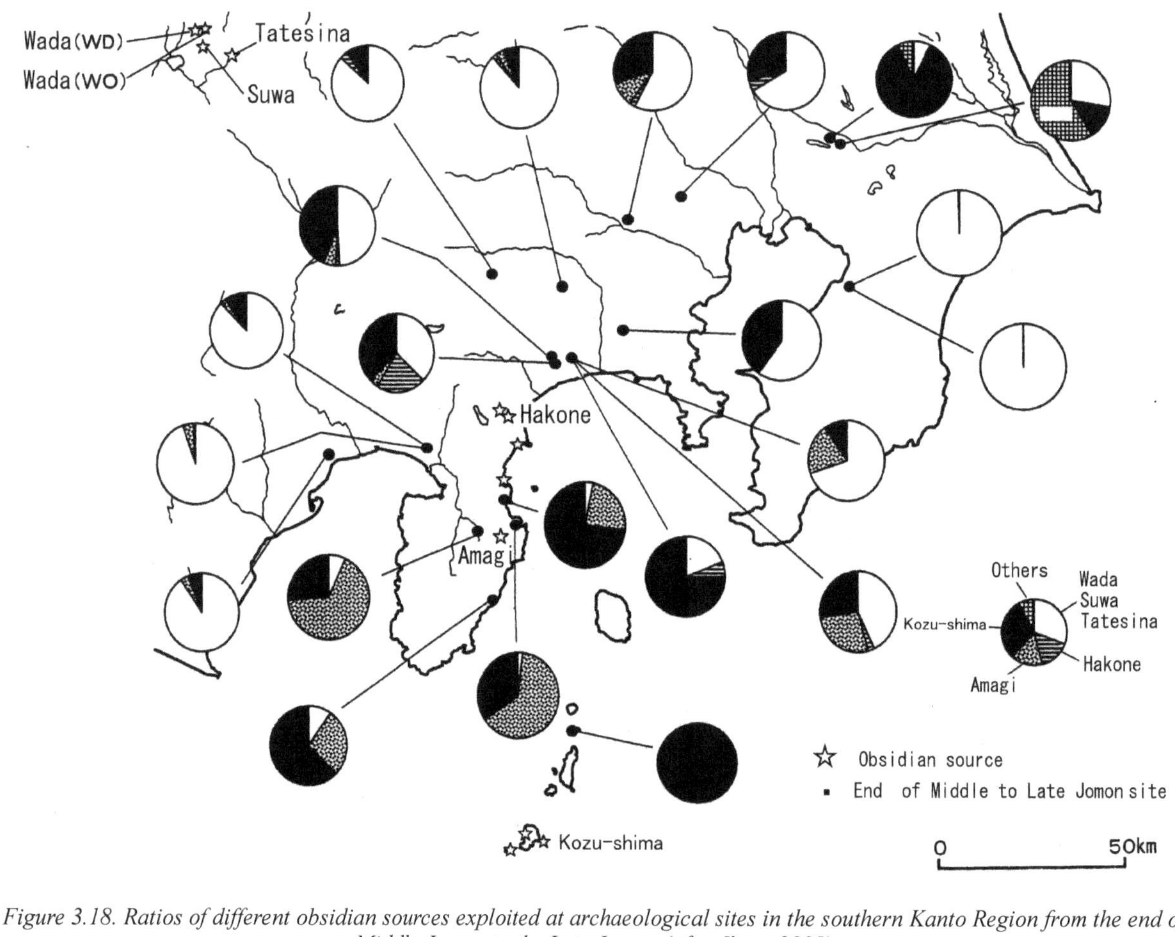

Figure 3.18. Ratios of different obsidian sources exploited at archaeological sites in the southern Kanto Region from the end of Middle Jomon to the Late Jomon (after Ikeya 2005)

from the Early Jomon onward, one might infer that a local group is likely to have controlled the quarries and exploited them, and distributed the obsidian to other groups. Nevertheless, residential sites of miners have not been found near the quarries. It is known that obsidian sources are located in the cold highlands, with elevation up to 1500m. It is believed that these places were uninhabitable in winter, although there may have been camps in other seasons. It is inferred that permanent settlements were located in the river valleys down from the sources. A number of Early Jomon and later settlements have been found in the basins around the sources, and the residents of those settlements may have been engaged in quarrying. The fourth problem is the mode of distribution at residential sites on the supply routes. One hypothesis proposes that obsidian was once stored at the main settlement on the supply path and then distributed to satellite settlements in the neighbouring area (Kanayama 1998). This question needs to be further addressed in discussion of the entire Jomon trade system.

Conclusion

This chapter has discussed obsidian exploitation on Honshu Island in the Upper Palaeolithic and the Jomon. The obsidian sources of Shinshu, Hakone, Amagi, Takaharayama, and Kozu-shima in central Honshu are known to have been exploited since the beginning of the Upper Palaeolithic (ca. 35,000 BP). The transport of obsidian from Kozu-shima Island in the open sea to mainland Honshu indicates that early Upper Palaeolithic bearers were able to cross the sea. Most probably, the colonisation of the Japanese Archipelago by anatomically modern humans cannot be dated back to beyond ca. 40,000 BP. The Japanese Islands were not connected to the Asian continent at that time, according to palaeogeographical data. Therefore, modern humans must have migrated using some kind of maritime transport. The exploitation of obsidian from sources in the open sea constitutes good circumstantial evidence of voyages from the Asian mainland to the Japanese Archipelago.

The obsidian supply zone in central Honshu in the Upper Palaeolithic was generally confined to about 200km in radius, although it varies according to the quality and size of a particular source. This economic sphere of ca. 200km is likely to represent a territorial range for regional society. However, within this space obsidian transported from remote sources was also found, suggesting the presence of a broad social network that spread across regional societies. On the other hand, obsidian exploitation of each source varies greatly through time. This can be attributed to changes in the accessibility of sources in the fluctuating natural environment, shifts in territories occupied by local human groups, and changes in the relationship between groups.

The most significant change in obsidian procurement was the transition from collecting to quarrying which occurs at the onset of Jomon. There are many quarries in the Wada – Suwa region corresponding to Early Jomon and later periods. Mining was likely undertaken by groups settled in areas neighbouring the source. The 200km range of obsidian distribution in the Jomon, almost the same as in the Upper Palaeolithic, is established. The large quantity of aquatic resources such as sea mollusc shells and fish bones found at Jomon sites clearly indicates a more advanced maritime adaptation compared to the Upper Palaeolithic. The procurement of Kozu-shima obsidian was made possible by improved seafaring techniques.

In future studies of the procurement and transport of prehistoric obsidian on Honshu Island, significantly more samples need to be studied to detect more easily minor sources of archaeological volcanic glass.

Acknowledgements

I am grateful to Drs Yaroslav V. Kuzmin and Michael D. Glascock for their invitation to participate in this volume, and for the editorial help they provided along with Dr Susan G. Keates and Ms Candace Lindsey.

References

Binford, L. R. 1979. Organization and Formation Processes: Looking at Curated Technologies. *Journal of Anthropological Research* 35, 255–273.

Daikuhara, Y. 2002. Kokuyoseki no Ryutsuu wo Meguru Shakai [Society Supervising Obsidian Distribution]. In *Jomon Shakairon Jokan,* edited by M. Anzai, 67–113. Tokyo, Douseisha Publishers.

Harunari, H. 1976. Sendoki, Jomon Jidai no Kakki ni Tsuite (1) [Epoch between the Pre-Ceramic and Jomon Periods (1)]. *Kokogaku Kenkyu* 22(4), 68–92.

Hashimoto, K. 2006. Kanjo Unit to Sekifu no Kakawari [The Relationship betwen Circular Unit and Axe]. *Kyusekki Kenkyu* 2, 35–46.

Higashimura, T., and T. Warashina. 1988. Sekki Genzai no Sanchi Bunseki [Source Analysis of Lithic Raw Materials]. In *Kokogaku to Kanren-Kagaku*, edited by H. Kobayashi, 447–491. Okayama, Japan, Kamaki Yoshimasa Sensei Koki Kinen Ronbunshu Kankou-kai.

Ikeya, N. 2005. *Kuroshio wo Watatta Kokuyouseki. Mitaka danma Iseki* [The Obsidian Transports Crossing the Japan Current: The Mitaka-danma Site]. Tokyo, Shinsensha Publishers.

Ikeya, N. 2006. Kan-Chubu-Kouchi Nantou-Iki ni Okeru Kokuyoseki Ryustuu to Gensanchi Kaihatsu [Circulation of Obsidian and its Quarry Development in the Southeast Area of Pan-Central Highlands]. *Kokuyoseki Bunka Kenkyu* 4, 161–171.

Kanayama, Y. 1989. Aira-Tn Kazanbai Koukaki Niokeru Kokuyouseki Sekkigun [Obsidian Tool Assemblages at the Time of the Aira-Tn Tephra Fall]. *Kokugakuin Daigaku Kokogaku Shiriyokan Kiyo* 6, 1–11.

Kanayama, Y. 1998. Shuraku-Kan no Koryu to Koeki [Inter-Settlement Exchanges and Trade]. *Kikan Kokogaku* 64, 59–63.

Kawazu Town Board of Education (ed.). 1980. *Mitaka-danma Iseki* [The Mitaka-danma Site]. Kawazu, Japan, Kawazu Town Board of Education.

Miyasaka, K., and S. Tanaka. 2001. *Kokuyouseki Gensanchi Iseki Bunpu-Chosa Houkokusho (Wada-toge,*

Kirigamine) I [Report of Obsidian Source Site Survey in Nagano Prefecture (Wada Pass and Kirigamine) I]. Shimosuwa, Japan, Shimosuwa Town Board of Education.

Mochizuki, A. 1997a. Keikou-X-Sen Bunseki ni Yoru Yokota Iseki Shutsudo no Kokuyousekisei Sekki no Sanchi-Suitei [Source Tracing of Obsidian Tools from the Yokota Site by X-ray Fluorescence Analysis]. *Saitama Koko Bessatsu* 5, 182–213.

Mochizuki, A. 1997b. Keikou-X-Sen Bunseki ni yoru Kashiwagaya-nagaosa Iseki Shutsudo no Kokuyousekisei Sekki no Sanchi-Suitei [Source Tracing of Obsidian Tools from the Kashiwagaya-nagaosa Site by X-ray Fluorescence Analysis]. In *Kashiwagaya-nagaosa Iseki,* edited by T. Tsutsumi, 411–439. Ebina, Japan, Kashiwagaya-nagaosa Iseki Chosadan.

Mochizuki, A. 1998. Chubu Chihou Kantou Chihou ni Okeru Naganoken-san Kokuyouseki no Ryutsuu [Distributions of Obsidian from Nagano Prefecture in the Chubu and Kanto Regions]. *Nagano-ken Kyusekki Bunka Kenkyu Kouryukai Happyou Shiryou* 10, 94–99.

Mochizuki, A. 2000a. Shimo-yanagisawa Iseki Shutsudo no Kokuyoseki wo Meguru Mondai [Problems with Obsidian from the Shimo-yanagisawa Site]. In *Shimo-yanagisawa Iseki,* edited by N. Kameda, 578–595. Tokyo, Waseda Daigaku Bunkazai Seirishitsu.

Mochizuki, A. 2000b. Kokuyouseki Sanchi Suitei Houkoku [Report of Obsidian Source Tracing]. In *Musashi Kokubunjiato Chousa Houkoku-Hokusei Chiiki (Tamaranzaka Iseki) no Chousa 4,* edited by S. Nakayama, 163–170. Tokyo, Fuchu City Board of Education.

Mochizuki, A. 2002a. Nojiri-ko Isekigun Shutsudo Kokuyouseki no Sanchi Suitei (I) [Source Tracing of Obsidian from the Nojiri-ko Cluster (I)]. In *Hinatabayashi B Iseki, Hinatabayashi A Iseki, Nanatsuguri Iseki, Ohira B Iseki,* edited by K. Tani, 233–241. Nagano, Nagano-ken Maizoubunkazai Centre.

Mochizuki, A. 2002b. Kokuyouseki Bunsekikagaku no Genjou to Tenbou [The Current State and Future Prospect of the Science of Obsidian Analysis]. *Kokuyouseki Bunka Kenkyu* 1, 95–102.

Mochizuki, A. 2004. Keikou-X-Sen Bunseki ni Yoru Tana-Shioda Isekigun Mukaihara Iseki No. 4 Chiten Shutsudo Kokuyousekisei Sekki no Sanchi Suitei [Source Tracing of Obsidian Tools from Locality 4 at the Mukaihara Site in the Tana-Shioda Cluster by X-ray Fluorescence Analysis]. In *Tana-mukaihara Iseki II,* edited by Sagamihara City Board of Education, 215–220. Sagamihara, Japan, Sagamihara City Board of Education.

Mochizuki, A., Y. Abe, and H. Ishikawa. 2003. Tsukimino Isekigun Dai I, II Iseki Shutsudo Kokuyouseki Sekkigun no Sanchi Suitei-Bunseki [Source Tracing Analysis of Obsidian Assemblages from the Tsukimino I and II Sites]. *Kokuyouseki Bunka Kenkyu* 2, 97–124.

Mochizuki, A., N. Ikeya, K. Kobayashi, and Y. Muto. 1994. Isekinai ni Okeru Kokuyosekisei Sekki no Gensanchi Bunseki ni Tsuite – Numazu-shi Doteue Iseki BB5 Sou no Gensanchi Suitei Kara [Intersite Distribution of Obsidian from Different Sources – Source Tracing at Layer BB5 at the Doteue Site, Numazu City]. *Shizuoka-ken Kokogaku Kenkyu* 26, 64–71.

Mochizuki, A., and T. Tsutsumi. 1997. Sagamino-Daichi no Saisekijin-Sekkigun no Kokuyouseki Riyou ni Kansuru Kenkyu [Studies on Obsidian Exploitation in the Microblade Industry on the Sagamino Upland]. *Yamato-Shishi Kenkyu* 23, 1–36.

Nagano City Board of Education (ed.). 1999. *Takayama Isekigun* [The Takayama Sites]. Nagano, Nagano City Board of Education.

Ninomiya, S., and K. Shimadate. 2001. Shizenkagakuteki Shuhou ni Yoru Bunseki [Analysis by the Methods of Natural Sciences]. *Kenkyu Kiyou* 22, 65–100.

Ono, A. 1975. Sendoki-Jidai Sekizai-Unpanron Note [A Note on Lithic Raw Material Transport in the Pre-Ceramic Period]. *Kokogaku Kenkyu* 21(4), 9–22.

Osawa, M., and S. Ninomiya (eds). 1991. *Kokuyoseki no Kagaku-Sosei* [Chemical Compositions of Obsidian]. Tokyo, Tokyo Gakugei Daigaku.

Sano City Board of Education (ed.). 2004. *Kambayashi Iseki* [The Kambayashi Site]. Sano, Japan, Sano City Board of Education.

Sato, H. 1996. Shakai-Kouzou [Social Structure]. *Sekki Bunka Kenkyu* 5, 329–340.

Sato, H., and T. Tsutsumi. 2007. The Japanese Microblade Industries: Technology, Raw Material Procurement, and Adaptations. In *Origin and Spread of Microblade Technology in Northern Asia and North America*, edited by Y. V. Kuzmin, S. G. Keates and C. Shen, 53–78. Burnaby, B.C., Canada, Archaeology Press.

Serizawa, S. 2006. Takaharayama-San Kokuyouseki no Ryutsuu ni Tsuite [On the Distribution of the Takaharayama Obsidian]. In *Takaharayama-San Kokuyouseki Chousa-jigyou Houkokusho,* edited by Yaita City Board of Education, 43–63. Yaita, Japan, Yaita City Board of Education.

Suwama, J. 2006. Sagamino-Daichi ni Okeru Kokuyoseki Riyo no Hensen [Historical Changes of Obsidian Use at Sagamino Upland]. *Kokuyoseki Bunka Kenkyu* 4, 151–160.

Suzuki, M. 1973. Chronology of Prehistoric Human Activity in Kanto, Japan. Part 1: Framework for Reconstructing Prehistoric Human Activity in Obsidian. *Journal of the Faculty of Science, University of Tokyo. Section 5 (Anthropology)* 4(3), 241–318.

Tamura, T. 1992. Toi Yama, Kuroi Ishi – Musashino II Ki Sekkigun no Shakaiseitai Gakuteki Ichi Kosatsu [Distant Mountains, Black Stones – Socio-Ecological Examination of the Musashino Phase II Lithic Assemblages]. *Senshi Kokogaku Ronshu* 2, 1–46.

Tamura, T., S. Kunitake, and M. Oya. 2006. Tochigi-Ken Takaharayama Kokuyoseki Gensanchi-Isekigun no Hakken to Sono Hyoka [New Discoveries of Palaeolithic Quarry Sites for Obsidian Resources: Mount Takahara and it's Meaning]. *Nihon Kokogaku* 22, 147–165.

Tani, K (ed.). 2000. *Hinatabayashi B Iseki; Hinatabayahsi A Iseki; Nanatsuguri Iseki; Ohira B Iseki* [The Hinatabayashi B Site; the Hinatabayashi A Site; Nanatsuguri Site; and the Ohira B Site]. Nagano, Nagano-ken Maizoubunkazai Centre.

Tsutsumi, T. (ed.). 1997a. *Kashiwagaya-nagaosa Iseki* [The

Kashiwagaya-nagaosa Site]. Ebina, Japan, Kashiwagaya-nagaosa Iseki Chousadan,.

Tsutsumi, T. (ed.). 1997b. *Kawarada Iseki* [The Kawarada Site]. Miyota, Japan, Miyota City Board of Education.

Tsutsumi, T. 2000. Souki no Kinou to Kanreitekio to Shiteno Hikaku-Riyo System [Environmental Adaptation by the Analysis of Endscrapers in the Upper Palaeolithic of Japan]. *Kokogaku Kenkyu* 47, 66–84.

Tsutsumi, T. 2002. Shinshu Kokuyouseki Gensanchi wo Meguru Shigen-Kaihatsu to Shigen-Jukyu [Obsidian Resource Exploitation in the Upper Palaeolithic in Japanese Central Highlands]. *Kokugakuin Daigaku Kokogaku Shiryokan Kiyo* 18, 1–21.

Tsutsumi, T. 2004. Nihon Rettou no Saisekijin Iseki to Saisekijin Shiryo [Microblade Sites and Assemblages in the Japanese Archipelago]. In *Nihon no Saisekijin Bunka III*, edited by Yatsugatake Palaeolithic Research Group, 86–87. Nagano, Yatsugatake Palaeolithic Research Group.

Tsutsumi, T. 2006. Yuri Shigenari Collection ni Miru Yadegawa Iseki no Saisekijin Sekkigun [Microblade Assemblage at the Yadegawa Site in Shigenari Yuri's Collection]. *Kokuyoseki Bunka Kenkyu* 4, 97–124.

Tsutsumi, T. 2007. The Dynamics of Obsidian Use by the Microblade Industries of the Terminal Late Palaeolithic. *Daiyonki Kenkyu* 46, 179–186.

Tsutsumi, T., M. Morishima, and K. Moriyama (eds). 1993. *Nagano-ken Kokuyouseki Gensanchi Iseki Bunpu Chousa Houkokusho (Wada-toge, Omegura) III* [Report of Obsidian Source Site Survey in Nagano Prefecture (Wada Pass and Omegura) III]. Wada, Japan, Wada Town Board of Education.

Warashina, T. 2004. Musashi-kokubunji-ato Kanren Iseki (Musashidai-nishi Chiku) Shutsudo Kokuyoseki no Gensanchi-Dotei [Source Tracing of Obsidian from the Musashi-kokubunji-kanren Site (Musashidai West Area)]. In *Musashi-Kokubunji-ato Kanren Iseki (Musashidai-nishi Chiku)*, edited by S. Nakayama, 310–324. Tokyo, Tokyo-to Maizoubunkazai Centre.

Warashina, T., and T. Higashimura. 1996. Tarukuchi Iseki Shutsudo Kokuyouseki, Anzangansei Sekki no Sekizai Sanchi Bunseki [Source Analysis of Obsidian and Andesite Tools from the Tarukuchi Site]. In *Tarukuchi Iseki*, edited by Asahi Town Board of Education, 176–185. Asahi, Japan, Asahi Town Board of Education.

Watanabe, H. 1990. *Jomon-shiki Kaisoka Shakai* [The Hierarchical Society of Jomon]. Tokyo, Rokko Shuppan Publishers.

Wiessner, P. 1982. Beyond Willow Smoke and Dog's Tails: A Comment on Binford's Analysis of Hunter-Gatherer Settlement Systems. *American Antiquity* 47, 171–178.

Yaita City Board of Education (ed.). 2006. *Takaharayama-San Kokuyouseki Chousa Jigyou Houkokusho* [Takaharayama Obsidian Research Project Report]. Yaita, Japan, Yaita City Board of Education.

Yamamoto, K. 1990. Space-Time Analysis of Raw Material Utilization for Stone Implements of the Jomon Culture in Japan. *Antiquity* 64, 868–889.

Chapter 4
Obsidian Trade Between Sources on Northwestern Kyushu Island and the Ryukyu Archipelago (Japan) During the Jomon Period

Hiroki Obata, Isao Morimoto, and Susumu Kakubuchi

Abstract: *The Ryukyu Archipelago is located about 100–700km south of Kyushu Island. While there are about 30 obsidian sources known on the Kyushu, there are none on the Ryukyus. Obsidian artefacts on the Ryukyus are recognised as one of the most persuasive pieces of evidence for prehistoric contacts. Until recently, the data remained vague because only limited geochemical analysis had been carried out. More than 140 obsidian artefacts from 39 sites were examined by X-ray Fluorescence analysis in this study. Obsidian from the Koshidake source on Kyushu Island comprises 94% of the total sample. The transport of Kyushu obsidian did not begin until the mid-Late Jomon which corresponds to the appearance of relatively strong cultural influences on the southern Ryukyus from the Kyushu Island. Observations of stone tools clearly show frequent use-wear or retouches on the edges, reduction processes, flakes manufactured by bipolar techniques, as well as some imitations of arrowheads imported from Kyushu Island. This suggests a limitation in obsidian availability which is more common in the southern than in the northern part of the Ryukyus. The total amount of obsidian in prehistoric assemblages from the Ryukyus is small, and it seems that obsidian trade with the Kyushu Island was not continuous.*

Keywords: *Obsidian, Sourcing, Jomon, Ryukyu Archipelago, Kyushu Island, Japan*

Introduction

About 30 sources of obsidian are known on Kyushu Island, and most of them were exploited in prehistoric times. The source with the best quality obsidian is the Koshidake Mount located near Imari City in northwestern Kyushu (Saga Prefecture). People began to utilise Koshidake obsidian at ca. 30,000 BP, and it was widely spread around the Kyushu Island by about 9000 BP. Several archaeologists have suggested that during the second half of the Jomon period Koshidake obsidian was transported by boats to the southern coast of the Korean Peninsula and to the Ryukyu island chain. Recent geochemical analyses support this (see below). Obsidian tools from northwestern Kyushu Island found in those regions may demonstrate that there was cultural exchange between the Kyushu, Korea, and the Ryukyu Archipelago. However, without identifying the origins of obsidian artefacts through geochemical analysis or without information about characteristics of obsidian tools in archeological context, we cannot address questions about whether obsidian was brought by boats of colonising populations or as an item of trade.

Geological Background of the Ryukyu Archipelago

No obsidian sources exist on the Ryukyus which lies south of Kyushu Island (Figure 4.1). The bedrock of the principal islands of this archipelago consists of Mesozoic and Palaeozoic formations, and the region from the northern part of Okinawa Island to Amami Ō-shima Island (Kagoshima Prefecture) is the southwestern prolongation of the Shimanto Supergroup. While Tertiary granites and other plutonic rocks are known on the Ryukyus, obsidian deposits are not found (NCKHI 1998). It seems most unlikely that such discoveries will be made south of the Amami island group in the future. Obsidian stone tools corresponding to the Jomon and Yayoi periods found on the Amami and Okinawa island groups in the northern part of Ryukyu region may have been brought from the main Kyushu Island. The distance from the nearest obsidian source on the southern Kyushu to the westernmost archaeological site in the Ryukyus where obsidian tools were discovered is about 700km. The site is located on the Kerama island group (Figure 4.1). There are several straits between this site and Kyushu Island which are difficult to cross, and make the Okinawa island group geographically extremely 'far' from the Kyushu.

Research History and Issues

Kamimura (1998) and Oda (2000, 2001, 2003) studied obsidian artefacts from the Ryukyu Archipelago. Kamimura (1998) documented 22 sites with obsidian tools and made the important statement that obsidian may have been brought from the Kyushu sources, most probably from the Koshidake. In addition, he assumed that some of the obsidian artefacts on the Ryukyus may have come from other sources than Koshidake, and stressed the need for geochemical study. However, he did not conduct chemical analysis but only made observations by naked eye. Following T. Kamimura's initial research, Oda (2000, 2001, 2003) collected obsidian flakes from 26 sites on the Ryukyu Islands. He pointed out that physical and chemical analyses (X-ray Fluorescence) were carried out on samples from only three sites: Nakadomari, Izena shell mound, and Ufujika. Observations by naked eye showed that some items are different in colour and texture from the Koshidake obsidian, thus highlighting the importance of additional geochemical examination.

As T. Kamimura and S. Oda stated, instrumental geochemical analysis of obsidian tools from the Ryukyu region has been conducted infrequently, and assumptions about the origins of the raw material have little scientific support. Furthermore, reports on obsidian tools do not form

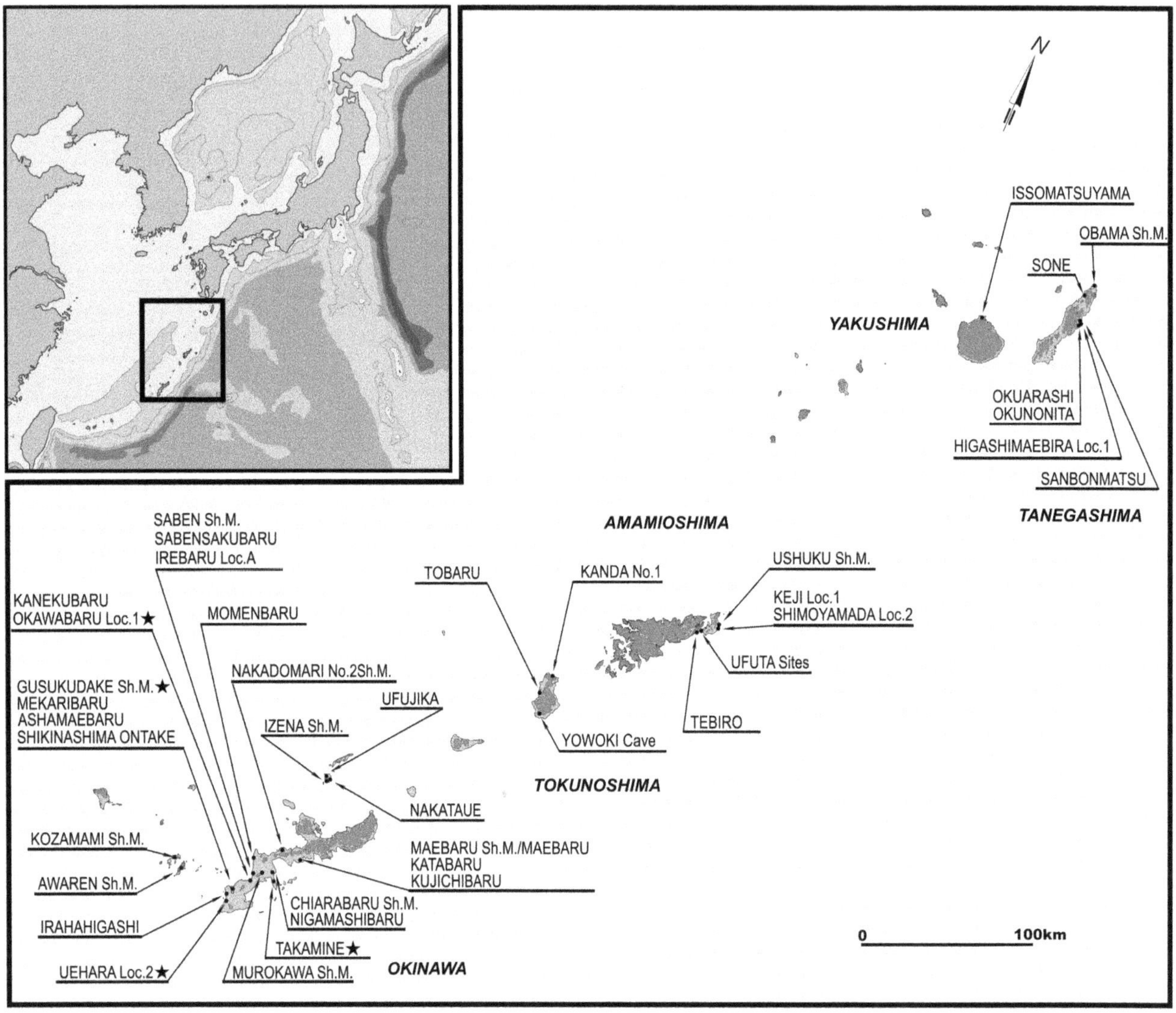

Figure 4.1. Location of archaeological sites with obsidian tools on the Ryukyu Islands; stars indicate non-analysed sites

a reliable dataset because there are inadequacies in terms of recognition and description of artefacts. It is a common rule to pay attention mainly to the archaeological aspects of quality and quantity of obsidian tools when it comes to getting into the discussion of obsidian trade. Although Oda (2003) took notice of that, his "double patina" hypothesis based on the existence of clearly different degrees of patinisation on the surfaces of obsidian tools is insufficient in that respect.

Our research programme took into account these problems; analysis was attempted on as many samples as possible to present observations from the archaeological point of view as well as qualitative and quantitative data. Sixty two samples from the Okinawa island group representing 22 archaeological sites, 20 samples from Tokuno-shima Island (three sites), 36 samples from Amami Ō-shima Island (seven sites), as well as 22 reference samples from Tanega-shima Island (six sites), and 6 reference samples from Yaku-shima Island (one site) were analysed. Altogether, 145 specimens from 39 sites were studied (Obata *et al.* 2004). The total number of obsidian finds on the Okinawa and Amami island groups is now 385 items from 36 sites (Figure 4.1).

Methods

For identification of the obsidian raw material's origin, Energy Dispersive X-ray Fluorescence (hereafter – EDXRF) analysis (e.g., Higashimura 1986; Mochizuki 1996) was performed; this allows non-destructive geochemical study (Kakubuchi and Utsunomiya 2002, 2003). Ten elements were measured: zirconium (Zr), yttrium (Y), strontium (Sr), rubidium (Rb), zinc (Zn), nickel (Ni), iron (Fe), manganese (Mn), and titanium (Ti). A reference sample was made from Japanese Standard Rock Sample JB-1 using the fusing agent Spectromelt A12 (Merk Co.). The glass tablet with a ratio between rock and fusing agent of 1:2 was made to measure the background intensity at three points: the Kα position, and the high and low angles, to calculate the real intensity (the measurement intensity or background intensity of the Kα line). Thus, the ratio of the X-ray intensity between the standard sample and the analysed specimen was determined. The reason for using a standard sample during the measurement was to avoid deterioration of precision due to the variation of X-ray intensity. The ratio of X-ray intensity was converted into X-ray counts, thus the intensity ratio between all measured elements in the samples could be determined.

Analyses were carried out with EDXRF equipment (Nihon Denshisei JSX-60S7) at the Department of Culture and Education, Saga University. Measurements were conducted using a rhodium (Rh) X-ray tube with 50kV voltage and 40mA current. A spectrum crystal LiF (200) was used, and scintillation counters were employed as detectors. After an obsidian artefact was set in a sample holder, it was covered with 25μm film to fix the sample into position. To adjust for irregularities on the sample's surface, measurements were carried out with the option "Sample spin ON".

Results

Geochemical data are shown in Tables 4.1–4.2. Determination of the origin of obsidian was achieved by comparing the X-ray intensity ratio of the measured elements with the values already obtained from obsidian sources throughout the Kyushu region. Basically, every element must be taken into account for determination of the origin; however, the Rb/Sr–Sr/Zr and Rb/Sr–Fe/Ze plots with the greatest diversity were finally used. The results for each island are presented below (see also Figures 4.2–4.3).

Okinawa, Tokuno-shima, and Amami Ō-shima Islands

Among the specimens from the main island of Okinawa, sample OK0057 is obviously not obsidian but artificial glass; samples OK0051 and OK1001 are also glass rather than obsidian. Sample OK0012 is identified as from the Hariojimakita (Hodaiyama) source, and other samples originate from the Koshidake source. All obsidian tools on the Tokuno-shima Island belong to the Koshidake source. Among the obsidian excavated at Amami Ō-shima

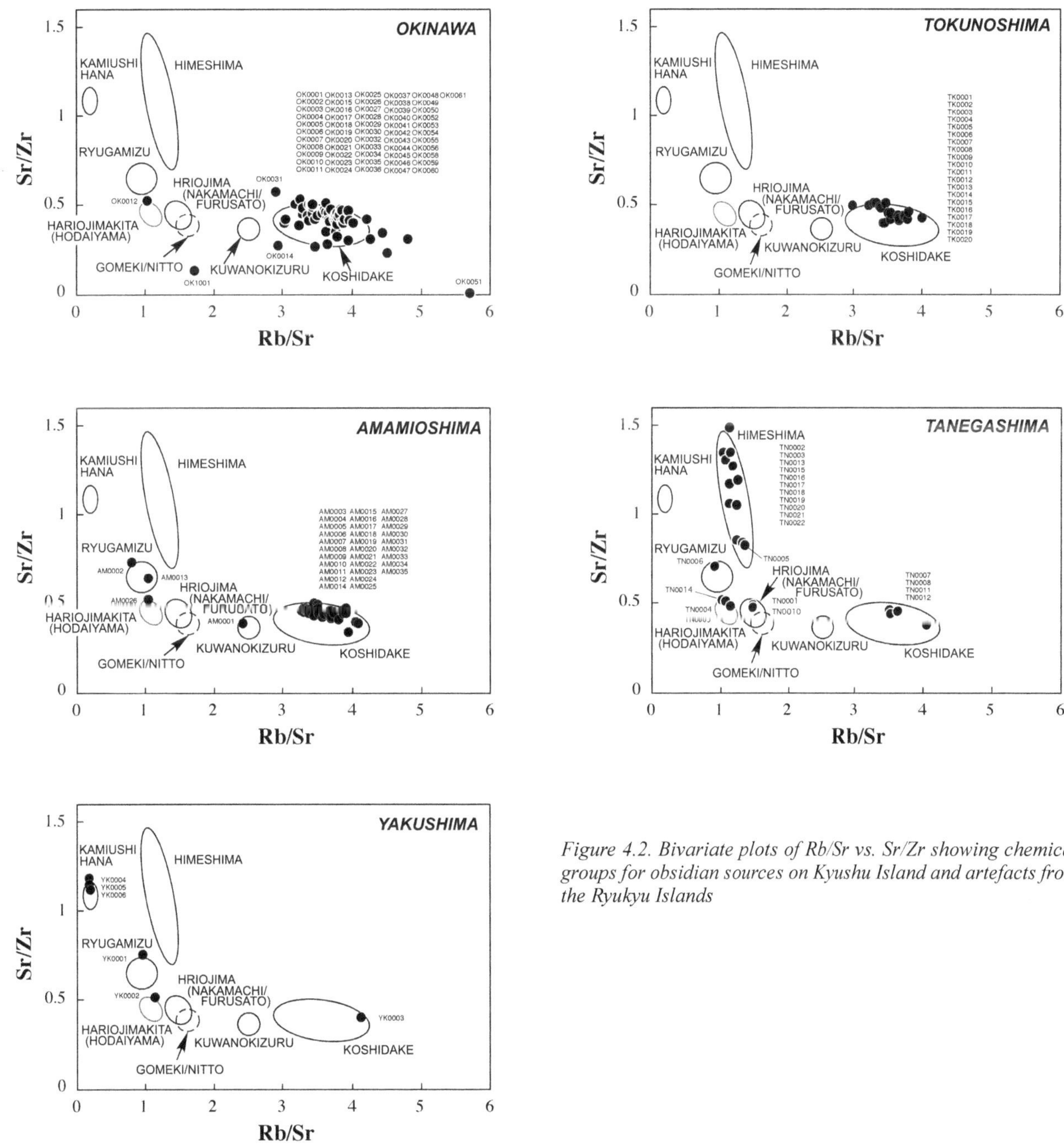

Figure 4.2. Bivariate plots of Rb/Sr vs. Sr/Zr showing chemical groups for obsidian sources on Kyushu Island and artefacts from the Ryukyu Islands

Table 4.1. Elemental composition ratios for obsidian from the Ryukyu Archipelago

Okinawa Island

Sample No.	OK0001	OK0002	OK0003	OK0004	OK0005	OK0006	OK0007	OK0008	OK0009	OK0010	OK0011	OK0012	OK0013	OK0014	OK0015	OK0016
Nb/Zr	0.194	0.212	0.274	0.259	0.220	0.234	0.195	0.212	0.308	0.241	0.237	0.116	0.205	0.262	0.257	0.218
Y/Zr	0.619	0.641	0.568	0.560	0.556	0.620	0.504	0.608	0.663	0.679	0.598	0.240	0.413	0.489	0.611	0.619
Sr/Zr	0.425	0.497	0.469	0.512	0.432	0.522	0.415	0.492	0.432	0.492	0.403	0.538	0.359	0.405	0.486	0.464
Rb/Zr	1.546	1.698	1.691	1.716	1.484	1.769	1.552	1.612	1.607	1.614	1.514	0.543	1.293	1.215	1.851	1.652
Zn/Zr	0.635	0.400	0.533	0.197	0.294	0.211	0.147	0.669	0.277	0.249	0.378	0.187	1.477	0.185	0.534	0.169
Ni/Zr	0.767	0.507	0.640	0.403	0.585	0.895	0.202	0.820	0.476	0.307	0.359	0.259	1.913	0.477	0.417	0.433
Fe/Zr	5.972	7.552	6.569	6.534	8.291	8.121	5.562	7.387	7.271	7.424	6.429	4.915	7.906	5.294	6.288	5.862
Mn/Zr	0.322	0.250	0.300	0.223	0.245	0.284	0.169	0.343	0.238	0.228	0.234	0.142	0.600	0.148	0.220	0.173
Ti/Zr	0.035	0.028	0.014	0.019	0.022	0.024	0.019	0.033	0.028	0.025	0.027	0.021	0.028	0.014	0.038	0.015
Rb/Sr	3.638	3.416	3.607	3.355	3.440	3.389	3.743	3.274	3.723	3.280	3.762	1.009	3.602	2.999	3.809	3.562
Estimated Source*	KOSHI	KOSHI	KOSHI	KOSHI	KOSHI	KOSHI	KOSHI	KOSHI	KOSHI	KOSHI	KOSHI	HARI-K	KOSHI	KOSHI	KOSHI	KOSHI

Sample No.	OK0017	OK0018	OK0019	OK0020	OK0021	OK0022	OK0023	OK0024	OK0025	OK0026	OK0027	OK0028	OK0029	OK0030	OK0031	OK0032
Nb/Zr	0.247	0.255	0.271	0.254	0.194	0.242	0.237	0.251	0.265	0.226	0.256	0.197	0.254	0.243	0.231	0.238
Y/Zr	0.591	0.639	0.627	0.707	0.488	0.616	0.567	0.590	0.614	0.634	0.715	0.636	0.691	0.608	0.642	0.639
Sr/Zr	0.425	0.471	0.459	0.458	0.419	0.480	0.443	0.477	0.519	0.510	0.431	0.410	0.485	0.513	0.585	0.510
Rb/Zr	1.784	1.843	1.747	1.719	1.651	1.841	1.674	1.872	1.875	1.617	1.672	1.581	1.780	1.754	1.684	1.734
Zn/Zr	0.165	0.228	0.196	0.169	0.740	0.217	0.259	0.131	0.140	0.290	0.216	0.372	0.215	0.177	0.139	0.138
Ni/Zr	0.484	0.293	0.254	0.257	0.985	0.198	0.457	0.045	0.276	0.577	0.342	0.339	0.225	0.102	0.086	0.072
Fe/Zr	6.792	6.215	5.836	6.998	6.587	6.136	7.533	5.095	6.078	8.474	7.311	6.446	6.165	6.075	5.697	6.430
Mn/Zr	0.197	0.208	0.186	0.214	0.360	0.204	0.238	0.159	0.215	0.258	0.194	0.248	0.205	0.186	0.167	0.181
Ti/Zr	0.019	0.028	0.015	0.042	0.021	0.015	0.031	0.019	0.023	0.050	0.026	0.034	0.021	0.012	0.029	0.026
Rb/Sr	4.195	3.916	3.807	3.753	3.944	3.835	3.779	3.920	3.610	3.171	3.876	3.855	3.671	3.419	2.878	3.398
Estimated Source	KOSHI	KOSHI	KOSHI	KOSHI	KOSHI	KOSHI	KOSHI	KOSHI	KOSHI	KOSHI	KOSHI	KOSHI	KOSHI	KOSHI	KOSHI	KOSHI

Sample No.	OK0033	OK0034	OK0035	OK0036	OK0037	OK0038	OK0039	OK0040	OK0041	OK0042	OK0043	OK0044	OK0045	OK0046	OK0047	OK0048
Nb/Zr	0.271	0.235	0.231	0.171	0.226	0.239	0.210	0.075	0.290	0.169	0.049	0.000	0.184	0.240	0.266	0.100
Y/Zr	0.605	0.698	0.645	0.601	0.706	0.682	0.683	0.481	0.598	0.565	0.356	0.278	0.356	0.626	0.636	0.309
Sr/Zr	0.450	0.443	0.462	0.449	0.507	0.467	0.539	0.429	0.455	0.317	0.317	0.284	0.290	0.350	0.407	0.274
Rb/Zr	1.566	1.657	1.678	1.500	1.725	1.759	1.745	1.301	1.487	1.518	1.345	0.826	1.053	1.549	1.627	0.945
Zn/Zr	0.767	0.147	0.139	0.808	0.187	0.179	0.530	0.629	0.246	0.844	2.779	1.412	0.671	0.552	0.561	3.171
Ni/Zr	1.050	0.149	0.118	0.994	0.405	0.203	0.367	0.840	0.244	0.591	3.613	1.434	0.694	0.681	0.514	4.018
Fe/Zr	7.050	5.234	5.440	6.476	6.327	6.543	7.575	8.093	8.585	9.806	14.494	15.399	9.394	10.253	10.033	13.698

Mn/Zr	0.326	0.184	0.180	0.371	0.203	0.195	0.283	0.339	0.312	0.383	1.267	0.695	0.357	0.395	0.374	1.381
Ti/Zr	0.037	0.015	0.013	0.013	0.012	0.015	0.054	0.017	0.057	0.109	0.151	0.150	0.120	0.109	0.175	0.155
Rb/Sr	3.480	3.744	3.633	3.341	3.400	3.765	3.239	3.030	3.270	4.786	4.247	2.910	3.625	4.422	3.996	3.449
Estimated Source	KOSHI	KOSHI	KOSHI	KOSHI	KOSHI	KOSHI	KOSHI	KOSHI	KOSHI	KOSHI	KOSHI	KOSHI	KOSHI	KOSHI	KOSHI	KOSHI

Sample No.	OK0049	OK0050	OK0051	OK0052	OK0053	OK0054	OK0055	OK0056	OK0057	OK0058	OK0059	OK0060	OK0061	OK1001
Nb/Zr	0.120	0.205	0.023	0.235	0.176	0.168	0.192	0.167	0.032	0.238	0.251	0.220	0.249	0.119
Y/Zr	0.410	0.479	0.076	0.589	0.569	0.527	0.410	0.454	0.081	0.591	0.556	0.581	0.643	0.135
Sr/Zr	0.307	0.331	0.013	0.415	0.360	0.425	0.235	0.391	0.364	0.466	0.417	0.438	0.487	0.137
Rb/Zr	1.204	1.247	0.072	1.592	1.351	1.427	1.055	1.261	0.291	1.592	1.525	1.591	1.748	0.232
Zn/Zr	0.730	0.634	0.024	0.324	0.832	0.888	1.196	0.989	1.045	0.146	0.285	0.746	0.154	0.030
Ni/Zr	0.705	0.581	0.015	0.280	0.795	1.106	0.794	0.846	1.417	0.103	0.243	0.927	0.073	0.020
Fe/Zr	11.333	10.063	1.705	9.068	10.063	10.017	10.131	12.325	8.776	7.429	8.622	9.863	8.113	2.568
Mn/Zr	0.421	0.371	0.037	0.325	0.405	0.514	0.483	0.590	0.457	0.243	0.289	0.491	0.268	0.045
Ti/Zr	0.113	0.113	0.018	0.055	0.111	0.091	0.162	0.143	0.207	0.042	0.079	0.080	0.048	0.021
Rb/Sr	3.917	3.763	5.693	3.835	3.748	3.361	4.489	3.222	0.802	3.416	3.660	3.632	3.586	1.699
Estimated Source	KOSHI	KOSHI	Not obsidian	KOSHI	KOSHI	KOSHI	KOSHI	KOSHI	Not obsidian	KOSHI	KOSHI	KOSHI	KOSHI	Not obsidian

Tokuno-shima Island

Sample No.	TK0001	TK0002	TK0003	TK0004	TK0005	TK0006	TK0007	TK0008	TK0009	TK0010	TK0011	TK0012	TK0013	TK0014	TN0014	TK0015
Nb/Zr	0.237	0.160	0.259	0.274	0.248	0.228	0.231	0.240	0.254	0.214	0.247	0.215	0.239	0.218	0.181	0.233
Y/Zr	0.751	0.730	0.713	0.699	0.747	0.600	0.634	0.649	0.576	0.725	0.612	0.541	0.568	0.479	0.313	0.671
Sr/Zr	0.498	0.427	0.450	0.412	0.489	0.411	0.430	0.476	0.444	0.497	0.486	0.408	0.385	0.387	0.760	0.444
Rb/Zr	1.725	1.556	1.710	1.648	1.573	1.556	1.614	1.610	1.553	1.658	1.450	1.443	1.325	1.332	0.838	1.571
Zn/Zr	0.430	0.653	0.146	0.263	0.155	0.393	0.205	0.254	0.290	0.321	1.230	0.342	0.610	0.484	1.244	0.252
Ni/Zr	1.092	1.532	0.078	0.254	0.126	0.543	0.175	0.333	0.293	0.292	1.684	0.354	0.713	0.560	1.147	0.224
Fe/Zr	12.792	17.302	7.317	8.117	8.347	9.513	7.808	9.180	9.308	9.364	11.686	8.769	12.390	12.169	18.150	9.499
Mn/Zr	0.406	0.462	0.265	0.299	0.263	0.373	0.285	0.284	0.351	0.333	0.689	0.279	0.382	0.491	0.962	0.298
Ti/Zr	0.084	0.116	0.025	0.049	0.051	0.071	0.037	0.050	0.051	0.064	0.106	0.059	0.134	0.110	0.213	0.082
Rb/Sr	3.462	3.644	3.801	4.003	3.219	3.787	3.751	3.380	3.499	3.334	2.982	3.535	3.444	3.443	1.102	3.536
Estimated Source	KOSHI	KOSHI	KOSHI	KOSHI	KOSHI	KOSHI	KOSHI	KOSHI	KOSHI	KOSHI	KOSHI	KOSHI	KOSHI	KOSHI	KOSHI	KOSHI

Sample No.	TK0016	TK0017	TK0018	TK0019	TK0020
Nb/Zr	0.140	0.250	0.245	0.218	0.262
Y/Zr	0.429	0.592	0.621	0.609	0.601

Sr/Zr	0.401	0.473	0.499	0.415	0.431
Rb/Zr	1.468	1.608	1.643	1.520	1.608
Zn/Zr	1.633	0.184	0.302	0.280	0.207
Ni/Zr	2.433	0.162	0.357	0.431	0.204
Fe/Zr	11.540	9.034	9.920	11.521	8.276
Mn/Zr	0.915	0.278	0.380	0.371	0.277
Ti/Zr	0.050	0.054	0.060	0.096	0.056
Rb/Sr	3.657	3.400	3.292	3.659	3.731
Estimated Source	KOSHI	KOSHI	KOSHI	KOSHI	KOSHI

Amami Ō-shima Island

Sample No.	AM0001	AM0002	AM0003	AM0004	AM0005	AM0006	AM0007	AM0008	AM0009	AM0010	AM0011	AM0012	AM0013	AM0014	AM0015	AM0016
Nb/Zr	0.155	0.070	0.230	0.250	0.252	0.242	0.231	0.255	0.239	0.257	0.236	0.246	0.080	0.260	0.263	0.220
Y/Zr	0.350	0.245	0.528	0.702	0.705	0.661	0.652	0.639	0.599	0.603	0.588	0.624	0.334	0.791	0.767	0.634
Sr/Zr	0.397	0.739	0.462	0.467	0.501	0.459	0.463	0.477	0.498	0.458	0.450	0.504	0.651	0.470	0.478	0.465
Rb/Zr	0.956	0.579	1.570	1.712	1.751	1.703	1.578	1.623	1.735	1.508	1.608	1.729	0.678	1.569	1.578	1.518
Zn/Zr	0.084	0.103	1.466	0.152	0.150	0.150	0.198	0.504	0.441	0.332	0.346	0.184	1.074	0.160	0.356	0.277
Ni/Zr	0.049	0.140	1.905	0.116	0.109	0.108	0.140	0.688	0.759	0.695	0.529	0.178	1.510	0.085	0.432	0.207
Fe/Zr	4.029	3.742	8.752	5.827	5.903	5.798	6.065	9.941	12.181	12.260	10.228	8.546	6.539	8.278	11.920	9.277
Mn/Zr	0.142	0.178	0.597	0.198	0.188	0.198	0.226	0.452	0.427	0.398	0.345	0.296	0.622	0.272	0.383	0.355
Ti/Zr	0.013	0.034	0.034	0.015	0.015	0.022	0.024	0.044	0.081	0.306	0.061	0.046	0.089	0.052	0.091	0.075
Rb/Sr	2.407	0.784	3.402	3.664	3.492	3.715	3.405	3.405	3.484	3.293	3.570	3.431	1.041	3.338	3.299	3.265
Estimated Source	KUWA?	RYU	KOSHI	KOSHI	KOSHI	KOSHI	KOSHI	KOSHI	KOSHI	KOSHI	KOSHI	KOSHI	KOSHI	KOSHI	KOSHI	KOSHI

Sample No.	AM0017	AM0018	AM0019	AM0020	AM0021	AM0022	AM0023	AM0024	AM0025	AM0026	AM0027	AM0028	AM0029	AM0030	AM0031	AM0032
Nb/Zr	0.240	0.238	0.265	0.241	0.273	0.244	0.271	0.262	0.238	0.137	0.265	0.237	0.253	0.238	0.229	0.215
Y/Zr	0.669	0.632	0.660	0.682	0.660	0.543	0.397	0.472	0.565	0.199	0.609	0.595	0.586	0.587	0.504	0.655
Sr/Zr	0.478	0.470	0.462	0.448	0.423	0.477	0.346	0.439	0.429	0.531	0.478	0.451	0.449	0.465	0.391	0.449
Rb/Zr	1.672	1.561	1.568	1.568	1.555	1.576	1.357	1.506	1.530	0.549	1.856	1.752	1.733	1.789	1.592	1.514

Zn/Zr	0.190	0.311	0.489	0.946	0.367	0.000	0.768	0.953	0.571	0.106	0.158	0.192	0.280	0.333	0.343	0.263
Ni/Zr	0.180	0.211	0.384	1.207	0.252	0.275	0.933	0.595	0.587	0.068	0.061	0.122	0.232	0.276	0.266	0.288
Fe/Zr	8.497	8.670	10.043	9.736	9.002	8.488	13.714	11.063	8.202	5.134	8.044	8.765	9.155	8.381	8.627	8.736
Mn/Zr	0.312	0.304	0.340	0.607	0.315	0.292	0.480	0.399	0.383	0.154	0.278	0.334	0.342	0.347	0.325	0.299
Ti/Zr	0.033	0.048	0.060	0.056	0.067	0.051	0.146	0.093	0.044	0.034	0.040	0.101	0.054	0.038	0.066	0.058
Rb/Sr	3.497	3.325	3.391	3.498	3.678	3.305	3.922	3.432	3.567	1.034	3.885	3.885	3.859	3.847	4.073	3.372
Estimated Source	KOSHI	KOSHI	KOSHI	KOSHI	KOSHI	KOSHI	KOSHI	KOSHI	KOSHI	HARI-K	KOSHI	KOSHI	KOSHI	KOSHI	KOSHI	KOSHI

Sample No.	AM0033	AM0034	AM0035	AM0036
Nb/Zr	0.203	0.230	0.230	0.224
Y/Zr	0.650	0.684	0.655	0.634
Sr/Zr	0.413	0.450	0.401	0.439
Rb/Zr	1.560	1.640	1.617	1.618
Zn/Zr	0.274	0.169	0.334	0.228
Ni/Zr	0.209	0.119	0.388	0.[illegible]84
Fe/Zr	8.119	8.042	7.702	9.[illegible]00
Mn/Zr	0.288	0.268	0.315	0.317
Ti/Zr	0.050	0.043	0.058	0.068
Rb/Sr	3.780	3.644	4.032	3.684
Estimated Source	KOSHI	KOSHI	KOSHI	KOSHI

Tanega-shima Island

Sample No.	TN0001	TN0002	TN0003	TN0004	TN0005	TN0006	TN0007	TN0008	TN0009	TN0010	TN0011	TN0012	TN0013	TN0014	TN0015	TN0016
Nb/Zr	0.099	0.410	0.323	0.103	0.083	0.068	0.183	0.217	0.118	0.085	0.216	0.227	0.456	0.165	0.221	0.353
Y/Zr	0.287	0.506	0.424	0.219	0.423	0.246	0.611	0.569	0.194	0.251	0.570	0.540	0.500	0.194	0.582	0.481
Sr/Zr	0.491	1.195	1.057	0.514	0.830	0.709	0.465	0.455	0.488	0.481	0.442	0.380	1.065	0.517	0.847	0.858
Rb/Zr	0.721	1.488	1.305	0.549	1.128	0.644	1.634	1.649	0.554	0.712	1.559	1.539	1.206	0.523	1.116	1.053
Zn/Zr	0.116	0.461	0.676	0.222	0.074	0.086	0.265	0.181	0.172	0.827	0.408	0.378	0.925	0.840	3.114	0.803
Ni/Zr	0.061	0.279	0.718	0.190	0.057	0.084	0.221	0.099	0.133	1.035	0.306	0.500	0.969	0.637	4.478	0.656
Fe/Zr	5.904	16.924	17.654	6.219	5.375	4.328	9.111	7.498	6.008	7.703	9.496	11.411	21.890	12.291	21.532	15.688
Mn/Zr	0.189	1.116	1.026	0.185	0.214	0.215	0.305	0.266	0.167	0.421	0.332	0.382	1.199	0.686	1.869	0.938

Ti/Zr	0.068	0.098	0.107	0.061	0.086	0.070	0.057	0.032	0.068	0.111	0.063	0.114	0.241	0.153	0.117	0.066
Rb/Sr	1.470	1.246	1.234	1.068	1.358	0.908	3.514	3.625	1.135	1.481	3.523	4.055	1.132	1.011	1.318	1.227
Estimated Source	HARI-N	HIME	HIME	HARI-K	(RYU)	RYU	KOSHI	KOSHI	HARI-K	HARI-N	KOSHI	KOSHI	HIME	(HIME)	HIME	HIME

Sample No.	TN0017	TN0018	TN0019	TN0020	TN0021	TN0022
Nb/Zr	0.337	0.365	0.457	0.426	0.467	0.540
Y/Zr	0.536	0.552	0.564	0.627	0.530	0.667
Sr/Zr	1.310	1.180	1.352	1.496	1.351	1.276
Rb/Zr	1.403	1.322	1.535	1.691	1.404	1.513
Zn/Zr	0.647	0.675	0.511	2.378	0.488	0.763
Ni/Zr	0.597	0.557	0.496	2.808	0.522	0.721
Fe/Zr	22.191	20.079	19.362	26.131	18.062	24.181
Mn/Zr	1.401	1.256	1.325	2.116	1.244	1.375
Ti/Zr	0.137	0.141	0.105	0.165	0.079	0.198
Rb/Sr	1.071	1.121	1.136	1.130	1.039	1.185
Estimated Source	HIME	HIME	HIME	HIME	HIME	HIME

Yaku-shima Island

Sample No.	YK0001	YK0002	YK0003	YK0004	YK0005	YK0006
Nb/Zr	0.067	0.123	0.227	0.006	0.034	0.018
Y/Zr	0.265	0.207	0.576	0.127	0.136	0.153
Sr/Zr	0.756	0.516	0.401	1.119	1.182	1.148
Rb/Zr	0.724	0.588	1.652	0.220	0.201	0.203
Zn/Zr	0.069	0.079	0.211	0.171	0.083	0.083
Ni/Zr	0.048	0.043	0.253	0.180	0.059	0.066
Fe/Zr	4.310	5.165	8.064	11.102	9.829	9.425
Mn/Zr	0.212	0.150	0.282	0.202	0.200	0.196
Ti/Zr	0.073	0.033	0.041	0.278	0.228	0.223

Rb/Sr	0.958	1.139	4.118	0.197	0.170	0.177
Estimated Source	RYU	HARI-K	KOSHI	KAMI	KAMI	KAMI

*Source names: KOSHI – Koshidake; HARI-K – Hario-Kita; HARI-N – Hario-Nakamachi; HIME – Himeshima; KUWA – Kuwanokizuru; KAMI – Kamiushihana; RYU – Ryugamizu

Table 4.2. The number and ratio of obsidian tools from the Ryukyu Archipelago by identified source

Region Source	**Okinawa Island**	**Tokuno-shima Island**	**Amami Ō-shima Island**	**Yaku-shima Island**	**Tanega-shima Island**
Koshidake	58 (93%)	20 (100%)	32 (88%)	1 (17%)	4 (18%)
Hariojima	1 (2%)		1 (3%)	1 (17%)	2 (9%)
Harionakamachi					2 (9%)
Kuwanokizuru			1 (3%)		
Kamiushihana				3 (49%)	
Ryugamizu			2 (6%)	1 (17%)	1 (5%)
Himeshima					11 (50%)
Unknown or modern glass	3 (5%)				2 (9%)
Total amount of artefacts	62	20	36	6	22

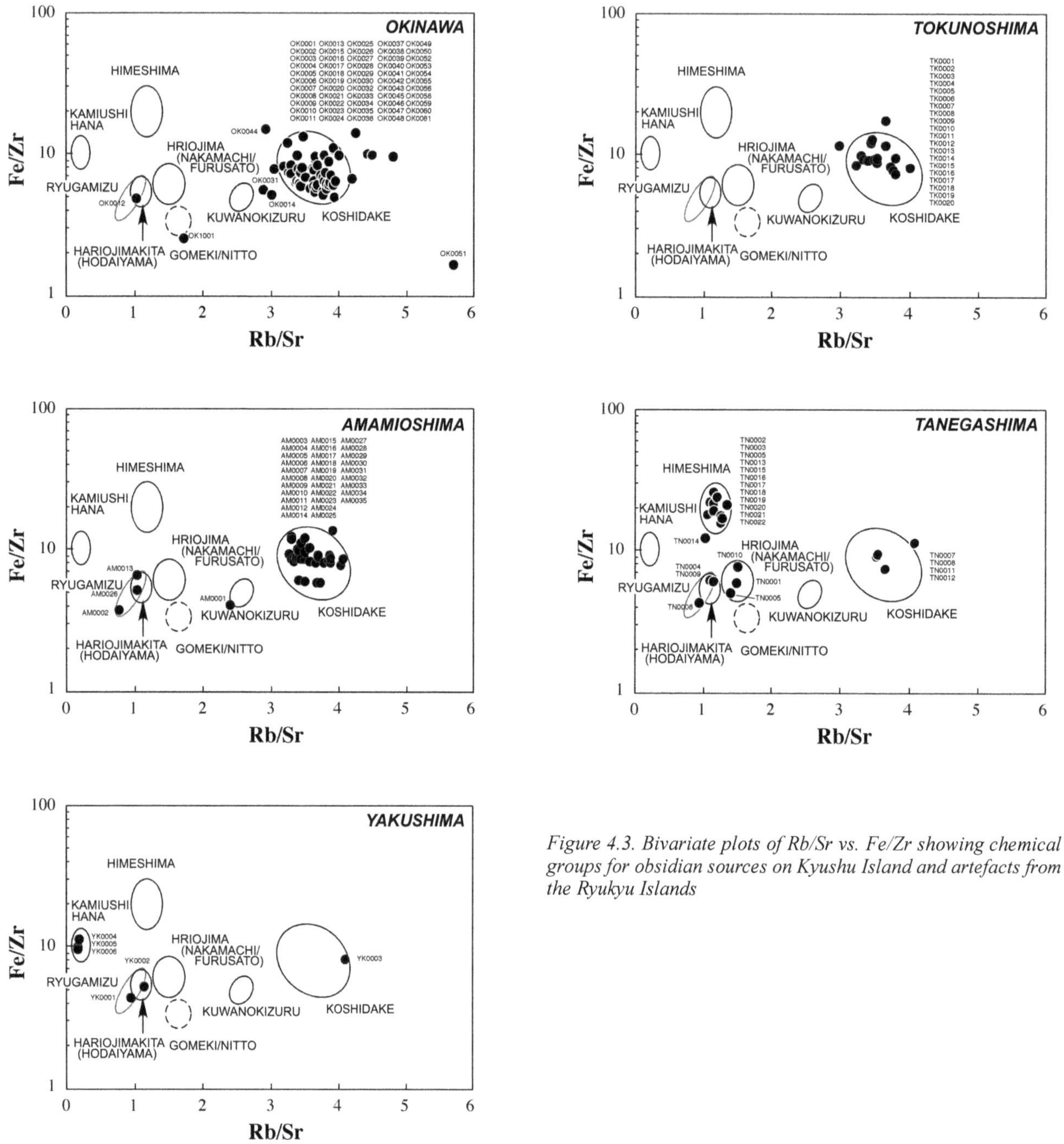

Figure 4.3. Bivariate plots of Rb/Sr vs. Fe/Zr showing chemical groups for obsidian sources on Kyushu Island and artefacts from the Ryukyu Islands

Island, one sample (AM0001) seems to originate from the Kuwanokizuru source; two artefacts (AM0002 and AM0013) are from the Ryugamizu source; one item (AM0026) is from the Hariojimakita (Hodaiyama) source; and the other 32 specimens are from the Koshidake.

Tanega-shima and Yaku-shima Islands

Eleven obsidian samples excavated on Tanega-shima Island (TN0002, TN0003, TN0013, and TN0015 through TN0022) originate from the Himeshima source; one item (TN0006) is from the Ryugamizu source; two samples (TN0004 and TN0009) are from the Hariojimakita (Hodaiyama) source; two specimens (TN0001 and TN0010) are from the Hariojima (Nakamachi/Furusato) source; and four samples (TN0007, T0008, TN0011, and TN0012) are from the Koshidake. In addition, on the Rb/Sr–Fe/Zr plot the TN0005 sample is close to the Hariojima (Nakamachi/ Furusato) source while on the Rb/Sr–Sr/Zr plot it is closer to the Himeshima source. The TN0014 sample on the Rb/Sr–Fe/Zr plot is more similar to the Himeshima source while on the Rb/Sr–Sr/Zr plot it is close to the Hariojimakita (Hodaiyama) source. Because it is not certain to which source these two artefacts belong, their origin is classified as "Unknown". Three obsidian samples excavated on Yaku-shima Island (YK0004, YK0005, and YK0006) are thought to be from the Kamiushihana source; one specimen (YK0001) is from the Ryugamizu source; one item (YK002) is from the Hariojimakita (Hodaiyama) source; and one specimen (YK0003) is from the Koshidake.

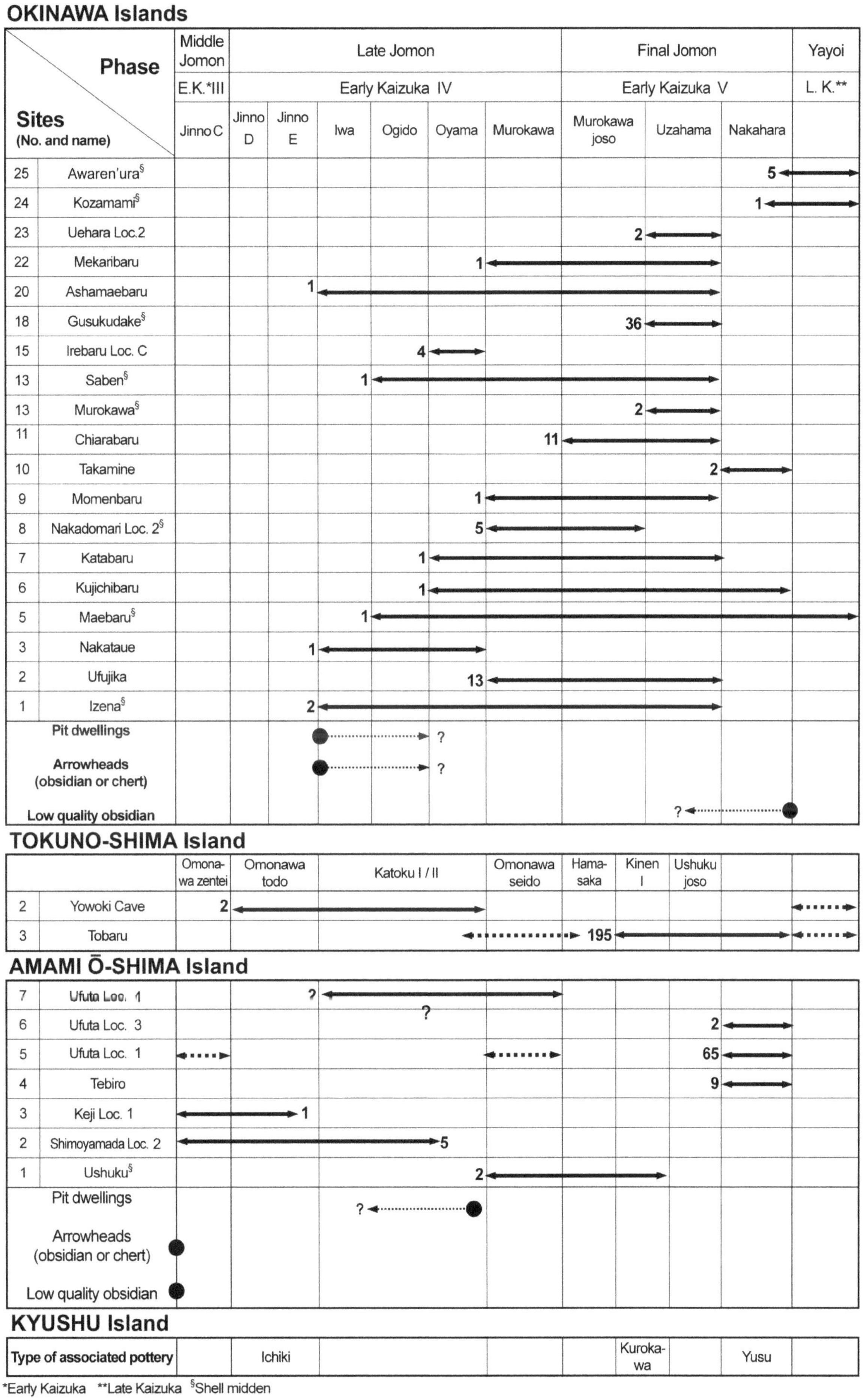

Figure 4.4. Chronologies of archaeological sites with obsidian tools on the Ryukyu Archipelago archipelago (values near the arrows mean the number of obsidian tools)

Discussion

Patterns of Obsidian Transportation from Kyushu Island to the Ryukyu Archipelago

As a result of chemical analyses, we can recognise a rapid increase in the quantity of obsidian imported from Kyushu Island sources to the Ryukyus from the beginning of the later half of the Late Jomon period (Figure 4.4). South of the Amami Ō-shima Island, cultural material with a strong influence of Jomon complex from the main island of Kyushu (including the remains of pit dwellings) is recognised during this phase. The advent of obsidian in the Ryukyus can be understood as part of the general influence from the Jomon cultural sphere that existed on the main islands of Japanese Archipelago (Houshu, Kyushu, and Shikoku).

For a discussion of the nature of the Watase Line (biological borderline) between Yaku-shima Island and Tokara-retto [Tokara Islands] as a possible barrier for the spread of obsidian and other goods, the key is the Tokara island group. Large-scale settlements such as Tachibana on the Kuchino-shima Island (30° N latitude) appeared from the Final Jomon onwards. No obsidian was found on that site; however, since remains of pit dwellings as well as pottery from Kyushu Island were unearthed, there is a clear indication of interaction between the two regions. There is no earlier evidence for exchange of goods, including obsidian, from the Kyushu, and it is highly probable that the Watase Line was a real cultural boundary.

In general, movement of obsidian can be observed during the Jomon between the Yaku-shima and Tanega-shima islands and the main island of Kyushu. Thus, these regions are thought to have interacted with each other throughout most of Jomon period. By contrast, it is most likely that obsidian appeared in the region south of Amami Ō-shima Island between the end of the Middle Jomon and the beginning of the Late Jomon at the earliest, and around the middle of the Late Jomon at the latest. A tendency of increasing quantities of imported obsidian until the Final Jomon can be observed. The appearance of obsidian seems to correlate with geographical position, since the time of the first appearance of volcanic glass has the tendency to become younger from the Amami Ō-shima and Tokuno-shima islands southward, to the Okinawa island group (Figure 4.5). In other words, the further away from the origin of raw material (i.e., the Kyushu Island obsidian), the later is its first appearance. According to this study, it becomes clear that no obsidian was imported to the Okinawa island group before the second half of the Late Jomon. From this time on, there was an increase in the quantity of obsidian import to all regions of the Ryukyus, particularly in the second half of the Final Jomon. This seems to have continued until the Early Yayoi period (Figure 4.4).

The present study makes it clear that obsidian artefacts unearthed on the Ryukyu Archipelago are all from seven obsidian sources located on Kyushu Island (Table 4.2). As for the Late and Final Jomon periods, obsidian from the Koshidake source constitutes 94% of the total amount. The share of obsidian from the Hariojima source (the Yodohime type) in northwestern Kyushu is 2.6%; the same 2.6% for the Ryugamizu source in southern Kyushu. However, the latter obsidian is distributed no further south than the Amami Ō-shima Island.

Assuming that the degree of superiority of raw material is reflected by its distribution (i.e., the distance from the source) and its quantity, one can ascribe a sequence of sources (top to bottom): Koshidake – Hariojima – Ryugamizu – Himeshima – Kamiushihana – Kuwanokizuru. The Koshidake source is the most distant one from the utilisation region, and although the cost of transportation was the highest compared to other raw materials, the fact is that it could compete with other materials and was the most preferred due to its excellent quality.

Late and Final Jomon Obsidian Artefacts from Okinawa and Amami Island Groups

Obsidian examined from the Amami and Okinawa island groups shows artefacts belonging to the Late and Final Jomon periods. Oda (2001) recognised tiny traces of flaking and striations on obsidian tools using the naked eye. He also observed that flakes and chips made of obsidian or chert were in part processed, and that edges of blades have traces of use. Oda (2001) called these artefacts "stone microtools". Among obsidian microtools at the Ufujika site and the Izena shell mound, many linear abrasions or tiny scratches called striations on the surface or ridges of primary flakes are noted (Oda 2001). Scars originated when artefacts were carried around, and together with different degrees of hydration (patina) they were used as evidence for a time lag between the production of a flake tool and its utilisation. The authors of this study, however, could not find any traces of patina on the samples which they examined. More than that, it is hard to imagine that this time lag was long enough to cause such a patina, and is it more likely that the striations on the fresh flake surfaces from Izena Island were caused by damage at the time of excavation. However, as Oda (2001) had already pointed out, it is a fact that even the smallest chips were processed on their edges by tiny flaking, and their frequency is extremely high. Use-wear analysis of these artefacts has not been done yet. However, there are clear tiny traces of flaking that occurred when the stone tools were used. Several linear abrasions or striations and concentrations of them as well as a combination of these phenomena can be observed. It is necessary to investigate the functional meaning of flake tools in the future by use-wear analysis.

The principles of stone tool reduction (i.e., theory of morphological transformation) are the key to interpretation of activity range for prehistoric groups and their interrelationships. It is therefore a highly meaningful theory. Several factors cause the morphological transformation of stone tools. Among the factors for Mesoamerican prismatic blade technique (Hirth and Andrew 2002) that occur through prehistory, one can mention the important factor of constraint by (A-2) condition of training and apprenticeship

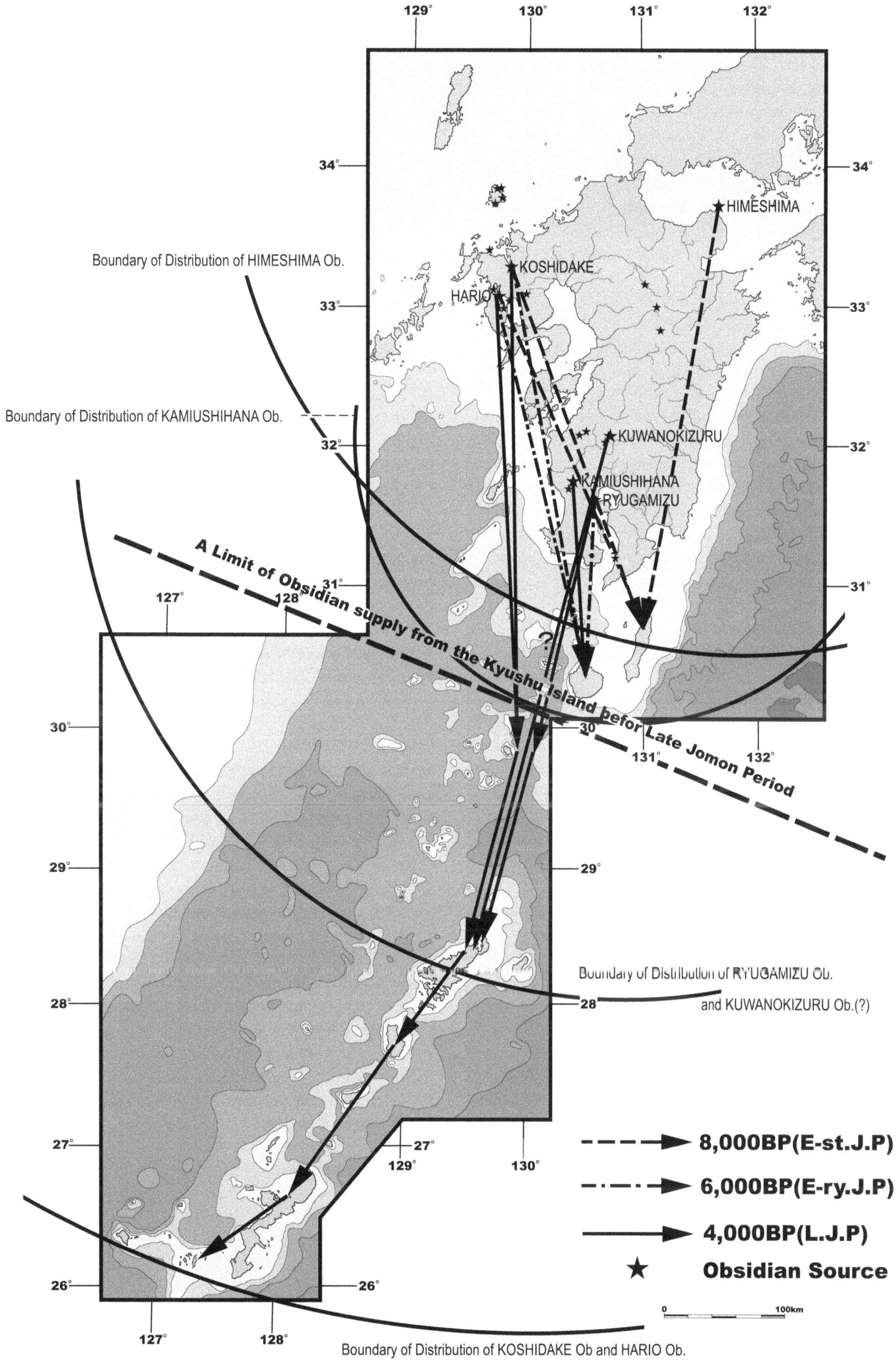

Figure 4.5. Probable obsidian transportation routes between the sources on Kyushu Island and the Ryukyu Archipelago

and (B-2) availability and distance from source. These phenomena appeared with remarkable similarity in regions located far from resources and information. In other words, in regions which have for topographic or social reasons an abundant supply of raw material and information for production of stone tools (in this research it is the area around the Koshidake source), this pattern is difficult to grasp as it is expected from theory. On the other hand, in regions with shortages of raw materials and information (islands of the Ryukyu Archipelago) this feature appears remarkably similar.

The study by Kajiwara (1991) of the reduction theory for Palaeolithic stone tools made on shale in the Tohoku region (Honshu Island), and research by Rickels and Cox (1993) of chert sources and sites of consumption in the Palaeoindian complexes of southwestern Texas may serve as reference cases (see Odell 2004). Considering the factors mentioned above, the characteristics of obsidian tools found on the Ryukyu Archipelago can be described as follows: 1) the cores are extremely small; 2) the flake tools are small; 3) many of the flake tools were produced from a wedge-shaped core using a bipolar technique; and 4) the amount of flakes actually used is high. These features indicate a situation where the necessary raw material was not continuously available.

Transport of Obsidian to the Ruykyu Archipelago: Current Problems of Study

Considering the patterns of obsidian exchange between Kyushu Island and the Ryukyus, Oda (2001, 214–215) thinks of that 'group of seafarers that added obsidian to a large number of exchange goods, transporting light, efficiently processed flakes rather than original nodules.' Thus, he stressed that flakes acted as exchange goods. However, on the Amami island group at Ufuta 1 site obsidian flakes including chips originating from three nodules were found, and on the Shimoyamada 2 site three small flake tools most probably made of the same core were discovered. Some flakes with cortex were also found at the Tebiro site (Amami Ō-shima Island). Furthermore, on Okinawa Island flakes with cortex were discovered at the Chiarabaru shell mound and Shikina-shimaontake site, and nodules (though being small) were excavated from the Aharen'ura shell mound and Kanekubaru 1 site (Figure 4.1). Judging from this, one can conclude that there was import of flaked raw material as Oda (2001) suggested. However, import of raw material in its natural, unprocessed form (as nodules) also took place.

From the point of view of size, quality, and quantity of obsidian, one cannot infer a conclusion about the frequent and secure import of obsidian. The estimated gross weight of obsidian raw material on the Amami island group and southwards during the Late–Final Jomon is around 360g. Only one or two pieces are similar to the average obsidian nodules scattered around the Koshidake source. The Suzuoke type blade cores typically distributed around northwestern Kyushu Island in the Late Jomon are found in extremely low quantities.

Conclusion

Two volcanic glass groups from sources on Kyushu Island are recognised on the Ryukyu Archipelago sites: 1) the group with the best quality of obsidian came from the northwestern slope of Koshidake Mount; and 2) the group with slightly less quality obsidian from clusters at the Umamodoshi site near Imari City (Obata 2004). The Group 2 type is known from the Okinawa island group, and it is particularly frequent on sites representing the later stages of obsidian exchange in distant regions like the Aharen'ura shell mound on Tokashiki-jima Island (Kerama island group). This situation with obsidian pebbles of low quality and obsidian arrowheads of high quality resembles archaeological sites on the southern coast of the Korean Peninsula where obsidian from northwestern Kyushu is also found (Takahashi *et al.* 2003). This is a typical situation for the outer peripheries of an obsidian supply zone. The difference between excellent products and inferior ones can be seen among stone arrowheads, not from the point of raw material constraint but from the difference in the level of craftsmanship. The elaborate arrowheads were either imported as processed goods or produced in the consumption area by people who learned from experience, and the inferior ones were "imitations" (Daikubara 2003) of these products made by domestic toolmakers. Depending upon the archaeological site, there may be no flakes or chips from production, and the whole assemblage may consist of excellent products only. These artefacts may either have been imported from settlements that served as production centres or may have been brought directly from the main island of Kyushu.

With time progressing to the Final Jomon, some sites – like Gusukudake, Tobaru, Hirota, and Ufuta – contain more obsidian compared to other localities, and traces of tool production can also be observed. These sites may be called 'centres', and one can interpret them as key settlements from where obsidian was redistributed to surrounding areas. To verify this hypothesis, it is necessary to discover deposits of obsidian raw material on the Ryukyu Archipelago and Kyushu Island, and to reconstruct the relationship between these trading points. However, as it was stated earlier, the amount of obsidian found on the Ryukyus could have easily been brought by canoes. Although there is no doubt that people were moving to the Ryukyu Islands during the Late–Final Jomon bringing obsidian, it was not a situation where the large migrating population was exchanging information and goods continuously.

Acknowledgements

We are grateful to Drs Yaroslav V. Kuzmin and Michael D. Glascock for the invitation to participate in SAA 2005 Symposium and this volume, and for editing the original version of this paper.

References

Daikubara, Y. 2003. Moho to Mozo – Koshitsuketsuganseiishisaji/ Ishiyari no Ryutsu to

Keishiki Henyo [Copying and Imitation – Distribution and Typological Transition of Stone Scrapers and Stone Spears Made of Hard Shale]. *Jomon Jidai* 14, 1–29.

Higashimura, T. 1986. *Sekki Sanchi Suiteiho* [Identification of Stone Sources]. Tokyo, Nyu Saiensusha Publisher.

Hirth, K., and B. Andrew (eds). 2002. *Path Way to Prismatic Blades*. Los Angeles, Costen Institure of Archaeology, University of California.

Kajiwara, H. 1991. Sekkigun Keisei ni Oyobosu Sekizai Kankyo no Igi [The Meaning of Stone Material in the Environment for the Formation of Assemblages of Stone Tools]. In *Kita Karano Shiten*, edited by Society for Japanese Archaeology, 51–62. Sendai, Society for Japanese Archaeology.

Kakubuchi, S., and M. Utsunomiya. 2002. Keiko X-sen Bunseki ni Yoru Kokuyoseki no Sanchi Dotei (1) [Determination of the Original Source of Obsidian with X-ray Fluorescence Analysis (1)]. *Saga Daigaku Bunka Kyoiku Gakubu Kenkyu Bunshu* 7(1), 49–52.

Kakubuchi, S., and M. Utsunomiya. 2003. Keiko X-sen Bunseki ni Yoru Kokuyoseki no Sanchi Dotei (2) [Determination of the Original Source of Obsidian with X-ray Fluorescence Analysis (2)]. *Saga Daigaku Bunka Kyoiku Gakubu Kenkyu Bunshu* 7(2), 47–58.

Kamimura, T. 1998. Ryukyu Retto no Sekizoku to Kokuyoseki – Sono Shusei to Igi [Stone Arrowheads and Obsidian Unearthed on the Ryukyu Archipelago – Corpus and Meaning]. *Jinruishi Kenkyu* 10, 200–213.

Mochizuki, A. 1996. Keiko X-sen Bunseki ni yoru Chubu/ Kanto Chiho no Kokuyuosekisannti no Hanbetsu [Identification of Obsidian Sources of Archaeological Artefacts from Chubu and Kanto Province by X-ray Fluorescence Analysis]. *X-sen Bunseki no Shinpo* 28, 157–168.

NCKHI (Nihon no Chisitsu "Kyushu Chiho" Henshu Iinkai) (ed.). 1998. *Nihon no Chisitsu 9 Kyushu Chiho* [Geology of Japan. Volume 9. The Kyushu Region]. Tokyo, Kyouritsu Shuppan Kabushikigaisha.

Obata, H. 2004. Sagaken Imarishi Ushimodoshi Iseki no Sekizai Shuseki Iko to Suzuokegata Sekijingiho [The Source of Stone Material at the Umamodoshi Site in Imari City, Saga Prefecture, and the Suzuoke Type Blade Technique]. *Sekki Gensanchi* 3, 107–128.

Obata, H., I. Morimoto, and S. Kakubuchi. 2004. Ryukyu Retto Shutsudo no Kokuyosekisei Sekki no Shutudo Yoso ni tsuite [Sources of Prehistoric Obsidian Tools on the Ryukyu Islands, Japan]. *Sekki Gensanchi* 4, 101–136.

Oda, S. 2000. Okinawa no Hakuhen Sekki ni tsuite [About Flake Tools from Okinawa]. In *Ryukyu Higashi Asia no Hito to Bunka 1,* edited by Committee for Publishing of Festschrift for 70th Birthday of Dr Hiroe Takamiya, 55–77. Nishihara-Machi, Okinawa, Japan, Committee for Publishing of Festschrift for 70th Birthday of Dr Hiroe Takamiya.

Oda, S. 2001. Izena Kaizuka no Sekki [Stone Tools from the Izena Shell Mound]. In *Izena Kaizuka*, edited by S. Oda, 106–234. Tokyo, Bensei Shuppan.

Oda, S. 2003. Okunawa Shoto no Hakuhen Sekki Shusei [Corpus of Flake Tools from the Okinawa Archipelago]. In *Okinawa-Ken Kayauchibanta Iseki: Nihon Jin Oyobi Nihon Jin no Kigen ni Kansuru Gakusaiteki Kenkyu Koko Shiryoshu 29,* edited by H. Harunari, 80–108. Sakura, Japan, National Museum of Japanese History.

Odell, G. H. 2004. *Lithic Analysis.* New York, Kluwer Academic/Plenum Press.

Takahashi, Y., I. Ha, and H. Obata. 2003. Hyong-goang X-son Bunsoku e Uihan Tongsam-dong/Bonbang Yujok Churto Hugyosok Sanji Chujong [Sources Identification of Obsidian Tools from Tongsamdong and Bonbang Sites by X-ray Fluorescence Analysis]. *Hanguk Shinsokki Yongu* 6, 83–99.

Chapter 5
Provenance Study of Obsidian Artefacts Excavated from Palaeolithic Sites on the Korean Peninsula

Nam-Chul Cho, Jong-Chan Kim, and Hyung-Tae Kang

Abstract: *The results from two case studies on obsidian sources for Palaeolithic sites in Korea are presented. While the initial research failed to recognise the source(s) of obsidian from two Upper Palaeolithic sites, Suyanggae and Sangmuyong-ri, a second attempt made by Kim et al. (2007) succeeded in the identification of Paektusan Volcano obsidian in the entire southern part of the Korean Peninsula. It also revealed the transport of obsidian from the Kyushu Island across the Korea (Tsushima) Strait to the Korean Peninsula, at ca. 25,000 BP as the earliest date. This is an important piece of solid evidence about the prehistoric contact and migration in Northeast Asia.*

Keywords: *Obsidian, Sources, Korean Peninsula, Upper Palaeolithic, Paektusan Volcano, Kyushu Island*

Introduction

Obsidian is a volcanic glassy rock formed by highly viscous and rapidly cooling lava with a high silicon dioxide (SiO_2) content. Most obsidian is black but some is red or green. Obsidian was used in prehistory as a raw material for making a variety of tools such as arrowheads and knives. In the earlier times, obsidian was exchanged for other items between distant regions by trade via land or sea. Since it is now excavated in large quantities at archaeological sites, obsidian is a valuable commodity for understanding the exchange relationship between regions during prehistory (e.g., Glascock *et al.* 1998).

In the modern Republic of Korea (hereafter – Korea), there are about 90 archaeological sites with obsidian artefacts (Kim *et al.* 2007, 122). Among the investigations that scientifically examined obsidian from Palaeolithic sites in Korea and thereby classified for provenance, one study analysed the trace elements Ba, Sr, and Zr using Neutron Activation Analysis (NAA). It was found that obsidian from the Sangmuyong-ri site in the Yanggu region (Kangwon Province; Figure 5.1) came from the Paektusan Volcano source (Shon 1989). Obsidian found at the Hahwage-ri site in the Hongcheon region (also Kangwon Province) (Figure 5.1) was subjected to X-ray Fluorescence (XRF) and microscopic and trace element analyses; but it was concluded that it is impossible to determine its provenance because the obsidian did not have any resemblance to the Paektusan source (Lee *et al.* 1992). Lee *et al.* (1990) tried to classify obsidian excavated from each region of the Korean Peninsula using multivariate statistical analysis after measuring the contents of trace elements through NAA. Though there was an additional study by Yi and Lee (1996), the provenancing of Korean obsidian still remained unsolved.

Except for research indicated above, there are no reported investigations that have systematically analysed and classified obsidian excavated from Korean Palaeolithic sites prior to the mid-2000s (see Kim *et al.* 2007). In addition, the studies mentioned above tried to establish obsidian provenance using only three or four trace elements for one or two samples, so it was not possible to obtain reliable results. Even though some researchers announced that most of the obsidian from Palaeolithic sites in Korea belongs to the Paektusan source, this was not sufficient evidence for verification.

The readers of this paper should bear in mind that this is an excerpt from the first author's PhD thesis compiled in February 2005 (Cho 2005), in which a thorough study was conducted on obsidian artefacts from two Palaeolithic sites, the Suyanggae and Sangmuyong-ri. Therefore, this paper reflects the status of Korean obsidian studies at that time. However, we felt that the up-to-date work on Korean obsidian provenancing by Kim *et al.* (2007) should be included in this paper. Thus, this paper consists of two parts: 1) results of the study of obsidian artefacts from the Suyanggae and Sangmuyong-ri sites; and 2) a summary of obsidian provenance analysis for other Korean Palaeolithic sites.

This study analyses the major chemical compounds using Scanning Electron Microscope (SEM) – Energy Dispersive Spectroscopy (EDS) on obsidian excavated at the Suyanggae site in the Danyang region (North Chungchong Province) and the Sangmuyong-ri site, and obsidian raw material obtained from the Paektusan source located on the border of North Korea and China (Figure 5.1). Trace elements were also examined using NAA. Subsequently, a definite correlation between the source and the sites was made. In addition, by observing the microcrystallites inside the obsidian and surveying the difference between the sites, a comparison of obsidian from Paektusan and Palaeolithic artefacts was made.

Sangmuyong-ri and Suyanggae Sites

The Sangmuyong-ri site is one of the key Palaeolithic sites in Korea, and stone tools such as choppers, points, scrapers, end-scrapers, and cutters, have been excavated here (Kangwon Provincial Government 1989). Most of the implements were made of quartzite, and some of obsidian

Figure 5.1. Location of Palaeolithic sites mentioned in this study and the Paektusan obsidian source on Korean Peninsula

and slate. It was assumed that during the Upper Palaeolithic people of this area used mainly quartzite which was easy to obtain. Since it was difficult to acquire obsidian and slate, inhabitants transported these rocks or exchanged them in cooperation with other Palaeolithic humans. Figure 5.2 shows the obsidian artefacts excavated at Sangmuyong-ri site (Chuncheon National Museum 2004).

Suyanggae is also one of the key Upper Palaeolithic sites in Korea. A total of 49 workshops where stone implements

Figure 5.2. Obsidian artefacts excavated at the Sangmuyong-ri site (after Chuncheon National Museum 2004)

were manufactured have been unearthed at this locality (Lee 1985); many chips and flakes were created as a by-product of the manufacture of stone implements. About 100 obsidian artefacts including choppers, tanged points, and grooving and dentate blades were excavated. Figure 5.3 shows the obsidian dentate and grooving blades from the Suyanggae site (Lee *et al.* 2004).

Material and Methods

In this study, 33 obsidian samples were analysed, including eight specimens from the Suyanggae site, 21 from the Sangmuyong-ri site, and four pieces found in the Paektusan Volcano area. Table 5.1 lists the samples used in this paper. Most of the obsidian samples are black; although some of them from the Sangmuyong-ri site are a mixture of red and black colours. Researchers were unable to distinguish any other features than those mentioned above with the naked eye; however, for some of the semi-translucent samples, the microcrystalline structure inside was observed with the help of microscope.

Major Chemical Compositions

To determine the chemical composition of the samples, small pieces were mounted using Epoxy resin and then polished with #200, 500, 1000, 2000, and 4000 grinding papers until there were no scratches on the surface. Then each obsidian sample was measured using SEM (Jeol Co., JSM-5910LV) – EDS (Oxford, INCA Energy) under the following analytic conditions: accelerating voltage – 20kV, measuring time – 120 seconds, and spot size – 50μm. This made it possible to analyse each sample under the same conditions by using 99.9% pure copper as a control.

Minor Elements

The sample surfaces were cleaned with acetone and dilute nitric acid, and then washed with distilled water. To remove the moisture left on the surface, the samples were dried for two hours at 110 °C; then sliced into 100–200mg pieces using a tungsten carbide blade and then crushed into powder. Forty milligrams of the powdered sample was measured and sealed in a polyethylene vial.

For the NAA sample preparation, a pneumatic tube was used with a rotating sample stage at the HANARO Reactor (Maximum Thermal Neutron Flux of HANARO is 5 – 10^{14} n/cm^{2}sec) of the Korean Atomic Energy Research Institute (Daejeon, Republic of Korea). The samples were exposed to neutron bombardment at a power of 20MW (1.7×10^{13} n/cm^{2}sec) for 10 hours. For analysis, a Gamma Counter (EG&G ORTEC, Dual amp 855) with 8000 channels connected to a high-purity germanium detector was used. Except for overlapping peaks due to neutron interference, the arrangement exhibited excellent detecting efficiency and high peak areas with ten elements measured.

Microcrystallites

To observe the microcrystalline structure inside the obsidian, the mounted samples were cut into slices 0.1mm thick using a diamond cutting wheel and then bonded to glass slides using Epoxy resin. After 24 hours, the samples were ground with #320, 800, 1200, 1500, 2400,

Table 5.1. List of samples used in this study

No.	Site	Origin and Age	Colour
1	Suyanggae	Artefact, Upper Palaeolithic	Black
2	Suyanggae	Artefact, Upper Palaeolithic	Black
3	Suyanggae	Artefact, Upper Palaeolithic	Black
4	Suyanggae	Artefact, Upper Palaeolithic	Black
5	Suyanggae	Artefact, Upper Palaeolithic	Black
6	Suyanggae	Artefact, Upper Palaeolithic	Black
7	Suyanggae	Artefact, Upper Palaeolithic	Black
8	Suyanggae	Artefact, Upper Palaeolithic	Black
9	Sangmuyong-ri	Artefact, Upper Palaeolithic	Black
10	Sangmuyong-ri	Artefact, Upper Palaeolithic	Black
11	Sangmuyong-ri	Artefact, Upper Palaeolithic	Black
12	Sangmuyong-ri	Artefact, Upper Palaeolithic	Black
13	Sangmuyong-ri	Artefact, Upper Palaeolithic	Black
14	Sangmuyong-ri	Artefact, Upper Palaeolithic	Black
15	Sangmuyong-ri	Artefact, Upper Palaeolithic	Black
16	Sangmuyong-ri	Artefact, Upper Palaeolithic	Black
17	Sangmuyong-ri	Artefact, Upper Palaeolithic	Khaki
18	Sangmuyong-ri	Artefact, Upper Palaeolithic	Black
19	Sangmuyong-ri	Artefact, Upper Palaeolithic	Khaki
20	Sangmuyong-ri	Artefact, Upper Palaeolithic	Black
21	Sangmuyong-ri	Artefact, Upper Palaeolithic	Black
22	Sangmuyong-ri	Artefact, Upper Palaeolithic	Black
23	Sangmuyong-ri	Artefact, Upper Palaeolithic	Black
24	Sangmuyong-ri	Artefact, Upper Palaeolithic	Black
25	Sangmuyong-ri	Artefact, Upper Palaeolithic	Black
26	Sangmuyong-ri	Artefact, Upper Palaeolithic	Red + Black
27	Sangmuyong-ri	Artefact, Upper Palaeolithic	Red + Black
28	Sangmuyong-ri	Artefact, Upper Palaeolithic	Red + Black
29	Sangmuyong-ri	Artefact, Upper Palaeolithic	Red + Black
30	Paektusan Volcano	Source	Black
31	Paektusan Volcano	Source	Black
32	Paektusan Volcano	Source	Black
33	Paektusan Volcano	Source	Black

and 4000 grinding papers to a thickness of 0.03mm. The microcrystalline structure of the obsidian was then observed using a polarising microscope (Leica, DMLP).

Results

Obsidian Classification by Major Chemical Compounds

Table 5.2 shows the composition of the major chemical elements for each sample. When igneous rocks are chemically classified based on their SiO_2 content, they are divided into acidic, intermediate, basic, and ultrabasic rocks (e.g., Raymond 2003). An acidic rock contains above 66% SiO_2 (by weight), and includes rhyolite and granite. An intermediate rock has SiO_2 content of 52–66%, and includes andesite and diorite. To be basic rock, SiO_2 content of 45–52% is required, and includes basalt and gabbro. The ultrabasic rock has SiO_2 content of less than 45%, and includes dunite and komatite (e.g., Raymond 2003). Because most of the samples used in this study have more than 66% SiO_2, they belong to the acidic rock group.

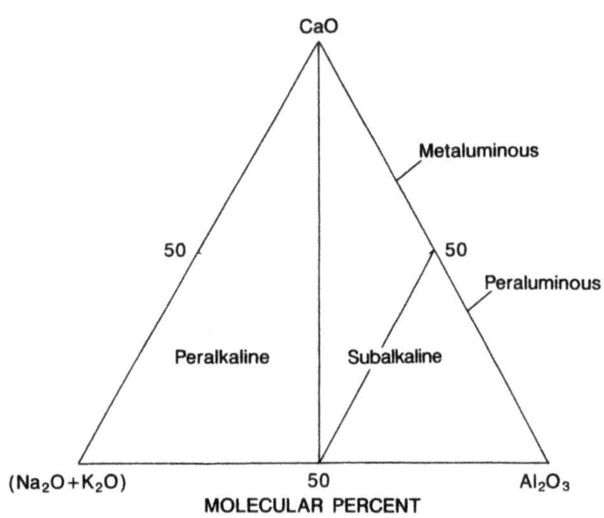

Figure 5.4. Diagram showing chemical classification of obsidians (after Williams-Thorpe 1995)

Figure 5.3. Obsidian artefacts excavated at the Suyanggae site (after Lee et al. 2004)

Table 5.2. Chemical composition of obsidian samples analysed for this study

		Concentration (%)									***Concentration (ppm)***									
No.	Site	Na_2O	MgO	Al_2O_3	SiO_2	K_2O	CaO	TiO_2	MnO	FeO	Sm	La	Ce	Sc	Cs	Hf	Lu	Rb	Tb	Sb
1	Suyanggae	3.43	0.25	12.3	76.7	5.04	0.47	0.21	0.02	1.54	6.96	61.9	96.3	1.04	3.36	7.60	0.47	252	1.48	0.00
2	Suyanggae	3.37	0.31	12.2	76.8	5.08	0.49	0.09	0.12	1.53	8.10	52.3	89.1	1.02	2.33	7.25	0.47	218	1.39	0.00
3	Suyanggae	3.30	0.20	12.3	76.8	4.80	0.55	0.28	0.32	1.43	6.73	55.1	92.7	0.93	2.88	7.62	0.40	253	1.22	0.00
4	Suyanggae	4.79	0.08	10.2	75.5	4.21	0.00	0.18	0.10	4.89	25.9	213	400	0.36	6.19	50.8	1.72	541	5.23	0.17
5	Suyanggae	2.98	0.19	12.4	76.8	5.02	0.63	0.14	0.22	1.66	6.54	60.3	116	1.07	3.20	8.44	0.55	235	1.51	0.53
6	Suyanggae	4.46	0.20	10.3	75.3	4.07	0.28	0.33	0.14	4.93	22.8	222	408	0.37	6.89	51.4	1.65	540	4.69	0.43
7	Suyanggae	3.06	0.12	12.1	77.0	5.19	0.61	0.24	0.00	1.78	7.22	62.7	89.6	1.01	2.73	7.78	0.45	228	1.29	0.00
8	Suyanggae	3.81	0.01	9.39	76.6	4.72	0.27	0.22	0.22	4.77	12.4	144	235	1.25	3.12	27.0	0.90	288	2.67	0.00
9	Sangmuyong-ri	3.20	0.13	12.6	76.7	5.03	0.55	0.31	0.00	1.66	9.53	75.3	141	1.01	3.37	6.13	0.52	314	0.00	0.00
10	Sangmuyong-ri	3.24	0.19	12.0	76.3	5.58	0.94	0.00	0.07	1.81	9.95	74.2	158	0.94	3.90	7.81	0.55	319	1.81	0.00
11	Sangmuyong-ri	3.30	0.45	12.3	76.5	5.05	0.39	0.19	0.25	1.67	9.61	72.0	153	1.15	3.26	7.80	0.51	313	1.84	0.00
12	Sangmuyong-ri	3.38	0.15	12.4	76.6	5.00	0.64	0.13	0.09	1.61	9.61	74.8	165	1.31	3.91	8.50	0.57	296	1.71	0.00
13	Sangmuyong-ri	3.18	0.35	12.2	77.0	5.12	0.62	0.07	0.00	1.62	9.77	73.0	155	1.20	3.96	8.17	0.55	306	1.74	0.55
14	Sangmuyong-ri	2.96	0.25	11.9	77.6	4.96	0.42	0.12	0.14	1.60	10.4	78.8	154	1.19	3.42	9.24	0.59	335	1.95	0.57
15	Sangmuyong-ri	2.99	0.21	12.2	76.9	5.00	0.66	0.20	0.11	1.74	9.28	68.9	158	1.21	3.02	8.00	0.50	302	1.69	0.70
16	Sangmuyong-ri	3.15	0.29	12.1	77.3	5.08	0.48	0.18	0.00	1.43	10.1	77.4	154	1.10	3.36	8.76	0.54	322	1.87	1.42
17	Sangmuyong-ri	0.58	2.30	14.0	80.0	0.14	0.06	0.55	0.07	2.37	1.63	10.3	17.4	7.94	0.00	3.30	0.20	0.00	0.36	0.16
18	Sangmuyong-ri	2.99	0.00	11.8	77.7	5.28	0.77	0.10	0.00	1.70	10.4	78.2	165	1.21	3.46	8.51	0.55	326	1.85	0.00
19	Sangmuyong-ri	0.57	3.27	15.7	77.4	0.00	0.31	0.22	0.16	2.37	3.39	1.16	5.30	6.56	0.00	3.42	0.22	0.00	0.09	0.00
20	Sangmuyong-ri	3.48	0.16	12.1	76.9	5.19	0.47	0.01	0.17	1.53	10.2	76.6	167	0.94	2.96	8.64	0.54	330	1.75	0.00
21	Sangmuyong-ri	3.27	0.02	12.4	76.9	5.27	0.54	0.08	0.07	1.49	11.8	90.5	124	1.80	3.87	10.3	0.65	395	2.09	0.00
22	Sangmuyong-ri	3.35	0.00	12.4	77.3	5.00	0.42	0.00	0.04	1.52	9.51	73.3	138	1.06	3.08	8.36	0.57	308	2.07	0.00
23	Sangmuyong-ri	3.24	0.52	12.5	76.7	4.82	0.57	0.00	0.09	1.60	10.4	78.1	132	0.98	3.36	8.55	0.55	321	1.72	0.22
24	Sangmuyong-ri	3.24	0.33	12.0	76.8	5.11	0.55	0.19	0.09	1.66	9.87	70.5	127	1.36	3.20	8.30	0.52	296	1.55	0.70
25	Sangmuyong-ri	3.28	0.16	12.6	76.6	5.11	0.55	0.04	0.08	1.63	7.76	58.7	98.3	0.81	2.49	7.07	0.42	244	1.27	0.00
26	Sangmuyong-ri	2.74	0.00	12.3	77.8	5.16	0.41	0.02	0.04	1.64	7.97	59.6	105	1.09	2.71	6.36	0.40	223	1.41	0.53
27	Sangmuyong-ri	3.45	0.19	12.3	76.4	5.18	0.64	0.19	0.00	1.72	8.23	61.3	104	1.05	2.58	6.91	0.42	245	1.40	0.18
28	Sangmuyong-ri	3.09	0.06	12.1	77.3	5.08	0.56	0.13	0.20	1.54	8.82	65.5	118	1.02	2.67	6.21	0.38	210	1.30	0.00
29	Sangmuyong-ri	3.07	0.23	12.2	76.6	5.35	0.43	0.09	0.15	1.87	9.88	73.0	138	0.87	2.69	8.06	0.56	308	1.47	0.00
30	Paektusan	4.50	0.28	14.8	67.3	5.89	1.33	0.44	0.22	5.29	11.0	81.0	157	4.27	1.10	12.4	0.90	115	1.21	0.00
31	Paektusan	4.83	0.24	15.3	66.1	5.92	1.43	0.84	0.00	5.43	12.9	95.1	182	4.58	1.51	14.4	1.02	121	1.52	0.00
32	Paektusan	5.22	0.00	15.4	70.1	5.99	0.21	0.25	0.09	2.90	19.6	152	289	1.39	2.72	35.1	2.01	219	2.68	0.46
33	Paektusan	4.88	0.51	14.7	67.0	5.72	1.28	0.50	0.03	5.40	11.8	77.5	145	5.17	1.21	12.1	0.95	119	1.33	4.14

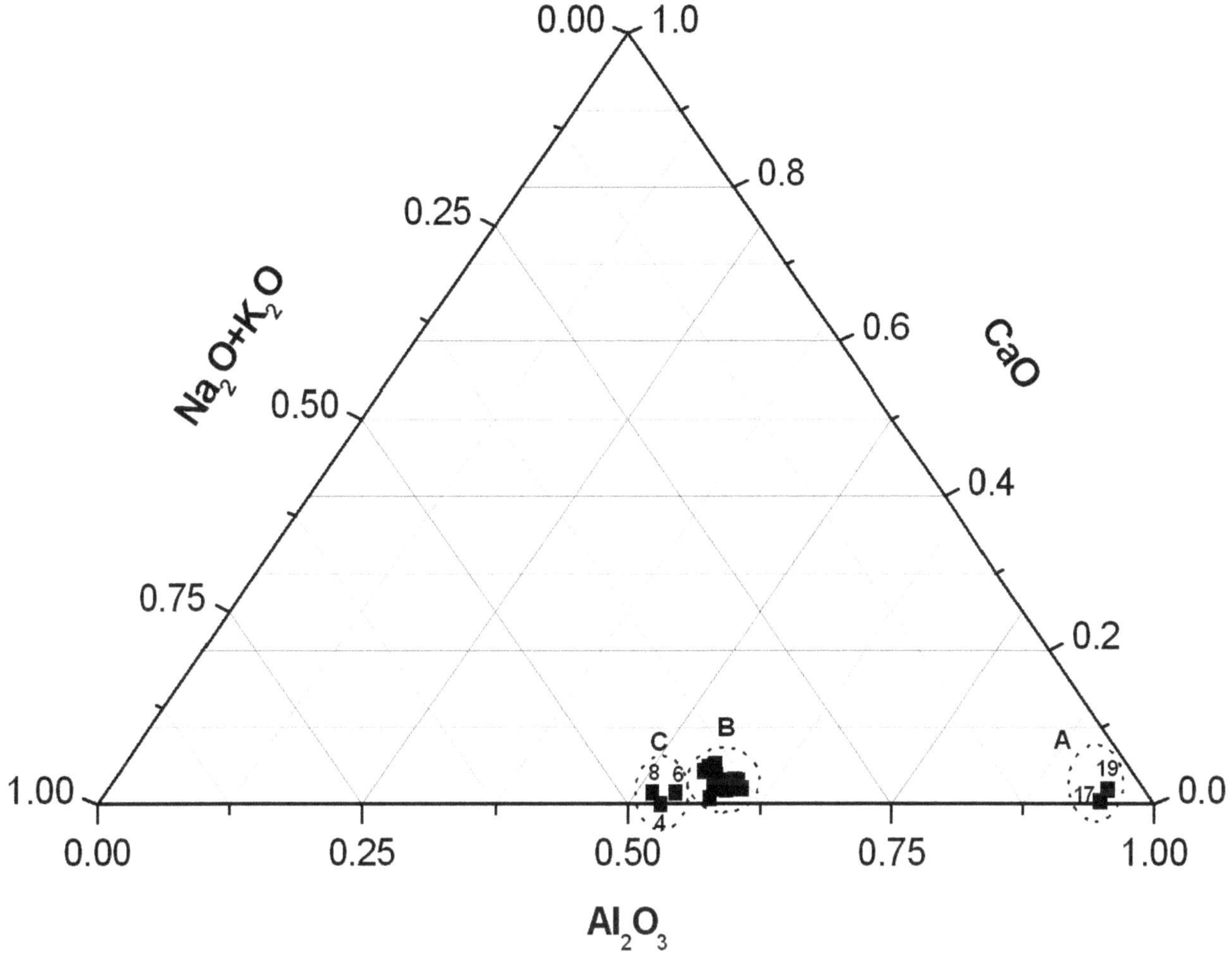

Figure 5.5. Chemical classification of obsidian excavated at two Korean Palaeolithic sites (see Table 5.2)

When classifying obsidian samples based on Al_2O_3 content, they can be subdivided into peralkaline, subalkaline, peraluminous, and metaluminous in accordance with the contents of Al_2O_3, CaO, and $Na_2O + K_2O$ (Williams-Thorpe 1995) (Figure 5.4). For peralkaline rock, the $Na_2O + K_2O$ content is more than that of Al_2O_3; for subalkaline rock, $Na_2O + K_2O$ is less than Al_2O_3 in content; for peraluminous rock, the Al_2O_3 content is more than CaO + $Na_2O + K_2O$; and for metaluminous rock, Al_2O_3 content is less than CaO + $Na_2O + K_2O$.

Most of the obsidian samples in this study belong to the subalkaline type which can be largely divided into three groups (Figure 5.5). Group A refers to samples Nos. 17 and 19 from the Sangmuyong-ri site. As the CaO + Na_2O + K_2O content is less than the amount of Al_2O_3, it belongs to the peraluminous series. Most of the obsidian samples (Group B) belong to the subalkaline series. Samples from the Suyanggae site constitute a discrete Group C because they have less Al_2O_3 and CaO compared to the other groups. In the case of groups A, B, and C, there is a big difference in the major chemical compounds, and they are divided into a separate series which means the obsidian in each group has an independent geological origin.

Principal component analysis (PCA) is one of the multivariate statistical approaches conducted to classify groups that are similar in characteristics and to check the relationships between them. In the case of inorganic artefacts like earthenware, ceramics, and obsidian, and

Table 5.3. Eigenvalues and proportions by principal components analysis for major elements

Component	*Eigenvalue*	*Proportion (%)*	*Cumulative (%)*
1	3.427	48.955	48.955
2	2.154	30.765	79.720
3	0.975	13.928	93.648
4	0.378	5.407	99.055
5	0.043	0.617	99.672
6	0.022	0.308	99.980
7	0.001	0.020	100.000

Table 5.4. Communalities for major elements

Compounds	*Communalities*	
	Initial	*Extraction*
Na_2O	1.000	0.869
MgO	1.000	0.970
Al_2O_3	1.000	0.736
SiO_2	1.000	0.985
K_2O	1.000	0.894
CaO	1.000	0.594
FeO	1.000	0.533

when their places of origin are not certain, this method is often used (Kim 2001). Therefore, in this study statistical analysis was used to correlate the content of major chemical compounds in the obsidian samples by the PCA method using the Statistical Package for Social Science (SPSS) (Chong and Choi 1998). As variables, seven components including Na_2O, MgO, Al_2O_3, SiO_2, K_2O, CaO, and FeO, were used, but not MnO or TiO_2 which were not detected in most of the samples when the PCA was conducted.

Table 5.3 shows the eigenvalues, the proportions, and the cumulative values. The eigenvalue is the amount of variance explained by each principal component. The principal component with the highest eigenvalue becomes the most important principal component, with "principal component 1 and 2" in the highest position. Then applying these principal components 1 and 2 to X and Y axes respectively, they can be displayed on a bivariate plot. The proportion is the value showing how much data is explained by each principal component. In the case of the Principal Component 1, as the proportion reaches 48.955%, it becomes the principal component covering 48.955% of the whole data variance. Meanwhile, the cumulative of the principal components 1 and 2 reaches 79.72% which means that it explains 79.72% of the whole data variance.

Table 5.4 provides the communalities of each composition. The communality is the sum of the variables explained through the extracted principal components 1 and 2. It indicates how much of the information contained in each variable can be displayed. The higher this value the greater the contribution it provides (Shin and Mun 1996). In Table 5.4, components such as SiO_2, MgO, K_2O, and Na_2O are making a larger contribution to each principal component than Al_2O_3, CaO, and FeO.

Figure 5.6 shows the values plotted using the principal components 1 and 2. Artefacts are divided into two groups. Group I is composed of obsidian from the Paektusan source, and Group II includes the obsidian excavated at the Suyanggae and Sangmuyong-ri sites. This means that the raw material (obsidian) of the artefacts from Suyanggae

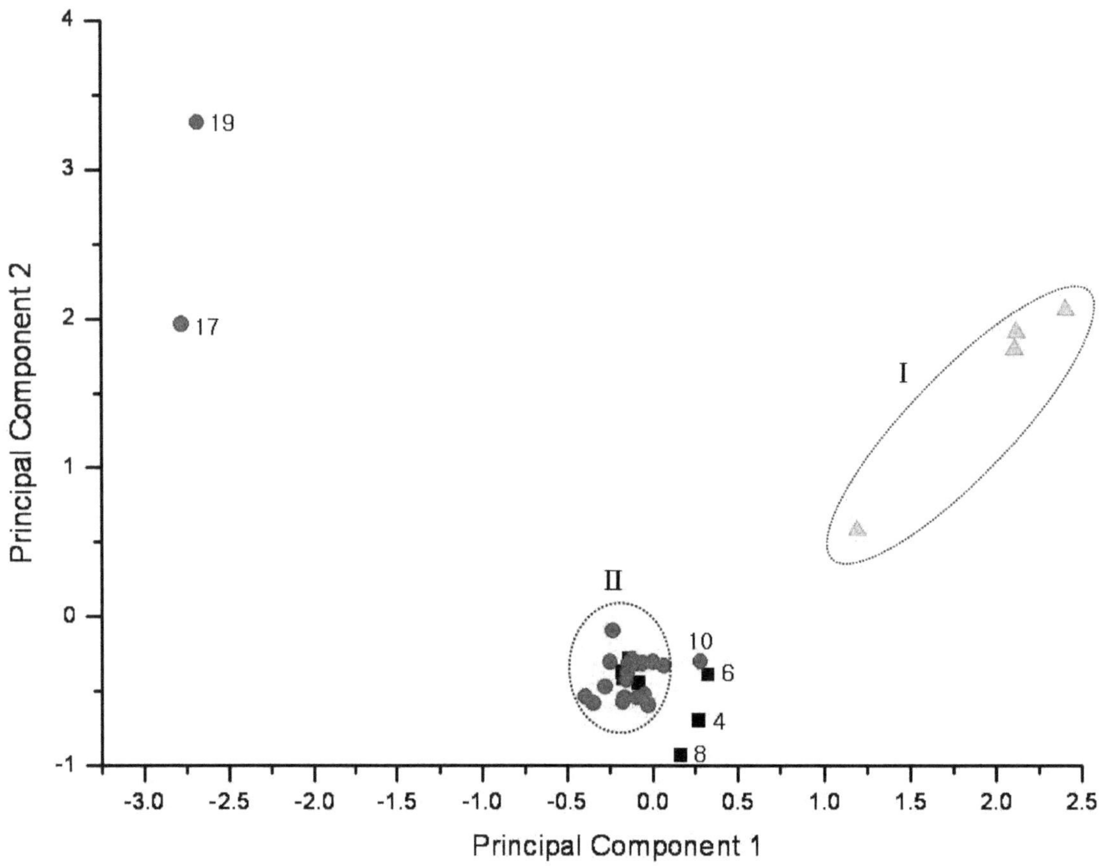

Figure 5.6. Bivariate plot of first and second principal components for seven major chemical compounds (squares are samples from the Suyanggae site, circles – from the Sangmuyong-ri site; and triangles – from the Paektusan source)

Table 5.5. Eigenvalues and proportions by principal component analysis for minor elements

Component	Eigenvalue	Proportion (%)	Cumulative (%)
1	6.918	76.863	76.863
2	1.569	17.438	94.301
3	0.236	2.624	96.925
4	0.107	1.193	98.118
5	0.067	0.744	98.863
6	0.043	0.478	99.341
7	0.032	0.353	99.694
8	0.015	0.172	99.866
9	0.012	0.134	100.000

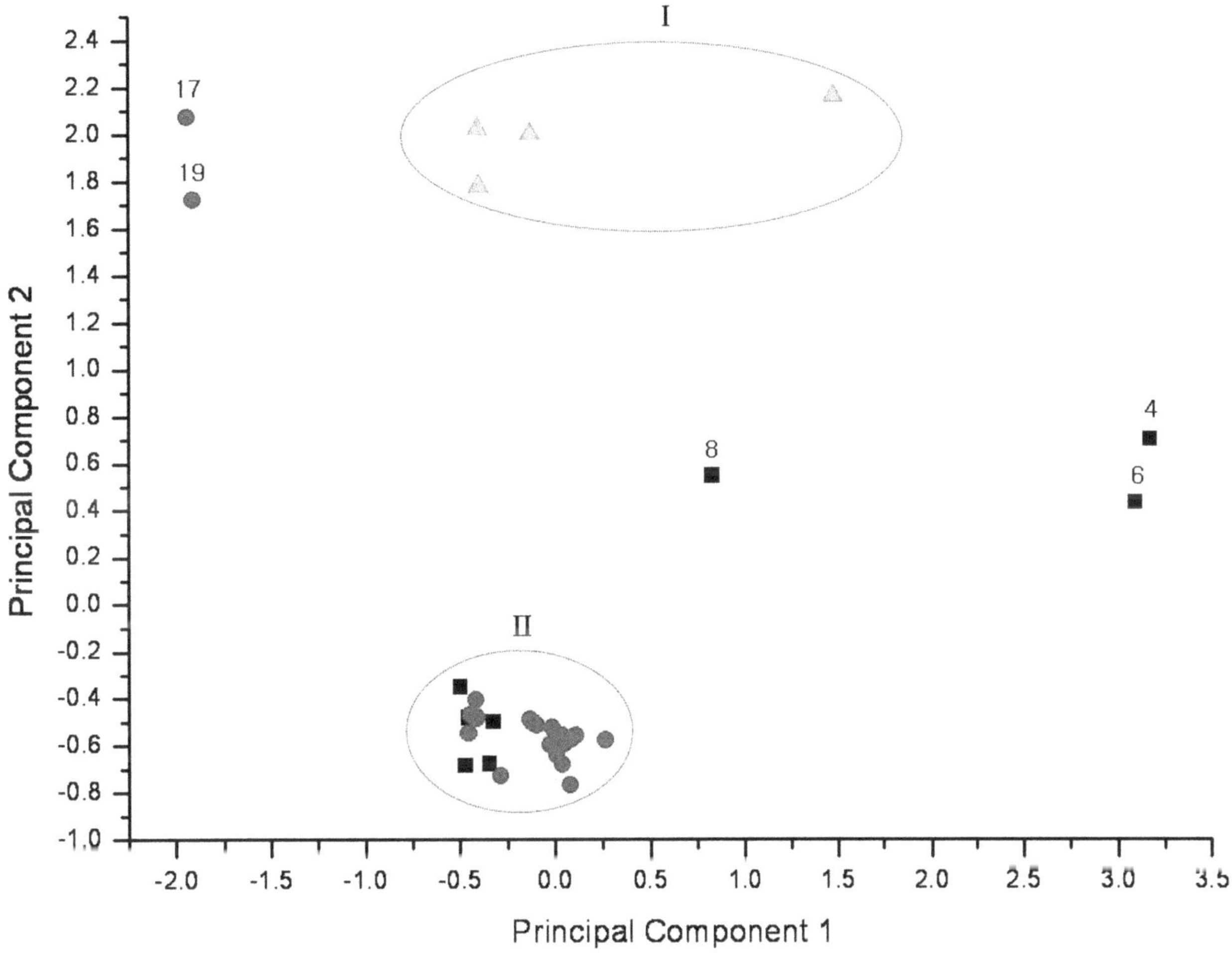

Figure 5.7. Bivariate plot of first and second principal components for nine minor elements (squares are samples from the Suyanggae site, circles – from the Sangmuyong-ri site; and triangles – from the Paektusan source)

and Sangmuyong-ri is of the same kind, and it is different from the Paektusan obsidian.

Some of the obsidian samples are distinct from the main groups. Samples Nos. 17 and 19 from the Sangmuyong-ri site and samples Nos. 4, 6, and 8 from the Suyanggae site have been separated from groups I and II. These samples potentially came from other sources. Although sample No. 10 from the Sangmuyong-ri site is apart from Group II, it should be compared using minor elements to determine its correlation to the group.

Obsidian Classification by Minor Elements

For determinating obsidian provenance using minor elements, the method used is the difference in element content according to the rock series and its ages. But in the case of Korea, there is no standard established to judge which elements are the most useful ones; therefore, it is necessary to undertake the provenance study by analysing several minor elements. The content of ten minor elements was measured by NAA, including Sm, La, Ce, Sc, Cs, Hf, Lu, Rb, Tb, and Sb (Table 5.2). Nine of the elements, except

Table 5.6. Communalities for minor elements

Elements	***Communalities***	
	Initial	***Extraction***
Sm	1.000	0.962
La	1.000	0.980
Ce	1.000	0.970
Sc	1.000	0.881
Cs	1.000	0.946
Hf	1.000	0.957
Lu	1.000	0.935
Rb	1.000	0.951
Tb	1.000	0.905

for Sb (which was not detected in many samples), were statistically processed by PCA using the SPSS software.

Table 5.5 shows the eigenvalues, the proportions, and the cumulatives for the samples being analysed. The cumulative of the principal components 1 and 2 with the highest eigenvalue reaches 94.301%. Table 5.6 has the communality values of minor elements, showing that all the elements make a significant contribution to the principal components.

Figure 5.7 is a chart where the values of the principal components 1 and 2 are plotted. It shows that they are divided into two large groups. The obsidian obtained from the Paektusan source constitutes Group I, and obsidian from the Suyanggae and Sangmuyong-ri sites comprise Group II. However, artefacts Nos. 17 and 19 from the Sangmuyong-ri site and Nos. 4, 6, and 8 from the Suyanggae site do not form a group but instead are scattered. This result corresponds to the classification of provenance using major chemical compounds (see above). Thus, obsidian excavated from two Palaeolithic sites on the Korean Peninsula (Suyanggae and Sangmuyong-ri) do not correlate with the Paektusan source. This suggests that the obsidian was brought to Korea from other regions.

Obsidian Classification by Microcrystallites

The texture of igneous rock can be largely categorised as holocrystalline, holohyaline, and hypocrystalline. The

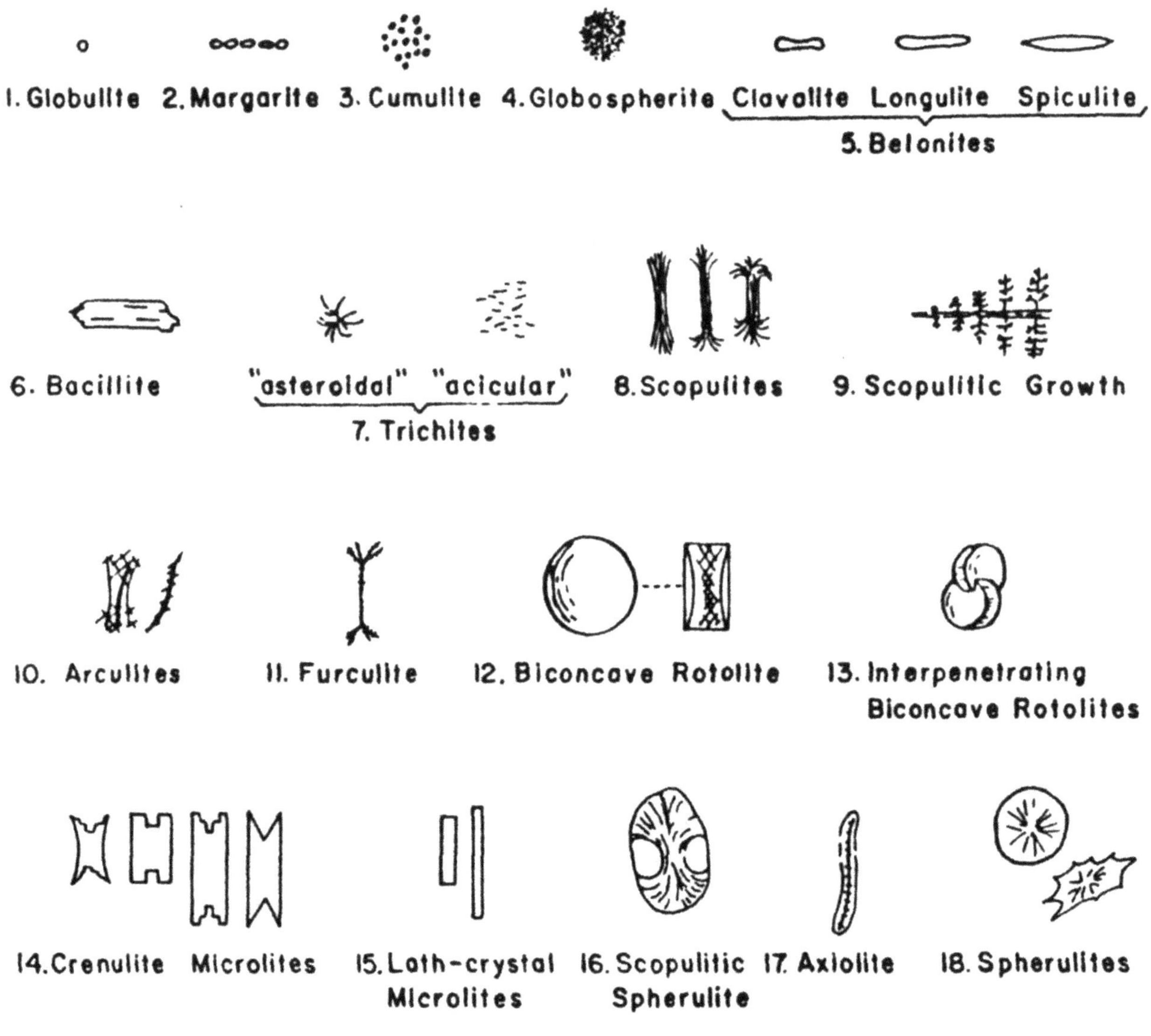

Figure 5.8. Types of crystallites and microlites in obsidian (after Michael and Ralph 1973)

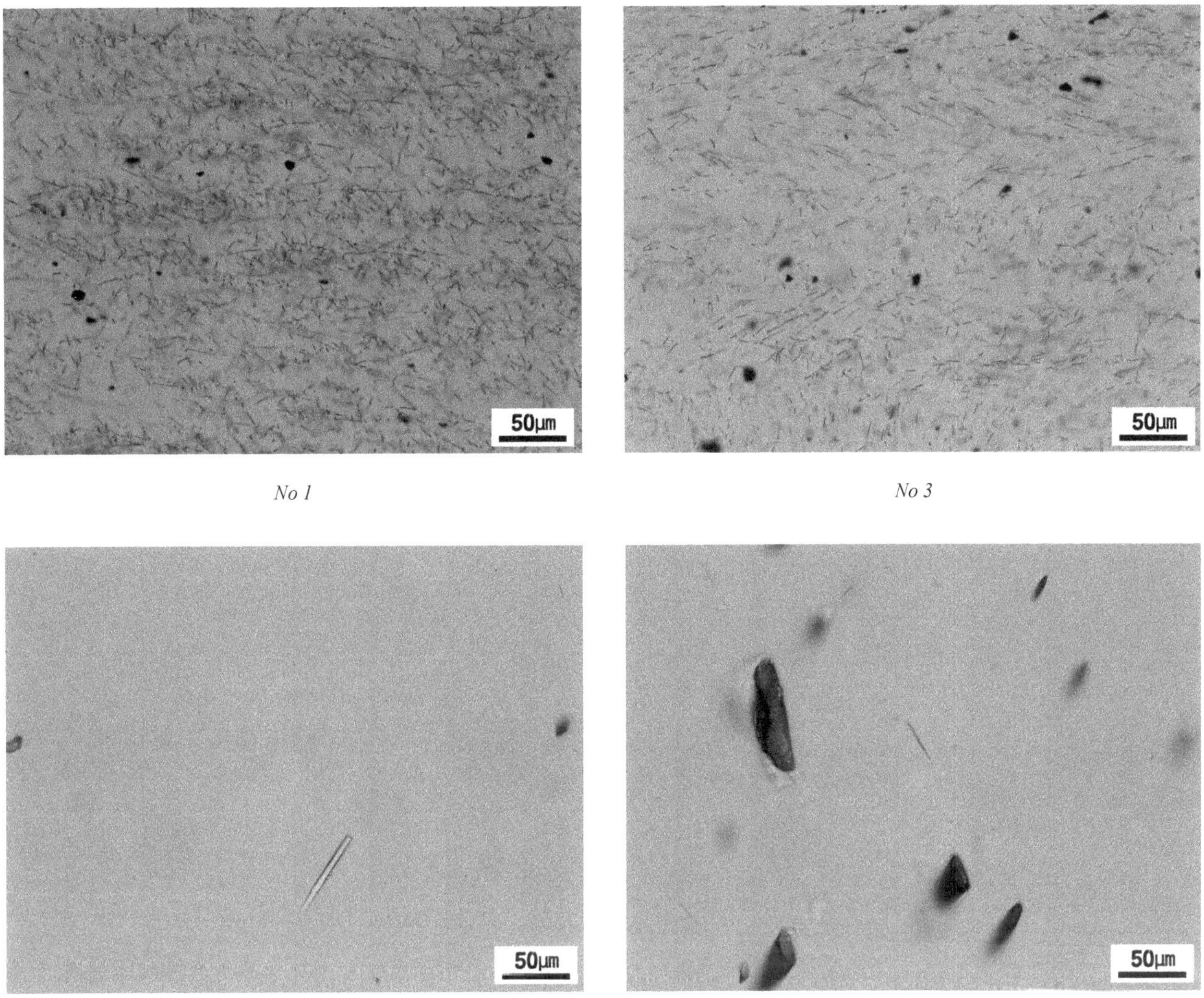

No 1 No 3

No 2 No 4

Figure 5.9. Textures of obsidian from the Suyanggae site (Nos. 1, 3 – acicular; Nos. 2, 4 – lath-crystal microlites)

holocrystalline texture is composed of crystalline minerals; the holohyaline one consists of glassy minerals; and the hypocrystalline texture is a mixture of crystalline and glassy minerals (e.g., Raymond 2003). In the case of obsidian, it is created when lava becomes solid through a rapid cooling process, and embryonic crystals are also formed at the stage between the amorphous and crystalline process (Kayani and McDonnel 1996; Michael and Ralph 1973; Shelley 1993). Figure 5.8 demonstrates a variety of crystallite and microlite forms (Michael and Ralph 1973). Most obsidian has a hypocrystalline texture in which plenty of crystallites or microlites exist. In this study, the minute textures of obsidian from the Suyanggae and Sangmuyong-ri sites were observed, and microcrystallines inside the obsidian were classified into several forms. Table 5.7 shows the arrangement of the microcrystallites.

The obsidian from Suyanggae site is divided into two groups. Most of the samples belong to the trichite-acicular type with a lot of microcrystallites inside. However, samples Nos. 4 and 6 are completely different in form and density. In terms of form, there are lath-crystals in samples but only a few of them are in a holohyaline texture state (Figure 5.9). This corresponds to provenance determination by chemical compounds. But even if samples Nos. 4 and 6 were excavated from the same place, they are geologically different in terms of their source from the rest of the artefacts at the Suyanggae site. This means that they have a different origin.

The obsidian from the Sangmuyong-ri site can be largely divided into three groups. Most of it is of the trichite-acicular type. Specimens Nos. 26 and 28 have a cumulite texture with black spots between a red, glassy matrix. Sample No. 19 has a feathery texture which is entirely different from the rest of the specimens at the Sangmuyong-ri site (Figure 5.10, No. 4; Table 5.7).

The obsidian from the Paektusan source is completely different in type, and is characterised by a flow texture. This type of texture can be observed in rhyolite and andesite, and is formed when minerals and glassy material are mixed such that the glassy material flows around the phenocrysts. Inside is found a variety of minute crystalline structures along with non-melted minerals (Figure 5.11).

When provenance was determined using the microcrystallites in the obsidian samples, it corresponded

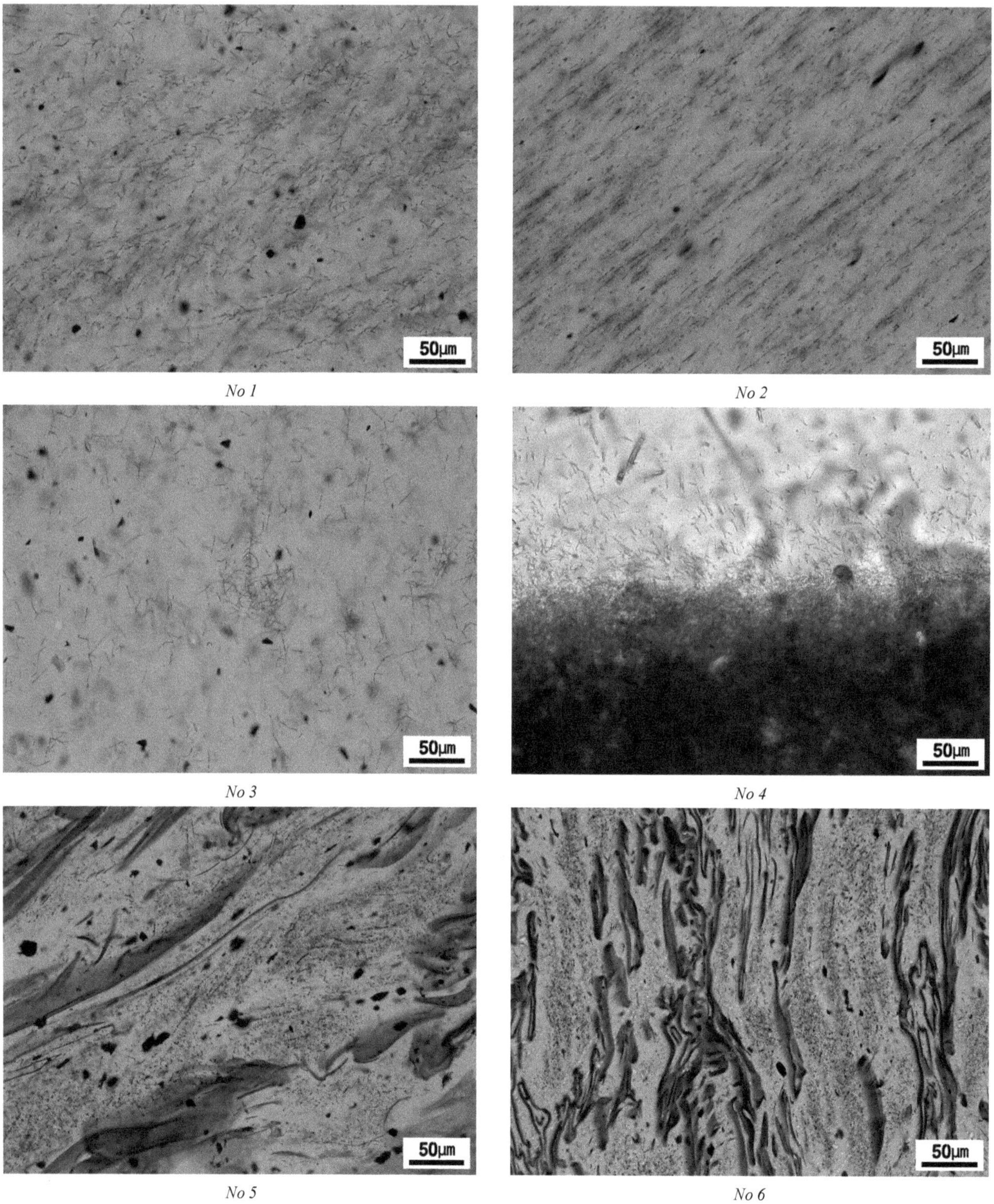

Figure 5.10. Textures of obsidian from the Sangmuyong-ri site (Nos. 1–3 – acicular; No. 4 – feather-shaped; Nos. 5–6 – cumulite)

to the results obtained by chemical analyses. Especially in the cases of samples Nos. 4 and 6 from Suyanggae and sample No. 19 from Sangmuyong-ri, it is clear they have different sources when compared to the majority of obsidian samples at these two sites. Obsidian from the Paektusan source has a completely different inner texture in relation to volcanic glass from these sites. Therefore, the results of this study show that prehistoric obsidian from the Suyanggae and Sangmuyong-ri sites has no relation to the Paektusan source. The Suyanggae and Sangmuyong-ri sites are representative of the Upper Palaeolithic epoch in Korea. The analytical results for four geologic obsidian samples from the Paektusan source show no resemblance to obsidian from these sites. This means that obsidian from the Suyanggae and Sangmuyong-ri is different from that at the Paektusan source and, therefore, archaeological obsidian did not originate at Paektusan (but see below, study by Kim *et al.* 2007).

Discussion

Obsidian samples were divided into two groups, Group

Table 5.7. Characteristics of microcrystallites and microlites inside obsidian from Korean sites

No.	***Site***	***Characterisation***
2	Suyanggae	Trichite: Acicular, irregular
4	Suyanggae	Lath-crystal microlite
5	Suyanggae	Trichite: Acicular, irregular
6	Suyanggae	Lath-crystal microlite
9	Sangmuyong-ri	Trichite: Acicular, irregular
10	Sangmuyong-ri	Trichite: Acicular, regular
11	Sangmuyong-ri	Trichite: Acicular, irregular
19	Sangmuyong-ri	Feather shape
26	Sangmuyong-ri	Cumulite, red glass
28	Sangmuyong-ri	Cumulite, red glass
30	Paektusan	Flow texture
31	Paektusan	Flow texture
32	Paektusan	Flow texture
33	Paektusan	Flow texture

I (Paektusan source) and Group II (Suyanggae and Sangmuyong-ri sites). In Korea, some locales including Myungju, Heiryung, and Gilju (all in Hamkeongbuk-do Province), Jeju Island, and Ullungdo Island, are known as sources of volcanic glass, but no pure obsidian has been collected. Thus, it has been assumed that most of the obsidian found at the Palaeolithic sites in the central part of Korea belongs to the Paektusan source.

However, since the collection details for the Paektusan source samples used in this study are not well-known and only four specimens were analysed, no definite conclusion can be made from the present study about the provenance of obsidian from the two Paleolithic sites mentioned above. Four specimens from the Paektusan source, which the authors analysed here, according to Popov *et al.* (2005) belong to Paektusan Type 2 [PNK2] and Type 3 [PNK3], while the Suyanggae and Sangmuyong-ri obsidians are mainly of Paektusan Type 1 [PNK1] (see Kim *et al.* 2007, 126). Thus, Cho (2005) failed to notice that obsidian from these two Palaeolithic sites is also from the Paektusan source. Likewise, Lee *et al.* (1992) analysed Hahwage-ri site obsidian which belong to PNK1 and PNK2 types and compared it to PNK3 samples brought from Paektusan Volcano, leading to conclusion that the source is non-identifiable.

Therefore, the previous failures of provenancing Paektusan obsidian on Korean archaeological sites should be attributed mainly to the difficulty of acquiring the PNK1 source material rather than a problem with geochemical analysis. More work related to the Paektusan source material is urgently needed. In previous reports, it was common that

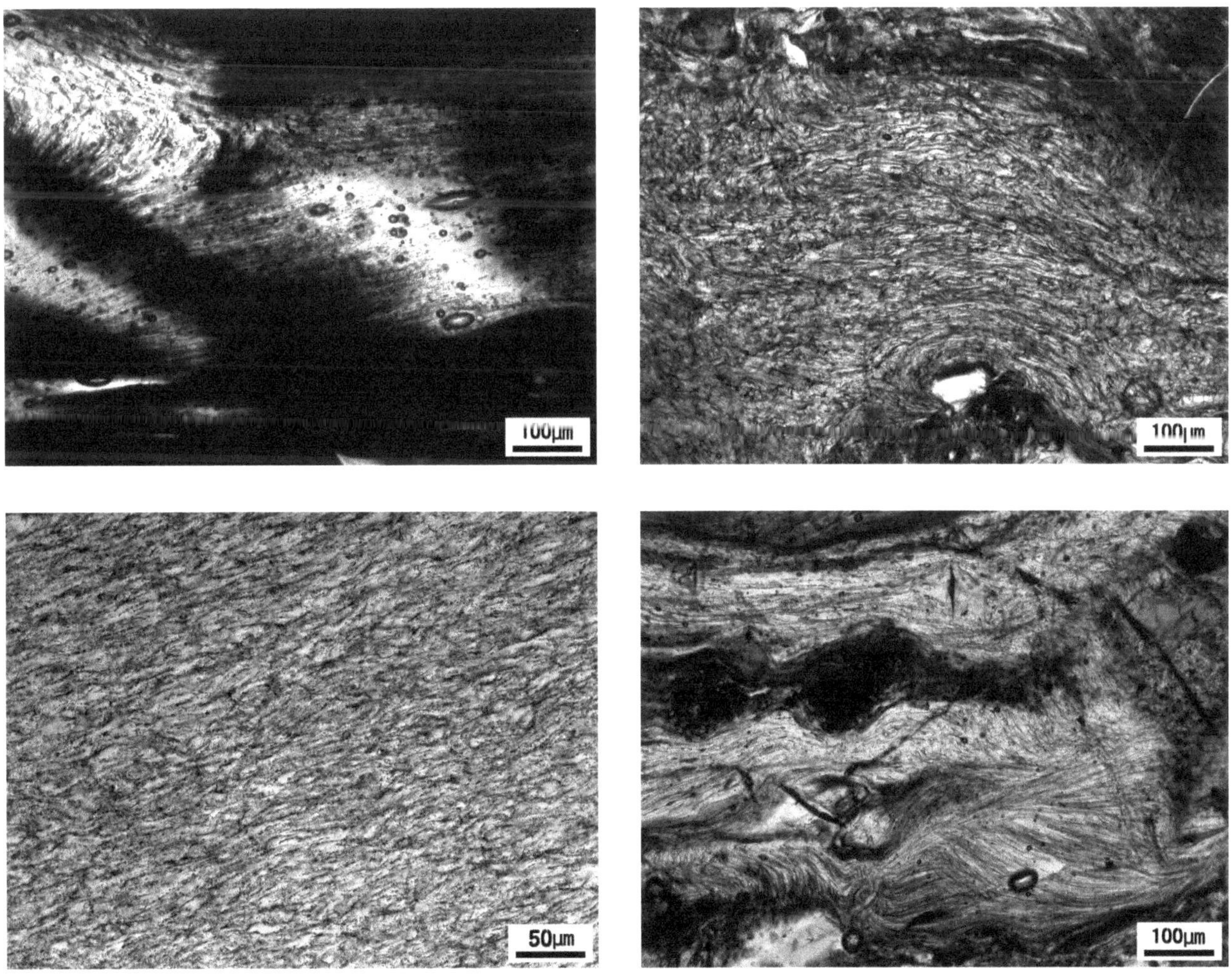

Figure 5.11. Textures of obsidian from the Paektusan source (flow type)

Table 5.8. Summary of source identifications for some Korean Palaeolithic sites (after Kim et al. 2007, modified)

Site	*No. of samples*	*Paektusan Type 1 Source*	*Paektusan Type 2 Source*	*Japanese sources*	*Unknown sources*
Hahwage–ri	6	1	5	–	–
Suyanggae	8	6	2	–	–
Sangmuyong–ri	21	19	–	–	2
Hopyung	20	11	7	–	2
Samri	10	10	–	–	–
Shinbuk	10	–	3	4	3
Totals (%)	75 (100%)	47 (62.7%)	17 (22.7%)	4 (5.3%)	7 (9.3%)

only one or two geochemically analysed samples were used for classification. However, to check the exact relationship between sources and utilisation sites, one needs a variety of analytical and statistical methods and a multitude of samples from both sources and archaeological sites.

Summary of Obsidian Sourcing by PIXE Method for Some Upper Palaeolithic Sites on Korean Peninsula

Recent progress in the identification of obsidian artefacts from the central and southern parts of the Korean Peninsula was achieved by Kim *et al.* (2007), and is briefly summarised here (see also Table 5.8). Obsidian from seven Upper Palaeolithic sites ranging in age from ca. 25,000 BP to ca. 15,000 BP was studied: Janghung-ri, Hopyung, Sam-ri, Hahwage-ri, Shinbuk, Syuanggae, and Sangmuyong-ri (Figure 5.1). Two analytical geochemical methods were employed, Proton-Induced X-ray Emission (PIXE) and NAA. Fifty flakes from the Hopyung, Sam-ri, and Shinbuk sites were examined by PIXE method, and three flakes from the Hahwage-ri and Janghung-ri sites were analysed by NAA. Also, a comparison between source samples from the Paektusan Volcano (Kuzmin *et al.* 2002; Popov *et al.* 2005) and 29 obsidian specimens from the Syuanggae and Sangmuyong-ri sites previously not assigned to a particular source (see above) was preformed.

The results of geochemical correlation between obsidian artefacts and sources from Paektusan and Kyushu Island (Kim *et al.* 2007) show that out of 75 samples of obsidian, 64 (85.4% of total set) belong to Paektusan and four samples (5.3%) are from Japanese sources at Kyushu Island. Seven specimens were not assigned to a particular source (Table 5.8). As for the provenance of obsidian from the Syuanggae and Sangmuyong-ri sites, the additional NAA analysis allowed correlation with the Paektusan source (Kim *et al.* 2007, 126). Two out of 21 samples from the Sangmuyong-ri site were unassigned.

At the Shinbuk site in the southernmost Korean Peninsula (Figure 5.1), obsidian from two sources was identified, Paektusan and Kyushu (Table 5.8). The distance between Paektusan and Shinbuk is about 800km in a straight line. This is similar to the distance between the source and prehistoric sites in the Primorye (Maritime) Province of the Russian Far East (Kuzmin *et al.* 2002). The direct evidence of contact between insular regions of Northeast Asia (such as the Japanese Archipelago) and the mainland at ca. 25,500–18,500 BP (this is the age of the Shinbuk site; e.g., Seong 2007, 2008), is extremely important for understanding prehistoric migrations and raw material exchange during the Upper Palaeolithic.

Conclusions

By identifying major chemical compounds, minor elements, and microcrystallites in 33 obsidians samples excavated at two Palaeolithic sites (Suyanggae and Sangmuyong-ri) on the Korean Peninsula, and four source obsidian samples from the Paektusan Volcano, the results are as follows. Concerning the major chemical compounds, most of the obsidian is rhyolite and belongs to the subalkaline series. Using PCA, samples are divided into two groups. Group I is obsidian from Paektusan source, and Group II represents obsidian from the archaeological sites of Suyanggae and Sangmuyong-ri. These groups have no relationship to each other in terms of origin. With identification of the minor elements, samples are also divided into two groups, except for some specimens based on the PCA approach. The result is essentially the same as the major compounds. Looking at microcrystallites in the volcanic glass matrix, it was found that some samples (Nos. 4 and 6 from the Suyanggae sites, and No. 19 from the Sangmuyong-ri site) have completely different textures and most probably different provenance compared to the rest of samples from these two sites. Obsidian from the Paektusan source also has a distinct texture compared to other obsidian items from the archaeological sites. All three methods give similar results, and it is clear that archaeological obsidian does not originate at the Paektusan source. There are some artefacts that are different from the majority of prehistoric obsidian and probably have a separate origin; however, they are also not from Paektusan.

The latest results allow the pinpointing of problematic obsidian samples from the Suyanggae and Sangmuyong-ri sites to the Paektusan source. This also supports a model of Paektusan as the major source of high quality volcanic glass for most of continental Northeast Asia, including the Primorye Province of Russia, Korean Peninsula, and probably Northeast China (Manchuria) (see also chapter by Y. V. Kuzmin in this volume).

Acknowledgements

We are grateful to Drs Yaroslav V. Kuzmin and Michael D. Glascock for their invitation to participate in this volume and for extensive editing of the original version of this paper.

References

Cho, N.-C. 2005. Classification of Obsidian Artifacts Found in Korean Peninsula Based on the Chemical Composition, Texture and Magnetic Property. Unpublished PhD thesis, Kangwon National University, Ch'unch'on, Korea (in Korean with English summary).

Chong, C. Y., and I. G. Choi. 1998. *Statistical Analysis by SPSSWIN*. Seoul, Muyeokgyeongyoung Corp.

Chuncheon National Museum. 2004. *Yunbo 2002–2003 Nyun Tongkwon* [Annual Report of 2002–2003] (Chuncheon Uglib Bagmulgwan 15). Chuncheon, Chuncheon National Museum.

Glascock, M. D., G. E. Braswell, and R. H. Cobean. 1998. A Systematic Approach to Obsidian Source Characterization. In *Archaeological Obsidian Studies: Method and Theory*, edited by M. S. Shackley, 15–65. New York and London, Plenum Press.

Kangwon Provincial Government and Central Museum of Kangwon National University. 1989. *Sangmuyongri Paroho Taesujiyuk Yujuk Balgul Josaboko* [Excavation Report of Sangmuyong-ri Palaeolithic Site in the Drainage basin of the Lake Paro]. Chuncheon, Kangwon University Press.

Kayani, P. I., and G. McDonnell. 1996. An Assessment of Back-Scattered Electron Petrography as a Method for Distinguishing Mediterranean Obsidians. *Archaeometry* 38, 43–58.

Kim, G. H. 2001. Hangugeseo Chultodoen Godaeyuliui Gogohwahagjeog Yeongu [The Study of Archaeological Chemistry on Ancient Glasses Found in Korea]. Unpublished PhD Dissertation, Chung-Ang University, Seoul.

Kim, J. C., D. K. Kim, M. Yoon, C. C. Yun, G. Park, H. J. Woo, M.-Y. Hong, and G. K. Lee. 2007. PIXE Provenancing of Obsidian Artefacts from Paleolithic Sites in Korea. *Bulletin of the Indo-Pacific Prehistory Association* 27, 122–128.

Kuzmin, Y. V., V. K. Popov, M. D. Glascock, and M. S. Shackley. 2002. Sources of Archaeological Volcanic Glass in the Primorye (Maritime) Province, Russian Far East. *Archaeometry* 44, 505–515.

Lee, C., M. Z. Czae, S. W. Kim , H. T. Kang, and J. D. Lee. 1990. A Classification of Obsidian Artifacts by Applying Pattern of Recognition to Trace Element Data. *Bulletin of Korean Chemical Society* 11, 450–455.

Lee, D. Y., J. Y. Kim, and C. G. Han. 1992. Hongcheon Hahwagyeriyujeogui Jihyeong Mit Jijil [Topography and Geology of Hahwage-ri Site in Hongcheon]. In *Munhwayujeok Balgul Josabogoseo,* edited by S. I. Chang, Y. H. Kim and Y. S. Son, 247–260. Chuncheon, Kangwon-do Press.

Lee, Y. J. 1985. Dannyang Suyanggae Guseokgi Yujeok Balguljosa Bogo [Report on the Excavation of Prehistoric Site in Suyanggae, Danyang]. In *Extended Research Report of Antiquities in Chungjudam Submerged Districts 16*. Cheongju, Chungbuk National University Museum.

Lee, Y. J., N.-C. Cho, and H. T. Kang. 2004. Dannyang Suyanggaeyujeok Heugyoseogui Teukseonghwa Yeongu [Characteristic Analysis on Obsidian Artefacts from the Suyanggae Site in Korea]. *Hanguk Guseokgi Hakbo* 10, 25–35.

Michael, H. N., and E. K. Ralph (eds). 1973. *Dating Techniques for the Archaeologist.* Cambridge, MA, MIT Press.

Popov, V. K., V. G. Sakhno, Y. V. Kuzmin, M. D. Glascock, and B.-K. Choi. 2005. Geochemistry of Volcanic Glasses of the Paektusan Volcano. *Doklady Earth Sciences* 403, 254–259.

Raymond, L. A. 2003. *Petrology of Igneous Rocks*. Seoul, Sigma Press.

Seong, C. 2007. Late Pleistocene Microlithic Assemblages in Korea. In *Origin and Spread of Microblade Technology in Northern Asia and North America*, edited by Y. V. Kuzmin, S. G. Keates and C. Shen, 103–114. Burnaby, B.C., Canada, Archaeology Press.

Seong, C. 2008. Tanged Points, Microblades and Late Palaeolithic Hunting in Korea. *Antiquity* 82, 871–883.

Shelley, P. H. 1993. A Geoarchaeological Approach to the Analysis of Secondary Lithic Deposits. *Geoarchaeology* 8, 59–72.

Shin, J. G., and S. H. Mun. 1996. *An Introduction to Multivariate Statistics*. Seoul, Jayu Academy.

Shon, B. K. 1989. Sangmuyongrieseo Balguldoen Heugyoseogui Gohyange Daehayeo [Provenance Study of Obsidians Excavated at Sangmuyong–ri]. In *Sangmuyong-ri: Paroho Taesujiyuk Yujuk Balgul Josaboko,* edited by B. K. Choi, 781–796. Chuncheon, Kangwon University Press.

Williams-Thorpe, O. 1995. Obsidian in the Mediterranean and Near East: A Provenancing Success Story. *Archaeometry* 37, 217–248.

Yi, S., and Lee X. X. 1996. Heugyoseok Seokgiui Jihwahakjeok Teukseonge Daehan Yebi Gochal [Preliminary Study of the Geochemical Property of Obsidian Stone Tools]. *Journal of the Korean Archaeological Society* 35, 173–187.

Chapter 6
Obsidian Provenance Studies on Kamchatka Peninsula (Far Eastern Russia): 2003–9 Results

Andrei V. Grebennikov, Vladimir K. Popov, Michael D. Glascock, Robert J. Speakman, Yaroslav V. Kuzmin, and Andrei V. Ptashinsky

Abstract: *The results of obsidian provenance research on the Kamchatka Peninsula based on extensive study of the chemical composition of volcanic glasses from both 'geological' sources and archaeological sites are presented. At least 16 geochemical groups reflecting different sources of obsidian have been identified for Kamchatka using Instrumental Neutron Activation Analysis. Seven sources of archaeological obsidian have been linked to specific geologic outcrops, with the distances between sites and obsidian sources up to 550km. At least seven geochemical groups based only on artefact analysis are also described. The use of multiple obsidian sources was a common pattern during the Palaeolithic, Neolithic, and Palaeometal periods of Kamchatkan prehistory.*

Keywords: *Obsidian, Source Identification, Palaeolithic, Neolithic, Kamchatka Peninsula, Russian Far East*

Introduction

Studies of the geochemistry of waterless volcanic glasses (i.e., obsidians) and sources of archaeological obsidian in the Russian Far East have been ongoing since the early 1990s, and the results were recently summarised (Kuzmin 2006; see also Kuzmin, this volume). Until the early 2000s, most areas under investigation were located in the southern part of the region, namely the Primorye (Maritime) Province, the Amur River basin, and Sakhalin Island. In the northern part of the Russian Far East, the Kamchatka Peninsula remains one of the most promising areas for obsidian provenance studies due to the abundance of volcanic glass sources and the extensive use of obsidian by prehistoric people. However, it was not until 2003 that systematic studies of obsidian geochemistry were initiated (Glascock *et al.* 2006a, 2006b; Kuzmin *et al.* 2006; Popov *et al.* 2005a; Speakman *et al.* 2005).

Summarised here are the results from the research undertaken on the geochemistry of obsidian from the Kamchatka Peninsula since 2003. Compared to earlier investigations (Glascock *et al.* 2006b; Speakman *et al.* 2005), data presented in this study are more comprehensive as a consequence of a larger dataset. Data described in this chapter contribute to a more detailed study of Kamchatkan prehistory, an area important in terms of its relationship with the peopling of the New World at the end of Pleistocene and in the Holocene, especially with regard to Alaska and the Aleutian Islands.

General Patterns of Geology and Archaeology of Kamchatka

The Kamchatka Peninsula, situated in the Northwestern Pacific, stretches approximately 1200km in a SSW–NNE direction with a maximum width of about 400km. The territory of Kamchatka covers about 370,000km^2 and is flanked by the Bering Sea in the east, the Sea of Okhotsk in the west, and the open Pacific Ocean in the south (Figure 6.1, A). The main geomorphic features of the Kamchatka Peninsula are two major mountain ranges, Central and Eastern, with a sedimentary basin between them occupied by the Kamchatka River drainage; mountains of the southern region; and lowlands on the western coast (Suslov 1961, 380–384; Ivanov 2002, 431–433) (Figure 6.1, A). The highest peak in the Central Range [*Sredinny Khrebet*, see The Times Atlas 1989] is Ichinsky Volcano [*Ichinskaya Sopka*] (3607m above sea level; hereafter – a.s.l.); and in the Eastern Range the highest point is the volcanic cone of Klyuchevskaya Sopka (4688m a.s.l.). It should be noted that the summits' heights and size of Kamchatka are to some extent different in Russian and US sources; for the purposes of this paper Russian dictionaries and atlases (Gorkin 1998; Sveshnikov and Krayukhin 2004; Topchiyan 1998) were used.

The Central Range includes a chain of extinct volcanoes located between 53° and 60° N. Among the highest summits are Khangar (2000m a.s.l.), Alnei (2598m a.s.l.), Shishel (2525m a.s.l.), Ostraya (2552m a.s.l.), Khuvkhoitun (2613m a.s.l.), Tylele (2234m a.s.l.), and Budakhanda (1707m a.s.l.). The Eastern Range consists of several ridges, Ganalsky, Valaginsky, Tumrok, Kumroch, and Gamchen; it also has separate high volcanic cones such as Avachinskaya Sopka (2741m a.s.l.), Koryakskaya Sopka (3456m a.s.l.), Zhupanovskaya Sopka (2923m a.s.l.), Krasheninnikov Volcano (1856m a.s.l.), Kronotskaya Sopka (3521m a.s.l.), Tolbachikskaya Sopka (3672m a.s.l.), and Shiveluch Volcano (3307m a.s.l.). The southern part of Kamchatka (south of 53° N) is represented by high plateaus and short ridges like Balaganchik, with volcanic cones situated between them. The highest summits are Mutnovskaya Sopka (2322m a.s.l.), Khodutka Volcano (2089m a.s.l.), Opala Volcano (2460m a.s.l.), Zheltovskaya Sopka (1957m a.s.l.), and Kambalnaya Sopka (2161m a.s.l.).

Tectonically, Kamchatka is located on the boundary between the Pacific and Eurasian plates (e.g., Khain 1994) which is one of the most active volcanic arcs in the world.

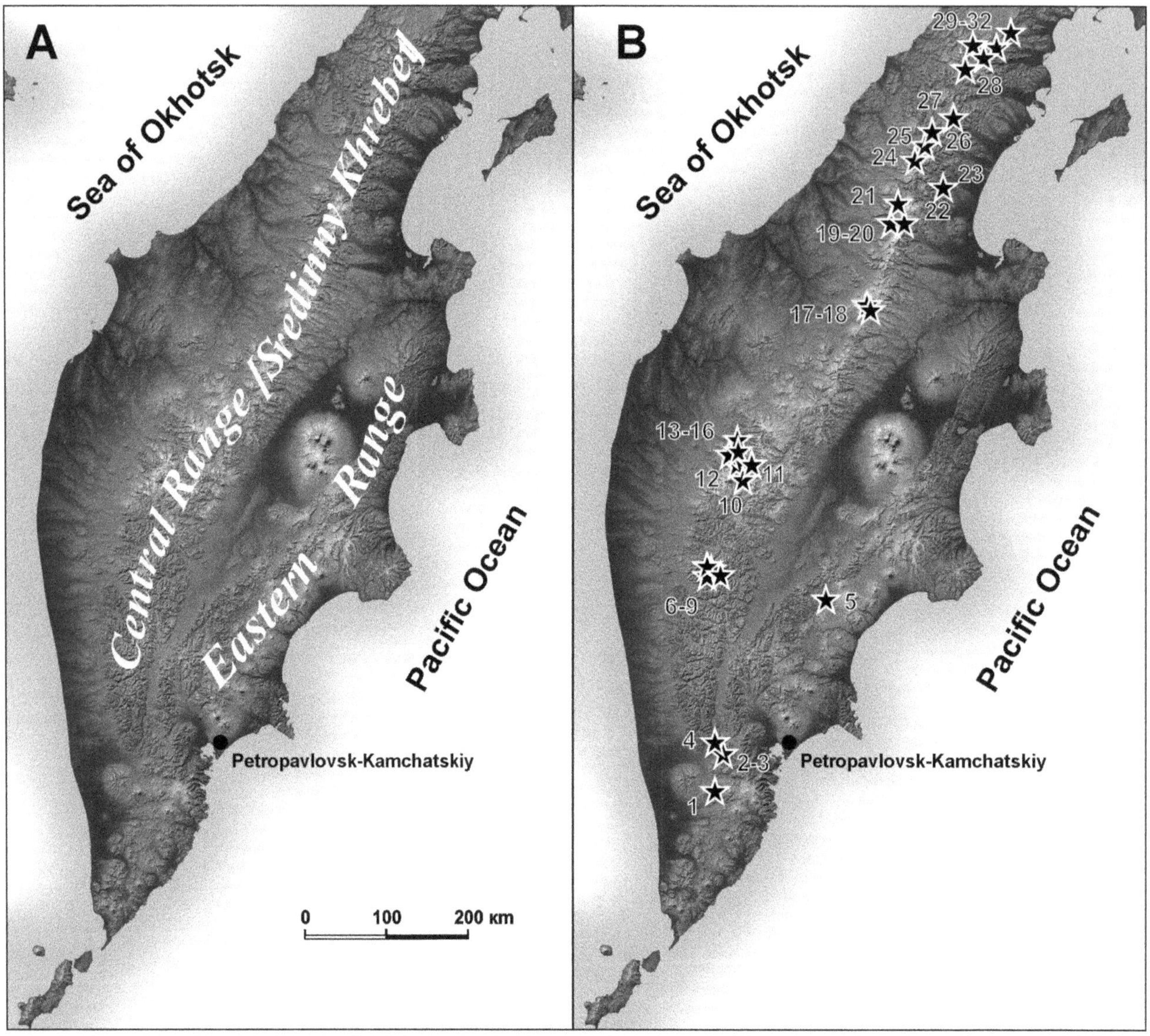

Figure 6.1. Major geomorphic features and obsidian sources at Kamchatka. A: Major mountain ranges. B: Main obsidian sources indicated by star signs (after Otchet 1992, modified). Sources: 1 – Tolmachev Dol (Chasha Maar); 2 – Nachiki Stream; 3 – Taburetka River; 4 – Nachiki (Shapochka Summit); 5 – Karimsky Volcano; 6 – Khangar Volcano (Southern); 7 – Khangar Volcano (Central); 8 – Khangar Volcano (Eastern); 9 – Khangar Volcano (Northern); 10 – Gigigilen; 11 – Payalpan; 12 – Maly Payalpan; 13 – Nosichan; 14 – Polyarnaya Summit; 15 – Tynya Summit; 16 – Belogolovaya Vtoraya River; 17 – Kunkhilok; 18 – Sedanka; 19 – Itkavayam (Southern); 20 – Itkavayam (Northern); 21 – Kevenei (Northern); 22 – Kevenei (Western); 23 – Levye Nachiki; 24 – Levoe Khailuli Plateau; 25 – Maryavaam; 26 – Palana (southern) and Korkavayam; 27 – Palana 1; 28 – Posledny Stream; 29 – Vanyavaam; 30 – Kichiga River, right side; 31 – Belaya River Headwaters; 32 – Kichiga River, left side

As a result, 28 active volcanoes are known on Kamchatka (Fedotov and Masurenkov 1991). Most of the Kamchatkan terrain consists of Cenozoic volcanic rocks, with some sedimentary and volcanic-sedimentary formations (Khain 1994). General petrological information about the volcanic rocks of Kamchatka is available from multiple sources (e.g., Leonov and Grib 2004; Popolitov and Volynets 1982; Volynets *et al.* 1990). Geochemical studies of acidic volcanic rocks on Kamchatka, related to investigations of the genesis of the island arc's acidic magmas, have been performed (Pampura *et al.* 1979; Popolitov and Volynets 1981). Volcanic glasses (obsidians and perlites) are widely distributed on Kamchatka, and they correspond to dacitic–rhyolitic volcanic complexes of the Neogene–Pleistocene age (Shevchuk 1981).

Today, about 30 sources of high and medium quality obsidian are known on Kamchatka (Otchet 1992; Rozenktrants 1981; Shevchuk 1981) (Figure 6.1, B). In the Central Kamchatkan volcanic belt, corresponding to the Central Range, obsidian-parent volcanic formations are dated to the Oligocene–Neogene. In the Eastern Kamchatkan volcanic belt (Eastern Range), obsidian is known mainly among the Pleistocene rocks (Khain 1994). In southern Kamchatka, obsidian-bearing rocks are dated to the Pliocene–Pleistocene. Volcanic glasses on Kamchatka occur in extrusive domes, lava and pyroclastic flows, and are also found in pyroclastic rocks (tephras and pumice tuffa) in the form of fragments. According to their chemical composition, the volcanic glasses correspond to dacites and rhyolites.

The largest number of obsidian sources on Kamchatka is found in the Central Range (Figure 6.1, B). Sources tend to be located at the higher elevations and occur as open-air scatters of colluvial origin on the mountain plateaus covered with tundra. Obsidian is present in blocks and big chunks. In the central part of the Eastern Range, several volcanic glass localities are found; and they are correlated with the Pleistocene phase of the acidic ignimbrite volcanism. South of the city of Petropavlovsk-Kamchatskiy, several volcanic glass sources of the Pliocene–Pleistocene age and rhyolitic composition were discovered.

The first natural science explorers on Kamchatka recorded obsidian sources in the mid-eighteenth century AD (e.g., Krasheninnikov 1972). To modern archaeologists, obsidian as a raw material was known in the prehistoric complexes of Kamchatka since at least the beginning of the twentieth century (e.g., Jochelson 1928), and its significance became more evident in the early 1960s when excavations of the Ushki site cluster began along with reconnaissance of the whole peninsula (Dikov 1965, 1968). Continued studies of the Kamchatkan archaeology (e.g., Dikova 1983; Ponomarenko 1985, 2000; Ptashinsky 2002) confirm the abundance of obsidian in the prehistoric assemblages. Since the early 1900s, obsidian artefacts have been discovered at more than 800 archaeological sites throughout Kamchatka. It was found that obsidian is prevalent in assemblages from the central and southern parts of Kamchatka in up to 96% of the total amount of raw material used (Glascock *et al.* 2006b). In northern Kamchatka, the use of obsidian for tool manufacture in prehistory was less common.

On the Kamchatka Peninsula, three main prehistoric stages, Palaeolithic, Neolithic, and Palaeometal (or Early Iron Age in some sources), have been established (e.g., Dikov 2003, 2004; Ponomarenko 2005). The Upper Palaeolithic is best represented by the two lowermost layers, 6 and 7, of the Ushki site cluster in the Kamchatka River Valley (Dikov 1996). The age of layer 7 of the Ushki, based on radiocarbon (hereafter – ^{14}C) dates is estimated as ca. 14,300–10,400 BP (Dikov 1996), or perhaps as ca. 11,300–10,000 BP according to the latest research (Goebel *et al.* 2003); and the age of layer 6 is ca. 11,100–10,000 BP. The newly discovered site Anavgai 2 with microblades is dated to ca. 10,900 BP (Ptashinsky 2009). The Neolithic period emerged on Kamchatka during the Middle Holocene, ca. 6000–5000 BP, and continued until ca. 1500 BP (Kuzmin 2000). A particular feature of Kamchatkan Neolithic is the very rare occurrence of pottery (e.g., Dikov 2003; Kuzmin 2000; Ponomarenko 2005). There are up to 100–150 Neolithic sites known on Kamchatka according to surveys (Ponomarenko 2005; Ptashinsky 2003), but only about ten of them are well-excavated and ^{14}C-dated. The Palaeometal stage (ca. 1500–300 BP) succeeds the Neolithic, although wide use of stone raw materials continued to exist until the contact with colonising Russian Cossacks during the late seventeenth – early eighteenth centuries AD. The number of sites attributed to the Palaeometal is in the several hundreds (Dikov 2003; Ponomarenko 1985, 1991, 1993, 1997, 2000; Ptashinsky 1989, 1999, 2002).

Materials and Methods

To achieve the primary aim of this study, 444 obsidian artefacts were collected from 45 archaeological sites and site clusters on Kamchatka (Table 6.1). The sites range in age from the late Upper Palaeolithic to the Palaeometal. Sixty-three samples of high quality volcanic glass also were obtained from 'geological' outcrops located throughout the Kamchatka Peninsula.

Geochemical studies of volcanic glasses were conducted using Instrumental Neutron Activation Analysis (hereafter – INAA; e.g., Glascock *et al.* 1998, 2007). INAA remains one of the most advanced methods to study the chemical composition of volcanic glasses, with sensitivity limits for most elements in the parts-per-million (ppm). The advantage of INAA is that one can use small samples (starting from a few milligrams) to measure more than 25 chemical elements, including the rare earth ones, to reveal a unique "geochemical fingerprint" for individual volcanic glass sources. For this study, INAA was performed at the University of Missouri Research Reactor (MURR).

A total of 507 obsidian samples from Kamchatka were submitted to MURR. Two analytical procedures were employed to measure the elemental concentrations. All of the samples first underwent a procedure that used a short irradiation and short decay to measure seven short-lived elements (Al, Ba, Cl, Dy, K, Mn, and Na). Based on these results, the preliminary geochemical groups were established. After that, a total of 162 samples were selected for the second procedure (i.e., a long irradiation). Samples selected for the long irradiation included those assigned to geochemical groups with small number of samples and also individual samples that failed to match any established geochemical group. The long irradiation procedure allows measurement of 22 medium and long-lived elements (Ba, La, Lu, Nd, Sm, U, Yb, Ce, Co, Cs, Eu, Fe, Hf, Rb, Sb, Sc, Sr, Ta, Tb, Th, Zn, and Zr). Statistical groupings, based on examination of bivariative and three-dimension plots, and cluster and discriminant classification analyses, were achieved with the help of GAUSS software (available from MURR) to indicate within a 95% degree of probability the major geochemical groups reflecting the sources of obsidian (Glascock *et al.* 1998).

Results and Discussion

Geochemistry of Kamchatkan Volcanic Glass: The 2009 State-of-the-Art

According to the chemical compositions determined in this study, the acidic volcanic glasses on Kamchatka belong to metaluminous and peraluminous rhyodacites and rhyolites of calc-alcali and subalcali types, and have a variable K/Na ratio. The SiO_2 content in volcanic glasses is from 72.65% to 75.84% weight. Volcanic glasses differ by content according to the amount of large-ion lithophilic elements (Rb, Ba, and Sr), high field strength elements (Y, Ta, Zr, Hf, Nb, and Th), and rare-earth elements, and also the K/Na, Rb/Sr, La/Yb, and Nb/Zr ratios. On the Hf–Rb–Ta diagram

Table 6.1. Archaeological sites and site clusters of the Kamchatka studied for the obsidian sources in 2003-7

Site No.	*Site/Cluster Name*	*Type of Site*	*Archaeological Age*
1	Lopatka Cape	Surface	Neolithic
2	Ozernovsky 1–4	Surface	Neolithic
3	Ozernaya River 1–2	Surface	Neolithic
4	Kurilskoe Lake	Surface	Neolithic
5	Kekhta River	Surface	Neolithic
6	Ust-Kovran	Cultural layer	Palaeometal
7	Kulki	Cultural layer	Neolithic
8	Palana-airport	Cultural layer	Neolithic
9	Anadyrka 1	Cultural layer	Palaeometal
10	Inchegitun 1	Cultural layer	Palaeometal
11	Chimei	Cultural layer	Palaeometal
12	Galgan 1	Cultural layer	Palaeometal
13	Zeleny Kholm	Cultural layer	Palaeometal
14	Pakhachi	Surface	Neolithic
15	Vaimintagin	Surface	Palaeometal
16	Nerpichye Lake	Surface	Palaeometal
17	Kozlov Cape	Surface	Palaeometal
18	Lisy	Surface	Palaeometal
19	Zhupanovo (Cape Pamyatnik)	Cultural layer	Palaeometal
20	Kopyto 1 (Zhupanovo River mouth)	Cultural layer	Palaeometal
21	Avacha; Avacha River, lower stream; Avacha River, animal farm	Cultural layer	Neolithic
22	ASK (Avacha 9); Severnye Koryaki Airport	Cultural layer	Neolithic
23	Plotnikova River (Nachiki Lake)	Surface	Neolithic
24	Lake Sokoch	Cultural layer	Neolithic
25	Viluchinsk 1–5; Sarannya Bay; Turpanka Bay	Surface	Palaeometal
26	Veselaya River (tributary of the Mutnaya R.)	Surface	Neolithic
27	Anavgai	Surface	Final Palaeolithic – Neolithic
28	Esso	Surface	Neolithic
29	Bolshoi Kamen	Cultural layer	Palaeometal
30	Karimshina River	Surface	Palaeometal
31	Elisovo 1–5; Nikolaevka	Surface	Palaeometal
32	Ilmagan	Surface	Neolithic
33	Kluchi	Cultural layer	Neolithic
34	Nikolka	Cultural layer	Neolithic
35	Siyushk	Cultural layer	Paleometal
36	Kozyrevsk	Cultural layer	Neolithic
37	Penzhina	Surface	Paleometal
38	Ushki 1, 2, 5	Cultural layer	Upper Palaeolithic – Neolithic
39	Kirpichnoe	Cultural layer	Neolithic
40	Zastoichik	Cultural layer	Neolithic
41	Lake Domashnee	Cultural layer	Neolithic
42	Kamaki	Cultural layer	Neolithic
43	Lopatka	Surface	Neolithic
44	Yavino 2	Surface	Palaeometal
45	Doyarki	Cultural layer	Neolithic

(Figure 6.2), the composition of volcanic glasses from the Central Range, Eastern Range, and the southern part of the peninsula are distinct; and the fields do not overlap.

The distribution of rare earth elements in volcanic glasses normalised by the primitive mantle provide evidence with regard to the depletion of all groups by heavy rare earth elements in relation to comparatively light elements, with a sharp decrease from La to Eu, and a steady increase from Eu to Lu. The clear Eu anomaly (in addition to obsidians from the Khangar Volcano, some of the Payalpan localities, and the Chasha Maar) is evident for volcanic glasses of Kamchatka. Based on Pierce's bivariant diagrams Nb–Y, Ta–Yb, Rb–(Y+Nb), and Rb–(Yb+Ta), these glasses belong to granitoids of volcanic arcs. The KAM-2 geochemical group of archaeological obsidian is the only exception; and

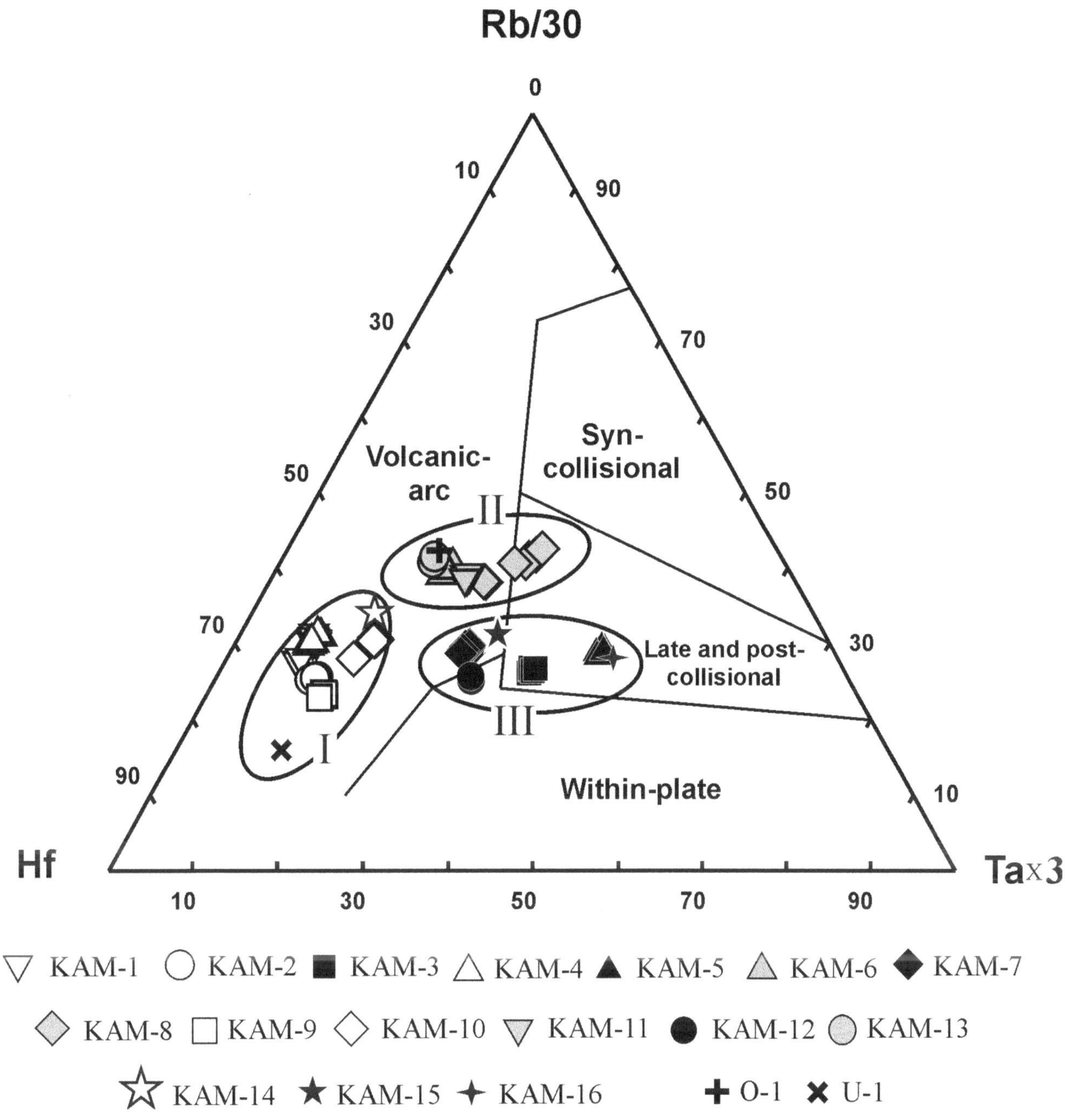

Figure 6.2. The Hf–Rb/30 – Ta×30 discrimination diagram for obsidian source samples and artefacts from Kamchatka (after Popov et al. 2007, with additions) showing the fields for volcanic arc granites, within-plate granites, syn-collisional granites, and late- and post-collisional granites. I – volcanic glasses from Eastern Kamchatka; II – volcanic glasses from Southern Kamchatka; III – volcanic glasses from Central Ridge; O–1 - obsidian from Opala Volcano; U–1 - perlite from Uzon Caldera. Site numbers correspond to those in Table 6.1. The plotting coordinates are from Harris et al. (1986)

it corresponds to an intraplate granite. For this group, the high concentrations of potassium (K), high field strength elements, rare earth elements, and low values of Nb/Zr ratio are considered typical. Volcanic glasses from the KAM-2 group correspond to the Eastern Volcanic Zone (Eastern Range). Volcanic glasses of Eastern Kamchatka (Uzon Caldera; Odnoboky Volcano in Karimsky Volcanic Centre) are characterised by minimal Nb/Zr values. The Nb/Zr ratio increases in the western part of Southern Kamchatkan Volcanic Zone (Shapochka and Opala volcanoes, Chasha Maar, and Yagodnaya Summit) and in Central Range (Payalpan group of volcanic glasses; Khangar and Obsidianovy volcanoes).

Based on the dataset, a total of 16 geochemical groups (KAM-01 through KAM-16) of obsidian were identified from the analysis of 407 specimens. An additional 37 obsidian samples were unassigned or ungrouped, and may be considered outliers and/or unknown sources. The groups and the number of samples in each are as follows:

Groups	Number of archaeological (geological) samples
KAM-01	113 (0)
KAM-02	46 (0)
KAM-03	34 (4)
KAM-04	28 (0)

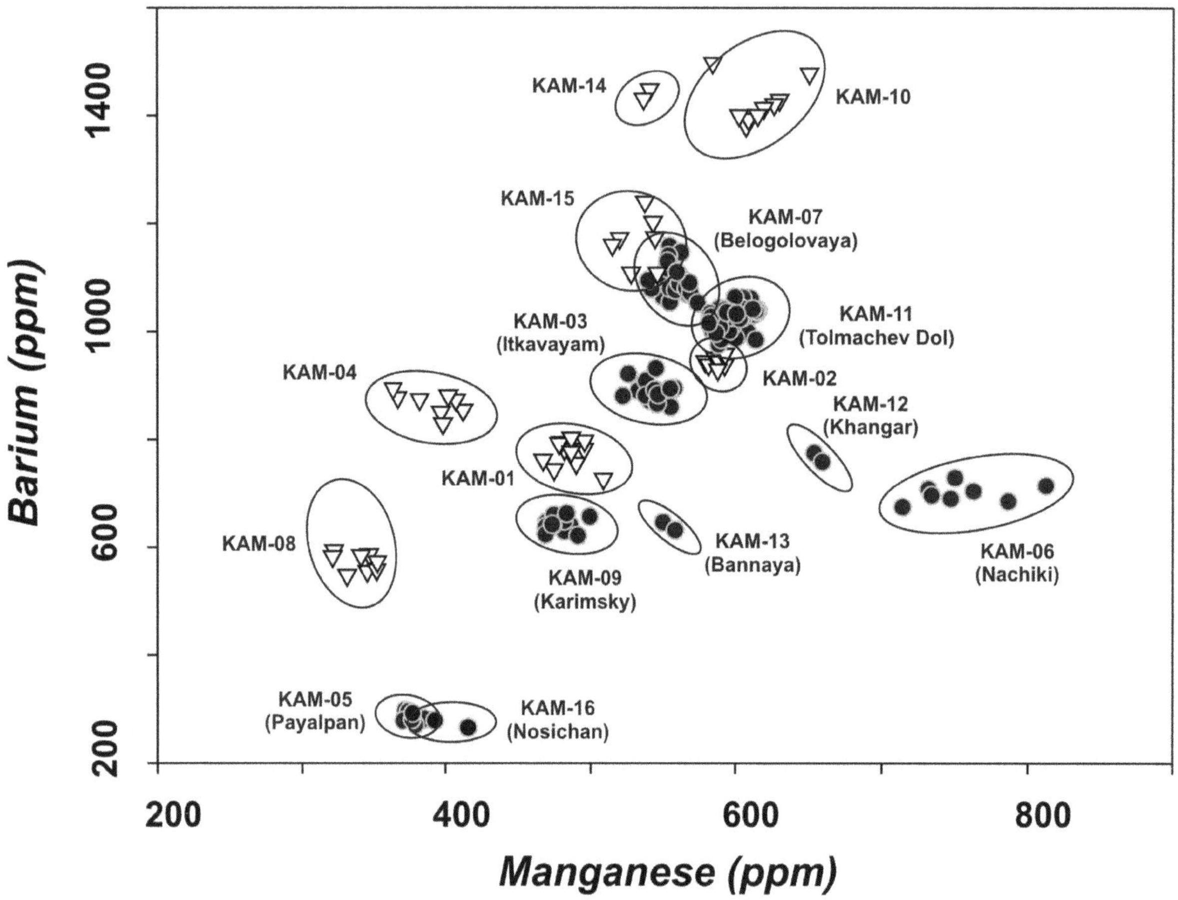

Figure 6.3. Bivariative plot of Mn vs. Ba concentrations for obsidian source samples and artefacts from Kamchatka analysed by INAA and geochemical group names (ellipses represent 95% confidence interval for group membership)

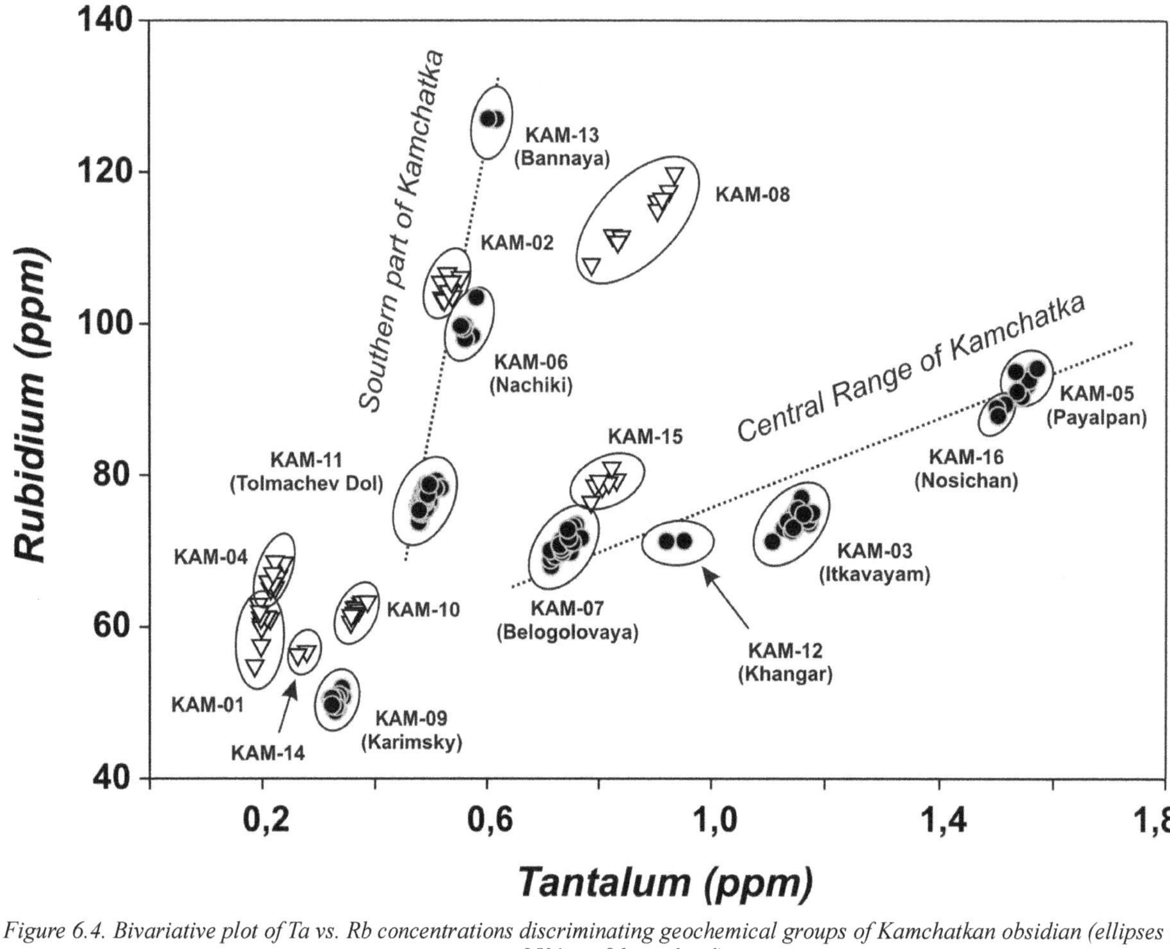

Figure 6.4. Bivariative plot of Ta vs. Rb concentrations discriminating geochemical groups of Kamchatkan obsidian (ellipses indicate 95% confidence level)

KAM-05	38 (2)
KAM-06	17 (2)
KAM-07	14 (0)
KAM-09	13 (11)
KAM-10	52 (0)
KAM-11	34 (30)
KAM-12	2 (2)
KAM-13	2 (2)
KAM-14	2 (0)
KAM-15	9 (0)
KAM-16	3 (2)
Unassigned/ungrouped	37 (10)
Total number of specimens	**444 (63)**

Bivariate plots of Ba versus Mn, and Rb versus Ta are shown in Figures 6.3 and 6.4, respectively. These plots illustrate the major geochemical groups in the dataset as of December 2008. Ba and Mn are particularly useful as discriminating elements because as large ions they are incompatible with crystallising solids; as magmas evolve the concentrations of incompatible elements will be different for each source. Figure 6.4 depicts the geographic position of obsidian sources in relationship to main volcanic belts of Kamchatka.

In Table 6.2, the means and standard deviations for the 16 source groups identified from the INAA data are summarised; unassigned/ungrouped samples are not included in Table 6.2 and are not depicted on Figures 6.3–6.4. The number of samples used to establish group statistics for Table 6.2 is much less than the total number of specimens analysed; this is due to the fact that only samples with determination of all 28 elements measured by the short and long irradiation were used to calculate mean values.

It was determined that (1) seven groups contained both archaeological and geological obsidians; (2) seven groups had only archaeological samples; and (3) two groups of geological obsidians had no matches with any archaeological specimens analysed thus far. Four compositional groups were directly attributed to 'geological' sources in the Central Range with precisely known locations: KAM-03 (Itkavayam), KAM-05 (Payalpan), KAM-07 (Belogolovaya Vtoraya River), and KAM-16 (Nosichan). One group, KAM-09 (Karimsky), was identified in the Akademii Nauk Caldera, part of the Karimsky Volcanic Centre in the Eastern Range. Two sources were identified in southern Kamchatka, groups KAM-06 (Nachiki) and KAM-11 (Tolmachev Dol) (Figure 6.4). Based on the results, the major sources of archaeological obsidian can be identified. This information is crucial to understanding the main patterns of raw material exploitation during the prehistory of Kamchatka.

Chemical data based on the variation of trace element compositions for volcanic glasses confirmed the geochemical zonal structure across the Kamchatkan Volcanic Arc which were established earlier (Churikova *et al.* 2001; Ishikawa *et al.* 2001; Popolitov and Volynets 1981; Volynets *et al.* 1987a, 1987b). The nature of this zonal pattern is related to a steep retreat of the subduction zone to the east, and the riftogenesis process and alkali volcanism of intraplate type in the western part of the Central Range. Rhyolites are characterised by the same geochemical peculiarities as basalts in the Neogene–Pleistocene volcanic series of Kamchatka.

Sources of Archaeological Obsidian at Kamchatka: The 2009 Status

Seven archaeological obsidian sources and the sites associated with them are presented in Table 6.3. In comparison to earlier studies (Glascock *et al.* 2006a, 2006b; Speakman *et al.* 2005), one new source is identified, Nosichan (KAM-16) in the Ichinsky Volcano region. Also, for several sources the number of related archaeological sites increased significantly. For example, the KAM-03 source now has 14 associated sites where before there were only eight sites where obsidian from this source was detected. The KAM-05 source has 17 associated sites versus ten in previous studies. The KAM-07 source obsidian is identified at ten localities compared to three sites as reported in Glascock *et al.* (2006b).

The KAM-03 source (Figure 6.5) is located near the headwaters of the Itkavayam River drainage basin on the western slope of the Central Range. The geographic coordinates are 58°05′ N and 160°46′ E. Volcanic glass from the KAM-03 source constitutes the cone of the small Obsidianovy Volcano which is probably not older than approximately 150,000 years. Obsidian is part of the lava flow, and occurs in layers of massive and striped volcanic glass of black and white colours about 0.4–15m thick. According to the chemical composition, the obsidian is rhyolitic.

In the vicinity of the Ichinsky Volcano, situated on the western slope of the Central Range, there are at least 11 distinct volcanic glass sources (Belogolovaya Vtoraya River, Tynya, Polyarnaya, Nosichan, Payalpan, Maly Payalpan, Galdavit, Studeny, Zemnoy Creek, Zemnoy Summit, and Gigigilen) based on data generated by the Geological Survey of Russia (Otchet 1992). Three of these—KAM-05 (Payalpan) (Figure 6.6), KAM-07 (Belogolovaya River) (Figure 6.7), and KAM-16 (Nosichan) (Figure 6.8) – are identified as sources of archaeological obsidian. Their coordinates are: 55°48′ N, 157°54′ E (KAM-05 and KAM-16); and 55°52′ N, 157°37′ E (KAM-07). The volcanic glass corresponds mainly to the dacite–rhyolite rocks of the upper part of Alnei Group (late Miocene) (Sheimovich and Patoka 2000). Some sources, such as Belogolovaya Vtoraya River, may have been created approximately 2,500,000 years ago.

The KAM-05 source is located on the western slope of Maly Payalpan Volcano and represents a lava flow originating from the intrusive dome. The flow is about 100m long and 5m thick. The obsidian has a black colour with dark-grey stripes, often subtranslucent. Based on chemical composition, the KAM-05 volcanic glass is rhyolitic. The KAM-07 source is situated in the headwaters of the Belogolovaya Vtoraya River, on the northern part of

Table 6.2. Concentration of elements (ppm) measured by INAA in obsidian samples from Kamchatka

Groups	KAM-01 (n = 12)		KAM-02 (n = 8)		KAM-03 (n = 15)		KAM-04 (n = 8)		KAM-05 (n = 9)		KAM-06 (n = 8)		KAM-07 (n = 24)		KAM-08 (n = 9)	
Element	**Mean**	**S.D.**	**Mean**	**S.D.**	**Mean**	**S.D.**	**Mean**	**S.D.**	**Mean**	**S.D.**	**Mean**	**S.D.**	**Mean**	**S.D.**	**Mean**	**S.D.**
Ba	773	24	945	10	890	19	867	21	289	18	700	17	1093	28	645	216
La	11.5	0.2	27.0	0.4	16.8	0.4	12.9	0.2	24.2	0.5	22.8	0.3	18.7	0.4	15.9	2.5
Lu	0.290	0.016	0.778	0.006	0.320	0.026	0.341	0.007	0.265	0.030	0.294	0.005	0.194	0.018	0.277	0.017
Nd	9.5	1.2	30.9	1.5	14.9	6.5	11.6	1.2	14.2	0.7	17.8	1.5	12.6	1.2	14.1	2.3
Sm	2.24	0.40	7.50	0.08	2.78	0.10	2.78	0.56	2.48	0.05	3.57	0.72	2.16	0.08	3.20	0.23
U	1.52	0.35	2.95	0.18	4.10	0.28	1.72	0.52	4.47	0.31	2.77	0.11	2.54	0.22	3.05	0.29
Yb	1.77	0.14	5.18	0.14	1.80	0.11	2.11	0.07	1.35	0.03	1.85	0.12	1.09	0.06	1.51	0.09
Ce	23.4	0.7	61.6	1.2	34.1	0.9	26.9	0.6	44.0	1.5	45.4	0.9	34.0	0.8	33.8	4.9
Co	1.290	0.118	0.595	0.016	0.351	0.012	1.029	0.081	0.163	0.031	0.223	0.024	0.954	0.110	0.909	0.158
Cs	3.54	0.09	4.74	0.07	3.23	0.08	4.40	0.15	2.26	0.05	4.58	0.07	1.26	0.02	10.39	0.81
Eu	0.473	0.044	1.006	0.015	0.455	0.017	0.491	0.011	0.296	0.007	0.499	0.010	0.513	0.011	0.502	0.041
Fe	10,777	367	13,489	181	5761	151	9586	404	4146	99	5272	128	8548	418	9285	810
Hf	4.06	0.11	8.66	0.14	3.50	0.08	4.41	0.09	2.95	0.08	3.39	0.11	3.44	0.08	3.01	0.36
Rb	60.0	2.2	104.8	1.3	74.2	1.5	66.6	1.4	92.2	1.5	99.8	1.6	70.8	1.4	114.0	3.8
Sb	1.28	0.13	1.01	0.10	0.41	0.03	1.73	0.22	0.41	0.02	0.50	0.05	0.12	0.01	0.24	0.02
Sc	3.03	0.52	7.48	0.11	1.99	0.04	3.26	0.14	1.72	0.04	2.02	0.03	1.82	0.05	3.26	0.07
Sr	206	20	84	41	111	17	157	7	53	6	77	10	354	54	153	18
Ta	0.20	0.01	0.53	0.01	1.15	0.02	0.22	0.01	1.56	0.04	0.56	0.01	0.74	0.02	0.87	0.05
Tb	0.31	0.03	1.23	0.03	0.37	0.01	0.39	0.01	0.29	0.01	0.39	0.01	0.22	0.01	0.47	0.03
Th	3.98	0.14	7.43	0.11	7.62	0.17	4.71	0.13	9.27	0.05	7.14	0.13	4.65	0.11	5.72	0.79
Zn	34.8	2.4	65.4	3.1	34.1	2.1	35.1	2.7	24.5	3.7	32.3	1.0	34.3	4.4	44.1	5.6
Zr	131	9	282	10	126	6	145	8	97	5	114	6	133	10	106	16
Al	70,756	2369	71,637	2724	70,995	2881	67,649	4003	66,722	2640	69,736	2802	75,867	2000	75,412	2186
Cl	376	73	686	105	112	23	356	24	239	42	409	29	238	79	37	20
Dy	1.84	0.38	7.78	0.47	2.24	0.28	2.71	0.22	1.64	0.16	2.25	0.21	1.09	0.26	2.68	0.35
K	26,219	3215	41,885	2051	31,992	1708	27,542	1536	39,407	1594	37,643	2011	32,521	1987	31,665	437
Mn	486	11	587	5	542	10	391	18	377	5	755	32	558	8	339	12
Na	30,867	971	32,292	357	32,134	411	29,473	707	28,142	773	30,858	338	31,966	703	30,404	690

Table 6.2. Concentration of elements (ppm) measured by INAA in obsidian samples from Kamchatka (end)

Groups	KAM-09		KAM-10		KAM-11		KAM-12		KAM-13		KAM-14		KAM-15		KAM-16	
	(n = 18)		(n = 11)		(n = 34)		(n = 2)		(n = 2)		(n = 2)		(n = 7)		(n = 3)	
Element	Mean	S.D.	Mean	S.D.	Mean	S.D.	Mean	S.D.	Mean	S.D.	Mean	S.D.	Mean	S.D.	Mean	S.D.
Ba	650	16	1391	100	1024	24	767	11	639	10	1439	12	1167	48	279	13
La	15.3	0.2	19.2	3.5	18.4	0.3	20.6	0.1	23.3	0.2	13.8	0.2	20.3	0.5	24.2	0.1
Lu	0.432	0.018	0.292	0.034	0.214	0.012	0.271	0.005	0.357	0.003	0.222	0.021	0.234	0.025	0.297	0.026
Nd	16.5	1.2	16.4	3.4	12.8	1.8	17.1	0.2	20.5	0.3	12.0	2.8	13.7	0.9	13.3	0.1
Sm	3.69	0.10	3.10	0.48	2.29	0.21	3.01	0.02	3.82	0.00	2.11	0.04	2.33	0.11	2.45	0.02
U	1.99	0.22	2.26	0.52	2.46	0.15	2.95	0.29	3.40	0.21	2.93	0.27	2.72	0.16	4.46	0.03
Yb	2.89	0.04	1.75	0.16	1.26	0.07	1.77	0.04	2.37	0.03	1.12	0.03	1.16	0.14	1.38	0.03
Ce	34.5	1.0	37.1	3.3	33.9	0.9	41.1	0.1	48.7	0.4	27.4	0.1	37.0	1.1	42.0	0.3
Co	0.724	0.150	0.563	0.101	0.264	0.098	0.235	0.004	0.197	0.038	0.336	0.036	0.361	0.029	0.171	0.034
Cs	1.74	0.03	1.68	0.13	2.66	0.06	1.69	0.00	5.67	0.03	2.80	0.01	1.44	0.02	2.20	0.00
Eu	0.594	0.013	0.611	0.037	0.458	0.012	0.661	0.008	0.46	0.001	0.4	0.002	0.498	0.011	0.294	0.004
Fe	9397	326	7758	291	5441	251	7358	256	5337	0	5491	76	6411	156	4069	96
Hf	4.63	0.07	3.68	0.22	2.53	0.06	4.19	0.08	4.24	0.10	2.59	0.30	3.28	0.12	2.90	0.06
Rb	50.4	0.8	63.1	2.9	76.9	1.4	71.3	0.0	127.0	0.0	56.5	0.3	78.9	1.4	88.7	0.8
Sb	0.39	0.05	0.41	0.11	0.36	0.02	0.24	0.00	0.71	0.01	2.16	0.02	0.14	0.01	0.36	0.02
Sc	3.10	0.12	2.11	0.11	1.52	0.10	1.55	0.03	2.38	0.01	1.72	0.01	1.49	0.04	1.71	0.04
Sr	119	20	282	21	216	24	205	4	39	0	219	25	276	25	47	5
Ta	0.33	0.01	0.40	0.11	0.49	0.01	0.94	0.02	0.61	0.01	0.27	0.01	0.81	0.02	1.51	0.01
Tb	0.60	0.02	0.37	0.04	0.26	0.01	0.36	0.01	0.46	0.00	0.22	0.02	0.24	0.01	0.28	0.01
Th	3.43	0.07	4.01	0.25	5.60	0.10	4.69	0.02	8.97	0.02	4.24	0.04	5.41	0.18	9.08	0.02
Zn	35.3	2.6	37.6	3.0	34.0	3.2	45.4	1.9	34.8	0.4	28.9	1.3	31.1	2.0	20.7	0.2
Zr	145	11	134	12	89	5	151	7	141	5	107	15	120	6	98	4
Al	69,299	2750	70,841	4079	70,320	2519	74,444	2946	66,621	257	66,800	1445	74,550	4352	72,595	3467
Cl	756	136	253	33	355	74	363	9	533	31	215	18	254	20	254	20
Dy	3.89	0.39	2.11	0.46	1.49	0.28	1.78	0.42	2.83	0.11	1.09	0.08	1.37	0.33	1.73	0.26
K	28,103	1774	31,381	1494	30,701	2051	28,277	681	39,995	481	29,760	97	33,121	1478	38,202	2921
Mn	481	10	610	22	599	10	657	4	554	6	539	3	534	12	395	19
Na	32,978	597	29,334	1893	29,489	393	36,401	314	29,709	100	27,945	112	30,424	699	29,830	1805

Table 6.3. Sources of archaeological obsidian at Kamchatka and distance from sources to sites

Source Name (Group No.)	***Site No.****	***Site Name***	***Number of samples***	***Distance from source (km)***
Itkavayam (KAM-03)	7	Kulki	1	170
	8	Palana-airport	3	100
	10	Inchegitun 1	5	120
	11	Chimei	2	140
	12	Galgan 1	7	90
	13	Zeleny Kholm	1	450
	16	Lake Nerpichye	1	220
	31	Elisovo 2	1	560
	33	Kluchi	1	210
	37	Penzhina	1	490
	38	Ushki 1	4	230
	38	Ushki 2	1	230
	41	Lake Domashnee	1	200
	45	Doyarki	1	220
Payalpan (KAM-05)	6	Ust-Kovran	1	170
	17	Kozlov Cape	1	280
	21	Avacha	1	315
	21	Avacha River, lower stream	1	310
	22	ASK (Avacha 9)	3	280
	23	Plotnikova River (Lake Nachiki)	1	310
	27	Anavgai	2	70
	31	Elisovo 2	1	290
	31	Elisovo 5	1	290
	32	Ilmagan	5	90
	34	Nikolka	1	140
	35	Siyushk	1	530
	36	Kozyrevsk	1	125
	38	Ushki 1	13	135
	38	Ushki 5	1	135
	41	Lake Domashnee	1	170
	45	Doyarki	1	160
Nachiki (KAM-06)	21	Avacha River, animal farm	1	50
	23	Plotnikova River (Lake Nachiki)	13	5
	24	Sokoch Lake	1	25
Belogolovaya Vtoraya River (KAM-07)	2	Ozernovsky 1	1	470
	2	Ozernovsky 2	1	470
	22	ASK (Avacha 9)	1	260
	33	Kluchi	1	190
	35	Siyushk	1	530
	36	Kozyrevsk	2	120
	38	Ushki 1	7	130
	40	Zastoichik	1	160
	41	Lake Domashnee	2	170
	42	Kamaki	1	230
Karimsky (KAM-09)	20	Kopyto 1 (Zhupanova River mouth)	2	40
Tolmachev Dol (KAM-11)	24	Lake Sokoch	2	60
	25	Viluchinsk 2	2	65
Nosichan (KAM-16)	22	ASK (Avacha 9)	1	260

*Site Nos. correspond to those on Figures 6.5–6.11.

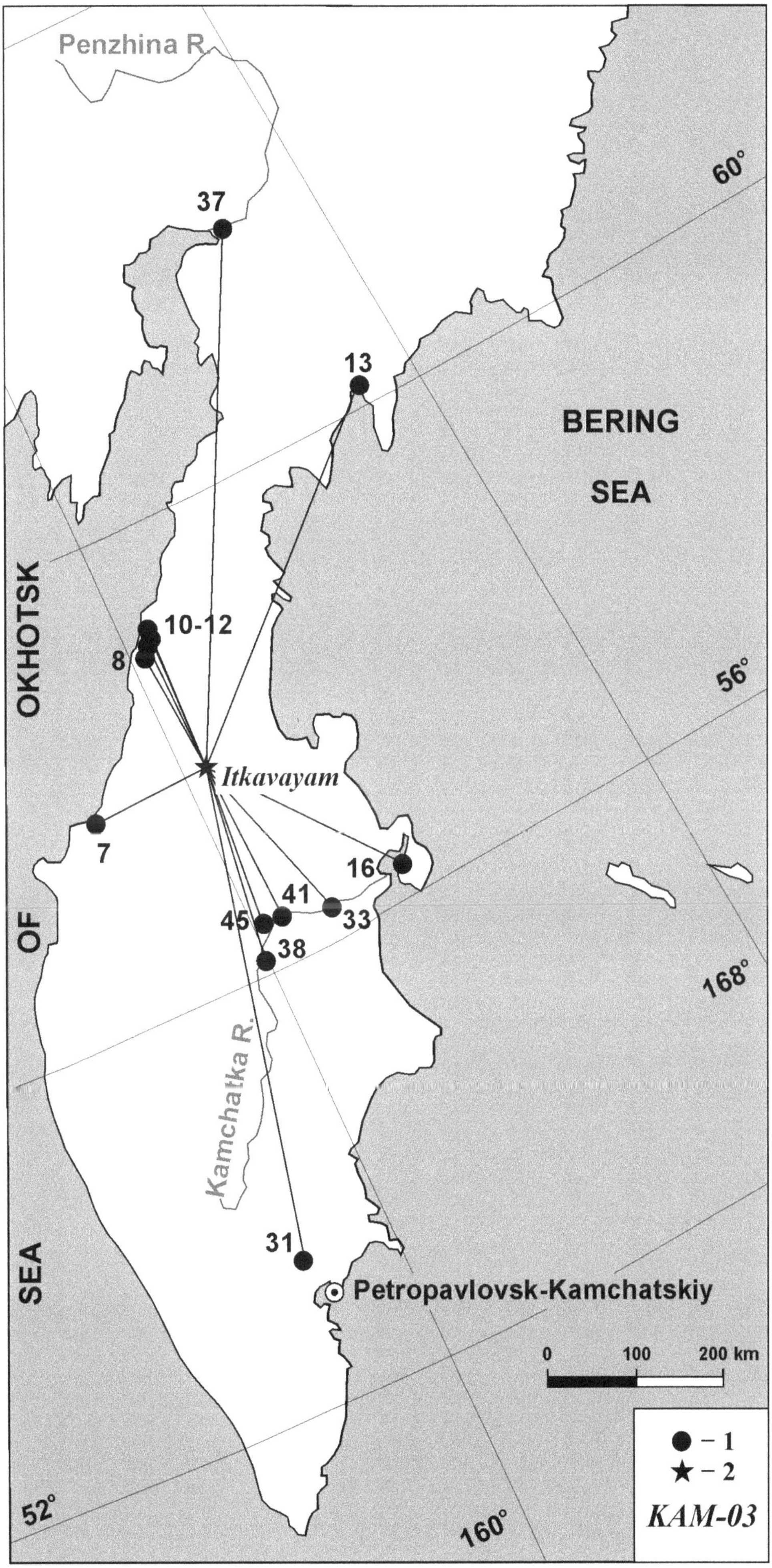

Figure 6.5. The KAM-03 (Itkavayam) obsidian source and associated archaeological sites (on Figures 6.5–6.11, site numbers correspond to those in Table 6.1)

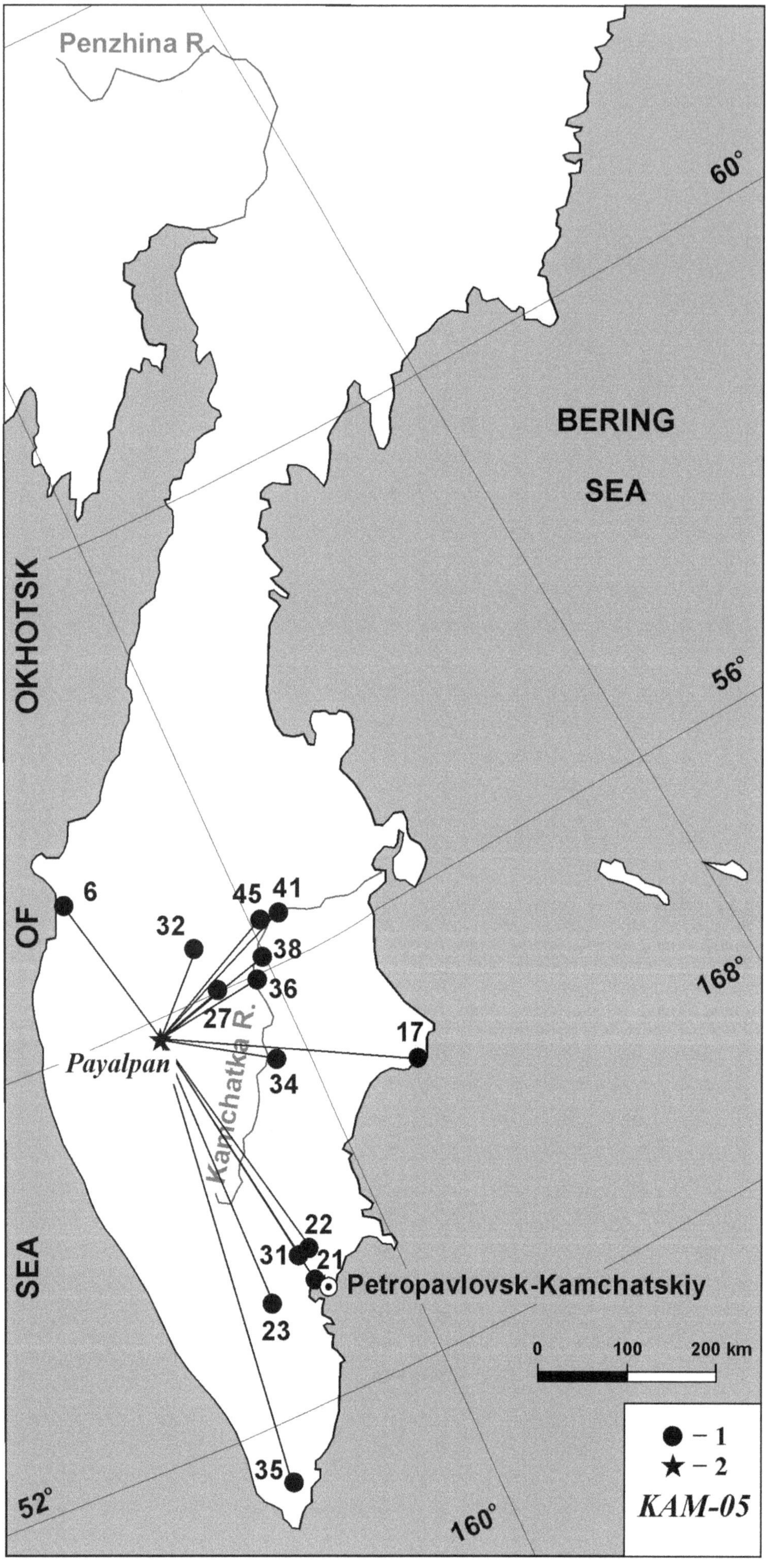

Figure 6.6. The KAM-05 (Payalpan) obsidian source and associated archaeological sites

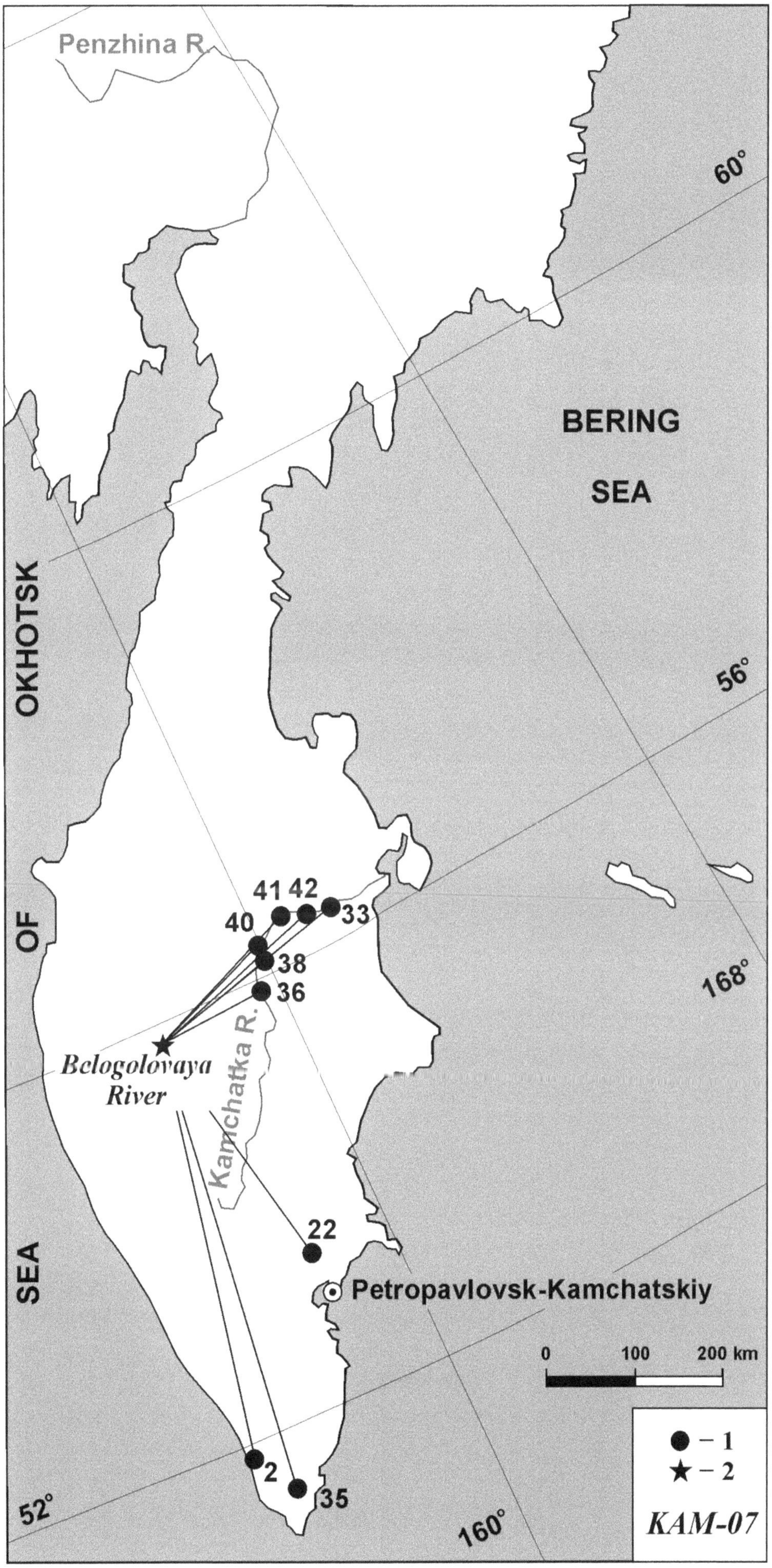

Figure 6.7. The KAM-07 (Belogolovaya Vtoraya River) obsidian source and associated archaeological sites

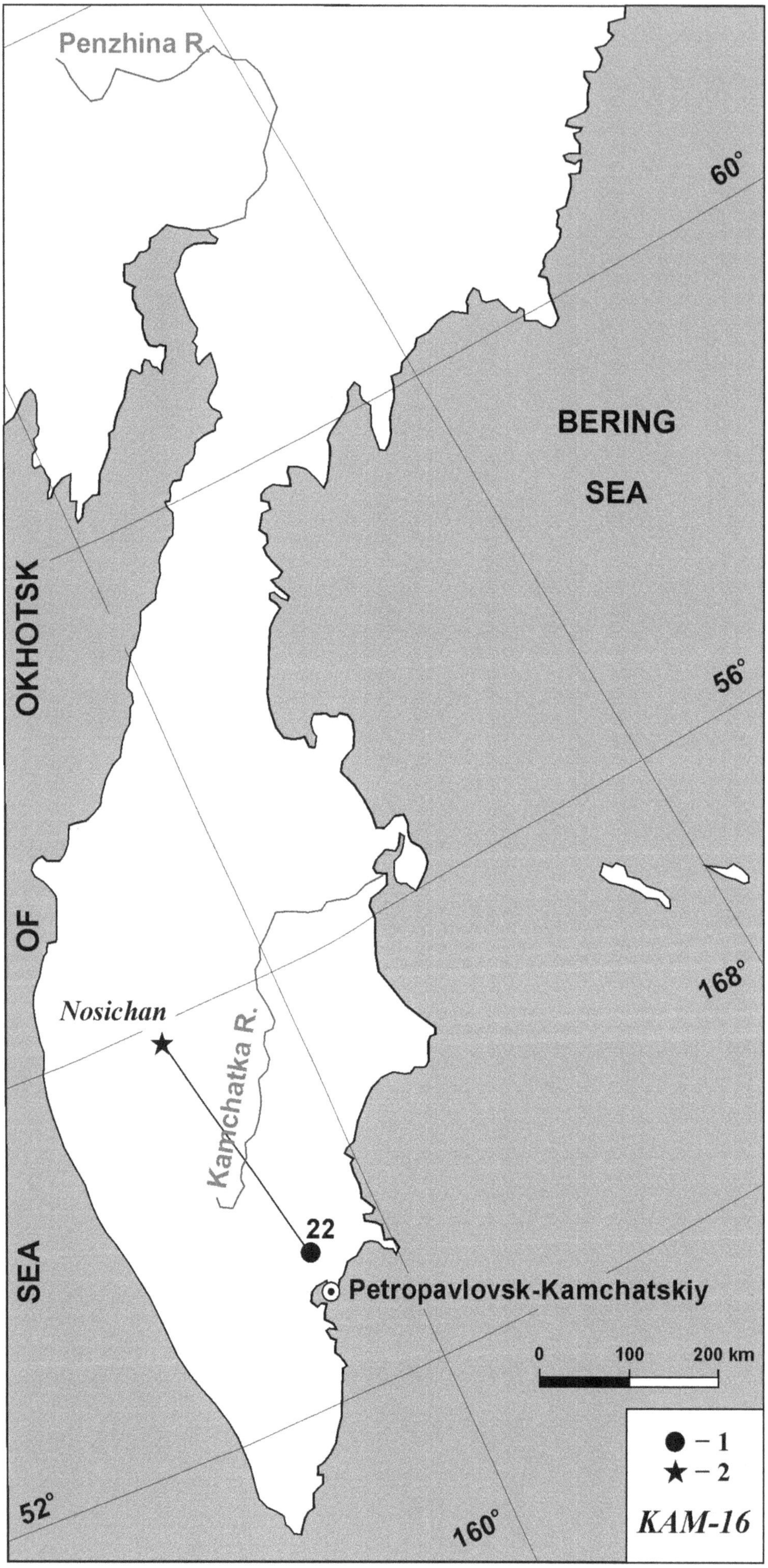

Figure 6.8. The KAM-16 (Nosichan) obsidian source and associated archaeological site

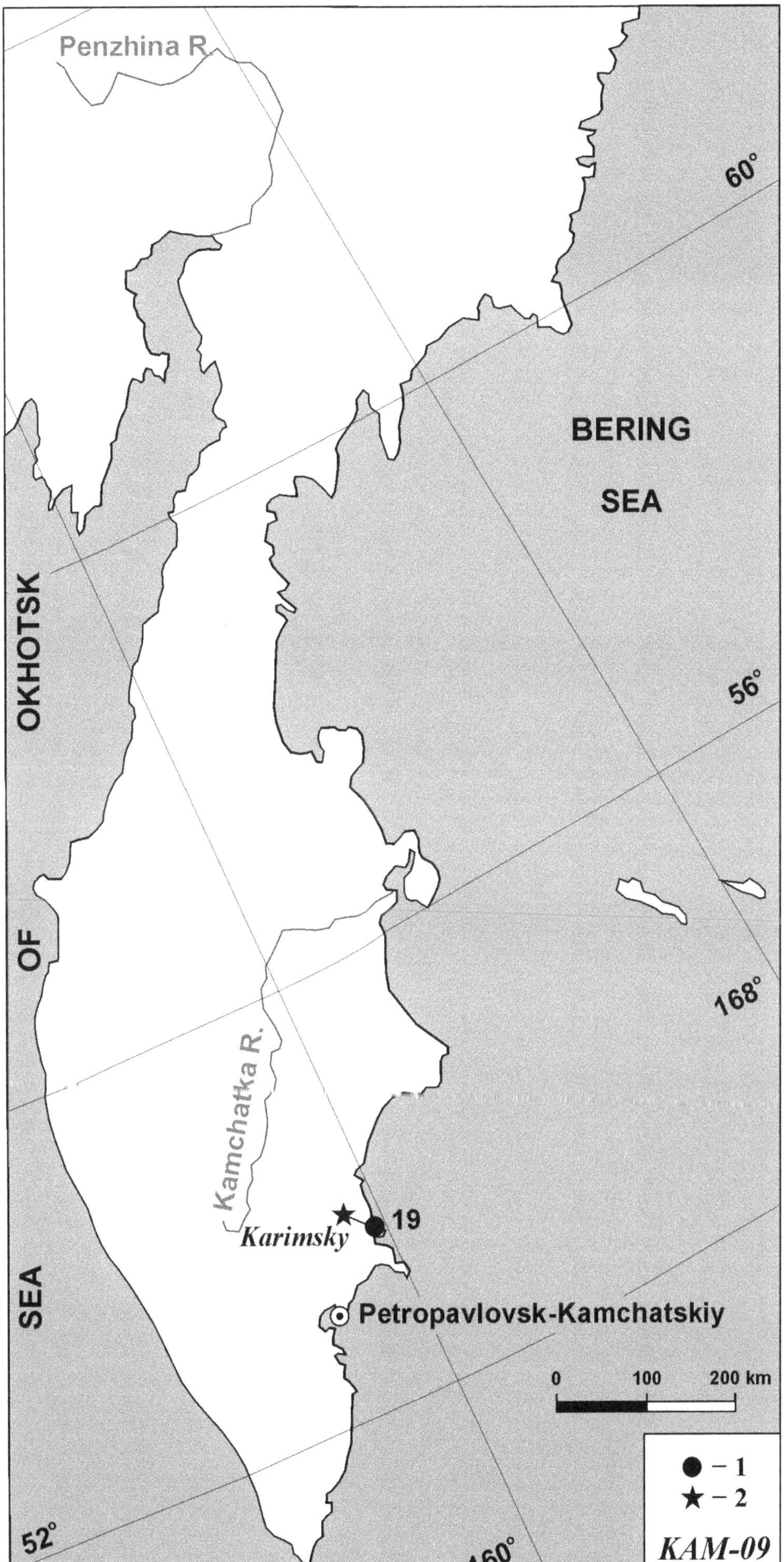

Figure 6.9. The KAM-09 (Karimsky Volcano) obsidian source and associated archaeological site

Tynya Ridge near the summits of Bystraya (1304m a.s.l.) and Tynya (1429m a.s.l.). Volcanic glass from this source is associated with rhyolitic extrusive domes and lava flows. The colours of the obsidian are black and dark-reddish, sometimes dark-gray and blue-grayish; the structures are massive and clastic. Some of the black obsidian pieces are semitransparent. The KAM-16 source is located near the Nosichan River, a tributary of the Belogolovaya Vtoraya River. Three small colluvial taluses and one bedrock outcrop were discovered. The KAM-16 source corresponds to the middle Pleistocene rhyolitic extrusive domes.

In the Eastern Range, the single obsidian source KAM-09 associated with prehistoric sites is in the Karimsky Volcanic Centre (coordinates 54°30′ N, 159°26′ E) (Figure 6.9). During the later stages of the development of this volcanic structure during the late Pleistocene and the Holocene, two calderas were created, Akademii Nauk and Karimskaya (e.g., Erlikh 1972). This source is composed of pumice tuffa, volcanic bombs, and lapilli, of rhyolithic and dacitic composition. High quality volcanic glass is found on the lake shore of the Akademii Nauk Caldera.

In the southern part of Kamchatka two sources of archaeological obsidian, KAM-06 and KAM-11, were identified. The KAM-06 source (Figure 6.10) originates from the eastern slope of Shapochka Summit in the Lake Nachiki area; geographic coordinates are 53°01′ N, 157°43′ E. Black obsidians are embedded into layers of volcanic glass, a marginal part of the rhyolitic extrusion. The KAM-11 source (coordinates 52°37′ N, 157°33′ E) is situated around the Chasha Maar in the central part of the Tolmachev Dol basaltic plateau (Figure 6.11). It was created by a catastrophic explosion around 5300 BP when tephra covered a territory about 15,000km^2, mainly toward the northeast of the Tolmachev Dol Plateau (Melekestsev *et al.* 1996; Dirksen *et al.* 2002). Volcanic products vary from pumice to waterless volcanic glass of grayish colour. The composition of volcanic glass from Tolmachev Dol source is rhyolitic.

Archaeological Obsidian Sources on Kamchatka Yet to Be Identified

In addition to 'geological' sources of archaeological obsidian which were identified, there are at least seven geochemical groups represented only by prehistoric artefacts; and the sources of these have yet to be located. Due to the large number of obsidian sources at Kamchatka, more research is needed to correlate the volcanic glass from archaeological assemblages with sources of obsidian. However, geochemical data and position of archeological sites suggests possible geographic locations for the unknown sources.

The KAM-01 group is widely distributed throughout all of the Kamchatka Peninsula (Figure 6.12) but concentrated mostly in the southeastern part. Artefacts belonging to this group are also identified in the northern Kurile Islands (Phillips and Speakman 2009; see also Phillips, this volume). The KAM-04 Group is situated mostly in the southeastern part of peninsula (Figure 6.13), and also in the northern Kuriles (Phillips and Speakman 2009; see also Phillips, this volume). The KAM-10 group comprises sites located in central and eastern parts of Kamchatka (Figure 6.14). The KAM-14 group consists of sites in central Kamchatka (Figure 6.15). All these groups have Nb/Zr ratios of around 0.04, which is characteristic for volcanic rocks of the East Kamchatkan Volcanic Belt (Ishikawa *et al.* 2001; Münker *et al.* 2004). It should be noted that sites belonging to these groups are in the vicinity of the Karimsky source (KAM-09) which has similar Nb/Zr ratio. Additionally, the chemical composition of this group is very similar to the KAM-09 group (Figure 6.4). Based on current understanding, it is assumed that the sources of the KAM-01, KAM-04, KAM-10, and KAM-14 groups are situated somewhere in the eastern part of Kamchatka.

The KAM-15 group comprises samples from the Ushki site cluster in central Kamchatka (Figure 6.16) (Kuzmin *et al.* 2008). According to the chemical data, this group has some resemblance to the Belogolovaya Vtoraya River source (KAM-07) (Figure 6.4). This fact can be used to tentatively place the KAM-15 source in the Ichinsky Volcano region. The KAM-02 group is represented by sites located mainly in the southern part of Kamchatka (Figure 6.17) and in the northern Kurile Islands (Phillips and Speakman 2009; see also Phillips, this volume). It has some chemical similarities with the Nachiki source (KAM-06) (Figure 6.4). This suggests a location for the KAM-02 source in southern Kamchatka.

Artefacts from sites in the northern part of Kamchatka constitute the KAM-08 group (Figure 6.18). Chemically, this group is intermediate relative to obsidian from the eastern and southern parts of Kamchatka (Figure 6.4). Therefore, it is difficult at this stage of research to assume a location for the obsidian source associated with this group. However, based on the overall distribution of artefacts assigned to this source, it seems very likely that source is located north of, or on, the Kamchatka Isthmus (narrow part of the peninsula in the north).

In addition to the groups that were comprised only of 'archaeological' obsidian, two geochemical groups contained only geological samples: KAM-12 (Khangar Volcano) and KAM-13 (Bannaya) (Figure 6.19). They have no matches with any artefacts from Kamchatka (Figures 6.3–6.4). It is possible that in the future at some archaeological sites obsidian from these sources will be identified.

Spatial-Temporal Patterns of Obsidian Use in Prehistory of Kamchatka (Preliminary Results)

The spatial distribution of obsidian from the seven known sources (Figures 6.5–6.11; Table 6.3) allows us to establish major patterns of transportation for raw material during Kamchatkan prehistory. The KAM-03 source (Figure 6.5) was widely used by ancient populations, with distance between source and sites up to 560km in a straight line. Similar features are typical for the KAM-05 and KAM-07

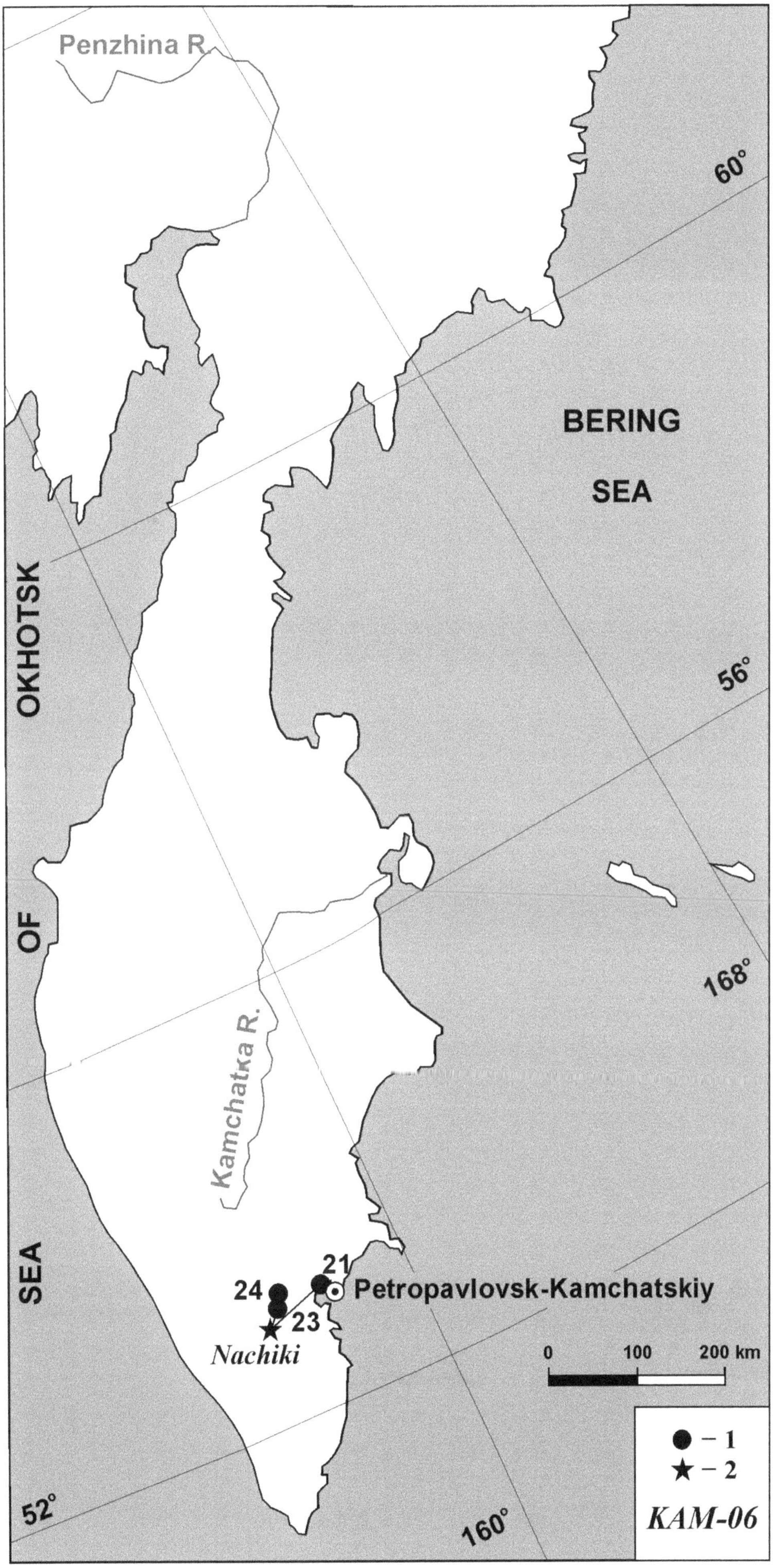

Figure 6.10. The KAM-06 (Nachiki) obsidian source and associated archaeological sites

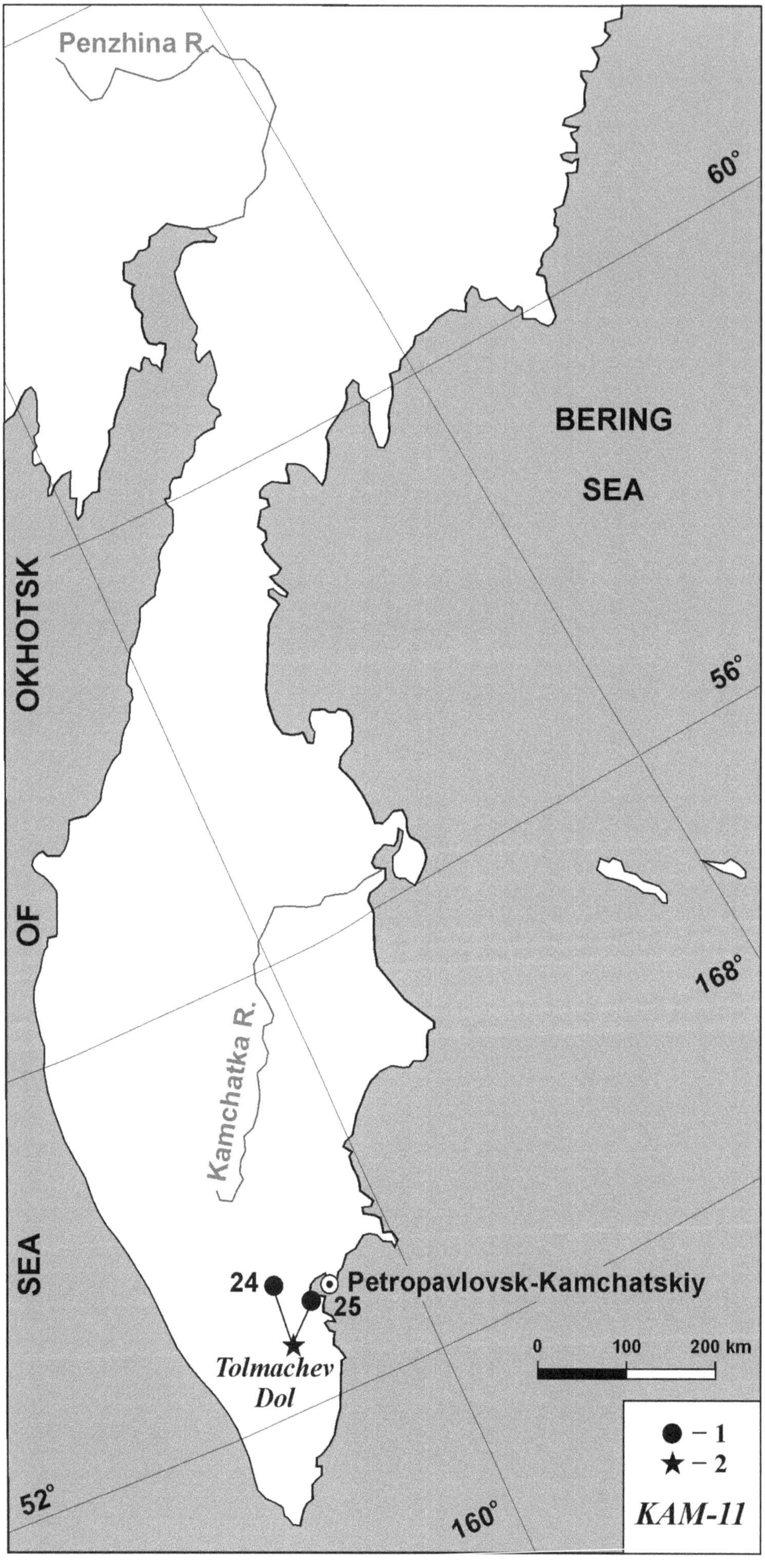

Figure 6.11. The KAM-11 (Tolmachev Dol) obsidian source and associated archaeological sites

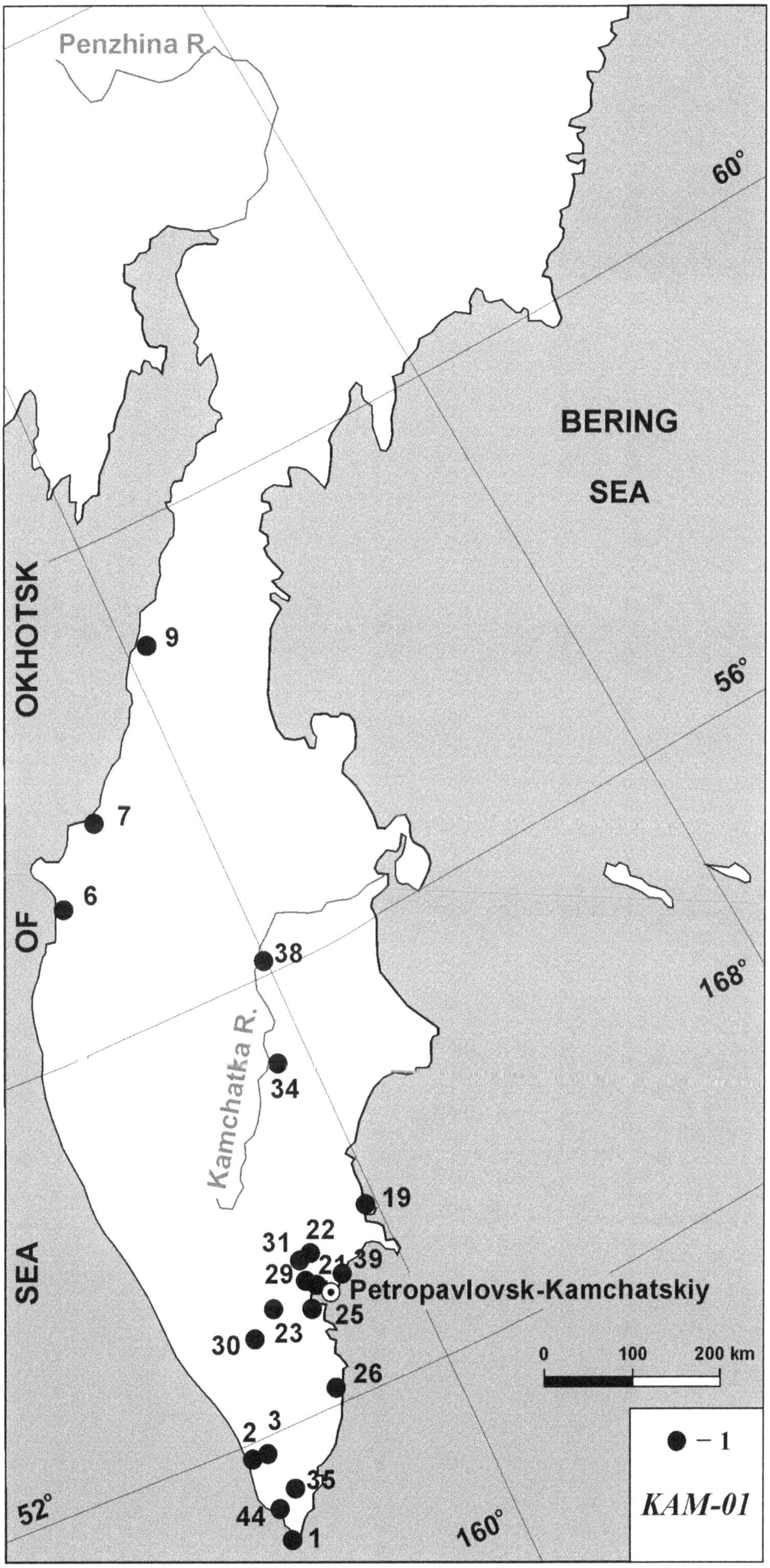

Figure 6.12. The distribution of archaeological sites with KAM-01 group obsidian artefacts (on Figures 6.12–6.18, site numbers correspond to those in Table 6.1)

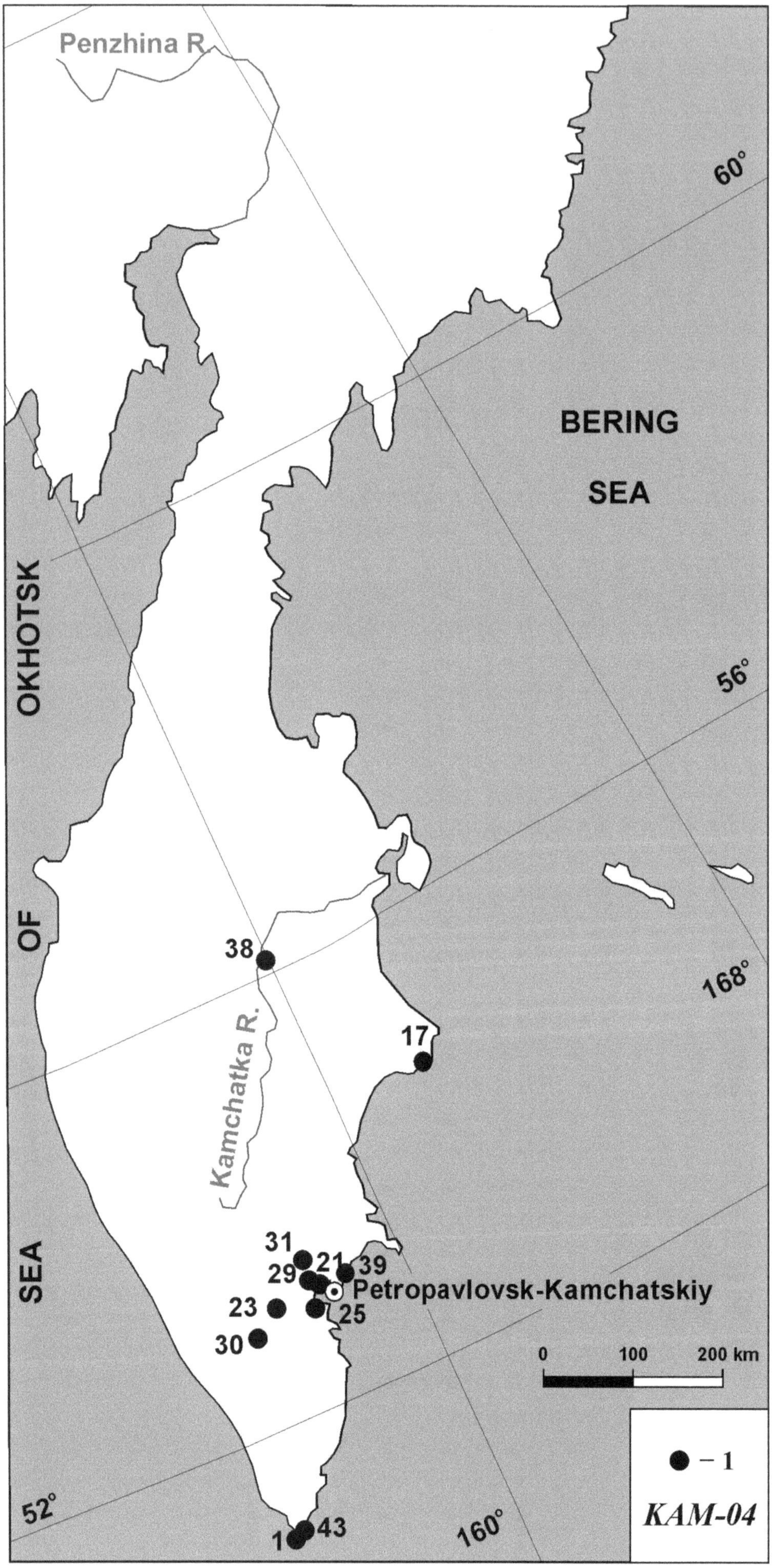

Figure 6.13. The distribution of archaeological sites with KAM-04 group obsidian artefacts

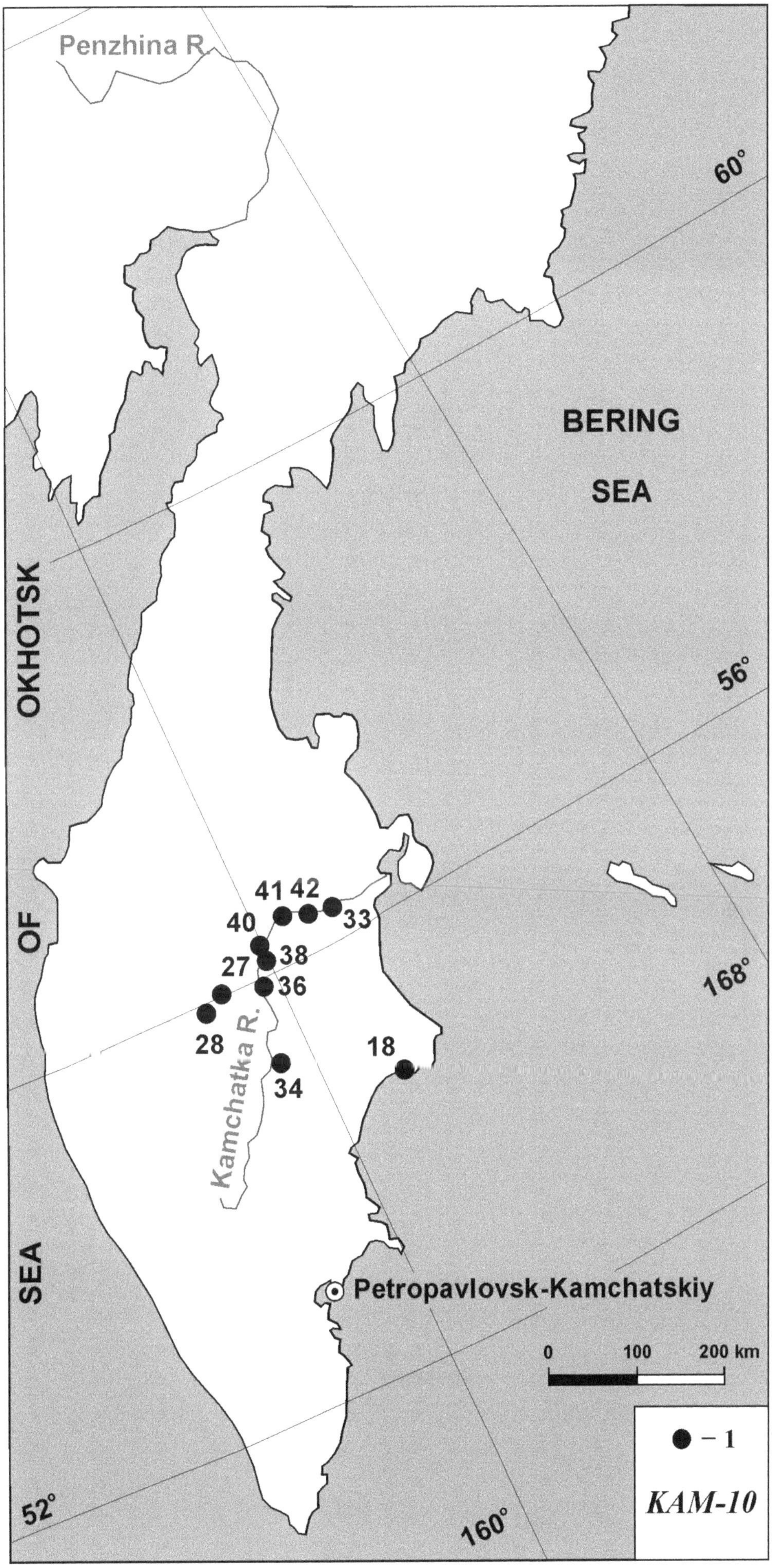

Figure 6.14. The distribution of archaeological sites with KAM-10 group obsidian artefacts

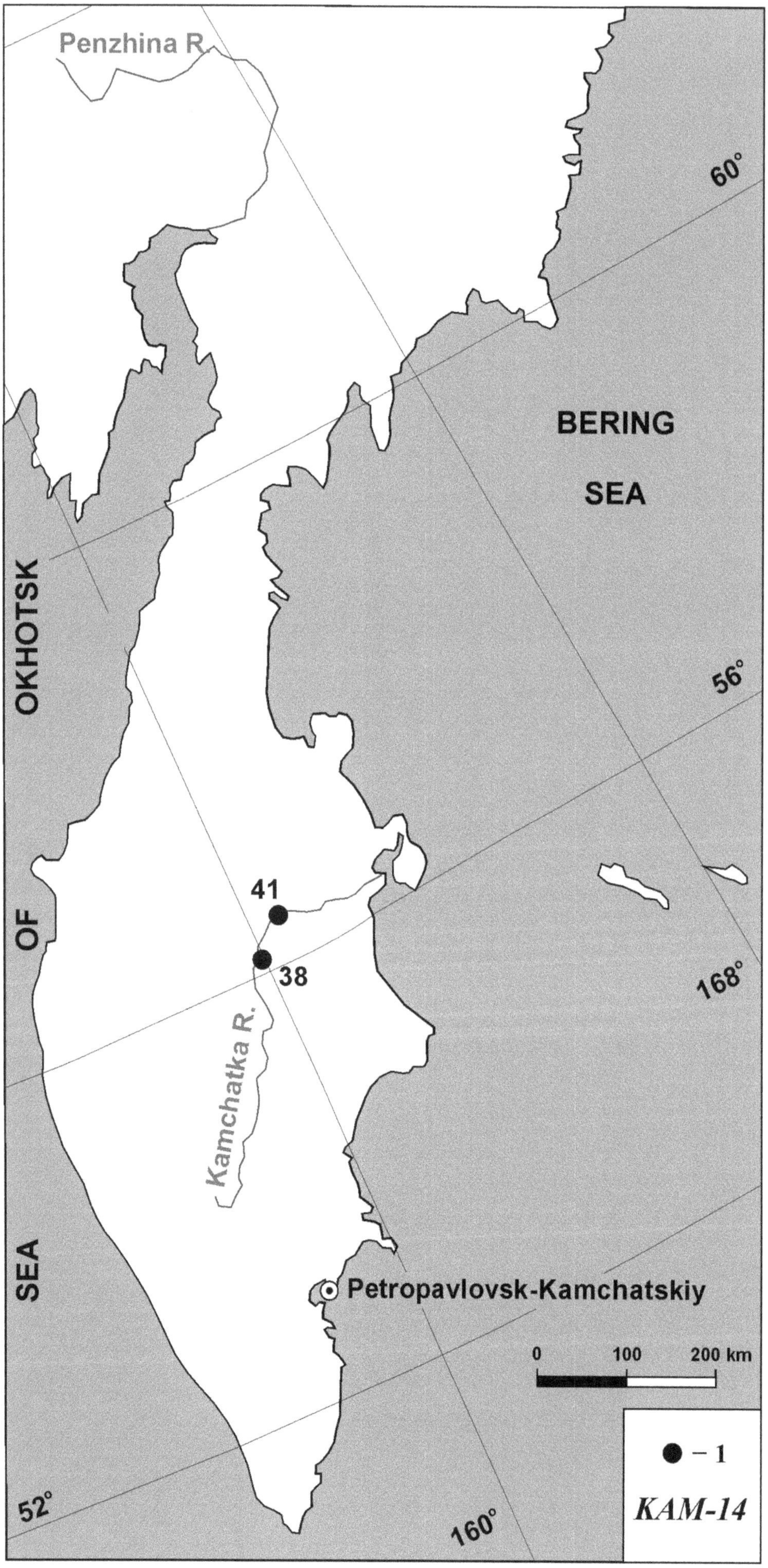

Figure 6.15. The distribution of archaeological sites with KAM-14 group obsidian artefacts

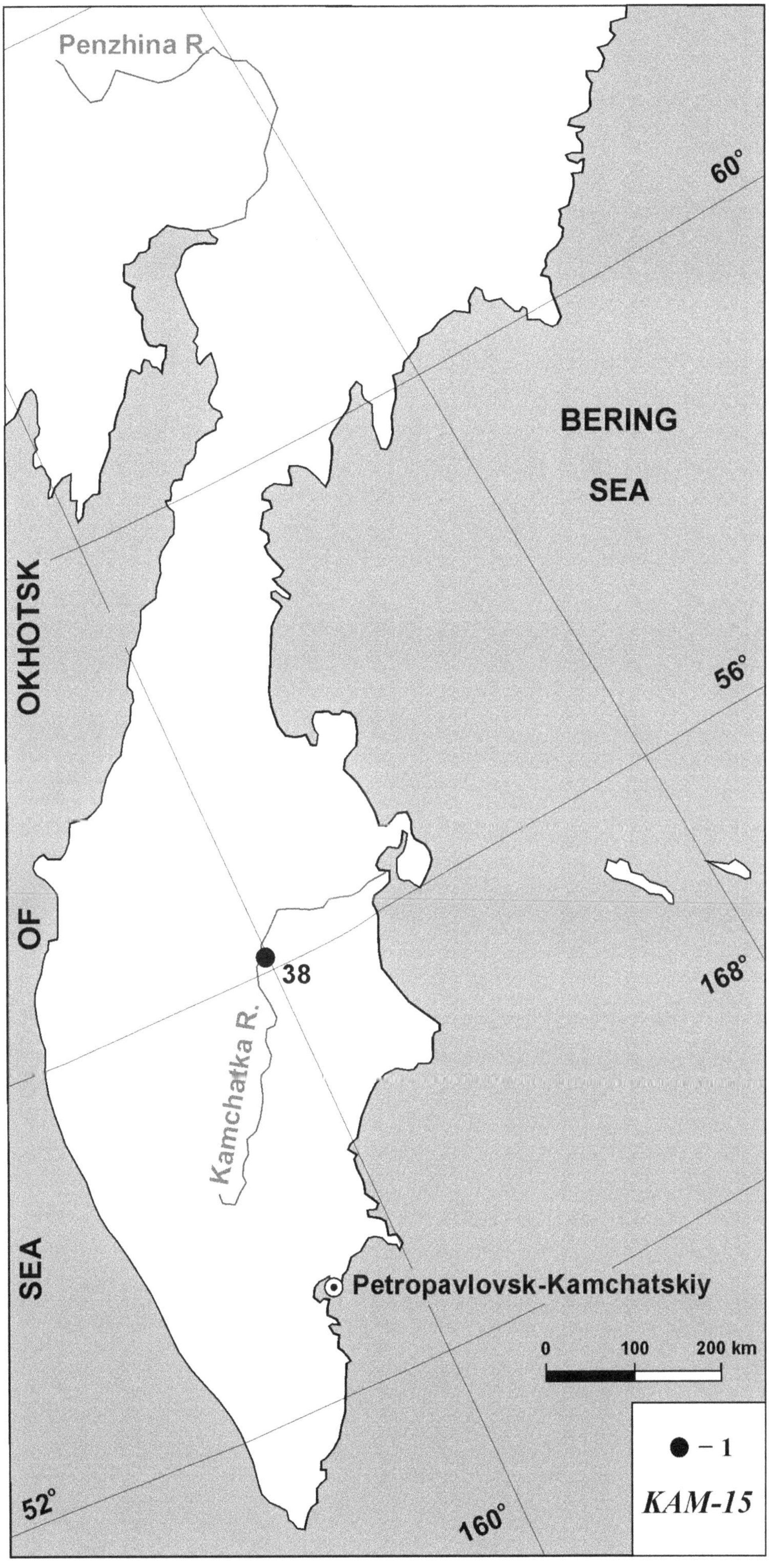

Figure 6.16. The distribution of archaeological sites with KAM-15 group obsidian artefacts

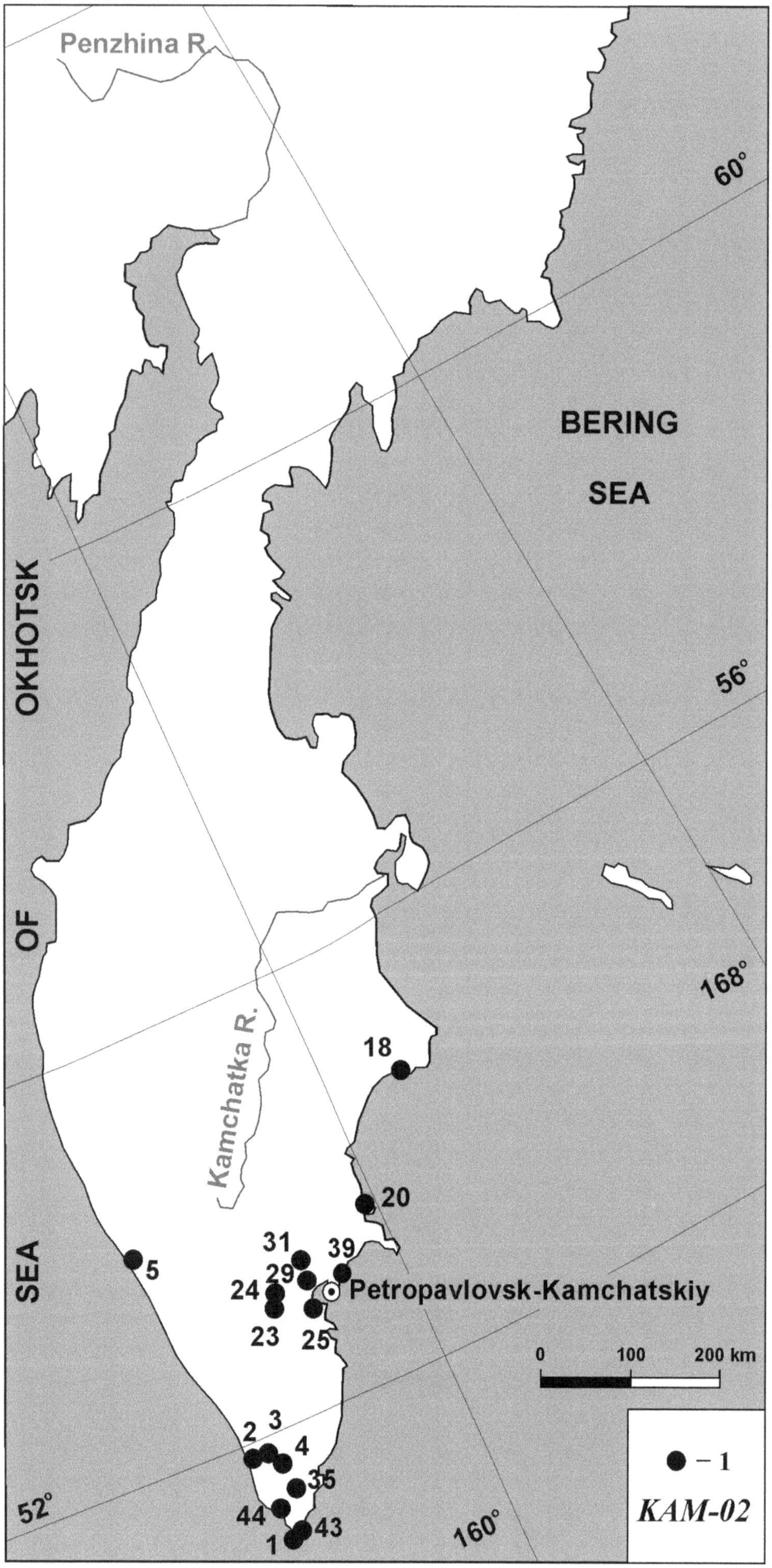

Figure 6.17. The distribution of archaeological sites with KAM-02 group obsidian artefacts

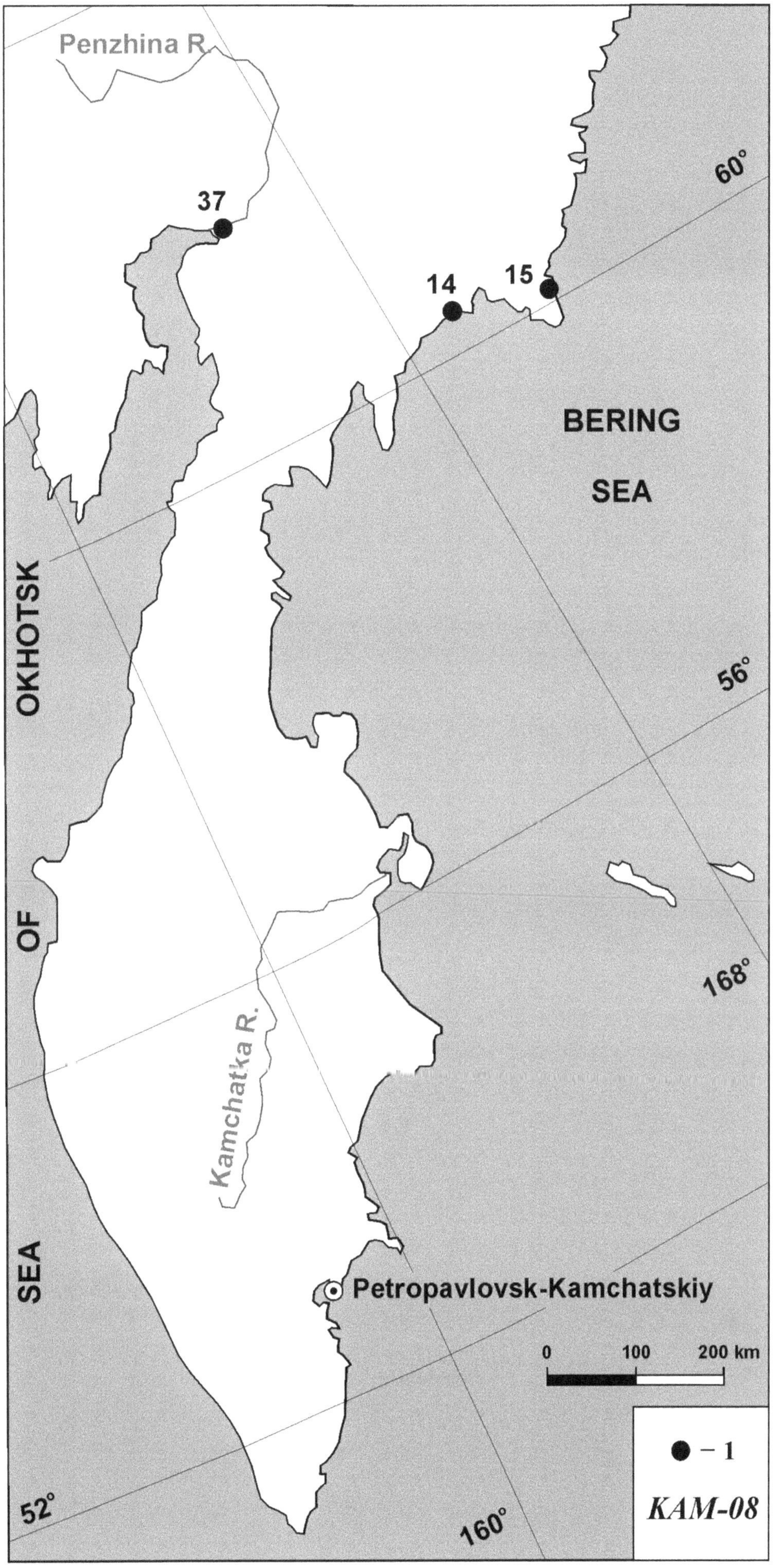

Figure 6.18. The distribution of archaeological sites with KAM-08 group obsidian artefacts

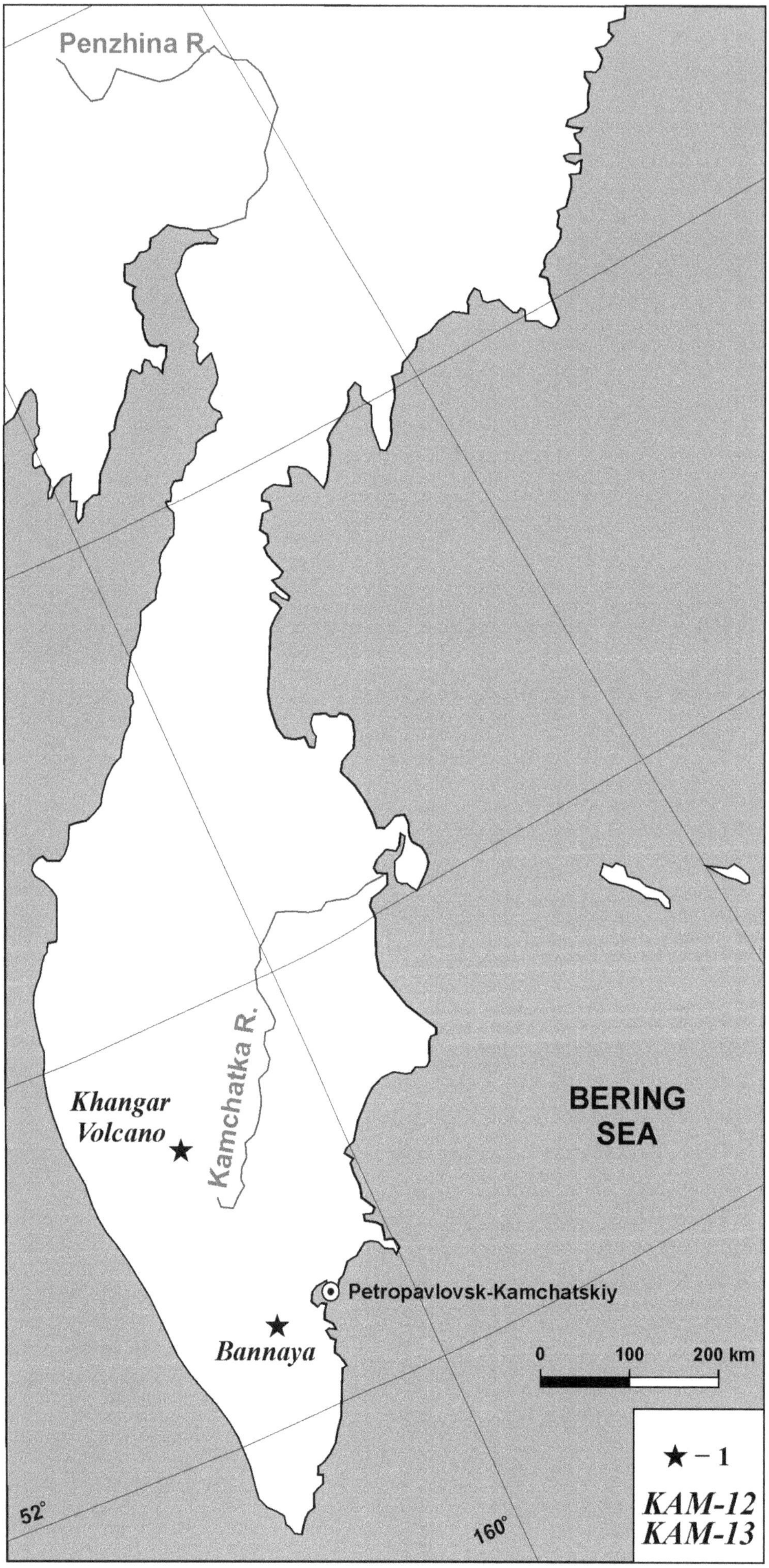

Figure 6.19. The position of KAM-12 (Khangar Volcano) and KAM-13 (Bannaya) obsidian sources

sources (Figures 6.6–6.7). The distance from KAM-16 to sites is about 260km (Figure 6.8). The sources KAM-09 (Figure 6.9), KAM-06 (Figure 6.10), and KAM-11 (Figure 6.11) were used more 'locally', with distances between sites and sources less than 100km.

As for transportation length in different stages of Kamchatkan prehistory, the following features can be tentatively established. The distance of obsidian movement during the late Upper Palaeolithic, ca. 14,300–10,200 BP (Ushki cluster), was up to 200–300km (Kuzmin *et al.* 2008). During the Neolithic, ca. 6000–1500 BP, the distance from sites to sources was from 90km up to 470km (obsidian from the KAM-07 source at Osernovsky 1–2 sites; Table 6.3). For some sites, such as Avacha River (animal farm), Plotnikova River, and Lake Sokoch (artefacts associated with the KAM-06 source); and the Lake Sokoch (specimens associated with the KAM-11 source), obsidian quarries were located only 5–60km away. Definite long-distance exchange and/or trade of obsidian from the KAM-05 and the KAM-07 sources existed in the Neolithic, with distances up to 315–470km. During the Palaeometal period, the long-distance transportation of the KAM-03 source obsidian was detected, with distances up to 450–560km (Table 6.3). Also, the extensive use of the KAM-05 source is noteworthy, with distances up to 280–315km. Some sources situated not far away from the sites also were used; for example, the KAM-09.

As for neighbouring regions of Northeast Asia where obsidian was widely used in prehistory, the distances for raw material acquisition are similar to Kamchatka. For example, during the Upper Palaeolithic obsidian from the Paektusan source on the modern North Korean–Chinese border is found at sites in the Primorye Province and the Korean Peninsula up to 500–800km away (Kim *et al.* 2007; Kuzmin *et al.* 2002a; Popov *et al.* 2005b; see also Kuzmin, this volume). The distance between sources on Hokkaido Island and Upper Palaeolithic sites on Sakhalin Island are up to 250km (Kuzmin and Glascock 2007; Kuzmin *et al.* 2002b), and on the Honshu Island up to 300–400km, sometimes even 600km (Tsutsumi 2002; see also Tsutsumi, this volume). During the Neolithic, the maximum range of obsidian exchange in Northeast Asia was up to 1000km (e.g., Kuzmin 2006; see also Kuzmin, this volume).

Of special interest is the use of multiple obsidian sources by prehistoric people of Kamchatka. It is common that at a particular site, obsidian from different sources is found (Table 6.4). For example, in the Upper Palaeolithic layers 7 and 6 of the Ushki cluster obsidian from six sources was identified (Kuzmin *et al.* 2008). In some Neolithic cultural complexes (Ushki and Plotnikova River), obsidian from five to seven sources was detected, although usually the number of sources was between one and three (average around two sources). During the Palaeometal, obsidian from four sources was found at some sites (Elisovo 1–5); the average number of sources is approximately two (Table 6.4).

A similar feature is observed in the southern part of the Russian Far East. Despite a limited number of obsidian sources in the Primorye Province, the Amur River basin, and on Sakhalin Island (Kuzmin 2006; Kuzmin and Glascock 2007; Kuzmin *et al.* 2002a, 2002b), volcanic glass from at least two sources was identified on several sites. This suggests that the strategy for obsidian acquisition in the Russian Far East has been quite complex since the late Upper Palaeolithic, ca. 14,000–12,000 BP.

Conclusion

It is obvious that obsidian provenance studies on the Kamchatka Peninsula are still in their infancy. Due to large numbers of 'geological' sources of good quality volcanic glass and vast terrain with a very low density of roads, it will take perhaps years to obtain reference samples from all major obsidian outcrops. Only in Mesoamerica has such a high concentration of obsidian sources been observed in such a restricted area (e.g., Glascock *et al.* 1998, 45–57). Even in more intensively surveyed regions compared to the Kamchatka Peninsula, the search for unknown obsidian sources takes time. For example, in a relatively well-studied part of western Arizona State (US Southwest) it took almost ten years to find the AZ Unknown B source (Shackley 2005, 37–39).

Nevertheless, after working on Kamchatka for four years multiple sources of archaeological obsidian have been established. Seven of them are securely associated with 'geological' occurrences of volcanic glass from Cenozoic volcanoes. At least seven other sources, unknown so far, have been detected in the central and southern parts of the peninsula, and determination of their exact location is one of the major tasks for the future. The long-distance (more than 200km from source to site) procurement and exchange of obsidian on the Kamchatka Peninsula started quite early, in the late Upper Palaeolithic, ca. 14,000–10,000 BP. It continued throughout Kamchatkan prehistory, until the contact with Russians and other European and Asian powers. The exchange distances in the Neolithic and Palaeometal were up to 450–550km.

Kamchatka and the neighbouring region of Chukotka have good potential for volcanic glass source studies because of two factors: 1) a large number of prehistoric sites with obsidian (e.g., Dikov 1997, 2003; Kiryak 2010), and 2) their location on the suggested path of the final Pleistocene colonising populations that moved from Siberia and Northeast Asia toward North America. The identification of long-distance obsidian trade and/or exchange in Northeastern Siberia and adjacent Northwestern North America can provide unequivocal data about the direction and time of prehistoric human contacts and migrations between the Asian and American continents.

Acknowledgements

This research, initiated in 2003, is the logical continuation of earlier studies of obsidian sources in the Russian Far East which began in 1992. We are grateful to Drs A. B. Perepelov, V. M. Okrugin, E. Y. Baluev, A. A. Gorbach, V.

Table 6.4. Number of obsidian sources used at Kamchatkan archaeological complexes

Site No.	Site/Cluster Name*	Obsidian Source(s)	Number of Sources Used
1	Lopatka Cape	KAM-01, 02, 04	3
2	Ozernovsky 1-4	KAM-01, 02, 07	3
3	Ozernaya River 1-2	KAM-01, 02	2
4	Lake Kurilskoe	KAM-02	1
5	Kekhta River	KAM-02	1
6	Ust-Kovran	KAM-01, 05	2
7	Kulki	KAM-01, 03	2
8	Palana-airport	KAM-03	1
9	Anadyrka 1	KAM-01	1
10	Inchegitun 1	KAM-03	1
11	Chimei	KAM-03	1
12	Galgan 1	KAM-03	1
13	Zeleny Kholm	KAM-03	1
14	Pakhachi	KAM-08	1
15	Vaimintagin	KAM-08	1
16	Lake Nerpichye	KAM-03	1
17	Kozlov Cape	KAM-04, 05	2
18	Lisy	KAM-02, 10	2
19	Zhupanovo (Cape Pamyatnik)	KAM-01, 09	2
20	Kopyto 1 (Zhupanovo River mouth)	KAM-01, 02, 09	3
21	Avacha Avacha River, lower stream Avacha River, animal farm	KAM-01, 03, 04, 05 KAM-01, 04 KAM-01, 06	4 2 2
22	ASK (Avacha 9) Severnye Koryaki airport	KAM-05 KAM-01, 07, 16	1 3
23	Plotnikova River (Lake Nachiki)	KAM-01, 02, 04, 05, 06	5
24	Lake Sokoch	KAM-02, 06, 11	3
25	Viluchinsk 1-5 Sarannya Bay Turpanka Bay	KAM-01, 02, 04, 11 KAM-01 KAM-01	4 1 1
26	Veselaya River	KAM-01	1
27	Anavgai	KAM-05, 10	2
28	Esso	KAM-10	1
29	Bolshoi Kamen	KAM-01, 02, 04	3
30	Karimshina River	KAM-01, 04	2
31	Elisovo 1-5 Nikolaevka	KAM-01, 02, 04, 05 KAM-01, 02	4 2
32	Ilmagan	KAM-05	1
33	Kluchi	KAM-03, 07, 10	3
34	Nikolka	KAM-01, 05, 10	3
35	Siyushk	KAM-01, 02, 07	3
36	Kozyrevsk	KAM-05, 07, 10	3
37	Penzhina	KAM-03	1
38	Ushki 1, 2, 5 (layers 6–7, Palaeolithic) Ushki 1, 2, 5 (layers 1–5, Neolithic)	KAM-01, 03, 05, 07, 10, 15 KAM-03, 04, 05, 07, 10, 14, 15	6 7
39	Kirpichnoe	KAM-01, 02, 04	3
40	Zastoichik	KAM-07, 10	2
41	Lake Domashnee	KAM-03, 05, 07, 10, 14	5
42	Kamaki	KAM-07, 10	2
43	Lopatka	KAM-02, 04	2
44	Yavino 2	KAM-01, 02	2
45	Doyarki	KAM-03, 05	2

*Site Nos. correspond to those on Figures 6.5–6.18.

L. Leonov, E. N. Grib, V. I. Andreev, S. M. Moskaleva, and M. M. Pevzner for supplying us with samples from obsidian sources of Kamchatka and for valuable discussions. We thank Prof. M. A. Kiryak (Dikova) and Dr A. A. Lebedintsev for providing access to late Prof. N. N. Dikov's obsidian collections from Kamchatka. We thank Dr T. Goebel for the review of the first draft and useful suggestions. This study was supported in 1992–2007 by grants from several foundations, including the US NSF (DBS-9205506, BNS-9102016, SBR-9503035 and 9802366) and CRDF (RG1-2538-VL-03); Russian RFFI (96-06-80688, 99-06-80348, 02-06-80282 and 06-06-80258), RFFI-DVO (06-08-96012) and DVO RAN (06-III-A-08-319); Japanese Ministry of Education, Science, Culture and Sport (Mombu Kagakusho) and the Japan Foundation; and Korean Brain Korea 21 Fund and the Korea Foundation. Finally, we are grateful to the staff members of the University of Missouri Research Reactor for providing assistance with sampling and processing of obsidian from Kamchatka.

The core for this paper is the presentation given at the Symposium "Crossing the Straits: Prehistoric Obsidian Source Exploitation in the Pacific Rim" as part of the 70th Annual Meeting of the Society for American Archaeology (SAA), Salt Lake City, Utah (USA), on 3 April 2005. This is the expanded version of the SAA 2005 presentation, with some new data and especially archaeological interpretations.

References

Churikova, T. S., F. Dorrendorf, and G. Werner. 2001. Priroda Geokhimicheskoi Zonalnosti Vkrest Prostiraniya Kamchatskoi Ostrovnoi Dugi [The Nature of Geochemical Zonal Pattern Across Kamchatka Island Arc]. In *Geodinamika i Vulkanizm Kurilo-Kamchatskoi Ostrovoduzhnoi Sistemy*, edited by B. V. Ivanov, 173–190. Petropavlovsk-Kamchatskiy, Institut Vulkanicheskoy Geologii i Geokhimii Dalnevostochnogo Otdeleniya Rossiiskoi Akademii Nauk.

Dikov, N. N. 1965. The Stone Age of Kamchatka and the Chukchi Peninsula in the Light of New Archaeological Data. *Arctic Anthropology* 3(1), 10–25.

Dikov, N. N. 1968. The Discovery of the Palaeolithic in Kamchatka and the Problem of the Initial Occupation of America. *Arctic Anthropology* 5(1), 191–203.

Dikov, N. N. 1996. The Ushki Sites, Kamchatka Peninsula. In *American Beginnings: The Prehistory and Palaeoecology of Beringia*, edited by F. H. West, 244–250. Chicago and London, University of Chicago Press.

Dikov, N. N. 1997. *Asia at the Juncture with America in Antiquity (The Stone Age of the Chukchi Peninsula)*. Anchorage, AK, Shared Beringian Heritage Program.

Dikov, N. N. 2003. *Archaeological Sites of Kamchatka, Chukotka, and the Upper Kolyma*. Anchorage, AK, Shared Beringian Heritage Program.

Dikov, N. N. 2004. *Early Cultures of Northeastern Asia*. Anchorage, AK, Shared Beringian Heritage Program.

Dikova, T. M. 1983. *Arkheologiya Yuzhnoi Kamchatki v Svyazi s Problemoi Rasseleniya Ainov* [Archaeology of South Kamchatka in Connection with Ainu Occupation Problem]. Moscow, Nauka Publishers.

Dirksen, O. V., V. V. Ponomareva, and L. D. Sulerzhitsky. 2002. The Chasha Crater (South Kamchatka) – Unique Example of Mass Explosion of Acidic Pyroclastics in the Field of Areas Basalt Volcanism. *Volcanology and Seismology* 24(5), 3–10.

Erlikh, E. N. (ed.). 1972. *Vulkany i Chetvertichniy Vulkanizm Sredinnogo Khrebta Kamchatki* [Volcanoes and Quaternary Volcanism of the Sredinniy Ridge of Kamchatka]. Moscow, Nauka Publishers.

Fedotov, S. A., and Y. P. Masurenkov (eds). 1991. *Active Volcanoes of Kamchatka. Volume 1.* Moscow, Nauka Publishers.

Glascock M. D., G. E. Braswell, and R. H. Cobean. 1998. A Systematic Approach to Obsidian Source Characterization. In *Archaeological Obsidian Studies: Method and Theory*, edited by M. S. Shackley, 15–65. New York and London, Plenum Press.

Glascock, M. D., V. K. Popov, Y. V. Kuzmin, R. J. Speakman, A. V. Ptashinsky, and A. V. Grebennikov. 2006b. Obsidian Sources and Prehistoric Obsidian Use on the Kamchatka Peninsula: Initial Results of Research. In *Archaeology in Northeast Asia: On the Pathway to Bering Strait* (University of Oregon Anthropological Papers 65), edited by D. E. Dumond and R. L. Bland, 73–88. Eugene, University of Oregon Press.

Glascock, M. D., R. J. Speakman, and H. Neff. 2007. Archaeometry at the University of Missouri Research Reactor and the Provenance of Obsidian Artefacts in North America. *Archaeometry* 49, 343–357.

Glascock, M. D., R. J. Speakman, V. K. Popov, and Y. V. Kuzmin. 2006a. Geochemistry and Provenance Research on Obsidian from the Kamchatka Peninsula. *Transactions of the American Nuclear Society* 95, 483–484.

Goebel, T., M. R. Waters, and M. Dikova. 2003. The Archaeology of Ushki Lake, Kamchatka, and the Pleistocene Peopling of the Americas. *Science* 301, 501–505.

Gorkin, A. P. (ed.-in-chief). 1998. *Geografiya Rossii: Entsiklopedichesky Slovar* [Geography of Russia: The Encyclopedic Dictionary]. Moscow, Bolshaya Rossiiskaya Entsiklopediya Publishers.

Harris, N. B. W., J. A. Pearce, and A. G. Tindle. 1986. Geochemical Characteristics of Collision-Zone Magmatism. In *Collision Tectonics* (Special Publication of the Geological Society of London 19), edited by M. P. Coward and A. C. Reis, 67–81. Oxford, Blackwell Scientific.

Ishikawa, T., F. Tera, and T. Nakazawa. 2001. Boron Isotope and Trace Element Systematics of the Three Volcanic Zones in the Kamchatka Arc. *Geochimica et Cosmochimica Acta* 65, 4523–4537.

Ivanov, A. 2002. The Far East. In *The Physical Geography of Northern Eurasia*, edited by M. Shahgedanova, 422–447. Oxford, Oxford University Press.

Jochelson, W. 1928. *Archaeological Investigations in Kamchatka*. Washington, D.C., Carnegie Institution of Washington.

Khain, V. E. 1994. *Geology of Northern Eurasia (Ex-USSR). Part 2. Phanerozoic Fold Belts and Young Platforms*. Berlin and Stuttgart, Gebrüder Borntraeger.

Kim, J. C., D. K. Kim, M. Yoon, C. C. Yun, G. Park, H. J. Woo, M.-Y. Hong, and G. K. Lee. 2007. PIXE Provenancing of Obsidian Artefacts from Paleolithic Sites in Korea. *Bulletin of the Indo-Pacific Prehistory Association* 27, 122–128.

Kiryak (Dikova), M. A. 2010. *The Stone Age of Chukotka, Northeastern Siberia (New Materials)* (B.A.R. International Series 2099). Oxford, BAR Publishing.

Krasheninnikov, S. P. 1972. *Explorations of Kamchatka, North Pacific Scimitar; Report of a Journey Made to Explore Eastern Siberia in 1735–1741, by Order of the Russian Imperial Government.* Portland, Oregon Historical Society.

Kuzmin, Y. V. 2000. Radiocarbon Chronology of the Stone Age Cultures on the Pacific Coast of Northeastern Siberia. *Arctic Anthropology* 37(1), 120–131.

Kuzmin, Y. V. 2006. Recent Studies of Obsidian Exchange Networks in Prehistoric Northeast Asia. In *Archaeology in Northeast Asia: On the Pathway to Bering Strait* (University of Oregon Anthropological Papers 65), edited by D. E. Dumond and R. L. Bland, 61–71. Eugene, University of Oregon Press.

Kuzmin, Y. V., and M. D. Glascock. 2007. Two Islands in the Ocean: Prehistoric Obsidian Exchange between Sakhalin and Hokkaido, Northeast Asia. *Journal of Island and Coastal Archaeology* 2, 99–120.

Kuzmin, Y. V., M. D. Glascock, and H. Sato. 2002b. Sources of Archaeological Obsidian on Sakhalin Island. *Journal of Archaeological Science* 29, 741–749.

Kuzmin, Y. V., V. K. Popov, M. D. Glascock, and M. S. Shackley. 2002a. Sources of Archaeological Volcanic Glass in the Primorye (Maritime) Province, Russian Far East. *Archaeometry* 44, 505–515.

Kuzmin, Y. V., V. K. Popov, A. V. Ptashinsky, R. J. Speakman, and M. D. Glascock. 2006. Obsidiany Sredinnogo Khrebta Kamchatki i ikh Ispolzovanie Drevnim Chelovekom [Obsidians of the Central Range of Kamchatka and their Use by Ancient Humans]. In *IV Dikovskie Chteniya*, edited by A. I. Lebedintsev, 81–83. Magadan, Severo-Vostochny Kompleksny Nauchno-Issledovatelsky Institut Dalnevostochnogo Otdeleniya Rossiiskoi Akademii Nauk.

Kuzmin, Y. V., R. J. Speakman, M. D. Glascock, V. K. Popov, A. V. Grebennikov, M. A. Dikova, and A. V. Ptashinsky. 2008. Obsidian Use at the Ushki Lake Complex, Kamchatka Peninsula (Northeastern Siberia): Implications for Terminal Pleistocene and Early Holocene Human Migrations in Beringia. *Journal of Archaeological Science* 35, 2179–2187.

Leonov, V. A., and E. N. Grib. 2004. *Strukturnaya Pozitsiya i Vulkanizm Chetvertychnykh Kalder, Kamchatka, Rossiya* [The Structural Position and Volcanism of the Quaternary Calderas, Kamchatka, Russia]. Vladivostok, Dalnauka Press.

Melekestsev, I. V., O. A. Braitseva, L. I. Bazanova, V. V. Ponomareva, and L. D. Sulerzhitsky. 1996. A Particular Type of Catastrophic Explosive Eruptions with Reference to the Holocene Subcaldera Eruptions at Khangar, Khodutka Maar, and Baraniy Amfiteatr Volcanoes in Kamchatka. *Volcanology and Seismology* 18(2), 3–24.

Münker, C., G. Wörner, G. Yogodzinsky, and T. Churikova. 2004. Behaviour of High Field Strength Elements in Subduction Zones: Constraints from Kamchatka–Aleutian Arc Lavas. *Earth and Planetary Science Letters* 224, 275–293.

Otchet. 1992. Otchet po Revizionno-Poiskovym Rabotam na Proyavleniakh Obsidiana Gory Skalistoi, Ichinskogo Vulkana i Predvaritalnaya Geologo-Ekonomicheskaya Otsenka Proyavleny Obsidiana Kamchatki za 1991–1992 Gody [Report on Revision and Survey Study of the Obsidian Localities of the Skalistaya Mount, Ichinsky Volcano, and Preliminary Geological-Economical Estimate of the Obsidian Localities at the Kamchatka in 1991–2]. Unpublished Manuscript on File. Petropavlovsk-Kamchatskiy, Kamchatgeologiya.

Pampura, V. D., E. I. Popolitov, O. N. Volynets, N. V. Ogorodov, A. A. Vageevskaya, N. E. Litasov, and N. N. Kozhemiaka. 1979. Geokhimicheskie Osobennosti Kislykh Vulkanitov Yuzhno-Kamchatskogo Vulkanicheskogo Poyasa [Geochemical Peculiarities of the Acidic Volcanites of Southern Kamchatkan Volcanic Belt]. In *Geokhimiya Endogennykh Processov*, edited by L. V. Tauson, 70–75. Irkutsk, Institut Geokhimii Sibirskogo Otdeleniya Akademii Nauk SSSR.

Phillips, S. C., and R. J. Speakman. 2009. Initial Source Evaluation of Archaeological Obsidian from the Kuril Islands of the Russian Far East Using Portable XRF. *Journal of Archaeological Science* 36, 1256–1263.

Ponomarenko, A. K. 1985. *Drevnya Kultura Itelmenov Vostochnoi Kamchatki* [The Ancient Culture of the Itelmens of Eastern Kamchatka]. Moscow, Nauka Publishers.

Ponomarenko, A. K. 1991. Arkheologicheskie Pamyatniki Doliny Reki Tigil i ee Pritokov – Rek Napana i Kulki [Archaeological Sites of the Tigil River Valley and its Tributaries, Napana and Kulki Rivers]. *Kraevedcheskie Zapiski* 7, 16–41.

Ponomarenko, A. K. 1993. Novye Arkheologicheskie Pamyaniki Yuzhnoi Kamchatki i Poluostrova Lopatka (k Voprosu ob Obitanii Ainov va Yuzhnoi Kamchatke) [New Archeological Sites of Southern Kamchatka (to the Problem of Ainu Inhabitance of Southern Kamchatka)]. *Kraevedcheskie Zapiski* 8, 3–136.

Ponomarenko, A. K. 1997. Arkheologicheskie Pamyatniki Dolin Rek Zapadnoi Kamchatki – Utkholok, Kovran, Tikhoi, Khairiuzova, Belogolovoi, Moroshechnoi, Sopochnoi, Kisun, Saichik [Archaeological Sites of the River Valleys of Western Kamchatka – Utkholok, Kovran, Tikhaya, Khairiuzova, Belogolovaya, Moroshechnaya, Sopochnaya, Kisun, and Saichik]. *Kraevedcheskie Zapiski* 10, 5–172.

Ponomarenko, A. K. 2000. *Drevnya Kultura Itelmenov Kamchatki* [The Ancient Culture of the Itelmens of Kamchatka]. Petropavlovsk-Kamchatskiy, Kamchatsky Gosudarstvenny Muzei.

Ponomarenko, A. K. 2005. Neolit Kamchatki: Periodizatsiya i Osnovnye Osobennosti Evolutsii Drevnikh Kultur [The Neolithic of Kamchatka: Periodisation and Main Peculiarities of the Evolution of Ancient Cultures]. In *Rossiisky Dalny Vostok v Drevnosti i Srednevekovye: Otkrytiya, Problemy, Gipotezy*, edited by Z. V. Andreeva,

268–291. Vladivostok, Dalnauka Press.

Popolitov, E. I., and O. N. Volynets. 1981. *Geokhimicheskie Osobennosti Chetvertichnogo Vulkanizma Kurilo-Kamchatskoi Ostrovnoi Dugi i Nekotorye Voprosy Petrogenezisa* [Geochemical Peculiarities of the Quaternary Volcanism of Kurile-Kamchatka Island Arc and Some Problems of Petrogenesis]. Novosibirsk, Nauka Publishers.

Popolitov, E. I., and O. N. Volynets. 1982. Geochemistry of Quaternary Volcanic Rocks from the Kurile-Kamchatka Island Arc. *Journal of Volcanology and Geothermal Research* 12, 299–316.

Popov, V. K., A. V. Grebennikov, A. B. Perepelov, Y. V. Kuzmin, M. D. Glascock, and R. J. Speakman. 2007. Geokhimicheskaya Tipizatsiya Kislykh Vulkanicheskikh Stekol Kamchatki [Geochemical Description of the Acidic Volcanic Glasses from Kamchatka Peninsula]. In *Problemy Geokhimii Endogennykh Protsessov i Okruzhayuschei Sredy. Tom 1. Geokhimiya Magmaticheskikh, Metamorficheskikh i Metasomaticheskikh Protsessov*, edited by A. Y. Medvedev, 201–206. Irkutsk, Institut Geokhimii Sibirskogo Otdeleniya Rossiiskoi Akademii Nauk.

Popov, V. K., A. V. Ptashinsky, Y. V. Kuzmin, M. D. Glascock, R. J. Speakman, V. L. Leonov, E. N. Grib, and A. A. Gorbach. 2005a. Geokhimiya Vulkanicheskikh Stekol i Istochniki Arkheologicheskogo Obsidian na Kamchatke (Dalny Vostok Rossii) [Geochemistry of Volcanic Glasses and Sources of Archaeological Obsidian on Kamchatka (Russian Far East). In *Severnaya Patsifika – Kulturnye Adaptatsii v Kontse Pleistotsena i Golotsene*, edited by N. A. Goryachev, E. M. Kokorev and Y. I. Muromtsev, 106–111. Magadan, Severny Mezhdunarodny Universitet.

Popov, V. K., V. G. Sakhno, Y. V. Kuzmin, M. D. Glascock, and B.-K. Choi. 2005b. Geochemistry of Volcanic Glasses of the Paektusan Volcano. *Doklady Earth Sciences* 403, 254–259.

Ptashinsky, A. V. 1989. Novye Pamyatniki Poberezhya Penzhinskoi Guby [New Sites on the Coast of Penzhina Bay]. *Kraevedcheskie Zapiski* 6, 91–116.

Ptashinsky, A. V. 1999. Kultura Morskikh Zveroboev Severo-Zapada Kamchatki [The Culture of Maritime Hunters of Northwestern Kamchatka]. In *Issledovaniya po Arkheologii Severa Dalnego Vostoka*, edited by R. S. Vasilievsky and A. I. Lebedintsev, 80–97. Magadan, Severo-Vostochny Kompleksny Nauchno-Issledovatelsky Institut Dalnevostochnogo Otdeleniya Rossiiskoi Akademii Nauk.

Ptashinsky, A. V. 2002. *Kultura Okhotnikov na Morskogo Zverya Severo-Vostochnogo Poberezhya Okhotskogo Morya (I–II Tysyacheletiya N.E.)* [The Culture of Maritime Hunters of the Northeastern Coast of Okhotsk Sea (Ist–IInd Millennia AD)]. Synopsis of the Candidate of Historical Sciences (PhD-equivalent) Dissertation. Moscow, Institut Arkheologii Rossiiskoi Akademii Nauk.

Ptashinsky, A. V. 2003. Novye Nakhodki Neolita Kamchatki [The New Finds of the Neolithic at Kamchatka]. *Vestnik Kamchatskoi Regionalnoi Assotsiatsii "Uchebno-Nauchny Tsentr" (KRAUNTS). Series "Gumanitarnye Nauki"* 2, 28–37.

Ptashinsky, A. V. 2009. A New Upper Paleolithic Site in Kamchatka (Russia). *Current Research in the Pleistocene* 26, 22–24.

Rozenkrantz, A. A. 1981. Kharakteristika Alneiskogo Mestorozhdeniya Perlitov i Obsidianov na Kamchatke [Characterisation of the Alnei Deposit of Perlites and Obsidians at Kamchatka]. In *Perlity*, edited by V. V. Nasedkin and V. P. Petrov, 123–127. Moscow, Nauka Publishers.

Shackley, M. S. 2005. *Obsidian: Geology and Archaeology in the North American Southwest*. Tucson, University of Arizona Press.

Sheimovich, V. S., and M. G. Patoka. 2000. *Geologicheskoe Stroenie Zon Aktivnogo Kainozoiskogo Vulkanizma* [Geological Structure of Active Cenozoic Volcanism Zones]. Moscow, GEOS Publishers.

Shevchuk, V. D. 1981. Geneticheskie Tipy Mestorozhdeniy Kislykh Vulkanicheskikh Stekol na Kamchatke [Genetic Types of the Acidic Volcanic Glass Deposits in Kamchatka]. In *Perlity*, edited by V. V. Nasedkin and V. P. Petrov, 43–51. Moscow, Nauka Publishers.

Speakman, R. J., M. D. Glascock, V. K. Popov, Y. V. Kuzmin, A. V. Ptashinsky, and A. V. Grebennikov. 2005. Geochemistry of Volcanic Glasses and Sources of Archaeological Obsidian on the Kamchatka Peninsula (Russian Far East): First Results. *Current Research in the Pleistocene* 22, 11–13.

Sveshnikov, V. V., and A. N. Krayukhin (eds). 2004. *Natsionalny Atlas Rossii* [The National Atlas of Russia]. Moscow, Federalnaya Sluzhba Geodezii i Kartografii Rossii.

Suslov, S. P. 1961. *Physical Geography of Asiatic Russia*. San Francisco and London, W. H. Freeman.

The Times Atlas. 1998. *The Times Atlas of the World.* 7th comprehensive edit. London, Times Books.

Topchiyan, A. L. (ed.). 1998. *Geografichesky Atlas Rossii* [Geographical Atlas of Russia]. Moscow, Proizvodstvennoe Kartosostavitelskoe Obiedinenie "Kartografiya".

Tsutsumi, T. 2002. Origins of Pottery and Human Strategies for Adaptation during the Termination of the Last-Glacial Period in the Japanese Archipelago. In *The Origins of Pottery and Agriculture*, edited by Y. Yasuda, 241–262. New Delhi, Roli Books and Lustre Press.

Volynets, O. N., G. N. Anoshin, Y. M. Pusankov, and V. S. Parkhomenko. 1987a. Geokhimicheskaya Tipizatsiya Pozdnekainozoiskikh Bazaltov Kamchatki (po Dannym Neitronno-Aktivatsionnogo Analiza) [Geochemical Description of the Late Cenozoic Basalts of Kamchatka (Based on Results of Neutron Activation Analysis]. *Doklady Akademii Nauk SSSR* 293(3), 685–688.

Volynets, O. N., V. S. Antipin, G. N. Anoshin, and A. B. Perepelov. 1987b. Geokhimicheskie Tipy Vulkanicheskikh Seriy Ostrovoduzhnoi Systemy Kamchatki [Geochemical Types of the Volcanic Series of Island Arc Systems of Kamchatka]. In *Geokhimicheskaya Tipizatsiya Magmaticheskikh Porod i ikh Geodinamika*, edited by L. V. Tauson, 34–55. Irkutsk, Institut Geokhimii Sibirskogo Otdeleniya Akademii Nauk SSSR.

Volynets, O. N., V. S. Antipin, A. B. Perepelov, and G. N. Anoshin. 1990. Geochemistry of Volcanic Series from Island Arc Systems with Respect to Geodynamics (Kamchatka). *Russian Geology and Geophysics* 31(5), 3–13.

Chapter 7
Bridging the Gap Between Two Obsidian Source Areas in Northeast Asia: LA-ICP-MS Analysis of Obsidian Artefacts from the Kurile Islands of the Russian Far East

S. Colby Phillips

Abstract: *Recent archaeological excavations in the Kurile Islands of the Russian Far East have recovered almost 2000 obsidian artefacts in the form of finished stone tools and flake debitage. While artefacts made of obsidian are present throughout the island chain, obsidian native to the Kurile Islands is not known to have been used prehistorically. An initial source provenance study of Kurile Island artefacts indicated that obsidian raw material was brought into the islands at least 2500 years ago from sources located on the Japanese island of Hokkaido and from the Kamchatka Peninsula (Russian Far East). This chapter reports on a larger provenance study using Laser Ablation Inductively-Coupled-Plasma Mass Spectrometry (LA-ICP-MS) that expands the initial research and provides the largest sample to date of obsidian artefacts from the Kurile Islands that can be assigned to obsidian source groups located in Northeast Asia. Identifying the sources used to produce obsidian artefacts is a key element necessary for reconstructing prehistoric Kurile Island migrations, colonisation events, and social network structures.*

Keywords: *Obsidian, Sourcing, Archaeology, Kurile Islands, Russian Far East, Kamchatka Peninsula, Hokkaido Island*

Introduction

Imported items found in archaeological sites are often seen as evidence for transport of materials via the movement/ migration of people or through trade/exchange networks (Pires-Ferreira 1978). The directions and distances associated with materials that have been transported, especially over long distances, can provide insight into the social as well as utilitarian nature of material exchange among hunter-gatherer groups that are essentially independent and economically self-sufficient (Eriksen 2002; Whallon 2006). Non-local lithic raw material used for the production of stone tools is one type of resource that can represent hunter-gatherer intergroup networks (Eriksen 2002).

Over the last decade, regional studies in the northeast Asian portion of the Pacific Rim have developed information and ideas about the prehistoric use of obsidian as an important raw material for the manufacture of stone tools. Research conducted specifically on Hokkaido Island (Japan) and the Primorye and Kamchatka regions of the Russian Far East, have provided detailed accounts of the location of primary and secondary obsidian sources, the movement of obsidian over long distances, and the differential use of various sources based on location and quality (see Table 7.1 for a list of relevant obsidian studies in Northeast Asia).

Until recently, the Kurile Islands of the Russian Far East (Figure 7.1) have represented an archaeological blank spot in terms of the use of obsidian as a raw material for manufacturing stone tools. While still relatively little is known about the overall culture history and technological adaptations of Kurile Island inhabitants, it recently was established that non-local obsidian was used prehistorically throughout the archipelago (Phillips and Speakman 2009). The Kurile Islands are a geographic bridge between northern Japan and the northern Russian Far East. The distribution of obsidian from these areas across the Kuriles has great potential to inform us about the migration movements and exchange relationships, and in turn the larger social organisational structure of the islands' inhabitants at different times in prehistory. This chapter provides a more in-depth view of the distribution of obsidian artefacts across the island chain through Laser Ablation Inductively-Coupled-Plasma Mass Spectrometry (hereafter – LA-ICP-MS) analysis of a larger sample of obsidian artefacts, with the goal of developing a higher-resolution data set from which studies of lithic technology, social organisation, and migration can be made.

Geographical and Geological Background

The Kurile Archipelago is a 1150km long chain of islands that spans the Okhotsk Sea – Pacific Ocean boundary, and is situated on the central portion of the Kurile-Kamchatka Island Arc which includes eastern Hokkaido, the Kurile Islands, and southern Kamchatka (Cook *et al.* 1986; Gorshkov 1970). The Kurile Islands vary in size from 5km^2 to 3200km^2, and the islands at the extreme southern and northern ends of the archipelago tend to be larger and more ecologically diverse than the more isolated central Kuriles (Pietsch *et al.* 2003).

Tectonically, Kurile Islands are associated with the subduction of the Pacific Plate under the Okhotsk Plate and consist of two island arc ridge systems, the Lesser Kurile Ridge and the Greater Kurile Ridge, located in between the oceanic Kurile Trench and the Kurile Basin of the Okhotsk Sea. The Lesser Kurile Ridge is an older, inactive arc that includes the Nemuro Peninsula of Hokkaido, the Habomai island group, and Shikotan Island, and then continues to the northeast as the submarine Vityaz Ridge (Gorshkov 1970; Ishizuka 2001). The Greater Kurile Ridge includes

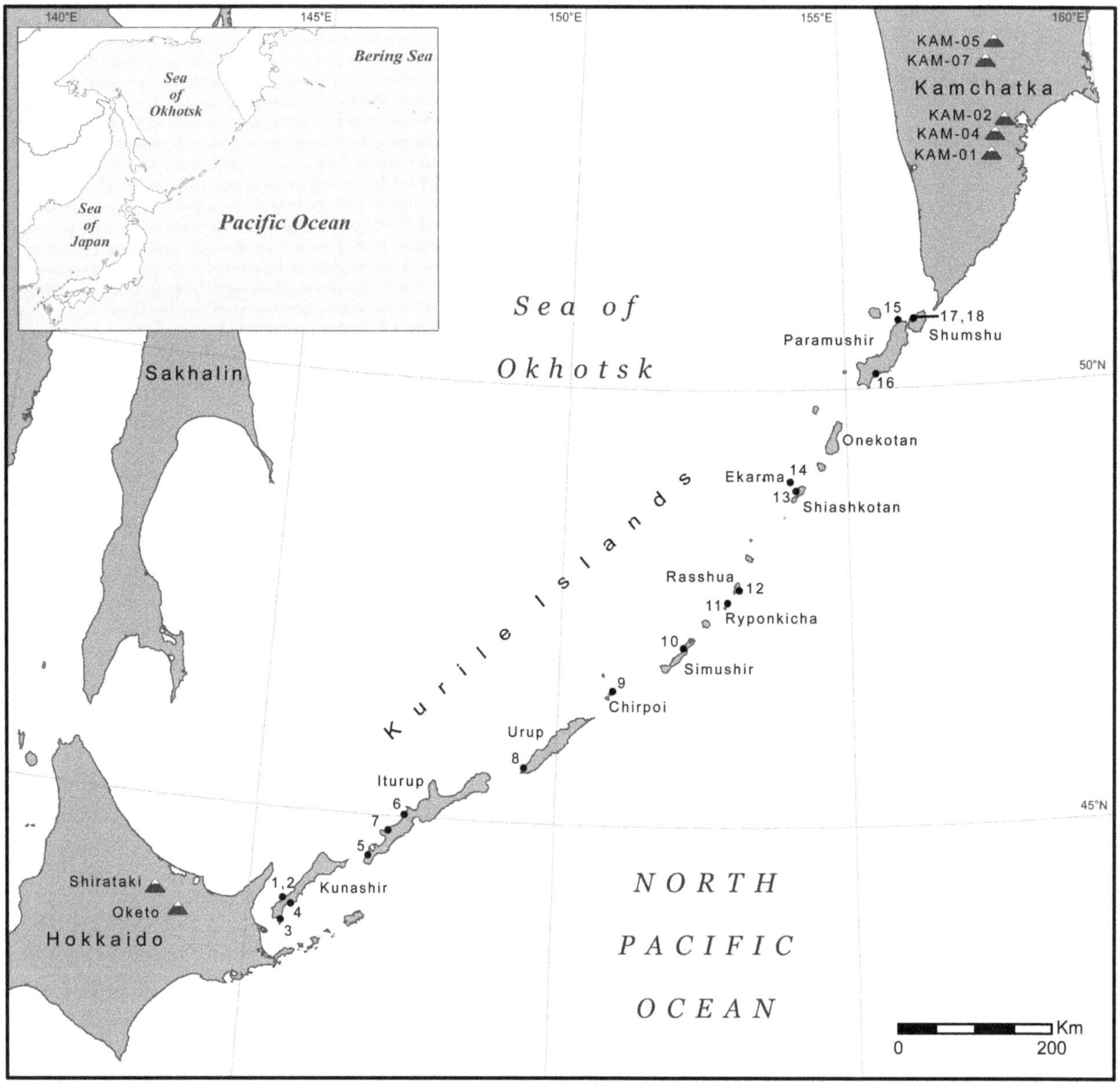

Figure 7.1. Map of Kurile Islands and surrounding region. Numbered points correspond to archaeological sites referenced in the text: 1) Alekhina; 2) Peschanaya 2; 3) Rikorda 1; 4) Sernovodsk 1; 5) Berezovka 1; 6) Kuibyushevskaya 1; 7) Tikhaya 1; 8) Ainu Creek 1; 9) Peschanaya Bay 1; 10) Vodopodnaya 2; 11) Ryponkicha 1; 12) Rasshua 1; 13) Drobnyye 1; 14) Ekarma 1; 15) Savushkina 1; 16) Tukharka River 1; 17) Baikova 1; 18) Bolshoi 1

the Shiretoko Peninsula of eastern Hokkaido, all of the remaining Kurile Islands from Kunashir north to Shumshu, and the southern tip of the Kamchatka Peninsula. The Greater Kurile Ridge was formed from submarine volcanic activity that began in the Miocene (about 23 million years ago) and is composed of 160 Quaternary terrestrial volcanoes and 89 submarine ones. The Greater Kurile Ridge is still tectonically and volcanically active, with 19 volcanoes that have erupted since AD 1945 (Gorshkov 1970; Simkin and Siebert 1994).

The Kurile chain is representative of the andesitic volcanism associated with most island arc systems, and is situated in between two centres of silicic volcanism, one located to the south in eastern Hokkaido and one to the north in southeastern Kamchatka (Erlich 1986). Since obsidian usually occurs in conjunction with silicic volcanism and is a product of magmas containing silicic melts with a SiO_2 content greater than 70% (Eichelberger 1995), Hokkaido and Kamchatka represent two potential areas with sources of obsidian that were used prehistorically as a raw material for stone tool production in the Kurile Islands.

Archaeological Background

Compared with Hokkaido, Kamchatka, and other parts of the larger Northeast Asia region, significantly less archaeological research has been conducted in the Kurile Islands in the last 50 years (Fitzhugh *et al.* 2002; Phillips and Speakman 2009; Vasilevsky and Shubina 2006; but see Prokofiev 2005 for a review of Japanese research in the Kuriles at the end of the 19th century and first half of the 20th century). More recently, the International Kurile Island Project (hereafter – IKIP) in 2000 and the

Table 7.1. Chronological listing of relevant obsidian studies in Northeast Asia

Year	*Author(s)*	*Title*
1995	Kimura, H.	Obsidian, Humans, and Technology
1996	Shackley, M. S., *et al.*	Geochemical Characterization of Archaeological Obsidian from the Russian Far East: A Pilot Study
1996	Glascock, M. D., *et al.*	Geochemical Characterization of Obsidian Artefacts from Prehistoric Sites in the Russian Far East: Initial Study
1999	Kuzmin, Y. V., *et al.*	Geochemical Source Analysis of Archaeological Obsidian in Primorye (Russian Far East)
2000	Glascock, M. D., *et al.*	Obsidian Geochemistry of the Archaeological Sites of the Sakhalin and its Sources
2000	Kuzmin, Y. V., *et al.*	The Sources of Archaeological Obsidian in Primorye and its Distribution in the Stone Age Cultures
2002	Hall, M., and Kimura, H.	Quantitative EDXRF Studies of Obsidian Sources in Northern Hokkaido
2002	Kimura, H.	A Prospect of the Obsidian Study in Hokkaido District
2002a	Kuzmin, Y. V., *et al.*	Sources of Archaeological Obsidian from Sakhalin Island
2002b	Kuzmin, Y. V., *et al.*	Sources of Archaeological Volcanic Glass in the Primorye (Maritime) Province, Russian Far East
2004	Doelman, T., *et al.*	Acquisition and Movement of Volcanic Glass in the Primorye Region of Far Eastern Russia
2004	Fitzhugh, B., *et al.*	Archaeological Paleobiogeography in the Russian Far East: The Kurile Islands and Sakhalin in Comparative Perspective
2004	Sato, H.	Prehistoric Obsidian Exploitation in the Russian Far East
2005	Naoe, Y., and Nagasaki, J.	Raw Material Consumption Strategy at Late Upper Paleolithic in Hokkaido
2005	Popov, V. K., *et al.*	Geochemistry of Volcanic Glasses of the Paektusan Volcano
2005	Speakman, R. J., *et al.*	Geochemistry of Volcanic Glasses and Sources of Archaeological Obsidian on the Kamchatka Peninsula (Russian Far East): First Results
2006	Glascock, M. D., *et al.*	Obsidian Sources and Prehistoric Obsidian Use on the Kamchatka Peninsula: Initial Results of Research
2006	Kimura, H.	The Shirataki Obsidian Mine Area and the Yubetsu–Horokazawa Technological Complex
2006b	Kuzmin, Y. V.	Recent Studies of Obsidian Exchange Networks in Prehistoric Northeast Asia
2006	Suzuki, H., and Naoe, Y.	The Shirataki Sites: An Overview of Upper Paleolithic Sites at an Obsidian Source in Hokkaido, Japan
2007	Izuho, M., and Sato, H.	Archaeological Obsidian Studies in Hokkaido, Japan: Retrospect and Prospects
2007	Kim, J. C., *et al.*	PIXE Provenancing of Obsidian Artefacts from Palaeolithic Sites in Korea
2007	Kluyev, N. A., and Slepstov, I. Y.	Late Pleistocene and Early Holocene Uses of Basaltic Glass in Primorye, Far East Russia: A New Perspective Based on Sites Near the Sources
2007	Kuzmin, Y. V., and Glascock, M. D.	Two Islands in the Ocean: Prehistoric Obsidian Exchange Between Sakhalin and Hokkaido, Northeast Asia
2007	Pantukhina, I.	The Role of Raw Material in Microblade Technology at Three Late Palaeolithic Sites, Russian Far East
2008	Doelman, T., *et al.*	Source Selectivity: An Assessment of Volcanic Glass in the Southern Primorye Region, Far Eastern Russia
2008	Kuzmin, Y. V., *et al.*	Obsidian Use at the Ushki Lake Complex, Kamchatka Peninsula (Northeastern Siberia): Implications for Terminal Pleistocene and Early Holocene Human Migrations in Beringia
2009	Phillips, S. C., and Speakman, R. J.	Initial Source Evaluation of Archaeological Obsidian from the Kurile Islands of the Russian Far East Using Portable XRF

Kurile Biocomplexity Project (hereafter – KBP) in 2006–7 have provided new data and the means to synthesise the archaeology of the island chain based on ceramic and lithic artefacts, faunal data, and radiocarbon (^{14}C) dating (Table 7.2).

During the Late Glacial period (ca. 18,000–15,000 BP) when global sea level is estimated to have been approximately 130m lower than today (e.g., Chappell and Shackleton 1986), the southern Kurile islands of Kunashir and Iturup and the northern islands of Paramushir and Shumshu were connected to Hokkaido and Kamchatka respectively. Confirmed early occupations in the southern Kurile Islands extend back to the Middle Jomon period, ca. 5000 BP, based on ^{14}C dates and pottery types from archaeological sites on Iturup Island (Golubev 1972; O. A. Shubina and I. A. Samarin, personal communication 2009), and people were potentially present in the southern

Kuriles as early as ca. 7000 BP (Zaitseva et al. 1993) [see also Yanshina and Kuzmin 2010 – *Editors*]. These earliest occupations probably represent Jomon hunter-gatherers who lived throughout Japan from ca. 12,000 to ca. 2500 BP (Imamura 1996; Kobayashi 2004; Yamaura and Ushiro 1999). Though little information currently exists for this period that some researchers have labeled the "Early Neolithic" of the southern Kurile Islands (Kuzmin *et al.* 1998; Vasilevsky and Shubina 2006; Zaitseva *et al.* 1993), the earliest occupants of the southern Kuriles likely lived as small and highly mobile populations subsisting primarily by terrestrial hunting and gathering, which was supplemented with fish and shellfish (Imamura 1996; Kikuchi 1999; Okada 1998), though direct evidence of subsistence activities at the earliest Kurile sites is currently lacking.

Consistent occupation in the southern Kurile Islands began ca. 4000 BP (Zaitseva *et al.* 1993), and at ca. 2500–1300 BP Epi-Jomon people with a developed maritime-adapted economy moved north out of Hokkaido and into the more

Table 7.2. Kurile Island culture history (after Fitzhugh and Dubreuil 1999; Fitzhugh et al. 2002, 2004; Kikuchi 1999; Ohnuki-Tierney 1976; Stephan 1976; Tezuka 1998; Vasilevsky and Shubina 2006; Zaitseva et al. 1993)

Culture period	***Dates (BP)***	***Presence in Kurile Islands***
Ainu	800–50	Southern to northern islands
Okhotsk	1400–800	Southern to northern islands
Epi-Jomon	2500–1400	Southern to north-central islands
Final Jomon	3200–2500	Southern islands
Late Jomon	4500–3200	?
Middle Jomon	5500–4500	Southern islands
Early Jomon	7300–5500	Southern islands?

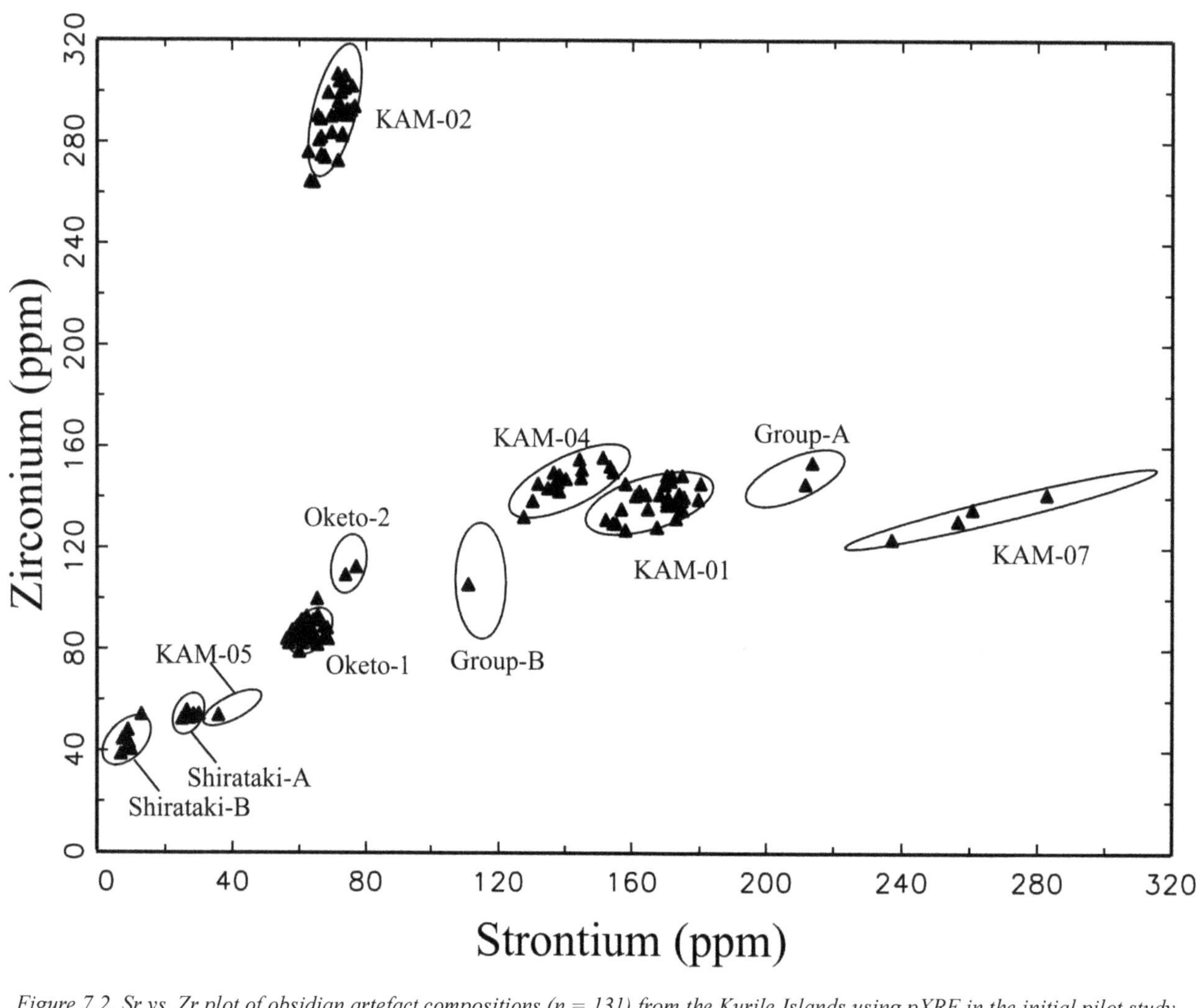

Figure 7.2. Sr vs. Zr plot of obsidian artefact compositions (n = 131) from the Kurile Islands using pXRF in the initial pilot study (Phillips and Speakman 2009). The ellipses surrounding each group are drawn at the 95% confidence level. Confidence ellipses were drawn using a minimum of four data points from a larger group of obsidian artefacts, though only the artefacts relevant to this paper are presented here

remote central Kuriles (Fitzhugh *et al.* 2002; Kikuchi 1999; Okada 1998; Vasilevsky and Shubina 2006; Yamaura 1998; Yamaura and Ushiro 1999). The Epi-Jomon period in the Kurile Islands represents a remnant continuation of the Jomon culture which had transitioned from a hunting and gathering economy to a more sedentary agriculture subsistence system in most of the Japanese Islands including southern Hokkaido. Epi-Jomon technology continued to rely on stone and bone tools with the introduction of a small amount of imported iron tools (Imamura 1996). Evidence of Epi-Jomon occupations based on recent ceramic identifications are found as far north as Shiashkotan Island in the north-central part of the Kurile chain (E. Gjesfjeld, personal communication 2009).

Around 1400 BP the intensively marine-oriented Okhotsk culture appeared on Sakhalin Island, potentially having migrated out of the Amur River basin on the Northeast Asian mainland (Okada 1998; Sato *et al.* 2007). Okhotsk people, identified by a complex mixture of ceramic styles from the Amur River basin and Sakhalin and Hokkaido islands, occupied the northern and eastern coasts of Hokkaido, and expanded throughout the entire Kurile Island chain and potentially into southern Kamchatka (Imamura 1996; Otaishi 1994; Yamaura 1998; Yamaura and Ushiro 1999). The Okhotsk reliance on marine resources is recognised through a technological tool kit that contained a variety of sea mammal hunting harpoon styles, composite fish hooks, and the predominance of bone tools made from sea mammal bones (Okada 1998).

After ca. 800 BP, the Okhotsk people were replaced on Hokkaido and in the Kurile Islands by the Ainu culture (Fitzhugh and Dubreuil 1999). The Ainu people are believed to be descendants of a combination of the Satsumon culture that established plant cultivation along with terrestrial hunting and riverine fishing in the interior of Hokkaido, and the remnant coastal Okhotsk culture (Okada 1998). In the Kurile Islands, the Ainu continued to employ a maritime-adapted resource economy, and also developed trade relationships with European and American explorers and trading companies (Ohnuki-Tierney 1976; Shubin 1994; Stephan 1974; Tezuka 1998; Vysokov 1994).

Based on previous analysis of small lithic assemblages from the Kuriles, it was proposed that the islands were sufficiently isolated to constrain the spread of non-local raw materials throughout the island chain via mobility or exchange (Fitzhugh *et al.* 2004). The limited surveys that have been conducted by the KBP for geological sources of obsidian in the Kurile Islands have only located sparse outcrops of fractured perlite (M. Nakagawa, personal communication 2007; A. V. Rybin, personal communication 2006). However, recent archaeological excavations by the KBP at sites in the southern, central, and northern parts of the island chain, and work by Russian archaeologist O. A. Shubina in the southern islands, have recovered obsidian artefacts from contexts associated with Epi-Jomon and Okhotsk site occupations spanning roughly 1750 years (ca. 2500–750 BP). An initial pilot project conducted by Phillips and Speakman (2009) analysed 131 obsidian flakes from 18 sites on eight islands using a portable X-ray Fluorescence (hereafter – pXRF) spectrometer. This study identified nine different source groups located in Hokkaido and Kamchatka (Figure 7.2) in this sample assemblage, indicating a long-term utilisation of non-local obsidian across the Kuriles. While this study provided a baseline of data for the region, only flakes approximately 5mm in diameter and larger and with one flat side were analysed due to the flake morphology and minimum size requirements inherent in pXRF instruments. The research presented here is an attempt to further refine these initial findings with a larger and more varied sample of Kurile obsidian artefacts.

Materials and Methods

More than 2000 obsidian artefacts, including formal tools, retouched flakes, and tool production debitage, were recovered from 26 archaeological sites across the southern, central, and northern Kurile Islands during the KBP 2006 and 2007 field seasons. Most of the archaeological sites investigated by the KBP were visited for only one or two days to conduct short surveys and small test excavations, and usually one to three 1 × 1 metre test pits were excavated at each site. The Ainu Creek, Vodopodnaya 2, and Drobnyye sites received more extensive investigation, with up to two weeks spent at the site, and the excavation of multiple 2 × 2m excavation units.

Because all formal lithic tools and retouched flakes are curated at the Sakhalin Regional Museum in Yuzhno-Sakhalinsk, Russia, only the flake debitage was available to analyse in the United States for the current study. The obsidian debitage assemblage is dominated by small flakes created through bifacial reduction and retouch with an average weight of 0.17g and average medial width of 4.14mm. The current study analysed 774 flakes from 18 sites (Figure 7.1) using LA-ICP-MS to build upon the previous study, and included flakes smaller than 5mm to potentially increase the diversity of the sources that occur in Kurile Island flake assemblages (Eerkens *et al.* 2007). Obsidian flake assemblages from six archaeological sites (Rikorda 1, Ainu Creek 1, Vodopodnaya 2, Drobnyye 1, Savushkina 1, and Baikova 1) distributed across the southern, central, and northern parts of the island chain make up 92.6% of the current study sample; the complete study sample represents all of the obsidian flakes in the Kurile assemblage that could not be analysed with pXRF.

In recent years, laser ablation (LA) systems used in tandem with inductively-coupled-plasma mass spectrometers (ICP-MS) have gained increasing popularity as a tool for chemical analyses of both organic and inorganic matrices. LA-ICP-MS offers several advantages over other analytical methods, including high accuracy and precision, low detection limits, rapid analytical time, low cost per sample, high sample throughput, and minimal damage to objects (Cochrane and Neff 2006; Speakman and Neff 2005; Speakman *et al.* 2002). Although LA-ICP-MS has been indispensable as a method of determining the chemical properties of cultural materials (Devos *et al.* 1999, 2000; Gratuze 1999; Gratuze *et al.* 2001; James and Dahlin 2005; Kennett *et al.* 2001;

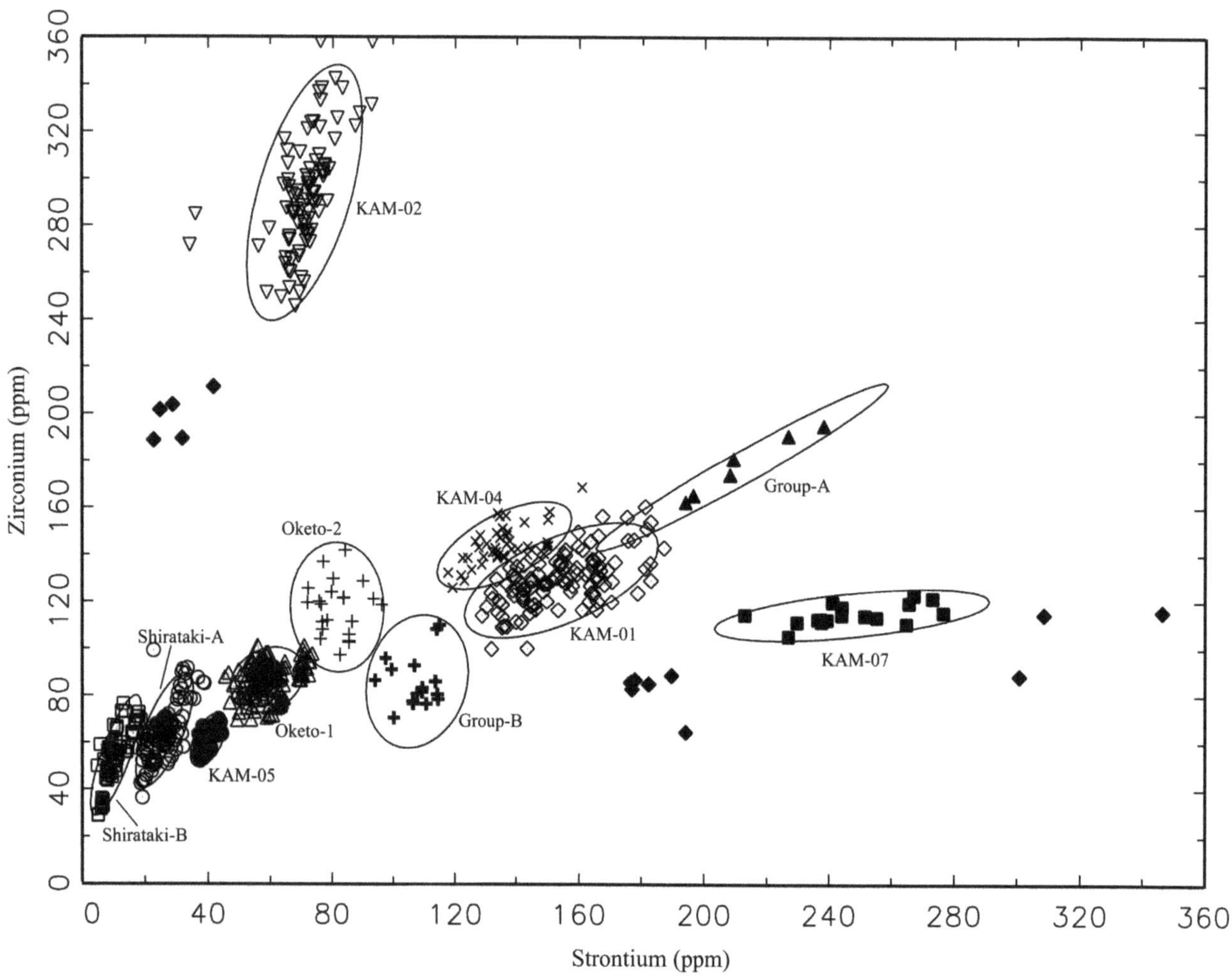

Figure 7.3. Sr vs. Zr plot of obsidian artefact compositions (n = 774) from the Kurile Islands using LA-ICP-MS. The ellipses surrounding each group are drawn at the 95% confidence level. Confidence ellipses were drawn using a minimum of four data points

Mallory-Greenough *et al.* 1999; Speakman and Neff 2005), its focus on small sample areas makes this technique ideally suited for the analysis of compositionally homogenous matrices, such as obsidian.

The current sample of obsidian flake debitage was analysed at the Smithsonian Institution's Museum Conservation Institute (Suitland, MD, USA) using a Perkin Elmer Elan 6000 ICP-MS and CETAC LSX 200nm laser ablation unit. Flakes were mounted so that the flattest face (dorsal or ventral) was exposed to the laser, and were pre-ablated to remove any surface contaminants with the laser operated at 70% power using a 200 micron diameter spot size running at a 20Hz pulse rate over a computer generated raster at a speed of 100 micron/sec. They were then ablated at 70% power using a 100 micron in diameter spot size operating at a 20Hz pulse rate over a computer generated raster at a speed of 30 micron/sec to generate elemental abundance data. Argon served as the transport gas from the laser ablation unit to the ICP-MS. At the beginning and end of each daily analysis session, a series of blanks, the NIST SRM 612 glass standard, and six well-characterised obsidian samples that previously had been analysed by Neutron Activation Analysis (NAA) were run to develop a set of calibration parameters and to monitor instrumental drift of the ICP-MS, which can affect count rates over several hours and/or days.

The ICP-MS analysis of obsidian artefacts generated data for 29 different elements; of these, Sr and Zr were used to discriminate between the primary Hokkaido and Kamchatkan sources and matched the elements used in the initial pilot project to differentiate obsidian source groups. Data generated from the analyses, expressed as counts per second, were ratioed to an internal standard (in this case Si) to normalise for different count rates between samples and standards. The normalisation approach used is modified from that suggested by Gratuze (1999) and others (e.g., Speakman and Neff 2005).

Results

The results of the LA-ICP-MS analysis of obsidian flakes indicate that 11 different source groups were identified (Figure 7.3) and which are in close agreement with the source groups determined in the original pXRF pilot study (Phillips and Speakman 2009). Although there are differences between pXRF and LA-ICP-MS in terms of their precision, specific source groups are accurately differentiated by each of the methods allowing for a comparison of the data sets created via both trace element analysis techniques.

Data for the source groups identified in the current artefact sample are summarised in Table 7.3. Four of the source groups are located in Hokkaido (Shirataki-A, Shirataki-B, Oketo-1 and Oketo-2), five are situated in Kamchatka (Kamchatka-1, Kamchatka-2, Kamchatka-4, Kamchatka-5, and Kamchatka-7 [named by primary investigators as KAM-01 through 07; see Kuzmin *et al.* (2008) – *Editors*]), while the locations of the Group-A and Group-B source groups are currently unknown. There were also 15 flakes that were not assigned to any source group. Obsidian from Hokkaido source groups represents 57.5% (n = 445) of the total Kurile Island sample assemblage that was analysed with LA-ICP-MS, with Kamchatkan source groups accounting for 37.7% (n = 292), Group-B – 2.1% (n = 16), flakes unassigned to any source group – 1.9% (n = 15), and Group-A – 0.8% (n = 6). The dominance of the Hokkaido source groups in the overall assemblage is less important due to the fact that most of those obsidian artefacts come from a single site, Ainu Creek 1, which had the largest volume of material excavated of any other archaeological site in the Kurile Islands during the KBP expeditions (Table 7.4). More interesting is the distribution of the different source groups across the island chain.

The initial study (Phillips and Speakman 2009) found that obsidian artefacts from the Hokkaido source groups were primarily distributed across the southern Kurile Islands, while Kamchatkan obsidian was concentrated in the central and northern islands, with very little crossover or mixing of Hokkaido and Kamchatkan sources. The current research found a similar pattern of distribution based on a sample size that is almost six times larger than the initial study.

The Hokkaido obsidian source groups are located in two major volcanic complexes in northeastern Hokkaido, the Shirataki (43°55′ N, 143°09′ E) and Oketo (43°42′ N, 143°32′ E) volcanoes (see Hall and Kimura 2002; Kuzmin *et al.* 2002a). Almost all of the obsidian from the Hokkaido source groups (97.9%; n = 436) is distributed across archaeological sites in the southern Kurile Islands, with only nine flakes from Hokkaido recovered from sites in the central and northern parts of the island chain (Table 7.5). Obsidian from the Oketo-1 source group represents 53% (n = 236) of the total Hokkaido obsidian-sourced artefact assemblage, and while the Oketo-1 source is the most abundant within all of the Hokkaido source groups, it is currently not necessarily the most widely distributed source in the southern islands. The Oketo-1 source is found in five archaeological sites across Kunashir, Iturup, and Urup islands, while the Shirataki-B source is found in seven sites, the Shirataki-A source in five sites, and the Oketo-2 source in four sites. However, the sites where Oketo-1 obsidian was not found suffer from small sample sizes (n ≤ 10), and it might be expected that obsidian from this source would be found in larger assemblages.

Small amounts of the Oketo-1, Oketo-2, and Shirataki-A source groups were found in the central and northern Kurile Islands. Three obsidian flakes from Oketo-1 were recovered from the Vodopodnaya 2 site on the central island of Simushir, and one flake from the Baikova 1 site on the northern Shumshu Island. Two flakes from the Oketo-2 source were found in two northern sites, one each at Baikova 1 and the Savushkina 1 site on Paramushir Island. Three flakes from the Shirataki-A source were excavated from the Vodopodnaya 2 site.

The Kamchatkan obsidian sources include five different source groups, KAM-01, KAM-02, KAM-04, KAM-05, and KAM-07. Only the KAM-05 and KAM-07 source groups have been confidently located geographically in central Kamchatka, with the KAM-05 group representing the Maly Payalpan Volcano and the KAM-07 group representing the Ichinsky Volcano near the headwaters of the Belogolovaya Vtoraya River (Kuzmin *et al.* 2008). The location of the KAM-01, KAM-02, and KAM-04 source groups have been loosely estimated based on the distribution of artefacts in archaeological sites in the southern and eastern parts of the Kamchatka Peninsula (Glascock *et al.* 2006; Kuzmin *et al.* 2008; see also Grebennikov *et al.*, this volume).

In the entire sample of Kamchatka-sourced obsidian, the KAM-01 (38.7%; n = 113) and KAM-02 (30.5%; n = 89) source groups are represented in the highest frequency in this study. Sixty-nine percent of the KAM-01 obsidian artefacts were found in the northern sites of Savushkina 1, Tukharka River 1 (both on Paramushir Island) and Baikova 1 (on Shumshu Island). The KAM-02 assemblage is concentrated in central Kurile sites, with 90% (n = 80) of the obsidian flakes from that group found in the Vodopodnaya 2, Ryponkicha 1, and Drobnyye 1 sites. Interestingly, almost 20% (n = 16) of the obsidian artefacts from the Rikorda 1 site on the southern island of Kunashir were made from KAM-05 obsidian, while only three other flakes from any Kamchatkan sources were found in the southern Kuriles. All of the KAM-05 obsidian found in the Rikorda 1 site came from two adjoining stratigraphic levels in the same test pit excavation, and may be the result of a single reduction event of a larger piece of KAM-05 obsidian. Confirmation of this hypothesis will require further lithic technological analysis currently in progress. Aside from one flake recovered from the southern Ainu Creek 1 site, the rest of the KAM-05 obsidian assemblage (39.3%; n = 11) was spread across four sites in the central Kuriles.

A total of 22 artefacts plotted together in two different groups that did not fit with any of the Hokkaido or Kamchatkan source groups were identified in the original pilot study. Group-A consisted of six artefacts that were all excavated from the Baikova 1 site on Shumshu Island in the northern Kuriles. Of the 16 artefacts assigned to the Group-B source group, 14 of them were recovered from sites in the central and northern Kuriles. The geographic distribution of Group-A and Group-B artefacts follows a similar pattern seen in the distribution of the Kamchatkan source groups; it is hypothesised that the Group-A and Group-B sources are located somewhere on the Kamchatka Peninsula. Finally, 15 obsidian flakes could not be assigned

Table 7.3. Means and standard deviations for elemental concentrations from obsidian artefacts analysed in this study; after Glascock et al. (2000, 2006); Kuzmin (2006b); Kuzmin and Glascock (2007); Kuzmin et al. (1999, 2000, 2002a, 2008); and Speakman et al. (2005)) for original obsidian source geochemical characterisation data. Values in ppm except where noted; S.D. – standard deviation

Source	**Shirataki-A**		**Shirataki-B**		**Oketo-1**		**Oketo-2**		**KAM-1**		**KAM-2**		**KAM-4**		**KAM-5**		**KAM-7**		**Group-A**		**Group-B**	
	***n* = 108**		***n* = 79**		***n* = 236**		***n* = 22**		***n* = 113**		***n* = 89**		***n* = 45**		***n* = 28**		***n* = 17**		***n* = 6**		***n* = 16**	
Element	**Mean**	**S.D.**	**Mean**	**S.D.**	**Mean**	**S.D.**	**Mean**	**S.D.**	**Mean**	**S.D.**	**Mean**	**S.D.**	**Mean**	**S.D.**	**Mean**	**S.D.**	**Mean**	**S.D.**	**Mean**	**S.D.**	**Mean**	**S.D.**
K, %	33.754	6.150	34.135	5.951	34.246	6.342	43.691	10.965	27.415	6.324	40.981	8.728	26.900	2.014	32.724	15.408	43.664	12.088	39.522	5.521	20.149	6.296
Mn	454	114	464	80	359	109	557	150	803	266	1224	430	596	192	343	102	892	253	1237	132	482	196
Fe, %	23.275	13.009	18.167	12.987	9.193	3.929	15.842	5.865	19.855	6.624	28.382	10.276	15.968	5.745	7.016	3.839	18.629	10.775	31.628	3.171	12.721	8.211
Zn	79	35	67	33	37	10	62	15	60	25	101	30	45	16	28	10	61	26	125	77	39	22
Ga	20	4	19	4	18	2	25	6	18	3	22	3	16	2	13	2	21	4	26	2	13	4
Rb	143	26	160	30	140	12	178	56	65	9	106	9	67	5	99	11	74	8	92	3	54	32
Sr	26	4	11	4	58	7	82	7	154	14	71	8	136	10	41	4	247	18	212	17	107	7
Y	29	6	33	6	24	3	32	7	18	4	62	9	21	3	16	4	11	2	22	2	14	8
Zr	64	10	54	12	86	6	117	11	129	11	294	25	144	8	61	6	115	4	178	13	86	11
Nb	5	1	5	1	4	1	6	2	2	0	7	1	2	1	6	4	7	1	2	0	2	1
Cs	9	2	11	2	7	1	9	2	4	1	5	0	4	0	4	1	1	0	5	0	3	2
Ba	1372	227	625	512	1590	180	2062	543	1131	165	1374	224	1218	130	890	388	1659	241	1516	173	812	232
La	21	3	16	6	22	2	28	7	13	2	35	5	14	1	19	4	20	2	18	1	10	5
Ce	43	6	35	9	45	4	58	15	28	5	79	10	29	2	36	6	37	3	38	2	21	9
Pr	5	1	4	2	5	1	6	2	4	1	11	3	4	1	4	1	4	1	5	1	3	1
Nd	17	3	14	5	17	2	23	6	12	2	33	5	13	2	12	2	13	2	16	2	8	5
Sm	4	1	4	1	4	0	5	1	3	1	12	2	4	0	3	0	3	0	4	0	2	1
Eu	0	0	0	0	0	0	1	0	1	0	2	1	1	0	0	0	1	0	1	0	0	0
Gd	4	1	4	1	3	1	4	1	3	1	10	2	3	1	2	0	2	0	3	1	2	1
Tb	1	0	1	0	1	0	1	0	0	0	1	0	0	0	0	0	0	0	0	0	0	0
Dy	4	1	5	1	4	0	5	1	3	1	10	2	3	0	2	1	2	0	3	0	2	1
Ho	1	0	1	0	1	0	1	0	1	0	2	1	1	0	0	0	0	0	1	0	0	0
Er	3	1	3	1	2	1	3	1	2	1	6	2	2	0	1	0	1	0	2	1	1	1
Tm	0	0	1	0	0	0	1	0	0	0	1	0	0	0	0	0	0	0	0	0	0	0
Yb	3	1	3	1	3	0	3	1	2	1	7	1	2	0	2	0	1	0	3	0	2	1
Lu	0	0	1	0	0	0	1	0	0	0	1	0	0	0	0	0	0	0	0	0	0	0
Hf	3	0	3	1	3	0	4	0	4	0	9	1	5	0	2	0	3	0	6	0	3	0
Ta	1	0	1	0	1	0	1	0	0	0	1	0	0	0	1	1	1	0	0	0	0	0
Th	11	2	10	3	12	1	15	5	4	1	8	1	5	0	9	1	5	0	6	0	4	3

Table 7.4. Distribution of obsidian source groups across Kurile Island archaeological sites; names of sources are in bold. Columns under the names of sources are (from left to right): Site No. and Name; No. of samples; % in total assemblage

Shirataki-A			***Shirataki-B***			***Oketo-1***			***Oketo-2***		
2 Peschanaya 2	1	0.9	1 Alekhina 1	1	1.3	3 Rikorda 1	37	15.7	2 Peschanaya 2	1	4.5
3 Rikorda 1	8	7.4	2 Peschanaya 2	2	2.5	4 Sernovodsk 1	6	2.5	3 Rikorda 1	3	13.6
5 Berezovka 1	1	0.9	3 Rikorda 1	20	25.3	5 Berezovka 1	1	0.4	4 Sernovodsk 1	3	13.6
6 Kuibyushevskaya 1	3	2.8	4 Sernovodsk 1	1	1.3	7 Tikhaya 1	1	0.4	8 Ainu Creek 1	13	59.3
8 Ainu Creek 1	92	85.2	5 Berezovka 1	2	2.5	8 Ainu Creek 1	187	79.3	15 Savushkina 1	1	4.5
10 Vodopodnaya 2	3	2.8	6 Kuibyushevskaya 1	7	8.9	10 Vodopodnaya 2	3	1.3	17 Baikova 1	1	4.5
			8 Ainu Creek 1	46	58.2	17 Baikova 1	1	0.4			
Total	**108**	**100%**	**Total**	**79**	**100%**	**Total**	**236**	**100%**	**Total**	**22**	**100%**
KAM-01			***KAM-02***			***KAM-04***			***KAM-05***		
3 Rikorda 1	1	0.9	8 Ainu Creek 1	1	1.1	10 Vodopodnaya 2	15	33.3	3 Rikorda 1	16	57.1
9 Peschanaya Bay 1	4	3.6	10 Vodopodnaya 2	28	31.5	13 Drobnyye 1	2	4.4	8 Ainu Creek 1	1	3.6
10 Vodopodnaya 2	5	4.4	11 Ryponkicha 1	3	3.4	15 Savushkina 1	27	60.1	10 Vodopodnaya 2	7	25.0
13 Drobnyye 1	25	22.1	13 Drobnyye 1	49	55.1	17 Baikova 1	1	2.2	12 Rasshua 1	1	3.6
15 Savushkina 1	30	26.5	15 Savushkina 1	1	1.1				13 Drobnyye 1	2	7.1
16 Tukharka River 1	1	0.9	17 Baikova 1	5	5.6				14 Ekarma 1	1	3.6
17 Baikova 1	47	41.6	18 Bolshoi 1	2	2.2						
Total	**113**	**100%**	**Total**	**89**	**100%**	**Total**	**45**	**100%**	**Total**	**28**	**100%**
KAM-07			***Group-A***			***Group-B***			***Unknown***		
10 Vodopodnaya 2	16	94.1	17 Baikova	6	100	4 Sernovodsk 1	1	6.3	8 Ainu Creek	1	6.7
13 Drobnyye 1	1	5.9				8 Ainu Creek 1	1	6.3	9 Peschanaya Bay 1	1	6.7
						9 Peschanaya Bay 1	6	37.3	10 Vodopodnaya 2	8	53.2
						10 Vodopodnaya 2	4	25.0	11 Ryponkicha 1	3	20.0
						14 Ekarma 1	1	6.3	12 Rasshua 1	1	6.7
						15 Savushkina 1	1	6.3	14 Ekarma 1	1	6.7
						17 Baikova 1	2	12.5			
Total	**17**	**100%**	**Total**	**6**	**100%**	**Total**	**16**	**100%**	**Total**	**15**	**100%**

to any of the 11 source groups currently identified through pXRF and LA-ICP-MS analyses.

Discussion

The distribution of non-local obsidian across the Kurile Islands can be placed into the broader regional context of obsidian movement and usage in Northeast Asia. It has been demonstrated that obsidian from sources on the Japanese island of Hokkaido was transported on distances of up to 1000km to Sakhalin Island of the Russian Far East for almost 20,000 years (Glascock *et al.* 2000, 2006; Kuzmin 2006a, 2006b; Kuzmin and Glascock 2007; Kuzmin *et al.* 2000, 2002a, 2008). Given the geographic proximity of the southern Kuriles to Hokkaido, it could be expected that the Hokkaido–Sakhalin obsidian trade/transport network would be extended into the Kurile Islands. The initial movement of obsidian onto the islands of Kunashir, Iturup, and Urup may have coincided with early colonisation of those islands by Early or Middle Jomon people.

Although less is known about the trade and transport of obsidian from Kamchatkan source groups, several recent studies have shown that obsidian was transported from sources to sites around the Kamchatka Peninsula at distances of up to 400km (Glascock *et al.* 2006; Kuzmin *et al.* 2008; Speakman *et al.* 2005). Based on the current work it is clear that obsidian from Kamchatkan sources was used extensively in the central and northern Kuriles. Only 19 out of a total of 292 flakes made from Kamchatkan obsidian are found in southern Kurile archaeological sites, 16 of those from the same test pit on the Rikorda 1 site.

The nature of obsidian source group distribution within the Kurile Islands represents a data set of archaeological materials that can be leveraged to explore issues related to human migration and social networking. Personal relationships between and among human groups provide a social means for circumventing the local subsistence and material resource constraints that are inherent to geographically isolated environments (Mackie 2001). At

Table 7.5. Obsidian source groups represented in Kurile Island archaeological sites. Site Nos. and Names are in bold. Columns under sites' names are (from left to right): Source Name; No. of samples; % in total assemblage

1			*2*			*3*			*4*			*5*		
Alekhina			*Peschanaya 2*			*Rikorda 1*			*Sernovodsk 1*			*Berezovka 1*		
Shirataki-B	1	100.0	Oketo-2	1	25.0	KAM-01	1	1.2	Group-B	1	9.1	Oketo-1	1	25.0
			Shirataki-A	1	25.0	KAM-05	16	18.8	Oketo-1	6	54.5	Shirataki-A	1	25.0
			Shirataki-B	2	50.0	Oketo-1	37	43.6	Oketo-2	3	27.3	Shirataki-B	2	50.0
						Oketo-2	3	3.5	Shirataki-B	1	9.1			
						Shirataki-A	8	9.4						
						Shirataki-B	20	23.5						
Total	1	100%	Total	4	100%	Total	85	100%	Total	11	100%	Total	4	100%
6			*7*			*8*			*9*			*10*		
Kuibyushevskaya 1			*Tikhaya 1*			*Ainu Creek 1*			*Peschanaya Bay 1*			*Vodopodnaya 2*		
Shirataki-A	3	30.0	Oketo-1	1	100.0	Group-B	1	0.3	Group-B	6	54.5	Group-B	4	4.5
Shirataki-B	7	70.0				KAM-02	1	0.3	KAM-01	4	36.4	KAM-01	5	5.6
						KAM-05	1	0.3	Unknown	1	9.1	KAM-02	28	31.4
						Oketo-1	187	54.8				KAM-04	15	16.9
						Oketo-2	12	3.5				KAM-05	7	7.9
						Shirataki-A	92	27.0				KAM-07	16	17.9
						Shirataki-B	46	13.5				Oketo-1	3	3.4
						Unknown	1	0.3				Shirataki-A	3	3.4
												Unknown	8	9.0
Total	10	100%	Total	1	100%	Total	341	100%	Total	11	100%	Total	89	100%
11			*12*			*13*			*14*					
Ryponkicha 1			*Rasshua 1*			*Drobnyye 1*			*Ekarma 1*					
KAM-02	3	50.0	KAM-05	1	50.0	KAM-01	25	31.6	Group-B	1	33.3			
Unknown	3	50.0	Unknown	1	50.0	KAM-02	49	62.1	KAM-05	1	33.3			
						KAM-04	2	2.5	Unknown	1	33.4			
						KAM-05	2	2.5						
						KAM-07	1	1.3						
Total	6	100%	Total	2	100%	Total	79	100%	Total	3	100%			
15			*16*			*17*			*18*					
Savushkina 1			*Tukharka River 1*			*Baikova 1*			*Bolshoi 1*					
Group-B	1	1.7	KAM-01	1	100.0	Group-A	6	9.5	KAM-02	2	100.0			
KAM-01	30	50.0				Group-B	2	3.2						
KAM-02	1	1.7				KAM-01	47	74.6						
KAM-04	27	44.9				KAM-02	5	7.9						
Oketo-2	1	1.7				KAM-04	1	1.6						
						Oketo-1	1	1.6						
						Oketo-2	1	1.6						
Total	60	100%	Total	1	100%	Total	63	100%	Total	2	100%			

the local scale, social networks are a way to exchange information related to the day-to-day extraction of subsistence and material resources that are unevenly distributed (Cashdan 1983). Mobility is another way of lessening the potential negative impact on resource acquisition success, especially when territories are large (Cashdan 1987, 1992). At a regional level, social networks are cooperative strategies that often form a 'safety net' of support that can be critical in times of local resource scarcity or failure as a result of environmental variability (e.g., natural catastrophic events) (Bender 1978; Kennett and Kennett 2000; Rautman 1993; Rensink *et al.* 1991; Whallon 1989, 2006).

A relative measure of the access to obsidian sources that inhabitants of the southern Kurile Islands maintained would be through a degree of the richness of sources found in southern Kurile archaeological sites. Obsidian source richness in sites is simply the number of different obsidian source groups found in a site's obsidian artefact assemblage. Before a richness measure can be considered for a site assemblage however, the potential effects of

Table 7.6. Rank Order of southern Kurile archaeological site assemblage sample size and source richness (Spearman's r = 0.926; p = 0.72)

Site	No. of samples	Rank (Samples)	Richness	Rank (Richness)
Alekhina	1	1.5	1	1.5
Peschanaya 2	4	3.5	3	4.5
Rikorda 1	85	7	6	7
Sernovodsk 1	11	6	4	6
Berezovka 1	4	3.5	3	4.5
Kuibyushevskaya 1	10	5	2	3
Tikhaya 1	1	1.5	1	1.5
Ainu Creek 1	342	8	7	8

sample size on richness must be considered (Grayson 1984; Grayson and Cole 1998). Table 7.6 displays the Spearman's Rank Order of southern Kurile Island archaeological sites in terms of site obsidian artefact assemblage size and in terms of the number of different obsidian sources that are represented in the site (see Larson and Farber 2003, 573 for explanation of Spearman's Rank Order; for examples of usage in archaeological research see Grayson 1984, 1989). These two ranks are highly correlated (Spearman's *rho*, $r_s = 0.926$; $p = 0.72$), demonstrating that the larger the assemblage for a site, the greater number of sources found in that assemblage.

An alternative to measuring source richness in terms of the number of sources present in a site, is to reverse the analysis and measure site richness in terms of the number of sites in which each source is present. Table 7.7 focuses on the Hokkaido obsidian source group assemblages and their distribution in southern Kurile archaeological sites. A rank order correlation shows that these two ranks are not significantly correlated (Spearman's *rho*, $r_s = 0.316$; $p = 1$), allowing for further analysis of the evenness of the distribution of Hokkaido sources across the southern Kurile Islands as a whole.

Evenness measures are used to quantify the distribution of species in a community (Beck 2008; Bobrowsky and Ball 1989, 7; Grayson and Cole 1998). The measure of evenness is given as:

$$E = H / \ln S$$

where H is a measure of source-site diversity calculated by the Shannon diversity index and represented as $H = p_i \ln p_i$, with p_i the proportion of the eight southern Kurile sites in which each source group is present, and $\ln S$ is the natural logarithm of the number of different Hokkaido source groups. The values for evenness are constrained between 1 and 0, with values closer to 1 indicating a more even distribution of specimens across classes. An evenness value of 0.985 indicates that the four Hokkaido obsidian source groups were evenly distributed in regards to their presence in archaeological sites across the southern Kuriles.

Human groups that are highly mobile and able to procure lithic raw material on the landscape as part of their regularly scheduled resource extraction activities or planned logistical trips to raw material sources should have access to a higher number of different sources. Groups that are less mobile and rely on exchange relationships with other groups to obtain lithic raw material may have access to fewer different sources (Binford 1979; Morrow and Jeffries 1989; Pecora 2001).

Based on these measures of Hokkaido source group distribution, hypotheses can be developed to direct further study of these assemblages in the southern Kuriles. Access to different obsidian sources should be dependent upon the nature of group mobility and raw material procurement. Occupants of Kunashir, Iturup, and Urup islands may have had more direct access to primary and secondary obsidian deposits located in northern and eastern Hokkaido, leading to a wider and more even circulation of Hokkaido source groups in the southern Kuriles. The geographic nature of the southern Kuriles may have supported frequent travel and direct access since the straits between islands that must be crossed average roughly 25km in width (compared to straits in other parts of the archipelago that are over 80km wide). Additionally, the southern Kurile Islands are larger, more ecologically diverse, and less isolated than the central and northern Kuriles. Social networks formed by inhabitants there may have focused on localised information exchange for resource extraction rather than on forming long-distance lines of support that may have been necessary for survival in the central and northern parts of the island chain.

The measure of source richness for central and northern Kurile archaeological sites suffers from the same sample size effects as the southern Kurile sites. A Spearman's Rank Order correlation for central and northern Kurile site obsidian artefact assemblage size and the number of sources that are found in those sites (Table 7.8) shows a high correlation between the two variables (Spearman's *rho*, $r_s = 0.941$; $p = 0.65$). Based on the relative abundance of Kamchatkan sources present in archaeological sites, there appears to be a difference in the concentration of some source groups in the central Kuriles versus in the

Table 7.7. Rank Order of Hokkaido source group assemblage sample size and site richness (Spearman's r = 0.316; p = 1.0)

Obsidian source group/assemblage	***No. of samples***	***Rank (Samples)***	***Rich-ness***	***Rank (Richness)***
Shirataki-A	105	2	5	4
Shirataki-B	79	3	7	1
Oketo-1	232	1	5	3
Oketo-2	20	4	4	2

Table 7.8. Rank Order of central and northern Kurile assemblage sample size and source richness (Spearman's r = 0.941; p = 0.65)

Site	***No. of samples***	***Rank (Samples)***	***Rich-ness***	***Rank (Richness)***
Peschanaya Bay 1	11	6	2	5
Vodopodnaya 2	89	10	8	10
Ryponkicha 1	6	5	2	5
Rasshua 1	1	1.5	1	2
Drobnyye 1	79	9	5	7.5
Ekarma 1	2	3.5	2	5
Savushkina 1	60	7	5	7.5
Tukharka River 1	1	1.5	1	2
Baikova 1	63	8	7	9
Bolshoi 1	2	3.5	1	2

Table 7.9. Rank Order of Kamchatka, Group-A, and Group-B source group assemblage sample size and site richness (Spearman's r = 0.873; p = 0.78)

Obsidian source group/assemblage	***No. of samples***	***Rank (Samples)***	***Rich-ness***	***Rank (Richness)***
KAM-01	112	1	6	1.5
KAM-02	88	2	6	1.5
KAM-04	45	3	4	4.5
KAM-05	11	6	4	4.5
KAM-07	17	5	2	6
Group-A	6	7	1	7
Group-B	19	4	5	3

northern islands. However, a rank order correlation between Kamchatkan source group assemblage size and site richness (Table 7.9) shows a similar correlation (Spearman's *rho*, r_s = 0.873; p = 0.78), precluding any further quantification of the evenness of the distribution of Kamchatkan source groups at this time.

Though further analysis of a larger sample will be required to make comparisons between the distributions of Hokkaido and Kamchatkan obsidian source groups, it appears that the Bussol Strait separating the southern and central Kuriles may have been a significant barrier to the transport of obsidian from both source areas. This strait is the widest one in the Kurile island chain, a combined 109km wide between Urup and Simushir islands (79km between Urup and Chirpoi islands and 30km between Chirpoi and Simushir islands). The Bussol Strait has a strong current flowing between the Pacific Ocean and the Sea of Okhotsk, and it is recognised as a biogeographic barrier to the movement of plants and animals from the southern to central part of the island chain (Pietsch *et al.* 2003). Human groups who moved from Hokkaido and the southern Kuriles into the central and northern part of the Kurile Archipelago may have found it too costly in terms of time, energy, and risk to maintain access to Hokkaido obsidian sources across this strait. Securing access to Kamchatka, obsidian sources would have provided a less costly alternative to Hokkaido obsidian, which may have been achieved by re-orienting seasonal or annual migration patterns towards the north and allowing for the direct procurement of Kamchatkan obsidian, or by developing trade and exchange relationships with the inhabitants of the southern Kamchatka Peninsula.

While a comprehensive ^{14}C chronology for the Kurile Islands is still being developed by the KBP and is currently unpublished, preliminary findings show that in the southern Kuriles obsidian from Hokkaido sources was accessed for over 1700 years beginning ca. 2500 BP, indicating long-term use of those sources. In the central and northern Kuriles, obsidian from Kamchatkan sources was used consistently for over 700 years spanning the Epi-Jomon and Okhotsk culture periods. Interestingly, small amounts of Hokkaido obsidian recovered from sites on Paramushir and Shumshu islands seem to come from Epi-Jomon contexts, which may indicate an early expansion by Epi-Jomon people throughout the entire length of the island chain (B. Fitzhugh, personal communication 2009).

Conclusions

Research on obsidian artefacts from the Kurile Islands demonstrates that the prehistoric inhabitants of the Kurile Archipelago had access to multiple sources of obsidian located on Hokkaido and the Kamchatka Peninsula. More interesting are the patterns of obsidian use in the southern, central, and northern Kurile Islands, which indicate that source access may have been influenced by geographic and social factors. In the Kuriles, changing patterns of obsidian distribution can be used to characterise hunter-gatherer social networks that were dynamic through time (Hofman *et al.* 2007), and to infer the role that the networks played in the overall organisation of small social groups living in unpredictable environments (Kirch 1988). Additional research on the provenance of obsidian artefacts found in Kurile archaeological sites, as well as the technological analysis of how obsidian was utilised vis-à-vis other lithic raw material types, will further explore these and other issues related to the human occupation of this region.

Acknowledgements

Jeff Speakman at the Smithsonian Institution provided invaluable assistance and input on the LA-ICP-MS analysis and interpretation. Many thanks to Nicole Little at the Smithsonian Institution for laboratory assistance and comments on the paper; Adam Freeburg at the University of Washington (Seattle, WA, USA) for help in creating maps; and Erik Gjesfjeld at the University of Washington for input and discussion regarding the distribution of Epi-Jomon and Okhotsk ceramics across the Kurile Islands. Drs Jeff Ferguson, Ana Steffen, and Ben Fitzhugh read a draft of this paper and provided valuable comments; their efforts are appreciated. Thanks also to Drs Michael D. Glascock and Yaroslav V. Kuzmin, and Jeff Speakman for the opportunity to contribute to this volume. Participation in the Kurile Biocomplexity Project was made possible in part by a grant from the US National Science Foundation (ARC-0508109) and various logistical and financial support from the University of Washington; the Hokkaido University Museum (Sapporo, Japan); the Historical Museum of Hokkaido (Sapporo, Japan); the Sakhalin Regional Museum (Yuzhno-Sakhalinsk, Russia); and the Far Eastern Branch of the Russian Academy of Sciences (Institute of Marine Geology and Geophysics, Yuzhno-Sakhalinsk; Institute of Volcanology and Seismology, Petropavlovsk-Kamchatskiy; and Northeastern Interdisciplinary Science Research Institute, Magadan).

References

Beck, R. K. 2008. Transport Distance and Debitage Assemblage Diversity: An Application of the Field Processing Model to Southern Utah Toolstone Procurement Sites. *American Antiquity* 73, 759–780.

Bender, B. 1978. Gatherer-Hunter to Farmer: A Social Perspective. *World Archaeology* 10, 204–222.

Binford, L. R. 1979. Organization and Formation Processes: Looking at Curated Technologies. *Journal of Anthropological Research* 35, 255–273.

Bobrowski, P. T., and B. F. Ball. 1989. The Theory and Mechanics of Ecological Diversity in Archaeology. In *Quantifying Diversity in Archaeology*, edited by R. D. Leonard and G. T. Jones, 4–12. New York, Cambridge University Press.

Cashdan, E. 1983. Territoriality among Human Foragers: Ecological Models and an Application to Four Bushman Groups [with Comments and Reply]. *Current Anthropology* 24, 47–66.

Cashdan, E. 1987. Trade and its Origins on the Botletli River, Botswana. *Journal of Anthropological Research* 43, 121–138.

Cashdan, E. 1992. Spatial Organization and Habitat Use. In *Evolutionary Ecology and Human Behavior*, edited by E. A. Smith and B. Winterhalder, 237–268. New York, Aldine de Gruyter.

Chappell, J., and N. J. Shackleton. 1986. Oxygen Isotopes and Sea-Level. *Nature* 324, 137–140.

Cochrane, E. E., and H. Neff. 2006. Investigating Compositional Diversity among Fijian Ceramics with Laser Ablation-Inductively Coupled Plasma Mass-Spectrometry (LA-ICP-MS): Implications for Interactions Studies on Geologically Similar Islands. *Journal of Archaeological Science* 33, 378–390.

Cook, D. B., K. Fujita, and C. A. McMullen. 1986. Present Day Plate Interactions in Northeast Asia – North-American, Eurasian, and Okhotsk Plates. *Journal of Geodynamics* 6, 33–51.

Devos, W., C. Moor, and P. Lienemann. 1999. Determination of Impurities in Antique Silver Objects for Authentication by Laser Ablation Inductively Coupled Plasma Mass Spectrometry (LA-ICP-MS). *Journal of Analytical Atomic Spectrometry* 14, 621–626.

Devos, W., M. Senn-Luder, C. Moor, and C. Salter. 2000. Laser Ablation-Inductively Coupled Plasma Mass-Spectrometry (LA-ICP-MS) for Spatially Resolved Trace Analysis of Early-Medieval Archaeological Iron Finds. *Fresenius Journal of Analytical Chemistry* 366, 873–880.

Doelman, T., N. Kononenko, V. Popov, G. Summerhayes, R. Torrence, R. Bonetti, A. Guglielmetti, A. Manzoni, and M. Oddone. 2004. Acquisition and Movement of Volcanic Glass in the Primorye Region of Far Eastern

Russia. *Rossiya i ATR* 4(46), 112–124.

Doelman, T., R. Torrence, V. Popov, M. Ionescu, N. Kluyev, I. Sleptsov, I. Pantyukhina, P. White, and M. Clements. 2008. Source Selectivity: An Assessment of Volcanic Glass Sources in the Southern Primorye Region, Far East Russia. *Geoarchaeology* 23, 243–273.

Eerkens, J. W., J. R. Ferguson, M. D. Glascock, C. E. Skinner, and S. A. Waechter. 2007. Reduction Strategies and Geochemical Characterization of Lithic Assemblages: A Comparison of Three Case Studies from Western North America. *American Antiquity* 72, 585–597.

Eichelberger, J. C. 1995. Silicic Volcanism: Ascent of Viscous Magmas from Crustal Reservoirs. *Annual Review of Earth and Planetary Sciences* 23, 41–63.

Eriksen, B. V. 2002. Fossil Mollusks and Exotic Raw Materials in Late Glacial and Early Postglacial Find Contexts: A Complement to Lithic Studies. In *Lithic Raw Material Economies in Late Glacial and Early Postglacial Europe* (B.A.R. International Series 1093), edited by L. E. Fisher and B. V. Eriksen, 27–52. Oxford, BAR Publishing.

Erlich, E. 1986. *Geology of the Calderas of Kamchatka and Kurile Islands with Comparisons to Calderas of Japan and the Aleutians*. Open File Report 86–291. Denver, CO, US Geological Survey.

Fitzhugh, B., S. Moore, C. Lockwood, and C. Boone. 2004. Archaeological Paleobiogeography in the Russian Far East: The Kurile Islands and Sakhalin in Comparative Perspective. *Asian Perspectives* 43, 92–122.

Fitzhugh, B., V. O. Shubin, K. Tezuka, Y. Ishizuka, and C. A. S. Mandryk. 2002. Archaeology in the Kurile Islands: Advances in the Study of Human Paleobiogeography and Northwest Pacific Prehistory. *Arctic Anthropology* 39(1–2), 69–94.

Fitzhugh, W. W., and C. O. Dubreuil (eds). 1999. *Ainu: Spirit of a Northern People*. Washington, D.C., Smithsonian Institution Press.

Glascock. M. D., A. A. Krupianko, Y. V. Kuzmin, M. S. Shackley, and A. V. Tabarev. 1996. Geochemical Characterization of Obsidian Artefacts from Prehistoric Sites in the Russian Far East: An Initial Study. In *Arkheologiya Severnoi Patsifiki*, edited by A. L. Ivliev, N. N. Kradin, and I. S. Zhushchikhovskaya, 406–410. Vladivostok, Dalnauka Press.

Glascock, M. D., V. K. Popov, Y. V. Kuzmin, R. J. Speakman, A. V. Ptashinksy, and A. V. Grebennikov. 2006. Obsidian Sources and Prehistoric Obsidian Use on the Kamchatka Peninsula: Initial Results of Research. In *Archaeology in Northeast Asia: On the Pathway to the Bering Strait* (University of Oregon Anthropological Papers 65), edited by D. E. Dumond and R. L. Bland, 73–88. Eugene, University of Oregon.

Glascock, M. D., M. S. Shackley, Y. V. Kuzmin, V. K. Popov, S. V. Gorbunov, A. A. Vasilevsky, and O. A. Shubina. 2000. Geokhimiya Obsidiana iz Arkheologicheskikh Pamyatnikov Sakhalina i Ego Istochniki [Obsidian Geochemistry of the Archaeological Sites of the Sakhalin Island and its Sources]. In *Vulkanicheskie Stekla Dalnego Vostoka Rossii: Geologicheskie i Arkheologicheskie Aspekty*, edited by Y. V. Kuzmin and V. K. Popov, 88–106. Vladivostok, Dalnevostochny Geologichesky Institut Dalnevostochnogo Otdeleniya Rossiiskoi Akademii Nauk.

Golubev, V. A. , 1972. *Arkheologiya Kurilskikh Ostrovov* [Archaeology of the Kurile Islands]. Synopsis of the Candidate of Historical Sciences [PhD-equivalent] Dissertation. Novosibirsk, Institut Istorii, Filologii i Filosofii Sibirskogo Otdeleniya Akademii Nauk SSSR.

Gorshkov, G. S. 1970. *Volcanism of the Upper Mantle. Investigations of the Kurile Island Arc*. New York, Plenum Press.

Gratuze, B. 1999. Obsidian Characterization by Laser Ablation ICP-MS and its Application to Prehistoric Trade in the Mediterranean and the Near East: Sources and Distribution of Obsidian within the Aegean and Anatolia. *Journal of Archaeological Science* 26, 869–882.

Gratuze, B., M. Blet-Lemarquand, and J. N. Barrandon. 2001. Mass Spectrometry with Laser Sampling: A New Tool to Characterize Archaeological Materials. *Journal of Radioanalytical and Nuclear Chemistry* 247, 645–656.

Grayson, D. K. 1984. *Quantitative Zooarchaeology: Topics in the Analysis of Archaeological Faunas*. New York, Academic Press.

Grayson, D. K. 1989. Sample Size and Relative Abundance in Archaeological Analysis: Illustrations from Spiral Fractures and Seriation. In *Quantifying Diversity in Archaeology*, edited R. D. Leonard and G. T. Jones, 79–84. New York, Cambridge University Press.

Grayson, D. K., and S. C. Cole. 1998. Stone Tool Assemblage Richness during the Middle and Early Upper Paleolithic in France. *Journal of Archaeological Science* 25, 927–938.

Hall, M., and H. Kimura. 2002. Quantitative EDXRF Studies of Obsidian Sources in Northern Hokkaido. *Journal of Archaeological Science* 29, 259–266.

Hofman, C. L., A. J. Bright, A. Boomert, and S. Knippenberg. 2007. Island Rhythms: The Web of Social Relationships and Interaction Networks in the Lesser Antillean Archipelago between 400 B.C. and 1942 A.D. *Latin American Antiquity* 18, 243–268.

Imamura, K. 1996. *Prehistoric Japan: New Perspectives on Insular East Asia*. London, University College London Press.

Ishizuka, Y. 2001. Volcanic Activity and Recent Tephras in the Kurile Islands: Field Result during the International Kurile Island Project (IKIP) 2000. Report on File with the Ben Fitzhugh (Project PI), University of Washington; available at: http://depts.washington.edu/ikip/Geology/volcanismindex.htm. Accessed 15 September 2009.

Izuho, M., and H. Sato. 2007. Archaeological Obsidian Studies in Hokkaido, Japan: Retrospect and Prospects. *Bulletin of the Indo-Pacific Prehistory Association* 27, 114–121.

James, W. D., and E. S. C. Dahlin. 2005. Chemical Compositional Studies of Archaeological Artifacts: Comparison of LA-ICP-MS to INAA Measurements. *Journal of Radioanalytical and Nuclear Chemistry* 263, 697–702.

Kennett, D. J., and J. P. Kennett. 2000. Competitive and Cooperative Responses to Climatic Instability in Coastal Southern California. *American Antiquity* 65, 379–395.

Kennett, D. J., H. Neff, M. D. Glascock, and A. Z. Mason. 2001. A Geochemical Revolution: Inductively Coupled Plasma-Mass Spectrometry. *The SAA Archaeological Record* 1, 22–36.

Kikuchi, T. 1999. Ainu Ties with Ancient Cultures of Northeast Asia. In *Ainu: Spirit of a Northern People*, edited by W. W. Fitzhugh and C. O. Dubreuil, 47–51. Washington, D.C., Smithsonian Institution Press.

Kim, J. C., D. K. Kim, M. Youn, C. C. Yun, G. Park, H. J. Woo, M.-Y. Hong, and G. K. Lee. 2007. PIXE Provenancing of Obsidian Artefacts from Paleolithic Sites in Korea. *Bulletin of the Indo-Pacific Prehistory Association* 27, 122–128.

Kimura, H. 1995. Kokuyouseki, Hito, Gijyutsu [Obsidian, Humans, and Technology]. *Hokkaido Kokogaku* 31, 3–63.

Kimura, H. 2002. Hokkaido Chiiki ni okeru Kokuyoseki Kenkyu no Tenbou [A Prospect of Obsidian Study in Hokkaido District]. *Kokuyouseki Bunka Kenkyu* 1, 69–82.

Kimura, H. 2006. The Shirataki Obsidian Mine Area and the Yubetsu–Horokazawa Technological Complex. *Current Research in the Pleistocene* 23, 9–11.

Kirch, P. V. 1988. Long-Distance Exchange and Island Colonization: The Lapita Case. *Norwegian Archaeological Review* 21, 103–117.

Kluyev, N. A., and I. Y. Sleptsov. 2007. Late Pleistocene and Early Holocene Uses of Basaltic Glass in Primorye, Far East Russia: A New Perspective Based on Sites near the Sources. *Bulletin of the Indo-Pacific Prehistory Association* 27, 129–134.

Kobayashi, T. 2004. *Jomon Reflections: Forager Life and Culture in the Prehistoric Japanese Archipelago*. Oxford, Oxbow Books.

Kuzmin, Y. V. 2006a. Palaeoenvironment and Chronology. In *Archaeology of the Russian Far East: Essays in Stone Age Prehistory* (B.A.R. International Series 1540), edited by S. M. Nelson, A. P. Derevianko, Y. V. Kuzmin, and R. L. Bland, 13–40. Oxford, BAR Publishing.

Kuzmin, Y. V. 2006b. Recent Studies of Obsidian Exchange Networks in Prehistoric Northeast Asia. In *Archaeology in Northeast Asia. On the Pathway to the Bering Strait* (University of Oregon Anthropological Papers 65), edited by D. E. Dumond and R. L. Bland, 61–71. Eugene, University of Oregon.

Kuzmin, Y. V., S. V. Alkin, A. Ono, H. Sato, T. Sakaki, S. Matsumoto, K. Orimo, and S. Ito. 1998. *Radiouglerodnaya Khronologiya Drevnikh Kultur Kamennogo Veka Severo-Vostochnoi Azii* [Radiocarbon Chronology of the Stone Age Cultures of Northeast Asia]. Vladivostok, Tikhookeansky Institut Geografii Dalnevostochnogo Otdeleniya Rossiiskoi Akademii Nauk.

Kuzmin, Y. V., and M. D. Glascock. 2007. Two Islands in the Ocean: Prehistoric Obsidian Exchange between Sakhalin and Hokkaido, Northeast Asia. *Journal of Island and Coastal Archaeology* 2, 99–120.

Kuzmin, Y. V., M. D. Glascock, and H. Sato. 2002a. Sources of Archaeological Obsidian on Sakhalin Island. *Journal of Archaeological Science* 29, 741–749.

Kuzmin, Y. V., V. K. Popov, M. D. Glascock, and M. S. Shackley. 2002b. Sources of Archaeological Volcanic Glass in the Primorye (Maritime) Province, Russian Far East. *Archaeometry* 44, 505–515.

Kuzmin, Y. V., V. K. Popov, A. V. Tabarev, M. D. Glascock, and M. S. Shackley. 2000. Istochniki Arkheologicheskovo Obsidiana Primorya i Ego Rasprostranenie v Kulturakh Kamennogo Veka [The Sources of Archaeological Obsidian in Primorye and its Distribution in the Stone Age Cultures]. In *Vulkanicheskie Stekla Dalnego Vostoka Rossii: Geologicheskie i Arkheologicheskie Aspekty*, edited by Y. V. Kuzmin and V. K. Popov, 83–88. Vladivostok, Dalnevostochny Geologichesky Institut Dalnevostochnogo Otdeleniya Rossiiskoi Akademii Nauk.

Kuzmin, Y. V., R. J. Speakman, M. D. Glascock, V. K. Popov, A. V. Grebennikov, M. A. Dikova, and A. V. Ptashinsky. 2008. Obsidian Use at the Ushki Lake Complex, Kamchatka Peninsula (Northeastern Siberia): Implications for Terminal Pleistocene and Early Holocene Human Migrations in Beringia. *Journal of Archaeological Science* 35, 2179–2187.

Kuzmin, Y. V., A. V. Tabarev, V. K. Popov, M. D. Glascock, and M. S. Shackley. 1999. Geochemical Source Analysis of Archaeological Obsidian in Primorye (Russian Far East). *Current Research in the Pleistocene* 16, 97–99.

Larson, R., and B. Farber. 2003. *Elementary Statistics* (2nd edit.). Upper Saddle River, NJ, Prentice Hall.

Mackie, Q. 2001. *Settlement Archaeology in a Fjordland Archipelago* (B.A.R. International Series 926). Oxford, BAR Publishing.

Mallory-Greenough, L. M., J. D. Greenough, G. Dobosi, and J. V. Owen. 1999. Fingerprinting Ancient Egyptian Quarries: Preliminary Results Using Laser Ablation Microprobe-Inductively Coupled Plasma-Mass Spectrometry. *Archaeometry* 41, 227–238.

Morrow, C. A., and R. W. Jeffries. 1989. Trade or Embedded Procurement? A Test Case from Southern Illinois. In *Time, Energy and Stone Tools*, edited by R. Torrence, 27–33. Cambridge, Cambridge University Press.

Naoe, Y., and J. Nagasaki. 2005. Hokkaido Koki Kyusekki Jidai Zenhan Ki no Sekizai Shohi Senryaku [Raw Material Consumption Strategy at Late Upper Palaeolithic on Hokkaido]. *Hokkaido Kyusekki Bunka Kenkyu* 10, 45–58.

Ohnuki-Tierney, E. 1976. Regional Variations in Ainu Culture. *American Ethnologist* 3, 297–329.

Okada, H. 1998. Maritime Adaptations in Northern Japan. *Arctic Anthropology* 35(1), 335–339.

Otaishi, N. 1994. Okhotsk Culture: Prehistoric Hunting and Fishing Reconstructed from the Sea of Okhotsk Paleofauna. *ArchaeoZoologia* 6, 71–78.

Pantukhina, I. 2007. The Role of Raw Material in Microblade Technology at Three Late Palaeolithic Sites, Russian Far East. *Bulletin of the Indo-Pacific Prehistory Association* 27, 144–153.

Pecora, A. M. 2001. Chipped Stone Tool Production Strategies and Lithic Debris Patterns. In *Lithic Debitage: Context, Form, Meaning*, edited by W. Andrefsky, Jr., 173–191. Salt Lake City, University of Utah Press.

Phillips, S. C., and R. J. Speakman. 2009. Initial Source Evaluation of Archaeological Obsidian from the Kurile

Islands of the Russian Far East Using Portable XRF. *Journal of Archaeological Science* 36, 1256–1263.

Pietsch, T. W., V. V. Bogatov, K. Amaoka, Y. N. Zhuravlev, V. Y. Barkalov, S. Gage, H. Takahashi, A. S. Lelej, S. Y. Storozhenko, N. Minakawa, D. J. Bennet, and T. R. Anderson. 2003. Biodiversity and Biogeography of the Islands of the Kurile Archipelago. *Journal of Biogeography* 30, 1297–1310.

Pires-Ferreira, J. W. 1978. Obsidian Exchange Networks: Inferences and Speculations on the Development of Social Organization in Formative Mesoamerica. In *Cultural Continuity in Mesoamerica*, edited by D. L. Browman, 49–78. The Hague, Mouton.

Popov, V. K., V. G. Sakhno, Y. V. Kuzmin, M. D. Glascock, and B.-K. Choi. 2005. Geochemistry of Volcanic Glasses of the Paektusan Volcano. *Doklady Earth Sciences* 403, 345–350.

Prokofiev, M. M. 2005. *Yaponskie Uchenye – Issledovateli Yuzhnogo Sakhalina i Kurilskikh Ostrovov (Konets XIX – Pervaya Polovina XX vv.). Biobibliograficheskie Ocherki* [Japanese Scholars – Researchers of the Southern Sakhalin and the Kurile Islands (End of the XIXth – First Half of the XXth Centuries). Biobibliographical Essays]. Yuzhno-Sakhalinsk, Sakhalinskoe Knizhnoe Izdatelstvo.

Rautman, A. E. 1993. Resource Variability, Risk, and the Structure of Social Networks: An Example from the Prehistoric Southwest. *American Antiquity* 58, 403–424.

Rensink, E., J. Kolen, and A. Spieksma. 1991. Patterns of Raw Material Distribution in the Upper Pleistocene of Northwestern and Central Europe. In *Raw Material Economies among Prehistoric Hunter-Gatherers* (Publications in Anthropology 19), edited by A. Montet-White and S. Holen, 142–159. Lawrence, University of Kansas.

Sato, H. 2004. Russia Kyokuto Niokeru Senshi Jidai no Kokuyoseki no Riyo [Prehistoric Obsidian Exploitation in the Russian Far East]. *Kokuyouseki Bunka Kenkyu* 3, 45–55.

Sato, T., T. Amano, H. Ono, H. Ishida, H. Kodera, H. Matsumura, M. Yoneda, and R. Masuda. 2007. Origins and Genetic Features of Okhotsk People, Revealed by Ancient Mitochondrial DNA Analysis. *Journal of Human Genetics* 52, 618–627.

Shackley, M. S., M. D. Glascock, Y. V. Kuzmin, and A. V. Tabarev. 1996. Geochemical Characterization of Archaeological Obsidian from the Russian Far East: A Pilot Study. *International Association of Obsidian Studies Bulletin* 17, 16–19.

Shubin, V. O. 1994. Aleut in the Kurile Islands. In *Anthropology of the North Pacific Rim*, edited by W. W. Fitzhugh and V. Chaussonet, 337–345. Washington, D.C., Smithsonian Institution Press.

Simkin, T., and L. Siebert. 1994. *Volcanoes of the World*. Tucson, AZ, Geoscience Press.

Speakman, R. J., M. D. Glascock, V. K. Popov, Y. V. Kuzmin, A. V. Ptashinsky, and A. V. Grebennikov. 2005. Geochemistry of Volcanic Glasses and Sources of Archaeological Obsidian on the Kamchatka Peninsula (Russian Far East): First Results. *Current Research in the Pleistocene* 22, 11–13.

Speakman, R. J., and H. Neff. 2005. The Application of Laser Ablation ICP-MS to the Study of Archaeological Materials—An Introduction. In *Laser Ablation ICP-MS in Archaeological Research*, edited by R. J. Speakman and H. Neff, 1–15. Albuquerque, University of New Mexico Press.

Speakman, R. J., H. Neff, M. D. Glascock, and B. J. Higgins. 2002. Characterization of Archaeological Materials by Laser Ablation-Inductively Coupled Plasma-Mass Spectrometry. In *Archaeological Chemistry: Materials, Methods, and Meaning,* edited by K. A. Jakes, 48–63. Washington, D.C., American Chemical Society.

Stephan, J. J. 1974. *The Kuril Islands: Russo-Japanese Frontier in the Pacific.* Oxford, Clarendon Press.

Suzuki, H., and Y. Naoe. 2006. The Shirataki Sites: An Overview of Upper Paleolithic Sites at an Obsidian Source in Hokkaido, Japan. *Current Research in the Pleistocene* 23, 31–33.

Tezuka, K. 1998. Long-Distance Trade Networks and Shipping in the Ezo Region. *Arctic Anthropology* 35(1), 350–360.

Vasilevsky, A. A., and O. A. Shubina. 2006. Neolithic of the Sakhalin and the Southern Kurile Islands. In *Archaeology of the Russian Far East: Essays in Stone Age Prehistory* (B.A.R. International Series 1540), edited by S. M. Nelson, A. P. Derevianko, Y. V. Kuzmin, and R. L. Bland, 151–166. Oxford, BAR Publishing.

Vysokov, M. S. 1994. *Istoryia Sakhalina i Kuril v Samom Kratkom Izlozhenii* [A Very Brief History of Sakhalin Island and the Kurile Islands]. Yuzhno-Sakhalinsk, Sakhalinskoe Knizhnoe Izdatelstvo.

Whallon, R. 1989. Elements of Cultural Change in the Later Paleolithic. In *The Human Revolution. Behavioral and Biological Perspectives on the Origins of Modern Humans*, edited by P. Mellars and C. Stringer, 433–454. Princeton, NJ, Princeton University Press.

Whallon, R. 2006. Social Networks and Information: Non-"Utilitarian" Mobility among Hunter-Gatherers. *Journal of Anthropological Archaeology* 25, 259–270.

Yamaura, K. 1998. The Sea Mammal Hunting Cultures of the Okhotsk Sea with Special Reference to Hokkaido Prehistory. *Arctic Anthropology* 35(1), 321–334.

Yamaura, K., and H. Ushiro. 1999. Prehistoric Hokkaido and Ainu Origins. In *Ainu: Spirit of a Northern People*, edited by W.W. Fitzhugh and C.O. Dubreuil, 39–46. Washington, D.C., Smithsonian Institution Press.

Yanshina, O. V., and Y. V. Kuzmin. 2010. The Earliest Evidence of Human Settlement in the Kurile Islands (Russian Far East): The Yankito Site Cluster, Iturup Island. *Journal of Island and Coastal Archaeology* 5, 179–184.

Zaitseva, G. I., S. G. Popov, A. P. Krylov, Y. V. Knorozov, and A. B. Spevakovskiy. 1993. Radiocarbon Chronology of Archaeological Sites of the Kurile Islands. *Radiocarbon* 35, 507–510.

Chapter 8
Crossing Mountains, Rivers, and Straits: A Review of the Current Evidence for Prehistoric Obsidian Exchange in Northeast Asia

Yaroslav V. Kuzmin

Abstract: *During the past decade, significant progress has been achieved in the study of obsidian exchange patterns in the Palaeolithic and Neolithic of Northeast Asia (Russia, Japan, and Korea). In the Russian Far East, research has confirmed earlier findings for several primary sources of archaeological volcanic glass (such as the Basaltic and Obluchie plateaus, Paektusan Volcano, and Shirataki and Oketo). The recent intensification of obsidian provenance research in Korea has brought to light new evidence regarding the extensive use of the Paektusan Volcano source by Upper Palaeolithic populations in the central and southern parts of the Korean Peninsula. Also, new data demonstrate that obsidian from another important source, Koshidake, located on Kyushu Island which lies across the modern Korea (Tsushima) Strait, was exploited by inhabitants of the Korean Peninsula since ca. 25,000 BP. Several long-distance obsidian exchange networks existed in Northeast Asia in the Palaeolithic and Neolithic (ca. 25,000–3000 BP), with the range of obsidian transport from sources to utilisation sites up to 1000km. People were able to cross natural obstacles, such as mountains, rivers, and even sea straits, to acquire this valuable raw material.*

Keywords: *Obsidian, Source Studies, Geochemistry, Russian Far East, Korean Peninsula, Hokkaido Island, Kyushu Island, Ryukyu Archipelago*

Introduction

In prehistory, especially throughout the Palaeolithic and Neolithic stages, obsidian was an important commodity that served mainly as raw material for tool manufacture. It was recognised in the 1960s that obsidian provenance studies could significantly contribute toward understanding patterns of ancient migration, social interaction, and exchange of raw material (e.g., Renfrew 1969; Renfrew *et al.* 1968; see review: Cann 1983). Because 'geological' sources of pure (waterless) volcanic glass are not very numerous, even in regions of modern active volcanism such as the Mediterranean and Mesoamerica (e.g., Glascock 2002; Williams-Thorpe 1995), the exchange and trade of obsidian was widely practiced in prehistoric societies in order to satisfy a high demand for good quality raw material. As a result, extensive exchange networks existed in many regions of the Old World, Oceania, and the Americas. There are many documented examples in which obsidian raw materials or tools were moved 1000km or more from their respective geologic sources (e.g., Barker *et al.* 2002; Brooks *et al.* 1997; Gratuze 1999; Kirch 1991; Molyneaux 2002; Pollmann 1993; Shackley 2005; Tykot 2002; Tykot and Ammerman 1997; Yacobaccio *et al.* 2002, 2004).

Prehistoric obsidian studies in Northeast Asia began in the 1960s in Japan, but became an important part of archaeological research in the continental part of the region, i.e. the Russian Far East and the Korean Peninsula, only in the 1990s. A brief information on obsidian from the far eastern Russian prehistoric complexes was published in the early 1980s (Butylina 1981), and major research programme based on internationally accepted methodological approaches began in the early 1990s. Results on obsidian source studies in Northeast Asia were described for Japan by Suzuki and Tomura (1983), Yamamoto (1990), Ono *et al.* (1992), Hall and Kimura (2002), and Tsutsumi (2002) (see résumé: Habu 2004, 221–224); and for the Russian Far East and Korea results have been presented by Kuzmin *et al.* (2002a, 2002b) and Popov *et al.* (2005). The major results of source identification for archaeological obsidian in Northeast Asia obtained before the mid-2000s currently were summarised by Kuzmin (2006a, 2006b). In this overview, the most recent developments in the field of prehistoric obsidian exchange in mainland and insular parts of Northeast Asia are presented.

Methods and Material

Obsidian provenance studies in the Russian Far East and Korea are conducted with the help of several geochemical methods, that is, silicate analysis (major oxides: SiO_2, TiO_2, Al_2O_3, Fe_2O_3, FeO, MnO, MgO, CaO, Na_2O, K_2O, and P_2O_5; and loss of ignition), X-ray Fluorescence (hereafter – XRF) analysis, and Neutron Activation Analysis (hereafter – NAA). The details of XRF and NAA techniques may be found in Glascock *et al.* (1998) and Pollard *et al.* (2007). Statistical grouping and discrimination of sources followed the approach described by Glascock *et al.* (1998). For the territory of the Korean Peninsula, Proton-Induced X-ray Emission (hereafter – PIXE) analysis (Kim *et al.* 2007a, 2007b) and $^{87}Sr/^{86}Sr$ isotope ratio (Cho 2005) were employed in some cases.

As for the reference 'geological' samples from obsidian sources, rock samples have been collected directly from the outcrops of volcanic glass in different regions of Northeast Asia, including Primorye [Maritime] and Amur provinces, Jewish Autonomous District (all in the Russian Federation); Hokkaido Island (Japan); and Jilin Province (People's Republic of China). Major sources examined by our group include: 1) Basaltic Plateau in southern Primorye Province; 2) Paektusan [Baitoushan] Volcano on Changbaishan Plateau in southeastern Jilin Province; 3) Obluchie Plateau

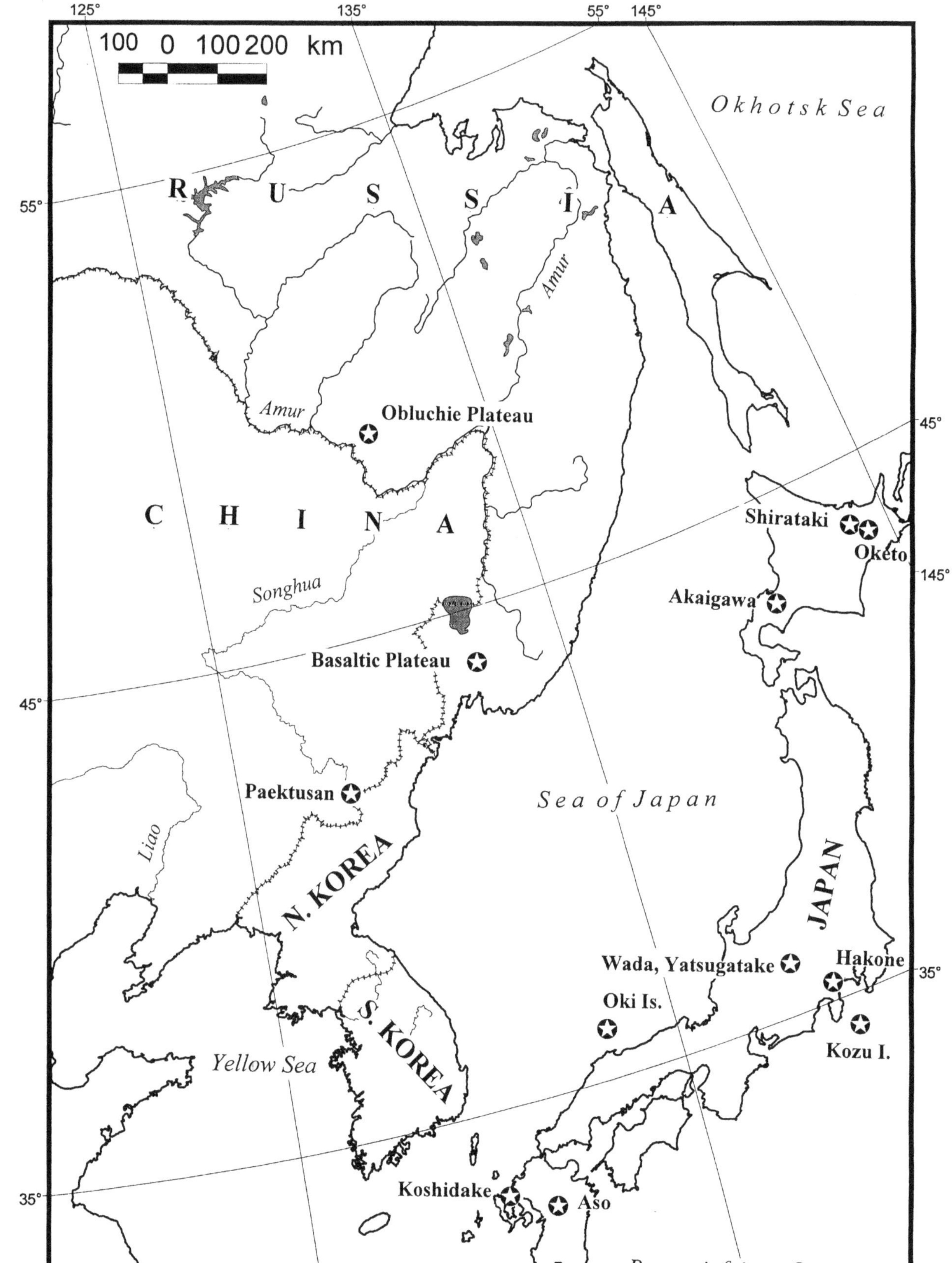

Figure 8.1. The major obsidian sources in Northeast Asia (after Habu 2004; Kuzmin 2006a; Ono et al. 1992; Tsutsumi 2002; modified)

in Amur Province and Jewish Autonomous District; and 4) Shirataki and Oketo on eastern Hokkaido (Figure 8.1).

Recent Progress in Archaeological Obsidian Source Studies in Northeast Asia

Russian Far East

Results of research conducted recently in the Primorye and Amur provinces were published by Popov *et al.* (2006a, 2006b). As for the Primorye Province, obsidian from two major sources was identified, i.e., the Basaltic Plateau and Paektusan; the minor source of Gladkaya River also was detected. All of these sources were originally identified by our team (Kuzmin *et al.* 2002a). At nine new sites, obsidian from the 'local' source of the Basaltic Plateau was established, with a distance of 20–130km (Figure 8.2). Obsidian from the 'remote' source of Paektusan was found at eight new sites. Most of them are situated in the southern part of the province, and only one site, Ustinovka 8, is located further north and east, at a distance of about 630km from the source (Figure 8.3). At two sites, volcanic glass from the Gladkaya River source with quite high water content (more than 4% weight) was discovered. These sites are in the immediate vicinity of the source, at a distance of 5–10km.

Doelman *et al.* (2008) conducted a joint Russian-Australian in-depth study of obsidian from the Basaltic Plateau and other sources of volcanic glass in southern Primorye, including issues of the raw material selection, acquisition, and intensity of use in prehistory. By using Proton-Induced X-ray Emission – Proton-Induced Gamma-ray Emission (PIXE–PIGME) method, Doelman *et al.* (2008) arrived at the same conclusion as Kuzmin *et al.* (1999, 2002a) in terms of identifying the sources of obsidian.

The research by Warashina *et al.* (1998) led to the identification of Paektusan obsidian at another site within the Ustinovka cluster in northeastern Primorye, the final Upper Palaeolithic complex of Ustinovka 6 dates to ca. 11,750–11,550 BP (e.g., Kuzmin 2006c). The amount of obsidian at the Ustinovka 6 site is limited; 14 obsidian artefacts were found, and they constitute 0.06% of the total assemblage (Kononenko *et al.* 2003, 59–61). Results by T. Warashina and coauthors support the conclusion that obsidian from the Paektusan source was brought to Primorye by at least ca. 10,000 BP (Kuzmin *et al.* 2002a, 513), and now this time frame can be revised to ca. 11,750 BP.

An error is noted in the work by Jia (2007, 147–148) where he states '…in the Upper Yinggeling (Xingcheng) culture, almost every site contains obsidian artefacts including the Sopohang and Odong sites located in the northeast corner of Korea (Li, Yungduo 1983:70), as well as the Zaisanovka site in the Primorye region of Russian Far East (Jilin Kaogusuo *et al.* 1998; Yanbian Museum 1991; Heilongjiang Kaogudui 1981).' This quotation appears to be from a Chinese excavation report, and it is unclear what it has to do with the presence of obsidian artefacts in the Zaisanovka cultural complex of Primorye, which is part of the Russian Far East. The citation of Chinese reports instead of the original Russian ones (Andreyev 1964; Andreev and Andreeva 1965, 129–132; Okladnikov 1965), which are readily available in English, is a serious oversight. In the original reports (e.g., Andreyev 1964, 267; Okladnikov 1965, 55–60), it is already stated that obsidian artefacts are known at several Zaisanovka sites.

In the Amur River basin, new data confirm that the major source of obsidian for prehistoric societies is situated in the middle stream of the river course, most probably in the Obluchie Plateau on the border between Amur Province and the Jewish Autonomous District (Popov *et al.* 2006b) (Figure 8.4). Alternatively, another Cenozoic basaltic plateau called Xunhe and located about 200km east of the Obluchie Plateau may have obsidian that was exploited by prehistoric people in the middle and lower streams of the Amur River (Figure 8.4). The straight-line distance between the source and utilisation sites is up to 700–800km.

At several Neolithic sites near the city of Khabarovsk in the lower course of the Amur River, the issue of obsidian sources is more complex compared to the middle part of the Amur River basin. At two sites, obsidian from the Basaltic Plateau was identified (Popov *et al.* 2006a) (Figure 8.2). This is the first finding of volcanic glass from Primorye sources in the Amur River basin, at a distance of 660km. At two other sites, obsidian artefacts do not match with any known sources, but their composition is identical to the geochemical group "Samarga" previously detected in the northern part of Primorye Province (Kuzmin *et al.* 2002a, 512). It seems that we have an unknown source of high quality volcanic glass in either northern Primorye Province or the southern part of Khabarovsk Province (Figure 8.5), probably within the watershed of the Svetlaya, Samarga, and Koppi rivers. This suggestion was initially put forward by our team (Kuzmin *et al.* 2002a, 512), and now we have more data in favour of this source's existence.

Concerning Sakhalin Island, some new results were obtained for the northern part of it (Kuzmin and Glascock 2007) compared with a previous study (e.g., Kimura 1998; Kuzmin 2006b; Kuzmin *et al.* 2002b). It was confirmed that most of the obsidian found in Sakhalin comes from two sources on Hokkaido Island, Shirataki and Oketo, at a distance of up to 1000km (Figure 8.6). The limited use of another source on Hokkaido, Akaigawa, was identified at some sites in southern Sakhalin.

The latest development in obsidian provenance research in the insular Russian Far East is the beginning of the source studies in the Kurile Islands (Phillips and Speakman 2009; see also Phillips, this volume) which are located between Hokkaido Island to the south and the Kamchatka Peninsula to the north. It was determined that in the southern (Kunashir and Urup islands) and some central (Chirpoi and Shiashkotan) Kuriles most obsidian was obtained from Hokkaido sources while in central (Chirpoi, Simushir, and Shiashkotan islands) and northern (Paramushir and

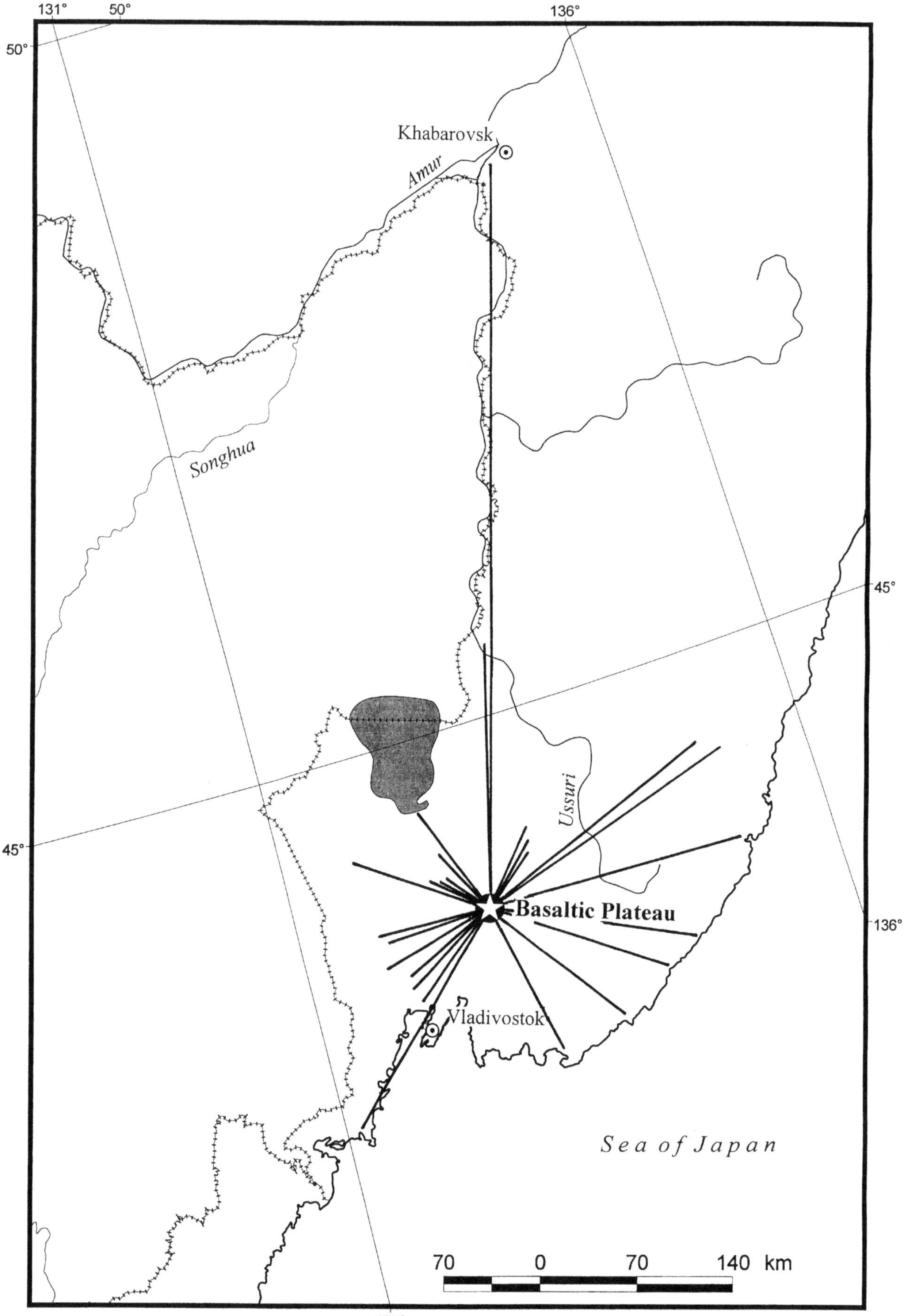

Figure 8.2. The distribution of obsidian from the Basaltic Plateau source in Primorye Province and the Amur River basin

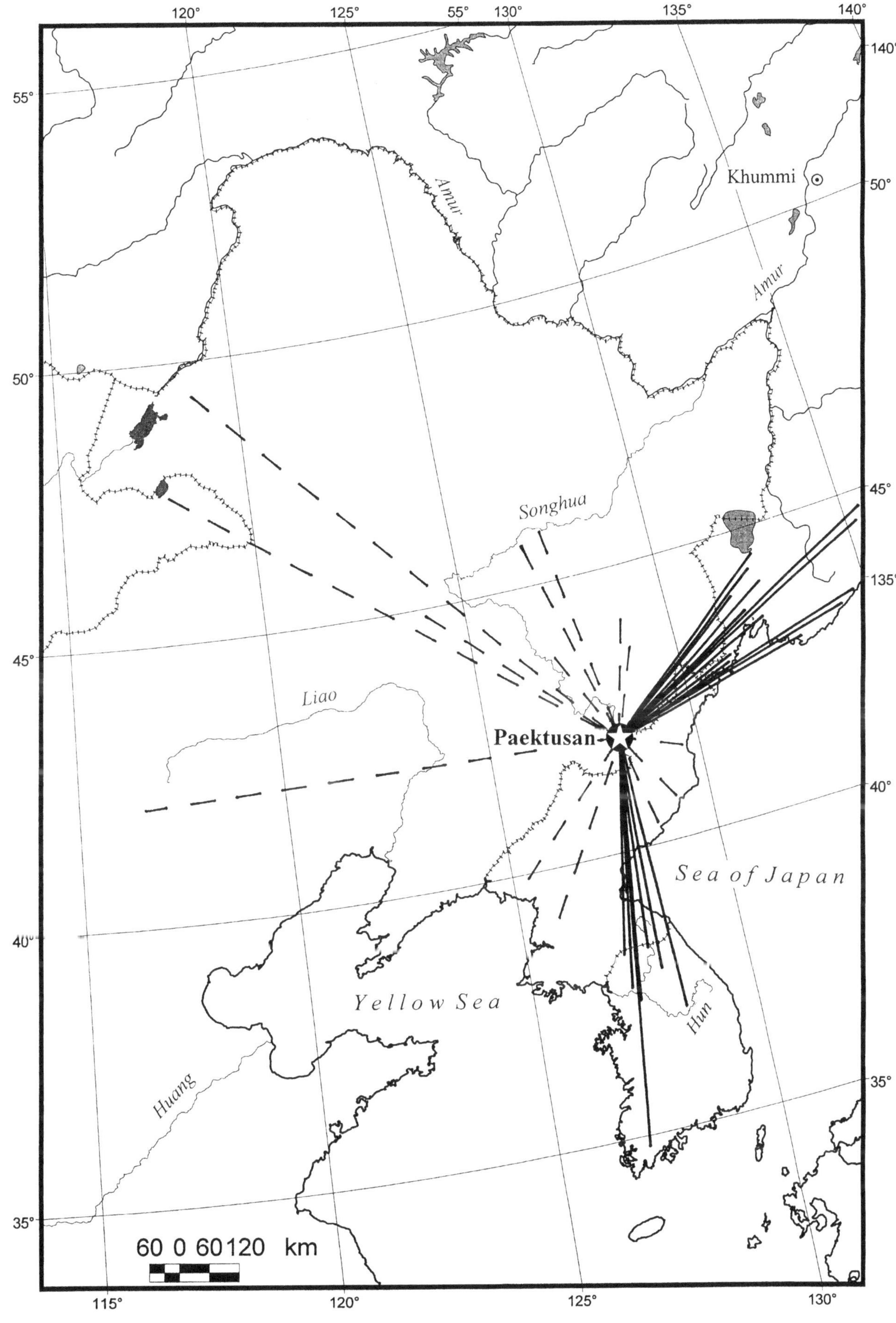

Figure 8.3. The distribution of obsidian from the Paektusan source to the Russian Far East, Korean Peninsula, and adjacent regions of Northeast Asia (solid lines – established routes; dashed lines – assumed routes)

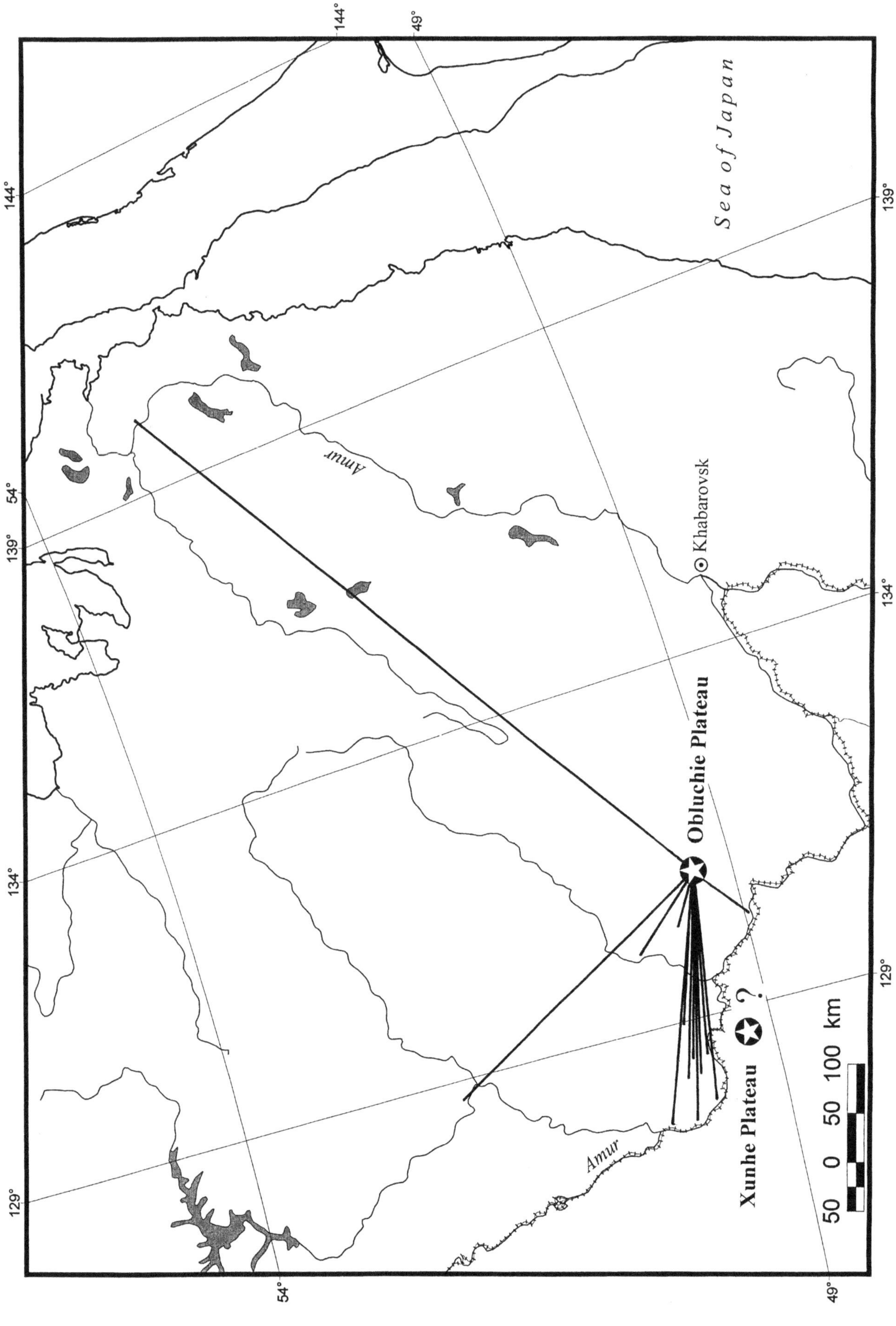

Figure 8.4. The distribution of obsidian from the Obluchie Plateau source in Amur River basin

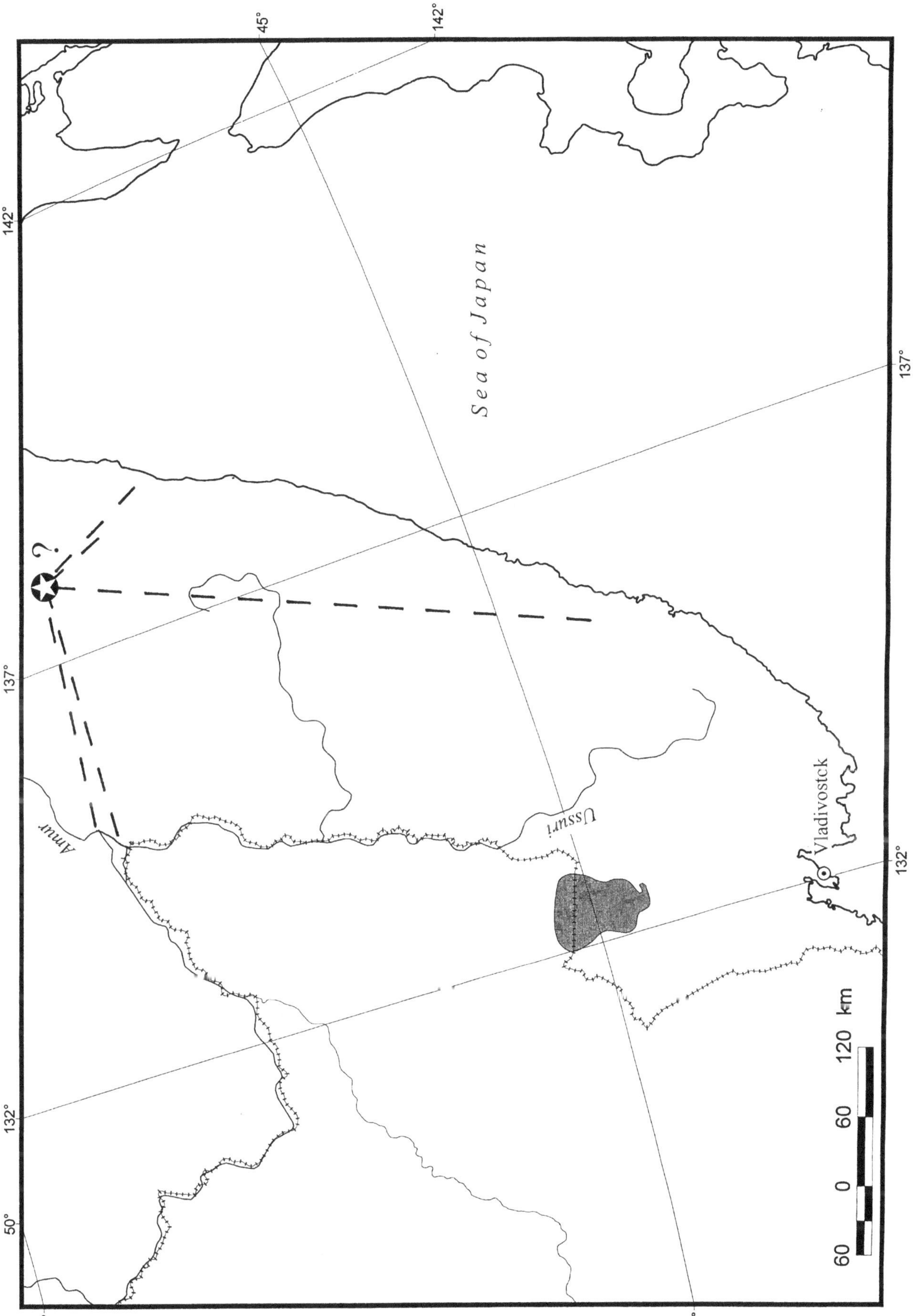

Figure 8.5. Location of archaeological sites with obsidian from the Samarga group

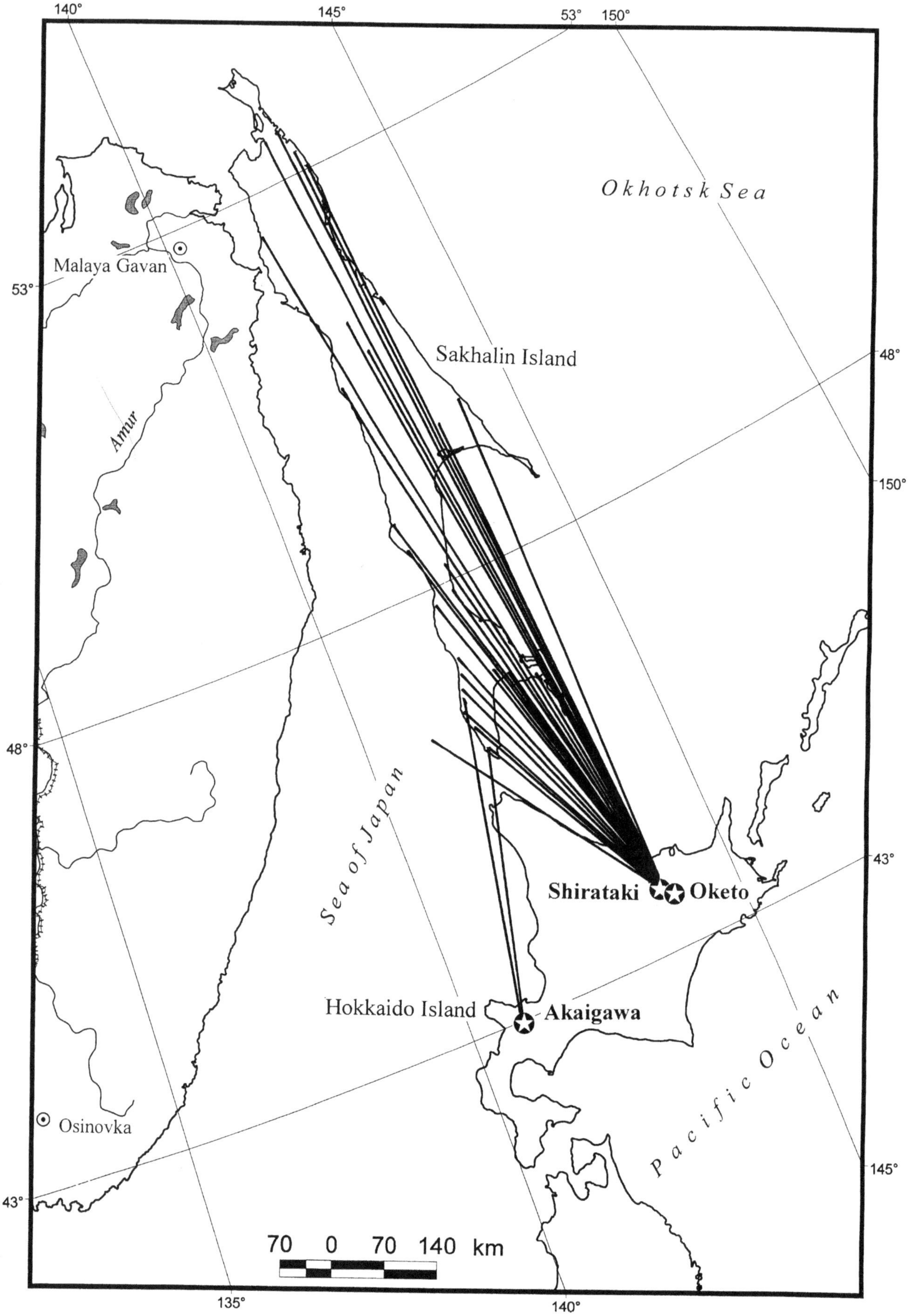

Figure 8.6. Prehistoric sites with obsidian artefacts at the Sakhalin Island and their obsidian sources

Shumshu islands) Kuriles the major sources of obsidian originated from the Kamchatka Peninsula.

This paper warrants some comments. Authors cite the obsidian source groups on Hokkaido Island, Oketo and Shirataki (Phillips and Speakman 2009, 1257: Figure 1; see also p. 1259) without reference to the original publications on this matter (Kuzmin and Glascock 2001, 2007; Kuzmin *et al.* 2002a). Also, authors stated that 'The movement of obsidian from Hokkaido is known to have covered large areas of Japan including the Sea of Japan rim area and into the Korean Peninsula (Izuho and Sato, 2007; Kim *et al.*, 2007).' (Phillips and Speakman 2009, 1261). In fact, Kim *et al.* (2007b) do not provide any data in favour of Hokkaido obsidian transportation to the Korean Peninsula. There is no other direct evidence of the import of Hokkaido obsidian to mainland Northeast Asia (e.g., Kuzmin 2006a; Kuzmin and Glascock 2007; also see below).

Korean Peninsula

Progress in obsidian source studies has been achieved during the last few years in Korea, in relation to previous attempts that were limited in terms of the approach and the amount of data collected (see review: Kuzmin 2006a). Geochemical analysis of obsidian artefacts was conducted at several Palaeolithic and Neolithic sites in the central and southern parts of the peninsula; politically these sites are in the territory of the Republic of Korea (South Korea) (Figure 8.7). The NAA and PIXE analyses for the Upper Palaeolithic sites of Hahwage-ri, Janghung-ri, Hopyung, and Sam-ri determined that Paektusan was the main source of obsidian artefacts (Kim *et al.* 2007a, 2007b; Popov *et al.* 2005). Analysis of obsidian at two other Palaeolithic sites, Suyanggae and Sangmuyong-ri, fails to match any of the known Korean and/or Japanese sources (Cho 2005).

The PIXE analyses of artefacts from other Upper Palaeolithic site, Shinbuk, shows that the second most important source of obsidian for this site is Koshidake located near the town of Imari, Saga Prefecture, Kyushu Island of Japan (Figure 8.8) (Kim *et al.* 2007b). This is an important discovery, because the ^{14}C dates for the Shinbuk site are ca. 18,500–25,500 BP (Kim *et al.* 2007b). In this case, we can establish obsidian transportation from the Koshidake source to the Korean Peninsula, not only during the Holocene as is testified by data for the Tongsamdong [Dongsamdong] site (Ono *et al.* 1992, 79) with an age range from ca. 6400 BP to ca. 3100 BP (Choe and Bale 2002), but also at the end of the Late Pleistocene.

Trace element analysis (Sm, La, Ce, Sc, Cs, Hf, Lu, Rb, Tb, and Sb) of artefacts from the Neolithic coastal sites in the southernmost part of the Korean Peninsula (Cho 2005), Yondaedo (ca. 6100 BP), Yokjido, Sangnodaedo (ca. 6600–4700 BP), Tongsamdong, and Songdo (ca. 5400 BP) (^{14}C dates are after Choe and Bale 2002), show that obsidian was transported across the Korea (Tsushima) Strait, where in northwestern Kyushu Island several obsidian sources are known; for example, Koshidake and Aso (Figures 8.7–8.8).

Along with progress in obsidian source studies on Korean Peninsula and neighbouring regions, misleading statements sometimes appeared in scientific literature. For example, Jia (2007, 148) associates the obsidian deposits at the Paektusan with a very large eruption that happened in the tenth millennium AD: '…obsidian in this area [Baitoushan Mountains – *Y.K.*] was formed by volcanic eruptions around 900 years ago (Jin, Bolu *et al.* 1994; Xu, Dongman *et al.* 1993). This obsidian resource was formed several thousand years after the Lower and Upper Yinggeling cultures and cannot be used for making their artefacts.' However, this statement is incorrect. The obsidian source at the Paektusan [Baitoushan] Volcano is about at least 65,000–100,000 years old (e.g., Popov *et al.* 2005; Sakhno 2007; Sakhno 2008, 23; Wang *et al.* 2003; Wei *et al.* 2003, 518–519; Wei *et al.* 2007), and it was readily available as raw material in the Upper Palaeolithic, ca. 25,000 BP (see above). The eruption around AD 970 or AD 1024 brought to surface mainly pumice and fine-grained tephra (e.g., Horn and Schmincke 2000; Wei *et al.* 2007; Y. V. Kuzmin, personal observations 2002, 2007), and some comendite and obsidian-like densely welded tuff (Wei *et al.* 2007, 321).

Discussion

The Spread of Obsidian from the Paektusan Source

Significant advances in the study of obsidian sources in the Russian Far East and the Korean Peninsula have permitted us to generate new data on the use of the Paektusan source in prehistory. For the Korean Peninsula it is now possible to establish the movement of obsidian from the Paektusan source toward the extreme south of the region (Figure 8.3), beginning in the Upper Palaeolithic at ca. 25,500 BP (Kim *et al.* 2007b). Previously, the earliest evidence of obsidian from Paektusan in the central Korean Peninsula was known from the Janghung-ri site, dated to ca. 24,000 BP (Popov *et al.* 2005).

As for the Russian Far East, the spread of obsidian from the Paektusan source to Primorye Province is now well-documented. Initially, data on the possible transportation of obsidian from Paektusan to the Upper Palaeolithic site of Ustinovka 1 in the eastern part of Primorye were published in the late 1980s (Vasilievsky and Gladyshev 1989, 101–102), with analytical details released later (Kuzmin and Popov 2000, 158–159). However, uncertainty about the stratigraphic position of obsidian samples from this site does not allow us to make any definite conclusion about the age of the artefacts, although they could be as old as ca. 20,000 BP (e.g., Kuzmin 2006c). Obsidian exchange between Primorye Province and Paektusan may have begun ca. 11,800 BP, as data for the Ustinovka 6 site suggest.

Some information has been published about the spread of Paektusan obsidian further north of Primorye Province to the Amur River basin where it was identified at the Khummi site (Warashina *et al.* 1998) (Figure 8.4). This conclusion deserves special attention. In the site's description, Lapshina (1999, 46–49) lists two obsidian cores. The number of stone artefacts in the lower layer of

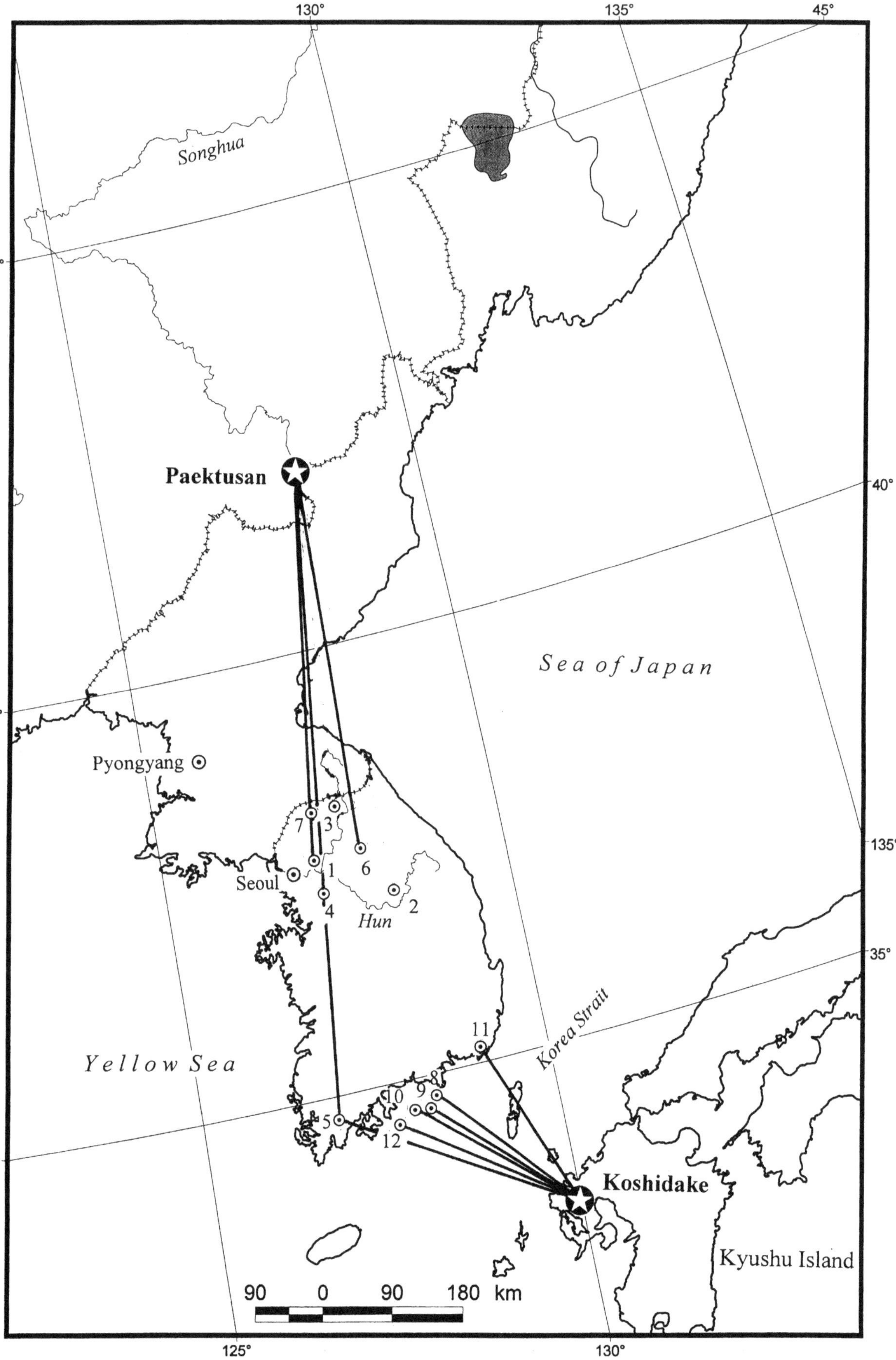

Figure 8.7. Archaeological sites in the central and southern Korean Peninsula with obsidian artefacts examined and the sources revealed (after Cho 2005; Kim et al. 2007b). 1 – Sites; 2 – Sources and their names. Sites: 1 – Hopyung; 2 – Suyanggae; 3 – Sangmuyong–ri; 4 – Sam–ri; 5 – Shinbuk; 6 – Hahwage–ri; 7 – Janghung–ri; 8 – Yondaedo; 9 – Yokjido; 10 – Sangnodaedo; 11 – Dongsamdong [Tongsamdong]; 12 – Songdo

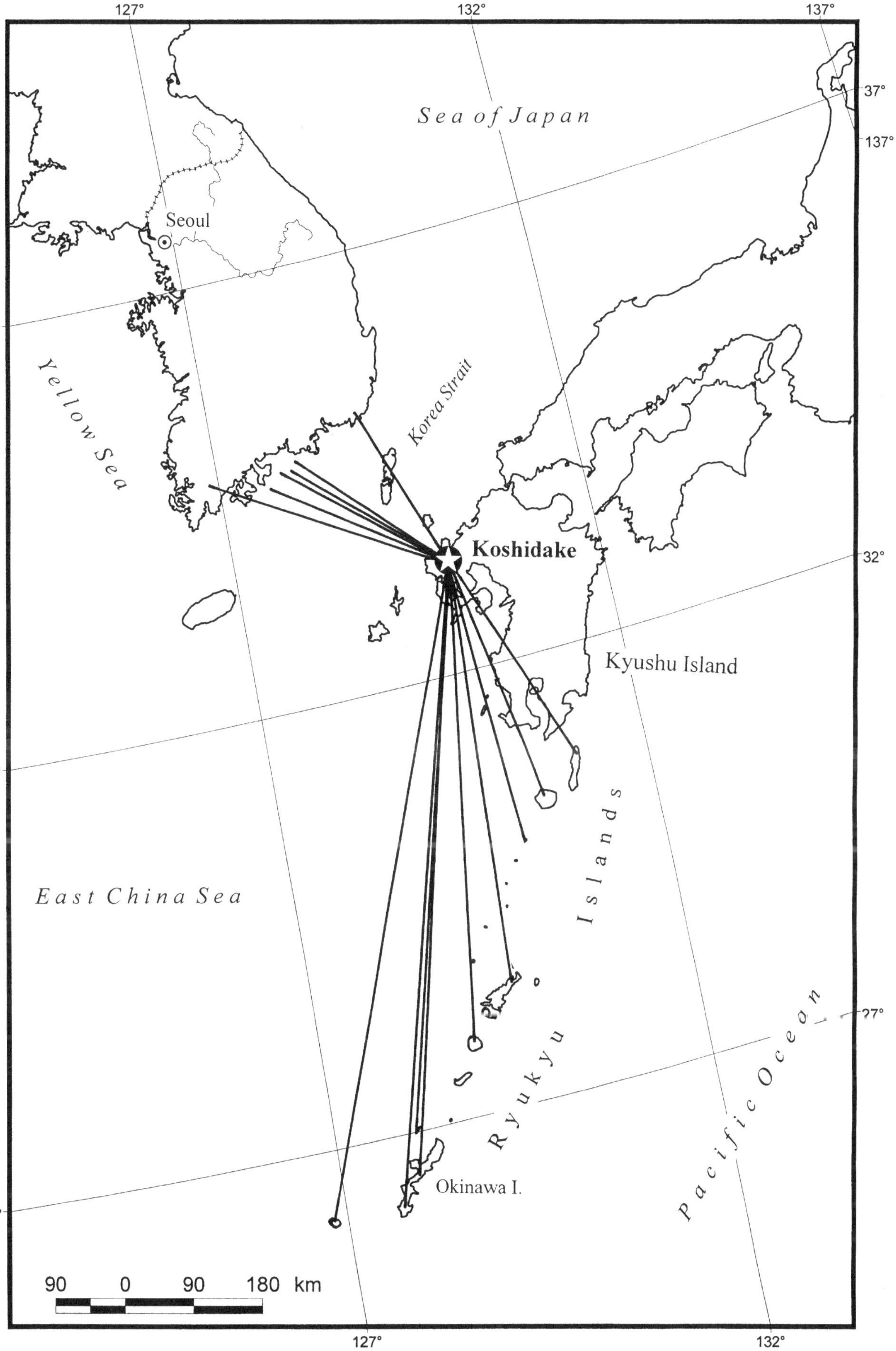

Figure 8.8. Spread of obsidian from the Koshidake source, Kyushu Island (Japan) (after Kim et al. 2007b; Obata et al. 2004, this volume; Ono et al. 1992; modified)

the Khummi site, associated with the Initial Neolithic, ca. 13,200–10,400 BP (Kuzmin 2006c), is 3772, and obsidian items constitute only 0.05% of the total assemblage. Unfortunately, the results of the obsidian source study were published without any details regarding the original chemical data (Warashina *et al.* 1998). Three artefacts from Khummi site were associated with two sources, Unknown A (one specimen) and Paektusan (two pieces). However, these results should not be accepted at face value. First, Warashina *et al.* (1998) provide no analytical data, and it is impossible to evaluate independently their conclusion. Second, as reference samples for the Paektusan source artefacts from the Kaineijodai site were used. This site is located near Onsong town in the northernmost part of North Hamgyong Province in the modern People's Democratic Republic of Korea (PDRK or North Korea). This artefact collection was obtained by Japanese archaeologists in the 1930s and is now stored at the Museum of Kyoto University (Matsushita 1998). Thus, the original 'geological' samples from the outcrops at the Paektusan Volcano were not used as reference material, in contrast to our research (Popov *et al.* 2005). It is therefore difficult to accept the conclusion by Warashina *et al.* (1998) on the spread of Paektusan obsidian to the lower course of the Amur River basin.

As for the distribution of obsidian into Northeast China, obsidian artefacts have been found in the Jilin and Heilongjiang provinces (sometimes referred to as Manchuria). Obsidian tools are known from some Palaeolithic sites (age estimate about 15,000–10,000 years ago; however, without ^{14}C dates) (Dong 1989; cited in Chen 2007, 17; Chen *et al.* 2009) and Neolithic settlements dated to ca. 5000–4000 BP (Liu 1995; Tan *et al.* 1995), and perhaps some later sites (Jia 2007, 147–148). These sites are close to the modern Chinese – North Korean border and in relatively close proximity to the Paektusan Volcano (distances up to 150–250km). Recently, an obsidian scraper was reported from the cultural layer of Xianrendong Upper Palaeolithic site in Jilin Province, dated to ca. 34,300 BP (Chen *et al.* 2007). If true, this is one of the earliest indications of obsidian use in Northeast Asia. The distance between this site and the Paektusan source is about 170km. A very large obsidian blade core (32 cm high) is described at the Xishan site (42°33′ N; 127°16′ E), with its age determined as the Upper Palaeolithic (Chen *et al.* 2009). The distance from site to possible source of obsidian at the Paektusan is about 90km. Recent progress in study of obsidian geochemistry at the Peking University (Beijing, China) provides hope that in the near future direct evidence for obsidian source usage in Manchuria will be available (Dr Xiaohong Wu and Ms Shuang Liu, personal communication 2008; Dr Peter W. M. Jia, presentation at the 19th Congress of the Indo-Pacific Prehistory Association, December 2009; see also Jia *et al.* 2010).

It is possible that obsidian artefacts are present in the prehistoric assemblages from the Songhua and Liao river basins and even further west, in Nei Monggol (Inner Mongolia) Province of China, as indicated by Obata (2003, 70) (Figure 8.3). The single solid evidence for obsidian artefacts in Mongolia proper is known from the Tamsagbulag cluster (approximate geographic coordinates are 47°15′ N, 117°30′ E), where two pieces of obsidian were found in a transitional Mesolithic–Neolithic context dated to ca. 5600 BP (Séfériadès 2004, 142–144). The distance between the Paektusan Volcano as the most probable source of obsidian in this part of Northeast Asia and Tamsagbulag is about 1000km as the crow flies (Figure 8.3). If true, it may represent one of the most remarkable examples of long-distance obsidian exchange in the later prehistory of continental Northeast Asia.

Obsidian Transportation from the Japanese Archipelago to Mainland Asia across Straits

Obsidian movement between southern part of Japanese Islands (Kyushu) and the Korean Peninsula through the Korea Strait originated at about 25,500 BP (Figure 8.8). Human migrations between the Korean Peninsula and Kyushu Island in the Upper Palaeolithic have also been suggested using archaeological data (e.g., Matsufuji 2003). Open water existed between Korea and Japan even at the height of the Last Glacial Maximum, ca. 20,000–18,000 BP, as marine geological research shows (e.g., Kim *et al.* 2000; Gorbarenko and Southon 2000; Lee and Nam 2003, 4; Lee *et al.* 2008; see coastline reconstructions: Kuzmin 1997; Lee *et al.* 2008). Some kind of water transport was probably used to cross the shrinking Korea Strait, about 15–20km wide. Previously, obsidian exchange between Korea and Japan was known only for the Holocene cultural complexes (e.g., Ono *et al.* 1992). Therefore, the exchange of obsidian between Japan and Korea was practiced for a long time, since at least the early Upper Palaeolithic (ca. 25,500 BP).

As for the northern part of the Japanese Archipelago, active exchange of obsidian between the sources on Hokkaido Island and the prehistoric complexes on neighbouring Sakhalin Island has been securely established (Kuzmin and Glascock 2007; Kuzmin *et al.* 2002b) (Figure 8.6). This movement of obsidian originated at ca. 19,400 BP and continued until ca. 800 BP. However, no obsidian from the Hokkaido sources was found at prehistoric sites in the Amur River basin (Figures 8.4 and 8.6). An "obsidian path" from Hokkaido through Sakhalin Island to the Amur River basin was suggested by Kimura (1995), who used obsidian from the Malaya Gavan site in the lower reaches of the Amur River as evidence of use of volcanic glass from the Hokkaido sources in the Asian mainland (Figure 8.6). There are some suggestions about the spread of obsidian in prehistory from Japanese sources to the mainland Russian Far East (see review: Sato 2004). These data need to be clarified because there is no reliable information about such movement of obsidian from Japan to the Russian Far East. For example, the chemical composition of obsidian from Malaya Gavan (samples KU279 and KU280, see Figure 8.9) in the Amur River basin does not match the Hokkaido sources (Kuzmin 2006a), and it is evident that Kimura's (1995, 1998) "obsidian path" hypothesis is not confirmed by chemical analyses. Furthermore, the suggestion about the possible correspondence of one of the obsidian artefacts from the Osinovka site in southern Primorye Province to the Hokkaido source (Kuzmin *et al.* 2002a, 512) was not

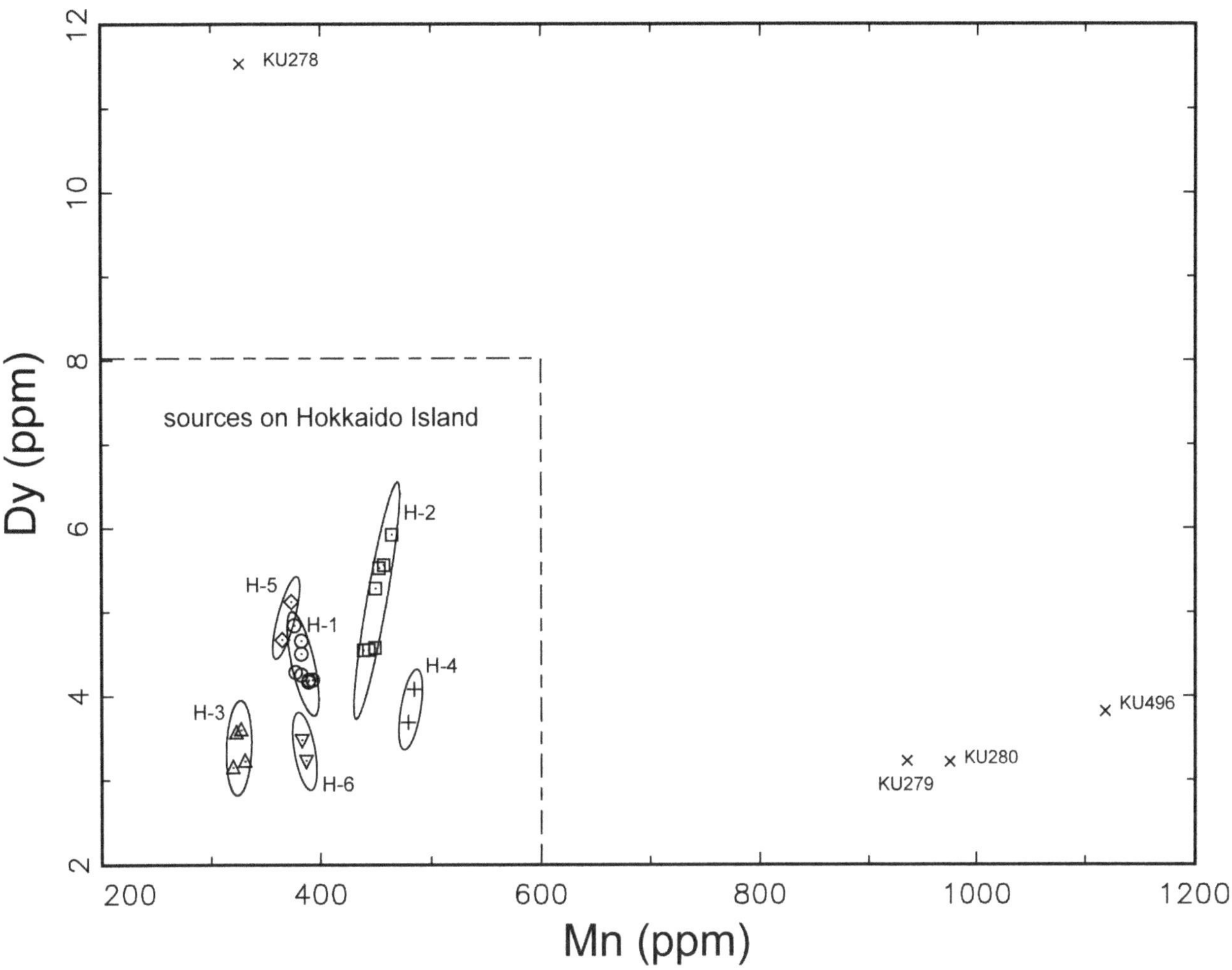

Figure 8.9. Chemical composition of obsidian from Malaya Gavan and Osinovka sites in comparison with the Hokkaido sources

supported by additional analyses of samples (KU278 and KU496; see Figure 8.9).

Data about the obsidian from Japanese sources (e.g., Oki, Shirataki, Akaigawa and others; see Sato 2004, 45) that was transported to the mainland Russian Far East cannot be supported by primary data, i.e. results of geochemical analyses. The first results from geochemical study of obsidian artefacts from some sites in Primorye and the Amur River basin, Ustinovka 1, Valentin-Peresheek, Kievka, Troitsa, Lva 2, Kalevala 1, Mustang, Rybak 1, and Malaya Gavan (Kuzmin and Popov 2000, 158–159), performed at the Institute for Atomic Energy, Rikkyo (St. Paul's) University (Tokyo, Japan) in 1987, were not properly interpreted and compared with known obsidian sources; at least this interpretation was never published (Prof. M. Suzuki, personal communication 2003). The conclusion about the spread of obsidian from Hokkaido sources to the Primorye (e.g., Kobayashi 2004, 59) and from other Japanese sources (see references: Sato 2004) is therefore based on insufficient data (see reviews: Kuzmin 2006a, 68; Sato 2004, 45–46).

Conclusion

Several long-distance obsidian exchange networks existed in prehistoric Northeast Asia. Some of them were located within particular regions, such as Honshu Island of Japan (see Tsutsumi, this volume). At least five networks, covering vast territories with distances to utilisation sites of up to 600–1000km, were centered around major obsidian sources: 1) the Basaltic Plateau in Primorye and Amur provinces, Russian Far East; 2) the Obluchie Plateau in Amur Province, Russian Far East; 3) Paektusan Volcano in continental Northeast Asia, including Primorye Province, the Korean Peninsula, and Manchuria; 4) Koshidake on Kyushu Island, spreading to the Korean Peninsula and the Ryukyu Archipelago; and 5) Shirataki–Oketo covering Hokkaido and Sakhalin Islands. In order to obtain obsidian, people must have travelled over mountains and crossed wide rivers and sea straits. The fact that ancient people began to move obsidian over long distances during the Upper Palaeolithic, at least at ca. 25,000 BP and possibly earlier, testifies that volcanic glass was a very important raw material. The identification of obsidian sources also serves as indisputable evidence of early human contacts and migrations within Northeast Asia. It is obvious that more research should be done in order to get better understanding of the creation and functioning of these exchange networks, and today we have good potential for it.

Acknowledgements

This paper is based on a presentation at the Symposium

"Crossing the Straits: Prehistoric Obsidian Exploitation in the Pacific Rim" that took place on 3 April 2005 as part of the Scientific Programme of the 70th Annual Meeting of the Society for American Archaeology, held at Salt Lake City, Utah (USA). I am grateful to my long-term colleagues, Drs Michael D. Glascock (University of Missouri, Columbia, MO, USA), Vladimir K. Popov (Far Eastern Geological Institute, Vladivostok, Russia), and M. Steven Shackley (University of California at Berkeley, CA, USA) for collaboration in data analysis and interpretation. The fieldwork assistance in Russia, China, and Japan was provided by Dr Hiroki Obata (Kumamoto University, Japan); Drs Vladimir K. Popov and Andrei V. Grebennikov (both from Far Eastern Geological Institute, Vladivostok); and Dr Masami Izuho (Tokyo Metropolitan University, Japan). Prof. Kim Jong Chan (Seoul National University, Korea) and Dr Cho Nam-Chul (Kongju University, Kongju, Korea) provided updated information on obsidian sources for Korean prehistoric sites. Drs Hiroyuki Sato (University of Tokyo, Japan), Hiroki Obata, Masami Izuho, and Mark E. Hall (Berkeley, CA, USA) helped with the translation of original Japanese publications. I am pleased to acknowledge Mr Makoto Tomii (Kyoto University, Japan) for the possibility to observe the obsidian collection of the Kaineijodai site. This research was supported by grants from the US Civil Research and Development Foundation (RG1-2538-VL-03) (2003–5); Korea Foundation (2002); Ministry of Education, Science, Culture and Sport of Japan (2003); US Fulbright Program (03-27672) (2004); and the Russian Foundation for Basic Sciences (RFFI, 99-06-80348, 02-06-80282, and 06-06-80258) (1999–2008).

References

Andreev, G. I., and Zh. V. Andreeva. 1965. 1959 Field Work of the Coastal Section of the Far Eastern Expedition in the Maritime Territory. *Arctic Anthropology* 3(1), 129–135.

Andreyev, G. I. 1964. Certain Problems Relating to the Shell Mound Culture. In *The Archaeology and Geomorphology of Northern Asia: Selected Works*, edited by H. N. Michael, 249–264. Toronto, University of Toronto Press.

Barker, A. W., C. E. Skinner, M. S. Shackley, M. D. Glascock, and J. D. Rogers. 2002. Mesoamerican Origin for an Obsidian Scraper from the Precolumbian Southwestern Untied States. *American Antiquity* 67, 103–108.

Brooks, S. O., M. D. Glascock, and M. Giesso. 1997. Source of Volcanic Glass for Ancient Andean Tools. *Nature* 386, 449–450.

Butylina, T. P. 1981. Some Aspects of the Stone Industry of the Ancient Inhabitants of South Primorye, USSR. In *Archaeological Studies of Pacific Stone Resources* (B.A.R. International Series 104), edited by F. Leach and J. Davidson, 45–54. Oxford, British Archaeological Reports.

Cann, J. R. 1983. Petrology of Obsidian Artefacts. In *The Petrology of Archaeological Artefacts*, edited by D. R. C. Kempe and A. P. Harvey, 227–255. Oxford, Clarendon Press.

Chen, C. 2007. Techno-Typological Comparison of Microblade Cores from East Asia and North America. In *Origin and Spread of Microblade Technology in Northern Asia and North America*, edited by Y. V. Kuzmin, S. G. Keates and C. Shen, 7–38. Burnaby, B.C., Canada, Archaeology Press.

Chen, Q., H. Zhao, and F. Wang. 2007. A Report of the 1993 Excavation of Xianrendong Paleolithic Site in Huadian, Jilin. *Renleixue Xuebao* 26(3), 222–236 (in Chinese with English Abstract).

Chen, Q., H. Zhao, and C. Wang. 2009. A Report on the Excavation at the Xishan Paleolithic Site, Xintunzi County, Fusong City. *Renleixue Xuebao* 28(2), 147–153 (in Chinese with English Abstract).

Cho, N.-C. 2005. Classification of Obsidian Artifacts Found in Korean Peninsula Based on the Chemical Composition, Texture and Magnetic Property. Unpublished PhD thesis, Kangwon National University, Ch'unch'on, Korea (in Korean with English summary).

Choe, C.-P., and M. T. Bale. 2002. Current Perspectives on Settlement, Subsistence, and Cultivation in Prehistoric Korea. *Arctic Anthropology* 39(1), 95–121.

Doelman, T., R. Torrence, V. Popov, M. Ionescu, N. Kluyev, I. Sleptsov, I. Pantyukhina, P. White, and M. Clements. 2008. Source Selectivity: An Assessment of Volcanic Glass Sources in the Southern Primorye Region, Far East Russia. *Geoarchaeology* 23, 243–273.

Dong, Z. A. 1989. Dabusu de Zishiqi [Microliths from Dabusu]. *Renleixue Xuebao* 8(1), 49–58.

Glascock, M. D. 2002. Obsidian Provenance Research in the Americas. *Accounts of Chemical Research* 35, 611–617.

Glascock, M. D., G. E. Braswell, and R. H. Cobean. 1998. A Systematic Approach to Obsidian Source Characterization. In *Archaeological Obsidian Studies: Method and Theory*, edited by M. S. Shackley, 15–65. New York and London, Plenum Press.

Gorbarenko, S. A., and J. P. Southon. 2000. Detailed Japan Sea Paleoceanography during the Last 25 Kyr: Constraints from AMS ^{14}C Dating and $\delta^{18}O$ of Planktonic Foraminifera. *Palaeogeography, Palaeoclimatology, Palaeoecology* 156, 177–193.

Gratuze, B. 1999. Obsidian Characterization by Laser Ablation ICP-MS and its Application to Prehistoric Trade in the Mediterranean and the Near East: Sources and Distribution of Obsidian within the Aegean and Anatolia. *Journal of Archaeological Science* 26, 869–881.

Habu, Y. 2004. *Ancient Jomon of Japan*. Cambridge, Cambridge University Press.

Hall, M. E., and H. Kimura. 2002. Quantitative EDXRF Studies of Obsidian Sources in Northern Hokkaido. *Journal of Archaeological Science* 29, 259–266.

Horn, S., and H.-U. Schmincke. 2000. Volatile Emission during the Eruption of Baitoushan Volcano (China/North Korea) ca. 969 AD. *Bulletin of Volcanology* 61, 537–555.

Jia, W. M. (P.). 2007. *Transition from Foraging to Farming in Northeast China* (B.A.R. International Series 1629). Oxford, BAR Publishing.

Jia, P. W., T. Doelman, C. Chen, H. Zhao, S. Lin, R. Torrence, and M. D. Glascock. 2010. Moving Sources: A Preliminary Study of Volcanic Glass Artifact

Distributions in Northeast China Using PXRF. *Journal of Archaeological Science* 37, 1670–1677.

Kim, J. C., M. Youn, S. C. Lee, C. C. Yun, Y. M. Song, J. Kang, H. R. Choi, M. Ashok, J. W. Kwak, and S. K. Kim. 2007a. Current Activities at the Seoul National University AMS Facility. *Nuclear Instruments and Methods in Physics Research B* 259, 57–61.

Kim, J. C., D. K. Kim, M. Yoon, C. C. Yun, G. Park, H. J. Woo, M.-Y. Hong, and G. K. Lee. 2007b. PIXE Provenancing of Obsidian Artefacts from Paleolithic Sites in Korea. *Bulletin of the Indo-Pacific Prehistory Association* 27, 122–128.

Kim, J.-M., J. P. Kenneth, B.-K. Park, D.-C. Kim, G.-Y. Kim, and B. Roark. 2000. Paleoceanographic Change during the Last Deglaciation, East Sea of Korea. *Paleoceanography* 15, 254–266.

Kimura, H. 1995. Kokuyouseki, Hito, Gijyutsu [Obsidian, Humans, and Technology]. *Hokkaido Kokogaku* 31, 3–63.

Kimura, H. 1998. Obsidian, Humans, Technology. In *Paleoekologiya Pleistotseha i Kultury Kamennogo Veka Severnoi Azii i Sopredelnyikh Territorii. Tom 2*, edited by A. P. Derevianko, 302–314. Novosibirsk, Izdatelstvo Instituta Arkheologii i Etnografii Sibirskogo Otdeleniya Rossiiskoi Akademii Nauk.

Kirch, P. V. 1991. Prehistoric Exchange in Western Melanesia. *Annual Review of Anthropology* 20, 141–165.

Kobayashi, T. 2004. *Jomon Reflections: Forager Life and Culture in the Prehistoric Japanese Archipelago*. Oxford, Oxbow Books.

Kononenko, N. A., A. V. Garkovik, A. M. Korotky, H. Kajiwara, and Y. Yokoyama. 2003. Stratigrafiya i Formirovanie Kulturnukh Otlozheny Stoyanok Ustinovka-7, 6 i 3 [Stratigraphy and Formation of Cultural Deposits at the Ustinovka 7, 6, and 3 Sites]. In *Okhotniki-Sobirateli Basseina Yaponskogo Morya na Rubezhe Pleistotsena–Golotsena*, edited by A. P. Derevianko and N. A. Kononenko, 49–73. Novosibirsk, Izdatelstvo Instituta Arkheologii i Etnografii Sibirskogo Otdeleniya Rossiiskoi Akademii Nauk.

Kuzmin, Y. V. 1997. Pleistocene Geoarchaeology of the Russian Far East: Updated Results. *Anthropologie* (Brno) [Special Volume 'Pleistocene Geoarchaeology', edited by J. Chlachula] 35, 131–136.

Kuzmin, Y. V. 2006a. Recent Studies of Obsidian Exchange Networks in Prehistoric Northeast Asia. In *Archaeology in Northeast Asia: On the Pathway to Bering Strait* (University of Oregon Anthropological Papers 65), edited by D. E. Dumond and R. L. Bland, 61–71. Eugene, University of Oregon.

Kuzmin, Y. V. 2006b. Palaeoeconomy of the Russian Far East (Stone Age Complexes). In *Archaeology of the Russian Far East: Essays in Stone Age Prehistory* (B.A.R. International Series 1540), edited by S. M. Nelson, A. P. Derevianko, Y. V. Kuzmin and R. L. Bland, 167–173. Oxford, BAR Publishing.

Kuzmin, Y. V. 2006c. Palaeoenvironment and Chronology. In *Archaeology of the Russian Far East: Essays in Stone Age Prehistory* (B.A.R. International Series 1540), edited by S. M. Nelson, A. P. Derevianko, Y. V. Kuzmin and R. L. Bland, 13–40. Oxford, BAR Publishing.

Kuzmin, Y. V., and M. D. Glascock. 2001. Source Analysis of Archaeological Obsidian on Sakhalin Island, Russian Far East. *Current Research in the Pleistocene* 18, 79–81.

Kuzmin, Y. V., and M. D. Glascock. 2007. Two Islands in the Ocean: Prehistoric Obsidian Exchange between Sakhalin and Hokkaido, Northeast Asia. *Journal of Island and Coastal Archaeology* 2, 99–120.

Kuzmin, Y. V., M. D. Glascock, and H. Sato. 2002b. Sources of Archaeological Obsidian on Sakhalin Island (Russian Far East). *Journal of Archaeological Science* 29, 741–749.

Kuzmin, Y. V., and V. K. Popov (eds). 2000. *Vulkanicheskie Stekla Dalnego Vostoka Rossii: Geologicheskie i Arkheologicheskie Aspekty* [Volcanic Glasses of the Russian Far East: Geological and Archaeological Aspects]. Vladivostok, Dalnevostochny Geologichesky Institut Dalnevostochnogo Otdeleniya Rossiiskoi Akademii Nauk.

Kuzmin, Y. V., V. K. Popov, M. D. Glascock, and M. S. Shackley. 2002a. Sources of Archaeological Volcanic Glass in the Primorye (Maritime) Province, Russian Far East. *Archaeometry* 44, 505–515.

Kuzmin, Y. V., A. V. Tabarev, V. K. Popov, M. D. Glascock, and M. S. Shackley. 1999. Geochemical Source Analysis of Archaeological Obsidian in Primorye (Russian Far East). *Current Research in the Pleistocene* 16, 97–99.

Lapshina, Z. S. 1999. *Drevnosti Ozera Khummi* [The Antiquities of the Lake Khummi]. Khabarovsk, Priamurskoe Geograficheskoe Obshchestvo.

Lee, E., S. Kim, and S. Nam. 2008. Paleo-Tsushima Water and its Effect on Surface Water Properties in the East Sea during the Last Glacial Maximum: Revisited. *Quaternary International* 176–177, 3–12.

Lee, E., and S. Nam. 2003. Freshwater Supply by Korean Rivers to the East Sea during the Last Glacial Maximum: A Review and New Evidence from the Korea Strait Region. *Geo-Marine Letters* 23, 1–6.

Liu, Z. 1995. Recent Neolithic Discoveries in Jilin Province. In *The Archaeology of Northeast China: Beyond the Great Wall*, edited by S. M. Nelson, 89–117. New York, Routledge.

Matsufiji, K. 2003. A Cultural Exchange between Kyushu Island and the Korean Peninsula in the Upper Palaeolithic Age. In *Palaeolithic Men's Lives and their Sites*, edited by Y.-J. Lee, 641–659. Seoul, Hakyeon Press.

Matsushita, H. 1998. Chosen Ikyo Hokudo Kainei Jyo Fukin no Sekki Nitsuite [Stone Tools near Kainei Castle in North Korea]. In *Toki no Kizuna Michi wo Tadoru*, edited by Ishizuki Kimio Sensei wo Shinobu, 439–464. Sapporo, Hon Kanko Iinkai.

Molyneaux, B. L. 2002. Exploring the Landscapes of Long-Distance Exchange: Evidence from Obsidian Cliffs and Devil Tower, Wyoming. In *Geochemical Evidence for Long-Distance Exchange*, edited by M. D. Glascock, 133–151. Westport, CT, Bergin and Garvey.

Obata, H. 2003. Kyoukutoni Okeru Kokuyouseki Shutsudoisekito Gensannchi Kenkyu [Study on the Prehistoric Obsidian Utilisation in Far East Asia – Review and Perspective]. *Sekki Gensanchi* 2, 67–88.

Obata, H., I. Morimoto, and S. Kakubuchi. 2004. Ryukyu Retto Shutsudo no Kokuyosekisei Sekki no Shutudo

Yousou ni Tuite [Sources of Prehistoric Obsidian Tools on the Ryukyu Islands, Japan]. *Sekki Gensanchi* 4, 101–136.

Okladnikov, A. P. 1965. *The Soviet Far East in Antiquity: An Archaeological and Historical Study of the Maritime Region of the USSR*. Toronto, University of Toronto Press.

Ono, A., H. Harunari, and S. Oda (eds). 1992. *Zukai: Nihon no Jinrui Iseki* [Atlas of Japanese Archaeology]. Tokyo, University of Tokyo Press.

Phillips, S. C., and R. J. Speakman. 2009. Initial Source Evaluation of Archaeological Obsidian from the Kuril Islands of the Russian Far East Using Portable XRF. *Journal of Archaeological Science* 36, 1256–1263.

Pollard, A. M., C. M. Batt, B. Stern, and S. M. M. Young. 2007. *Analytical Chemistry in Archaeology*. Cambridge, Cambridge University Press.

Pollmann, H.-O. 1993. *Obsidian im Nordwestmediterranen Raum. Seine Verbreitung und Nutzung im Neolithikum und Äneolithikum* (BAR International Series 585). Oxford, BAR Publishing.

Popov, V. K., Y. V. Kuzmin, M. D. Glascock, A. V. Grebennikov, N. A. Klyuev, A. N. Popov, I. Y. Sleptsov, S. Y. Batarshev, A. A. Krupianko, E. B. Krutykh, and S. S. Malkov. 2006a. Novye Dannye po Geokhimii Vulkanicheskikh Stekol iz Arkheologicheskikh Pamyatnikov Primorya (Rezultaty Rabot 2000–2003 gg.) [New Data on Geochemistry of Volcanic Glasses from Archaeological Sites in Primorye Province (Results of Research in 2000–3)]. In *Pyatye Grodekovskie Chteniya. Tom 1*, edited by S. V. Goncharova, 109–118. Khabarovsk, Khabarovsky Kraevedchesky Muzei.

Popov, V. K., Y. V. Kuzmin, I. Y. Shewkomud, A. V. Grebennikov, M. D. Glascock, N. N. Zaitsev, V. G. Petrov, B. V. Naumchenko, and A. K. Konopatsky. 2006b. Obsidian v Arkheologicheskikh Pamyatnikakh Srednego i Nizhnego Priamurya: Geokhimichesky Sostav i Istochniki [Obsidian in Archaeological Sites of the Middle and Lower Parts of Amur River Basin: Geochemical Composition and Sources]. In *Pyatye Grodekovskie Chteniya. Tom 1*, edited by S. V. Goncharova, 99–108. Khabarovsk, Khabarovsky Kraevedchesky Muzei.

Popov, V. K., V. G. Sakhno, Y. V. Kuzmin, M. D. Glascock, and B.-K. Choi. 2005. Geochemistry of Volcanic Glasses of the Paektusan Volcano. *Doklady Earth Sciences* 403, 254–259.

Renfrew, C. 1969. Trade and Culture Process in European Prehistory. *Current Anthropology* 10, 151–169.

Renfrew, C., J. E. Dixon, and J. R. Cann. 1968. Further Analysis of Near Eastern Obsidians. *Proceedings of the Prehistoric Society* 34, 319–331.

Sakhno, V. G. 2007. Chronology of Eruptions, Composition, and Magmatic Evolution of the Paektusan Volcano: Evidence from K-Ar, $^{87}Sr/^{86}Sr$, and $\delta^{18}O$ Isotope Data. *Doklady Earth Sciences* 412, 22–28.

Sakhno, V. G. 2008. *Noveishyi i Sovremenny Vulkanizm Yuga Dalnego Vostoka (Pozdnepleistotsen–Golotsenovy Etap)* [The Latest and Recent Volcanism in the Southern Far East (Late Pleistocene–Holocene Stage)]. Vladivostok, Dalnauka Press.

Sato, H. 2004. Prehistoric Obsidian Exploitation in the Russian Far East. In *Obsidian and Its Use in Stone Age of East Asia (Obsidian Summit International Workshop Meiji University Session)*, edited by M. Ambiru, K. Yajima, K. Sasaki, K. Shimada and A. Yamashina, 43–51. Tokyo, Meiji University Centre for Obsidian and Lithic Studies.

Séfériadès, M. L. 2004. An Aspect of Neolithisation in Mongolia: The Mesolithic–Neolithic Site of Tamsagbulag (Dornod District). *Documenta Praehistorica* 31, 139–149.

Shackley, M. S. 2005. *Obsidian: Geology and Archaeology in the North American Southwest*. Tucson, University of Arizona Press.

Suzuki, M., and K. Tomura. 1983. Basic Data for Identification the Source of Archaeological Obsidian by Activation Analysis and Discriminant Analysis. *St. Paul's Review of Science* 4(4), 99–110.

Tan, Y., X. Sun, H. Zhao, and Z. Gan. 1995. The Neolithic in Heilongjiang Province. In *The Archaeology of Northeast China: Beyond the Great Wall*, edited by S. M. Nelson, 118–144. New York, Routledge.

Tsutsumi, T. 2002. Origins of Pottery and Human Strategies for Adaptation during the Termination of the Last-Glacial Period in the Japanese Archipelago. In *The Origins of Pottery and Agriculture*, edited by Y. Yasuda, 241–262. New Delhi, Roli Books and Lustre Press.

Tykot, R. H. 2002. Chemical Fingerprinting and Source Tracing of Obsidian: The Central Mediterranean Trade in Black Gold. *Accounts of Chemical Research* 35, 618–627.

Tykot, R. H., and A. J. Ammerman. 1997. New Directions in Central Mediterranean Obsidian Studies. *Antiquity* 71, 1000–1006.

Vasilievsky, R. S., and S. A. Gladyshev. 1989. *Verkhny Paleolit Yuzhnogo Primorya* [The Upper Palaeolithic of Southern Primorye]. Novosibirsk, Nauka Publishers.

Wang, Y., C. Li, H. Wei, and X. Shan. 2003. Late Pliocene–Recent Tectonic Setting for the Tianchi Volcanic Zone, Changbai Mountains, Northeast China. *Journal of Asian Earth Sciences* 21, 1159–1170.

Warashina, T., T. Higashimura, H. Sato, and Z. S. Lapshina. 1998. Sekki Genzai no Sanchi Bunseki [Analysis of Lithic Artefact Sources]. In *Nihon Bunkazai Kagaku-kai dai 15 kai Taikai Kenkyu Happyo Yoshi*, edited by Nihon Bunkazai Kagaku-kai, 138–139. Nara, Nihon Bunkazai Kagaku-kai.

Wei, H., R. S. L. Sparks, R. Liu, Q. Fan, Y. Wang, H. Hong, H. Zhang, H. Chen, C. Jiang, J. Dong, Y. Zheng, and Y. Pan. 2003. Three Active Volcanoes in China and their Hazards. *Journal of Asian Earth Sciences* 21, 515–526.

Wei, H., Y. Wang, J. Jin, L. Gao, S.-H. Yun, and B. Jin. 2007. Timescale and Evolution of the Intracontinental Tianchi Volcanic Shield and Ignimbrite-Forming Eruption, Changbaishan, Northeast China. *Lithos* 96, 315–324.

Williams-Thorpe, O. 1995. Obsidian in the Mediterranean and Near East: A Provenancing Success Story. *Archaeometry* 37, 217–248.

Yacobaccio, H. D., P. S. Escola, M. Lazzari, and F. X. Pereyra. 2002. Long-Distance Obsidian Traffic in

Northwestern Argentina. In *Geochemical Evidence for Long-Distance Exchange*, edited by M. D. Glascock, 167–203. Westport, CT, Bergin and Garvey.

Yacobaccio, H. D., P. S. Escola, F. X. Pereyra, M. Lazzari, and M. D. Glascock. 2004. Quest for Ancient Routes: Obsidian Sourcing Research in Northwestern Argentina. *Journal of Archaeological Science* 31, 193–204.

Yamamoto, K. 1990. Space-Time Analysis of Raw Material Utilization for Stone Implements of the Jomon Culture in Japan. *Antiquity* 64, 868–889.

Chapter 9
Long-Distance Exchange of Western North American Obsidian

Carolyn D. Dillian

Abstract: *Preliminary analysis of a small sample of obsidian artefacts from archaeological contexts in the mid-Atlantic region of North America has yielded tentative trade connections to western US obsidian sources. This research has resulted in suggestions that cross-continental down-the-line exchange may have sporadically occurred in prehistory. However, the existing models of long-distance exchange are insufficient to explain the movement of materials over continental distances. Instead, casual, face-to-face, interactions between individuals may better illustrate the mechanisms operating to move obsidian from western sources to the eastern part of North America.*

Keywords: *Obsidian, Eastern North America, Provenance, Long-Distance Exchange*

Introduction

The movement of exotic materials over long distances is commonplace in our highly mobile society today. Yet we often assume that this exchange of material culture is a recent phenomenon, an assumption that may be erroneous. Given that human occupation of the New World has spanned more than 15,000 years, and that many Native American populations were mobile hunter-gatherers participating in extensive trade and exchange throughout much of prehistory, we must question whether long-distance movement of exotic items could have occurred.

The presence of western North American obsidian in Mississippian and Hopewellian contexts in the mid-continental United States has been clearly documented in the archaeological literature (Anderson *et al.* 1986; DeBoer 2004; Griffin 1965; Griffin *et al.* 1969; Hatch *et al.* 1990; Hughes 1992, 1995; Hughes and Fortier 1997; Lepper *et al.* 1998), and there is little doubt that long-distance exchange networks connecting unique raw material sources with population centres in these regions were in operation during late prehistory. However, the assertion that obsidian objects may have reached points further east has been largely anecdotal and is often received with well-deserved skepticism (Boulanger *et al.* 2007). This synthesis is an attempt to provide evidence for mid-Atlantic archaeological occurrences of artefacts from western US obsidian sources. Obsidian, though rare, was indeed a component of mid-Atlantic stone tool assemblages. Though several of the finds discussed in this paper were recovered in situations of less than ideal provenience, the quantity of obsidian artefacts from eastern United States archaeological sites suggests a pattern of volcanic glass movement across the continent. The exchange mechanism transporting obsidian to eastern US locales was likely one of casual face-to-face interaction resulting in the transfer of obsidian from one person to another. Through time, this served to move a small number of obsidian artefacts from western US sources to the eastern United States in prehistory.

Archaeological Theories of Exchange

Exchange can be defined as 'the spatial distribution of materials from hand to hand and from social group to social group' (Earle 1982, 2). However, exchange also reinforced and created social ties, promoted information sharing, and established and maintained positions of status. It commanded both economic and social roles within and between prehistoric societies. Exchange was a form of resource redistribution (Torrence 1986), provided a buffer against resource fluctuations (Arnold 1992, 77; Cohen 1981, 290), introduced and circulated prestige items (Appadurai 1986; Bennyhoff and Hughes 1987, 161; Hughes 1978, 53; Munn 1986), created communication and information networks, and served as a social tie between spatially and culturally distant peoples (Sahlins 1972, 186).

Archaeologically, exchange is visible through the spatial distribution of artefacts and stylistic patterns. Chemical characterisation techniques such as X-ray Fluorescence (hereafter – XRF), Proton-Induced X-ray Emission – Proton-Induced Gamma-ray Emission (hereafter – PIXE-PIGME), and Neutron Activation Analysis (hereafter – NAA), have been used to characterise and source lithic materials and ceramics (e.g., Deutchman 1980, 128–130; Ericson 1981; Summerhayes *et al.* 1998, 146–155; Tykot 1998, 76–79). Through chemical characterisation, it is possible to determine the geologic point of origin of materials such as obsidian, basalt, and ceramic tempers. Stylistic patterning also helps to identify exchange and culture areas in contact through the appearance of similar stylistic elements on locally manufactured objects (Fry 1980, 16; Plog 1978, 143–150).

Archaeologists study exchange to 'explain economic formations and their articulation with broader sociocultural contexts' (Earle 1982, 1). Exchange served to move objects and ideas through the landscape and across cultural boundaries. This movement was accomplished through mechanisms of reciprocal exchange, gifting, redistribution, feasting, marriage exchanges, and trade fairs. Exchange occurred as internal trade, between individuals *within* a social or geographic unit, or as external trade, between individuals of different social or geographic units (Renfrew 1984, 86). By recording patterns of the spatial distribution of archaeologically visible exchange objects, models have been developed to reconstruct prehistoric exchange networks. These models propose ways in which prehistoric people and societies interacted in the past. However, purely economic reasons are insufficient to fully comprehend exchange. According to Hodder (1982, 199), many models are 'inadequate because they fail to incorporate the symbolism of the artifacts exchanged'. As a result, alternative approaches that consider the social and cultural causes and effects of exchange provide additional insight into the many aspects of exchange in the past.

Exchange is often examined from an economic perspective; and the movement of obsidian in prehistoric times is frequently discussed within this context (e.g., Tykot 1992, 57). From this viewpoint, exchange served to move goods through space from producers to consumers. The specific properties of obsidian that made it a desirable raw material contributed to its economic importance (Tykot 1992, 57). For example, obsidian's sharp worked edge and perfect conchoidal fracture that is superior to chert or flint are often presented as reasons for the selection of obsidian over locally available materials.

Exchange provided a way of redistributing necessary or desired resources. In regions of patchy resources or where territorial circumscription limited travel to desired resources, exchange served to bring them to the consumer. In this manner, exchange functioned as a form of redistribution by moving goods throughout a region. According to Renfrew (1984, 91), 'in cases where there is also marked local diversity, with ecological variations within the region, a desire to obtain the products of a neighboring niche will inevitably promote exchange'. C. Renfrew also suggests that this will eventually lead to centralised exchange in the form of markets or trade fairs.

Exchange was an important method for mitigating resource fluctuations by serving as a form of social storage (O'Shea 1981, 173). Food resources during years of abundance were exchanged for more durable goods, which retained a culturally recognised value. During periods of resource stress, these valuables were in effect, 'cashed-in' for food resources from individuals or groups experiencing relative abundance. Obsidian may have been one form of durable social storage, and its use and exchange in many regions of the world where obsidian was available is testament to its importance as a valuable, yet often utilitarian material. However, obsidian's use as a prestige good would be more likely in geographic regions where it was not readily available, such as its documented occurrence in the mid-continental United States (Anderson *et al.* 1986; DeBoer 2004; Griffin 1965; Griffin *et al.* 1969; Hatch *et al.* 1990; Hughes 1992, 1995; Hughes and Fortier 1997; Lepper *et al.* 1998), and perhaps even further east (Dillian *et al.* 2007). These economic functions are extremely important; however, exchange did not occur in a vacuum and the social contexts of exchange are equally significant. Objects may also hold symbolic meanings that are separate from their utilitarian uses.

Many network studies of economics and prehistoric exchange have emphasised the trade of prestige items, rather than subsistence goods. Sahlins (1972, 218) believed that prestige items and subsistence goods circulated on different planes, in different exchange networks, and were not exchanged for each other except under dire circumstances. However, O'Shea (1981, 177) suggests that such exchange did occur, as prestige goods were one form of social storage. Rather than falling neatly into categories of subsistence goods and prestige goods, all potential objects for exchange existed on a continuum, in which all items could potentially be exchanged for any other item. Sahlins (1972) advocated a social model of exchange, and he suggested that exchange is both constrained and dictated by social boundaries and cultural etiquette.

Interestingly, these models of exchange suggest that the recipient of a product has at least some knowledge of its production. In other words, a consumer knows where that object is from and is familiar with the material. Yet in cases of continental-distance exchange, such as that proposed for obsidian found in the eastern United States, such familiarity may not have existed. Instead, obsidian may have been attractive simply because it was an unfamiliar material. In cases of exceptionally long-distance exchange, symbolic meanings of this raw material may in fact be more important than utilitarian properties. R. Torrence suggests that unmodified fragments of unusual lithic material 'are more easily linked to distant, unknown, unpeopled and mysterious places than are products that exhibit identifiable skills and/or forms which come from populated and possibly known places or individuals' (Torrence 2005, 366). In other words, there was an element of mystery and exoticism in objects from unknown locales.

Mid-Atlantic Obsidian Artefacts

More than one hundred years ago, archaeologists alleged that obsidian does appear, though rarely, in mid-Atlantic assemblages. In the late 1800s, Henry Chapman Mercer recorded in his diaries that 'A farmer near West Chester [Pennsylvania] showed me an arrowhead of volcanic glass or obsidian found in his field; and if his story is true and there was no trick, the Lenape must have got it from Mexico or the Yellowstone Park' (Mercer 1895, 2). In the early 1900s, Charles Conrad Abbott recorded the presence of rare obsidian artefacts in New Jersey sites. Abbott clearly stated that he discovered obsidian arrowheads, flakes, and scrapers in New Jersey that either came from Utah or Oregon, though he does not discuss how he arrived at this alleged geologic provenance (Abbott 1907, 57; Abbott 1908, 72–74; Abbott 1912, 28).

A review of Abbott's personal correspondence with Frederic Ward Putnam documented the recovery of obsidian specimens from archaeological contexts in New Jersey. Abbott was a well-known archaeologist who worked in New Jersey and adjoining states during the late 1800s. He published over 200 books and articles in leading scientific journals of the day, and was particularly known for his involvement in the debate on the origins of modern Native Americans. In addition to his writing between 1876 and 1889, Abbott worked for Putnam at the Peabody Museum of American Archaeology and Ethnology, Harvard University, though he was sporadically sending artefacts to the institution as early as 1872. Abbott's employment was unpaid; but he received money to purchase collections and for personal expenses incurred as part of his museum activities.

Dr Abbott's collections, which included more than 25,000 artefacts, were displayed at the Peabody Museum in the late 1800s; and he personally contributed to the arrangement and interpretation of the Delaware Valley material. Abbott's artefacts occupied the northern wall and long front table case of the third floor of the Peabody Museum (*Guide to the Peabody Museum,* Putnam 1898 as published in Williams 1973). The artefacts were arranged chronologically to show three periods of occupation including stone implements representing early Palaeolithic man from the glacial gravels at Trenton; argillite tools, flakes, and spear points, representing the end of the last Ice Age; and arrowheads, celts, pestles, axes and other objects showing more recent Delaware village occupations (Putnam 1898 as published in Williams 1973; Browman 2002).

Abbott was aware of the rarity of obsidian in eastern United States contexts. He wrote extensively of his work and collections to Putnam, who was Curator of the Peabody Museum. In his correspondence with Putnam, dated 11 April 1886, Abbott wrote: 'Now don't go on wild. I found an obsidian flake or piece of one, yesterday. The spot has a curious history, which I'll tell you if you stop over' (letter on file; Peabody Museum, Harvard University). Unfortunately, Abbott did not record the "curious history" of the site in his correspondence or personal diaries (diaries on file; Firestone Library, Princeton University). On 6 May 1886, Abbott further elaborated on his obsidian find in a letter accompanying a shipment of artefacts to the Peabody Museum. He wrote that 'I have also the pleasure of announcing the "surface find" of a fragment of obsidian, found on the brow of the plateau upon which my house stands and about 200 yards east of it. I believe it is the first incidence of this mineral being found in New Jersey.' (letter on file; Peabody Museum, Harvard University). Abbott's house was located in Hamilton Township, Mercer County, New Jersey.

Interestingly, Abbott was also aware of the danger of unprovenienced obsidian specimens being incorrectly attributed to archaeological sites in eastern United States. He stated in his correspondence that he received obsidian specimens from other collectors, but that they had little research value. He wrote to Putnam on 10 February 1887 that 'By the way. Berthoud has sent me some exquisite obsidian points, etc. (one mounted and poisoned) etc. etc. Do you want them, my dear boy? Or shall they be turned into scarf-pins, etc.' (letter on file; Peabody Museum, Harvard University).

Dr Abbott was later (in 1889) appointed the first Curator of the American Section at the University of Pennsylvania Museum of Archaeology and Anthropology, then called the Museum of Archaeology and Paleontology (Aiello 1976, Horan 1992). He assembled extensive Native American collections at the Museum through his own archaeological fieldwork and by supporting the work of other scholars, and created a Native American Exhibit in the Furness Building that rivaled those of other prominent museums of the time. Abbott resigned in 1894 over disagreements with Sara Yorke Stevenson, who served as Museum Secretary until 1905 (Kopytoff n.d.), although it has been noted that Abbott had little experience in museum curating and apparently neglected many of his duties (Hinsley 1985).

Abbott, writing in 1908 stated that obsidian artefacts found in the mid-Atlantic are from the far west and represent long-distance commerce, saying that: 'It is something more than barter. It is distinctly a feature of fixed conditions and that have been long-fixed and are generally known' (Abbott 1908, 73). He goes on to say 'That obsidian, that is not found east of the Mississippi, should find its way to the middle country, and from there occasionally to the Atlantic coast, is not an unthinkable proposition, however, improbable it may seem at first. The fact, however, remains that these foreign productions do occur here and that the Indian brought them. They have been found under circumstances that set aside all possibility of their presence being attributed to even the earliest European travelers ...' (Abbott 1908, 75).

Max Schrabisch, another mid-Atlantic archaeologist also working in the early 1900s, mentioned the occurrence of obsidian in private collections from south-central and south-western Sussex County, New Jersey (Schrabisch 1915, 8, 26). He reported a site on the Swartswood farm close to Trout Brook that contained obsidian artefacts, and a site referred to as the "Indian Spring" on the east bank of the Paulins Kill in the village of Lafayette as also yielding artefacts made out of obsidian (Schrabisch 1915, 47, 63).

More recently, other mid-Atlantic archaeologists have also published accounts of archaeological obsidian in the region. Dumont *et al.* (1974, 17) reported an obsidian biface in an Archaic context from the Rockelein I site, also in Sussex County, New Jersey. Kraft and Cavallo (1974, 17) have reported five temporally non diagnostic obsidian points discovered by avocational fossil hunters sifting stream sediments near Colts Neck, Monmouth County, New Jersey, and also noted the occurrence of other obsidian artefacts from the same county. Lenik (1985, 86–91) has published an obsidian fluted biface found at the Sheffield Playground site in Passaic County, New Jersey. This point, small and apparently unfinished, was discovered eroding out of the east bank of the Pompton River in Wayne, Passaic County. In discussing this specimen, Lenik (1985, 90) also mentioned the recovery of an obsidian core from a farm in Middlesex County, New Jersey. Another obsidian specimen has been reported at Hainesport, Burlington County, New Jersey (Stewart 1932, 50, 55, Plate 8).

These alleged occurrences of obsidian have not been examined; confirmation that these specimens are indeed obsidian and not a very high quality, glassy chert, has not been obtained. However, all of the occurrences mentioned above have been gleaned from published reports by professional archaeologists, and I hope that accurate identification of the raw material was provided [an example of misidentification of a chert as obsidian is published in Bello and Cresson (1995)]. Although reported by professionals, some of these pieces were excavated by avocational archaeologists. If necessary, geochemical characterisation can easily be used to distinguish between obsidian and other materials. However, the archaeological provenience of obsidian artefacts from the mid-Atlantic region of North America is, of course, also always questioned. Historic Euro-American transport of obsidian, perhaps as western curiosity or souvenirs, or poor museum records can result in erroneous assignment of obsidian specimens to mid-Atlantic archaeological sites. This danger was recognised as early as the late 1800s and early 1900s by Dr Abbott and others.

The occurrence of obsidian artefacts in archaeological sites in the middle United States has been well-documented; but prior to the initiation of this research (see Dillian *et al.* 2007), the proposition that obsidian may have reached points further east was frequently dismissed as exceptionally unlikely. The rare instances when obsidian has been documented in east coast of North America collections were usually attributed to errors in museum documentation, falsified information by collectors, or historic transport of obsidian artefacts. Yet new research suggests that obsidian was, though infrequently, but indeed *was* transported into the prehistoric northeast region (Dillian *et al.* 2007).

Figure 9.1. Obsidian projectile point from West Deptford Township, Gloucester County, New Jersey (for Figures 9.1–9.5 scale bars are in cm)

Figure 9.2. Two obsidian fragments from Neshanic Station, Hunterdon County, New Jersey

Geologic Provenance of Eastern United States Obsidian

In the United States of America and Canada, obsidian is readily available throughout a large portion of the west, including states and provinces of California, Oregon, Nevada, New Mexico, Arizona, Alaska, Idaho, Wyoming, South Dakota, Colorado, Utah, and British Columbia, and the Northwest Territories (Hughes 1986, Shackley 2005). Obsidian, as a non-crystalline substance, rapidly absorbs water and within a short geological period becomes perlite, a rock that cannot be used to produce chipped stone artefacts. Obsidian older than about 15,000,000 years is very rare. No obsidian sources have been found in the eastern USA, nor will they, since this region lacks late Tertiary or later volcanism. Yet obsidian from Obsidian Cliff in Yellowstone National Park (Wyoming State) has been recovered from Hopewell sites in Ohio State, illustrating its far-reaching importance in prehistory (Hatch *et al.* 1990).

A recent publication by Dillian *et al.* (2007) documented the geologic provenance of seven obsidian specimens alleged to have been recovered from eastern United States archaeological contexts (Figures 9.1–9.5). The study included obsidian specimens from states of New Jersey, New York, and Pennsylvania. Archaeological specimens were analysed at the Archaeological XRF Laboratory, University of California at Berkeley. XRF is one of the most commonly employed geochemical methods used for obsidian artefacts, and it provides accurate chemical characterisation assessments without requiring extensive sample preparation. A number of elements that XRF measures accurately just happens to be those incompatible elements that can be used to discriminate sources with a great degree of precision (Hughes and Smith 1993; Shackley 2005). Objects can be placed whole inside the sample chamber, providing they are small enough to fit within the closed chamber. Every effort is made to analyse a flat surface of the sample; yet irregular surface configurations do not hinder source assignments in most cases (Davis *et al.* 1998). Finally, XRF is a non-destructive technique, which is ideal for archaeological specimens. XRF can also easily distinguish between obsidian and

Figure 9.3. Obsidian secondary flake and projectile point from Monmouth County, New Jersey

Figure 9.4. Obsidian biface from Paterson, New Jersey

Figure 9.5. Obsidian retouched blade from Old Central Bridge, Schoharie County, New York

other cryptocrystalline silicates such as chert, jasper, or chalcedony, as well as between obsidian and synthetic glasses or other non-obsidian materials.

The trace element analyses were performed in the Archaeological XRF Laboratory, Department of Earth and Planetary Sciences, University of California at Berkeley, using a Spectrace/Thermo™ QuanX Energy Dispersive X-ray Fluorescence (hereafter – EDXRF) spectrometer. It is equipped with an air cooled Cu X-ray target with a 125 micron Be window, an X-ray generator that operates from 4–50 kV/0.02–2.0mA at 0.02 increments, using an IBM PC based microprocessor and WinTrace™ reduction software. The X-ray tube is operated at 30kV, 0.14mA, using a 0.05mm (medium) palladium (Pd) primary beam filter in an air path at 200 seconds livetime to generate X-ray intensity Kα-line data for elements titanium (Ti), manganese (Mn), iron (as Fe^T), thorium (Th), rubidium (Rb), strontium (Sr), yttrium (Y), zirconium (Zr), and niobium (Nb). Weight percent iron ($Fe_2O_3^T$) can be derived by multiplying part-per-million (hereafter – ppm) estimates by 1.4297×10^{-4}.

Trace element intensities were converted to concentration estimates by employing a least-squares calibration line established for each element from the analysis of international rock standards certified by the National Institute of Standards and Technology (NIST), the US Geological Survey (USGS), Canadian Centre for Mineral and Energy Technology, Japan Geological Survey, and the Centre de Recherches Pétrographiques et Géochimiques in France (Govindaraju 1994). Line fitting is linear for all elements but Fe where a derivative fitting is used to improve the fit for Fe and thus for all the other elements. Further details concerning the petrological choice of these elements in obsidian are available in Shackley (1995, 2005; see also Hughes and Smith 1993; Mahood and Stimac 1990). Specific standards used for the best fit regression calibration for elements Ti through Nb include G-2 (basalt), AGV-1 (andesite), GSP-1, SY-2 (syenite), BHVO-1 (hawaiite), STM-1 (syenite), QLO-1 (quartz latite), RGM-1 (obsidian), W-2 (diabase), BIR-1 (basalt), SDC-1 (mica schist), TLM-1 (tonalite), SCO-1 (shale), all US Geological Survey standards; BR-N (basalt) from the Centre de Recherché Pétrographiques et Géochimiques; and JR-1 and JR-2 (obsidian) from the Geological Survey of Japan (Govindaraju 1994). In addition to the reported values here, nickel (Ni), copper (Cu), zinc (Zn), and gallium (Ga) were measured; but these are rarely useful in discriminating glass sources and are not generally reported.

The data from the WinTrace™ software were translated directly into Excel for Windows software for manipulation and on into SPSS for Windows for statistical analyses. To evaluate these quantitative determinations, machine data were compared to measurements of known standards during each run. Table 9.1 shows a comparison between values recommended for RGM-1 as of 1 December 2002. RGM-1 is analysed during each sample run for obsidian artefacts to check machine calibration. BHVO-1 and/or AVG-1 is run when basalt or andesite artefacts are being analysed. XRF concentrations for selected trace elements of RGM-1 (n =

Table 9.1. Trace element values (in ppm) and geologic source assignments for obsidian specimens presented in the text

Specimen	*Ti*	*Mn*	*Fe*	*Zn*	*Ga*	*Rb*	*Sr*	*Y*	*Zr*	*Nb*	*Th*	*Ba*	*Source, State*
1 (a)	1220	602	8796	56	19	110	34	30	189	20	6	–	Blue Spring, California
2 (b)	861	468	7818	36	21	260	12	61	102	35	39	–	Black Rock, Utah
3 (b)	801	467	8040	47	21	267	8	59	101	23	34	–	Black Rock, Utah
4 (c)	1014	477	8219	66	22	453	12	50	140	63	81	–	Topaz Mountain, Utah
5 (c)	959	461	7709	41	20	424	13	49	133	64	88	–	Topaz Mountain, Utah
6 (d)	1005	214	9370	–	–	174	12	40	239	15	–	–	Reas Pass, Idaho (Yellowstone)
7 (e)	1266	361	10,214	–	–	82	101	23	122	9	–	1209	Malad, Idaho
RGM–1–H1	1500	306	13,274	34	18	154	113	27	223	8	14	–	Standard

Table 9.2. X-ray Fluorescence concentrations for selected trace elements for RGM-1 (in ppm)

Sample	*Ti*	*Mn*	*Fe*	*Rb*	*Sr*	*Y*	*Zr*	*Nb*
RGM-1 (Govindaraju 1994)	1600	279	12,998	149	108	25	219	9
RGM-1 (this study; n = 11)	1573 ± 67	313 ± 20	13,299 ± 82	148 ± 1	110 ± 2	22 ± 3	218 ± 4	6 ± 3

38 runs). The ± values represent first standard deviation computations for the group of measurements. All values are in ppm as reported in Govindaraju (1994) and this study (Table 9.2).

The results of this analysis showed that five different obsidian sources were represented in the published assemblage. These sources included Blue Spring obsidian from Modoc County, California; Black Rock obsidian from central Utah; Topaz Mountain obsidian from central Utah; Reas Pass obsidian from southeastern Idaho; and Malad obsidian from Idaho (Figure 9.6) (Dillian *et al.* 2007, 95). The variety of obsidian sources represented suggests a lack of patterning in obsidian procurement, yet questions certainly arise regarding the archaeological provenience of these alleged artefacts. The projectile point made from Blue Spring obsidian, for example, was recovered during construction monitoring in the summer of 1960 (Bello and Cresson 1998, 127–128). The alleged artefacts made of Utah obsidian sources are all from questionable archaeological provenience, including surface collection from an eroding slope and from donated collections at the Monmouth County Historical Association in New Jersey (Dillian *et al.* 2007, 96–97). The artefact made of Reas Pass obsidian has a better-documented archaeological provenience, having been found by Harold Arndt, the former Curator of Minerals in the Department of Geology at Bryn Mawr College (Bello 1997, 105; Dillian *et al.* 2007, 97). And, finally, the alleged artefact made from Malad obsidian was found during field walking of a surface scatter designated New York State site No. 9499 (Holloway 4) (Old Central Bridge; Figure 9.5). Its surface provenience is unfortunate, but recovery within a well-documented archaeological site suggests that it at least could have been associated with prehistoric occupation (Dillian *et al.* 2007, 98).

Two of these sources, Malad and Reas Pass, are known to have been traded across long distances during prehistory, suggesting that they could indeed represent prehistoric, continental-scale transport of lithic materials. Obsidian artefacts have been found in the middle United States, at sites near the Great Lakes and Mississippi River drainages. This connection between the middle United States and the Rocky Mountains has been well-documented in the archaeological literature (see, for example, DeBoer 2004). Furthermore, occurrences of copper, likely from the Great Lakes region, mica, unusual cherts, and other exotics found on archaeological sites in New Jersey provide evidence of trade and exchange networks that link the mid-Atlantic region to these larger western connections (Stewart 1989; Veit *et al.* 2004). Reas Pass obsidian, located in Idaho and part of the Yellowstone obsidian source group, was commonly used during Hopewell times and has been traded extensively into the middle United States. It would not be unlikely that this object could have been further exchanged into eastern United States contexts, given that exchange between Hopewell population centres and the eastern USA has been documented for other materials (Stewart 1989). The obsidian artefact with a Reas Pass provenance therefore fits with existing patterns of lithic material transport.

The Malad locality near Malad City in southern Idaho was the source for an obsidian retouched blade found in New York State. Malad obsidian was extensively traded during prehistory, and has been documented in sites as far south as Texas (Hester 2000), though no other apparent finds of this type of obsidian have been recorded for the eastern United States. The long-distance exchange of Malad obsidian also suggests that it could have been transported farther east as well and may fit with a model of long-distance exchange of lithic material.

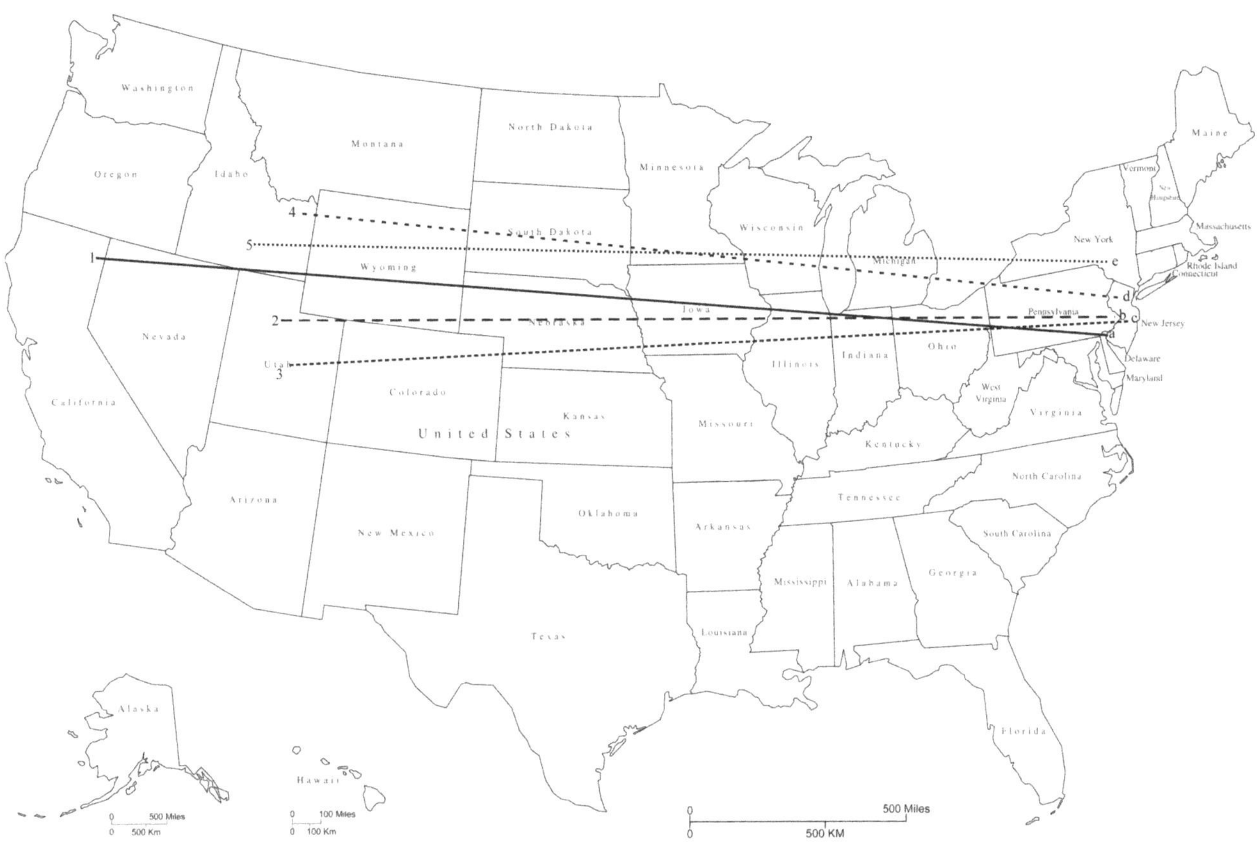

Figure 9.6. Map of obsidian findspots (letters) and source locations (numbers) for obsidian artefacts discussed in the text. Findspots: A – Site 28–Gl–309, West Deptford Township, New Jersey; B – Neshanic Station, New Jersey; C – Monmouth County, New Jersey; D – Catalogue No. 76–01–015A, Paterson, New Jersey; E – Site 9499 (Holloway 4), Old Central Bridge, Schoharie County, New York. Obsidian sources: 1 – Blue Spring; 2 – Black Rock; 3 – Topaz Mountain; 4 – Reas Pass; 5 – Malad

There have indeed been multiple published references to eastern United States obsidian finds that appear to have dubious archaeological provenience. A study of Hopewell obsidian artefacts was published by Gramly (2003). It focused primarily on a cache from Ohio State, but also included the chemical characterisation of four artefacts from the Genesee River valley in New York State, including three biface fragments and a single projectile point. Unfortunately, archaeological provenience for these artefacts is poor. They were collected by Howdie Lang (now deceased) and are allegedly Woodland Period in age, though no further provenience information is available (Gramly 2003). The artefacts were analysed by M. D. Glascock at the University of Missouri Research Reactor using XRF and NAA. The results of these analyses are compatible with the data obtained by EDXRF technique at the University of California at Berkeley (Glascock *et al.* 1998).

XRF and NAA revealed that four different obsidian sources are represented, including Blue Mountain obsidian in Modoc County, California; Newberry Crater (East Lake subsource) obsidian in central Oregon; Annadel obsidian north of San Francisco in California; and Bodie Hills obsidian located in eastern California, just south of Reno, Nevada (M. D. Glascock, personal communication 2005). All of these sources are close to 4800km from the Genesee River valley. It is important to note that M. D. Glascock strongly discounts the possibility that these artefacts could represent prehistoric transport of obsidian material. Specifically, M. D. Glascock (personal communication 2005) has emphasised the questionable archaeological provenience of the alleged obsidian artefacts.

Additionally from this region, one obsidian projectile point was recently published with provenience recorded as Livingston County, New York, also in the Genesee River valley (Saunders 2007). The projectile point morphologically resembled the Levanna type common to the Middle to Late Woodland period. However, XRF analysis of this artefact at the XRF Laboratory, University of California at Berkeley, revealed that it was made of Buck Mountain obsidian from the Warner Mountains of Modoc County, California (M. S. Shackley, personal communication 2007). This obsidian locality is a favourite of rockhounds and collectors as a beautiful mahogany source. Though it was indeed used in prehistory, the recovery of a projectile point of Buck Mountain obsidian in the eastern United States raises questions about more recent transport of volcanic glass specimens or recently knapped reproductions of prehistoric artefacts.

Analysis of an obsidian artefact purported to have been discovered in the Connecticut River valley of Vermont State has also been recently published (Boulanger *et al.* 2007); it was manufactured from the Double H obsidian source located near the border of Nevada and Oregon (Boulanger *et al.* 2007, 88). This biface, like those of the Genesee River valley in New York, has been discounted as more likely to have been the result of relatively recent artefact movement rather than representing prehistoric exchange. The relative obscurity of the obsidian source, lack of precise archaeological provenience, and western United States projectile point resembling the Elko Eared type, suggested that it may not be the product of prehistoric exchange (Boulanger *et al.* 2007, 88).

Models of Long-Distance Exchange

The lack of consistency in represented geological sources for eastern United States obsidian suggests that systematic exchange routes linking mid-Atlantic consumers to western sources were not in existence. Instead, periodic down-the-line exchange was likely the dominant mechanism for transporting obsidian material during prehistory. This type of exchange may perhaps best be termed *casual exchange,* in which objects were passed between individuals as gifts, unstructured trades, heirlooms, or even through petty theft. These kinds of interactions, however, imply face-to-face contact between the owner and recipient of the object. Such casual exchange interactions may result in very slow movement of obsidian or other exotics that may not be directional. In fact, objects may move into and out of communities in a relatively random pattern. Over time, however, this could result in the movement of objects across very large distances, including those of continental scale. Yet the time frame for such movement could be hundreds of years and multiple generations, implying little, if any, contact between the original procurer of an exotic item and the ultimate loss or discard of the object in an archaeological context.

Existing models of exchange in the mid-Atlantic region of North America do not sufficiently explain the patterns, or lack of patterning, apparent in obsidian exchange. Broad-based systems, which are as hand-to-hand, down-the-line systems of networked relationships, are typical of the informal exchange networks that operate between individuals, communities, and regions. These systems rarely contain specific production of items for exchange and can include trade of raw materials as well as finished goods (Stewart 1989, 1994, 2004). Focused networks frequently comprise the exchange of exotic items and imply the movement of mid-Atlantic peoples on long-distance, sporadic trading missions to the sources of these exotic goods. Focused networks, by definition, do not entail the personal contact and face-to-face interaction that is essential to a broad-based system (Stewart 1989, 1994, 2004).

The exchange networks implied by obsidian movement in the middle Atlantic region do not fit precisely with either broad-based or focused systems. As exotic items, obsidian artefacts should fall within a focused system, yet their movement across such a large geographic distance is not easily explained by direct trading missions to their source, as would be suggested in a focused system. This synthesis is not offering a systematic effort by inhabitants of northeastern North America to obtain obsidian. Instead, it proposes the exchange of exotic items as a much more informal process – a negotiation between individuals acting on personal impulse, rather than obtaining obsidian based on a culturally-determined system of value for this exotic material. Realistically, many people in the northeastern United States may not have ever even seen obsidian, which suggests that a fixed definition of worth and exchange was unlikely. Though the utility of such objects and the pragmatic use-value of such high quality raw material would have been apparent to any flintknapper, this synthesis suggests instead that the value of obsidian in the prehistoric northeastern North America was based on its foreign, exotic, and possibly even otherworldly aura, and the draw of that exotic appeal transcends time and space.

Acknowledgements

This research was supported in part by a Social Sciences Research Grant from Princeton University and a research grant from the Archaeological Society of New Jersey. I would like to thank Charles Bello and M. Steven Shackley for their work on the data collection and analysis (as published in Dillian *et al.* 2007); Jack Cresson; Megan Springate, Monmouth County Historical Association; Christina Rieth, New York State Museum; Joe Connelly; and Tamara Johnston, Bryn Mawr College, for providing samples for analysis. Thanks also go to Michael R. Gramly and Michael D. Glascock for contributing information on the alleged Genesee Valley obsidian artefacts. Any errors are my own.

References

Abbott, C. C. 1907. *Archaeologia Nova Caesarea. No. 1.* Trenton, NJ, MacCrellish and Quigley Printers.

Abbott, C. C. 1908. *Archaeologia Nova Caesarea. No. 2.* Trenton, NJ, MacCrellish and Quigley Printers.

Abbott, C. C. 1912. *Ten Years Diggings in Lenâpè Land 1901–1911.* Trenton, NJ, MacCrellish and Quigley Printers.

Aiello, L. 1976. A 19th-Century Odyssey: Dr. Charles Conrad Abbott's Search for Primitive Man in New Jersey. *The Journal of the Medical Society of New Jersey* 73, 23–26.

Anderson, D. C., J. A. Tiffany, and F. W. Nelson. 1986. Recent Research on Obsidian from Iowa Archaeological Sites. *American Antiquity* 51, 837–858.

Appadurai, A. 1986. Introduction: Commodities and the Politics of Value. In *The Social Life of Things: Commodities in Cultural Perspective,* edited by A. Appadurai, 3–63. Cambridge, Cambridge University Press.

Arnold, J. E. 1992. Complex Hunter-Gatherer-Fishers of Prehistoric California: Chiefs, Specialists, and Maritime Adaptations of the Channel Islands. *American Antiquity* 57, 60–84.

Bello, C. A. 1997. An Obsidian Biface Found "Near Paterson." *Bulletin of the Archaeological Society of New Jersey* 52, 105–106.

Bello, C. A., and J. H. Cresson. 1995. Island Beach Fluted Point Findspot (28Oc104). Archaeological Site Registration Form on file, Bureau of Archaeology/ Ethnology, New Jersey State Museum, Trenton, NJ.

Bello, C. A., and J. H. Cresson. 1998. An Obsidian Biface from the Lower Delaware Valley. *Bulletin of the Archaeological Society of New Jersey* 55, 127–128.

Bennyhoff, J. A., and R. E. Hughes. 1987. Shell Bead and Ornament Exchange Networks between California and the Great Basin. *Anthropological Papers of the American Museum of Natural History* 64(2), 79–175.

Boulanger, M., T. Jamison, C. Skinner, and M. Glascock. 2007. Analysis of an Obsidian Biface Reportedly Found in the Connecticut River Valley of Vermont. *Archaeology of Eastern North America* 35, 81–92.

Browman, D. L. 2002. The Peabody Museum, Frederic W. Putnam, and the Rise of U.S. Anthropology, 1866–1903. *American Anthropologist* 104, 508–519.

Cohen, M. N. 1981. Pacific Coast Foragers: Affluent or Overcrowded. In *Affluent Foragers: Pacific Coasts East and West* (Senri Ethnological Studies 9), edited by S. Koyama and D. H. Thomas, 275–295. Osaka, National Museum of Ethnology.

Davis, M. K., T. L. Jackson, M. S. Shackley, T. Teague, and J. H. Hampel. 1998. Factors Affecting the Energy-Dispersive X-Ray Fluorescence (EDXRF) Analysis of Archaeological Obsidian. In *Archaeological Obsidian Studies: Method and Theory,* edited by M. S. Shackley, 159–180. New York and London, Plenum Press.

DeBoer, W. R. 2004. Little Bighorn on the Scioto: The Rocky Mountain Connection to the Ohio Hopewell. *American Antiquity* 69, 85–107.

Deutchman, H. L. 1980. Chemical Evidence of Ceramic Exchange on Black Mesa. In *Models and Methods in Regional Exchange* (Society for American Archaeology Papers 1), edited by R. E. Fry, 119–134. Washington, D.C., Society for American Archaeology.

Dillian, C., C. Bello, and M. S. Shackley. 2007. Crossing the Delaware: Documenting Super-Long Distance Exchange in the Mid-Atlantic. *Archaeology of Eastern North America* 35, 93–104.

Dumont, E. M., J. Tolosky, and H. Tolosky. 1974. An Obsidian Biface from the Rockelein Site. *Bulletin of the Archaeological Society of New Jersey* 31, 17.

Earle, T. K. 1982. Prehistoric Economics and the Archaeology of Exchange. In *Contexts for Prehistoric Exchange,* edited by J. E. Ericson and T. K. Earle, 1–12. New York, Academic Press.

Ericson, J. E. 1981. *Exchange and Production Systems in Californian Prehistory: The Results of Hydration Dating and Chemical Characterization of Obsidian Sources* (B.A.R. International Series 110). Oxford, British Archaeological Reports.

Fry, R. E. 1980. Models of Exchange for Major Shape Classes of Lowland Maya Pottery. In *Models and Methods in Regional Exchange* (Society for American Archaeology Papers 1), edited by R. E. Fry, 3–18. Washington, D.C., Society for American Archaeology.

Glascock, M. D., G. E. Braswell, and R. H. Cobean. 1998. A Systematic Approach to Obsidian Source Characterization. In *Archaeological Obsidian Studies: Method and Theory,* edited by M. S. Shackley, 15–66. New York and London, Plenum Press.

Govindaraju, K. 1994. Compilation of Working Values and Sample Description for 383 Geostandards. *Geostandards Newsletter* 18 (Special Issue), 1–158.

Gramly, R. M. 2003. Obsidian Sourcing: An Example from an Hopewell Artifact Assemblage in Ohio. *The Amateur Archaeologist* 9(1), 35–39.

Griffin, J. B. 1965. Hopewell and the Dark Black Glass. *The Michigan Archaeologist* 11, 115–155.

Griffin, J. B., A. A. Gordus, and G. A. Wright. 1969. Identification of the Sources of Hopewellian Obsidian in the Middle West. *American Antiquity* 34, 1–14.

Hatch, J. W., J. W. Michels, C. M. Stevenson, B. E. Scheetz, and R. A. Geidel. 1990. Hopewell Obsidian Studies: Behavioral Implications of Recent Sourcing and Dating Research. *American Antiquity* 55, 461–479.

Hester, T. R. 2000. Archaeological Roots. *Discovery* 15(3). Available online at: http://www.utexas.edu/opa/pubs/discovery/disc2000v15n3/disc_archeology.html. Accessed 6 November 2005.

Hinsley, C. M. 1985. From Shell-Heaps to Stelae: Early Anthropology at the Peabody Museum. In *Objects and Others: Essays on Museums and Material Culture* (History of Anthropology. Vol. 3), edited by G. W. Stocking, Jr., 49–74. Madison, University of Wisconsin Press.

Hodder, I. 1982. Toward a Contextual Approach to Prehistoric Exchange. In *Contexts for Prehistoric Exchange,* edited by J. E. Ericson and T. K. Earle, 199–211. New York, Academic Press.

Horan, S. 1992. Charles Conrad Abbott Associations with the Peabody and the Museum of Archaeology and Paleontology. *Bulletin of the Archaeological Society of New Jersey* 47, 20–36.

Hughes, R. E. 1978. Aspects of Prehistoric Wiyot Exchange and Social Ranking. *Journal of California Anthropology* 5, 53–66.

Hughes, R. E. 1986. *Diachronic Variability in Obsidian Procurement Patterns in Northeastern California and South-Central Oregon* (University of California Publications in Anthropology 17). Berkeley, University of California.

Hughes, R. E. 1992. Another Look at Hopewell Obsidian Studies. *American Antiquity* 57, 515–523.

Hughes, R. E. 1995. Source Identification of Obsidian from the Trowbridge Site (14WY1), a Hopewellian Site in Kansas. *Midcontinental Journal of Archaeology* 20, 105–113.

Hughes, R. E., and A. C. Fortier. 1997. Identification of the Geologic Sources for Obsidian Artifacts from Three Middle Woodland Sites in the American Bottom, Illinois. *Illinois Archaeology* 9, 79–92.

Hughes, R. E., and R. L. Smith. 1993. Archaeology, Geology and Geochemistry in Obsidian Provenance Studies. In *Effects of Scale on Archaeological and Geoscientific Perspectives* (Geological Society of America Special Papers 283), edited by J. K. Stein and A. R. Linse, 79–91. Boulder, CO, Geological Society of America.

Kopytoff, I. n.d. The University of Pennsylvania Department of Anthropology: A Brief Historical Background. Electronic document, available online at: http://www.sas.upenn.edu/anthro/ppt_files/index.html. Accessed 9 March 2007.

Kraft, H. C., and J. Cavallo. 1974. Obsidian Points from New Jersey. *Bulletin of the Archaeological Society of New Jersey* 31, 17.

Lenik, E. J. 1985. *The Archaeology of Wayne.* Wayne, NJ, Wayne Township Historical Commission.

Lepper, B. T., C. E. Skinner, and C. M. Stevenson. 1998. Analysis of an Obsidian Biface Fragment from a Hopewell Occupation Associated with the Fort Hill (33HI1) Hilltop Enclosure in Southern Ohio. *Archaeology of Eastern North America* 26, 33–40.

Mahood, G., and J. A. Stimac. 1990. Trace-Element Partitioning in Pantellerites and Trachytes. *Geochimica et Cosmochimica Acta* 54, 2257–2276.

Mercer, H. C. 1895. The Red Man's Bucks County. Paper read by Henry Chapman Mercer before the Bucks County Historical Society, at Wolfe Rock, Buckingham, Pennsylvania, 16 July 1895. Manuscript on file; Bucks County Historical Society, Doylestown, PA (Mercer Archaeological Research Notes, Series 5, Folder 40).

Munn, N. 1986. *The Fame of Gawa: A Symbolic Study of Value Transformation in a Massim (Papua New Guinea) Society.* Durham, NC, Duke University Press.

O'Shea, J. 1981. Coping with Scarcity: Exchange and Social Storage. In *Economic Archaeology: Towards an Integration of Ecological and Social Approaches* (B.A.R. International Series 96), edited by A. Sheridan and G. Bailey, 167–183. Oxford, British Archaeological Reports.

Plog, S. 1978. Social Interaction and Stylistic Similarity: A Reanalysis. In *Advances in Archaeological Method and Theory. Volume 1*, edited by M. B. Schiffer, 143–182. New York, Academic Press.

Renfrew, C. 1984. *Approaches to Social Archaeology.* Cambridge, MA, Harvard University Press.

Sahlins, M. D. 1972. *Stone Age Economics.* Chicago, Aldine.

Saunders, G. 2007. I Found What? Where? *Indian Artifact Magazine* 26(3), 56–57.

Schrabisch, M. 1915. *Indian Habitations in Sussex County, New Jersey* (Geological Survey of New Jersey Bulletin 13). Union Hills, NJ, Dispatch Printing Company.

Shackley, M. S. 1995. Sources of Archaeological Obsidian in the Greater American Southwest: An Update and Quantitative Analysis. *American Antiquity* 60, 531–551.

Shackley, M. S. 2005. *Obsidian: Geology and Archaeology in the North American Southwest.* Tucson, University of Arizona Press.

Stewart, F. 1932. *Indians of Southern New Jersey.* Woodbury, NJ, Gloucester County Historical Society.

Stewart, R. M. 1989. Trade and Exchange in Middle Atlantic Prehistory. *Archaeology of Eastern North America* 17, 47–78.

Stewart, R. M. 1994. Late Archaic Through Late Woodland Exchange in the Middle Atlantic Region. In *Prehistoric Exchange Systems in North America*, edited by T. Baugh and J. Ericson, 73–98. New York, Plenum Press.

Stewart, R. M. 2004. Changing Patterns of Native American Trade in the Middle Atlantic Region and Chesapeake Watershed: A World Systems Perspective. *North American Archaeologist* 25, 337–356.

Summerhayes, G. R., J. R. Bird, R. Fullagar, C. Gosden, J. Specht, and R. Torrence. 1998. Application of PIXE-PIGME to Archaeological Analysis of Changing Patterns of Obsidian Use in West New Britain, Papua New Guinea. In *Archaeological Obsidian Studies: Method and Theory,* edited by M. S. Shackley, 129–158. New York and London, Plenum Press.

Torrence, R. 1986. *Production and Exchange of Stone Tools: Prehistoric Obsidian in the Aegean.* Cambridge, Cambridge University Press.

Torrence, R. 2005. Valued Stone – How So? In *Many Exchanges: Archaeology, History Community, and the Work of Isabel McBryde* (Aboriginal History Monographs 11), edited by I. MacFarlane, M. Mountain and R. Paton, 357–372. Canberra, Aboriginal History, Inc.

Tykot, R. H. 1992. Sources and Distribution of Sardinian Obsidian. In *Sardinia in the Mediterranean: A Footprint in the Sea* (Monographs in Mediterranean Archaeology 3), edited by R. Tykot and T. Andrews, 57–70. Sheffield, Sheffield Academic Press.

Tykot, R. H. 1998. Mediterranean Islands and Multiple Flows: The Sources and Exploitation of Sardinian Obsidian. In *Archaeological Obsidian Studies: Method and Theory,* edited by M. S. Shackley, 67–82. New York and London, Plenum Press.

Veit, R., G. Lattanzi, and C. Bello. 2004. More Precious than Gold: A Preliminary Study of the Varieties and Distribution of Pre-Contact Copper Artifacts in New Jersey. *Archaeology of Eastern North America* 32, 73–88.

Williams, S. (ed.) 1973. *The Selected Archaeological Papers of Frederic Ward Putnam.* New York, AMS Press, Inc.

Chapter 10
Trace Element Characterisation of Archaeologically Significant Volcanic Glasses from the Southern Great Basin of North America

Richard E. Hughes

Abstract: *From 1999–2003, this study collected and subjected to trace element analysis a number of archaeologically significant obsidians from southern Nevada (Great Basin of North America), focusing specifically on those glasses occurring within the Nellis Air Force Range (NAFR) and adjacent areas. This research resulted in the identification of four distinct varieties of obsidian contiguous and proximate to Obsidian Butte, as well as other chemically distinct varieties located to the east of the NAFR. The trace element fingerprints determined for these obsidians provided geologic counterparts for the many "unknown" varieties of obsidian heretofore identified in regional archaeological provenance studies. A brief example illustrates the utility and potential of employing these source-specific trace element signatures, in combination with analysis of time-sensitive projectile points, to reveal both change and continuity in prehistoric obsidian acquisition patterns.*

Keywords: *Obsidian, Provenance Study, Obsidian Butte Source, Great Basin of North America, Southern Nevada, Utah, California*

Introduction

Over the past 25 years, within the framework of sourcing the North American archaeological volcanic glasses many of the major obsidian sources of archaeological significance in California and parts of the Great Basin have been chemically characterised (e.g., Ambroz *et al.* 2001; Bowman *et al.* 1973; Ericson 1981; Hughes 1983, 1985, 1986a, 1986b, 1986c, 1988, 1989, 1990, 1994a, 2001a, 2005; Jack 1976; Jackson 1989; Lyons *et al.* 2001; Macdonald *et al.* 1992; Nelson 1984; Nelson and Holmes 1979; Shackley 1994; Stross *et al.* 1976), and archaeologists have employed these trace elements contrasts to investigate change and continuity in prehistoric obsidian conveyance over the past 10,000 years. Although many of the well-known and widely used obsidians have been analysed, there remain poorly studied regions – like the central and southern parts of the Great Basin of North America – where comparatively less is known about the inventory and trace element characteristics of archaeologically significant volcanic glasses.

Between 1999 and 2003 field reconnaissance, collection, and geochemical analysis using Energy Dispersive X-ray Fluorescence (hereafter – EDXRF) spectrometry of obsidian from one of these poorly known areas – Obsidian Butte, Nellis Air Force Range (hereafter – NAFR), State of Nevada (Figure 10.1), was conducted. The study also involved reconnaissance and collection at obsidian localities in areas adjacent to NAFR, such as the Nevada Test Site (hereafter – NTS). The goal of this research was not only to geochemically identify and distinguish among the different varieties of Obsidian Butte area volcanic glass (obsidian), but also to identify "new" chemical signatures of geologic obsidians located in surrounding areas so that archaeological research and management issues could be meaningfully addressed. This chapter describes the obsidian collection and processing methods, and laboratory analysis, interprets the results from samples obtained during 2001–3, synthesizes the results from an earlier study (Hughes 2001b), and provides a geochemical baseline for prehistoric obsidian source use studies at archaeological sites in the southern and central Great Basin.

Project Background

Since at least 1983 it has been known that geochemical variability exists among volcanic glasses (obsidians) within the Obsidian Butte area. As reviewed in Hughes (2001b, 29), X-ray Fluorescencc (hereafter – XRF) data generated by Thomas Jackson from cobbles/nodules collected from Tolicha Wash near the main highway adjacent to NAFR indicated the presence of perhaps three different varieties of obsidian. Subsequent research conducted later in 1983 indicated that two chemically distinct varieties of artefact-quality obsidian were present (which were termed Obsidian Butte Variety H-3 and Obsidian Butte Variety H-5), but a few nodules at both localities hinted at the existence of a third chemical variant.

In November 1999, reconnaissance was conducted and field collections made at Obsidian Butte with the goal of addressing the following research questions: 1) how many chemically distinct types of obsidian were present here; 2) what could be said about their spatial distribution; 3) how many of the geologic units contained artefact-quality obsidian; 4) what was the nature of utilisation of these obsidians; 5) were different obsidians used at different points in archaeological time; and 6) could the use-life histories of different chemical varieties of obsidian here be reconstructed. It was fully anticipated that all of these questions could not be answered on the basis of a single field and laboratory analysis (certainly not questions 5 and 6), but an excellent start was made addressing questions 1 through 4.

Fieldwork in 1999 included collection of samples from 14 geologic outcrops and quarrying loci from areas on

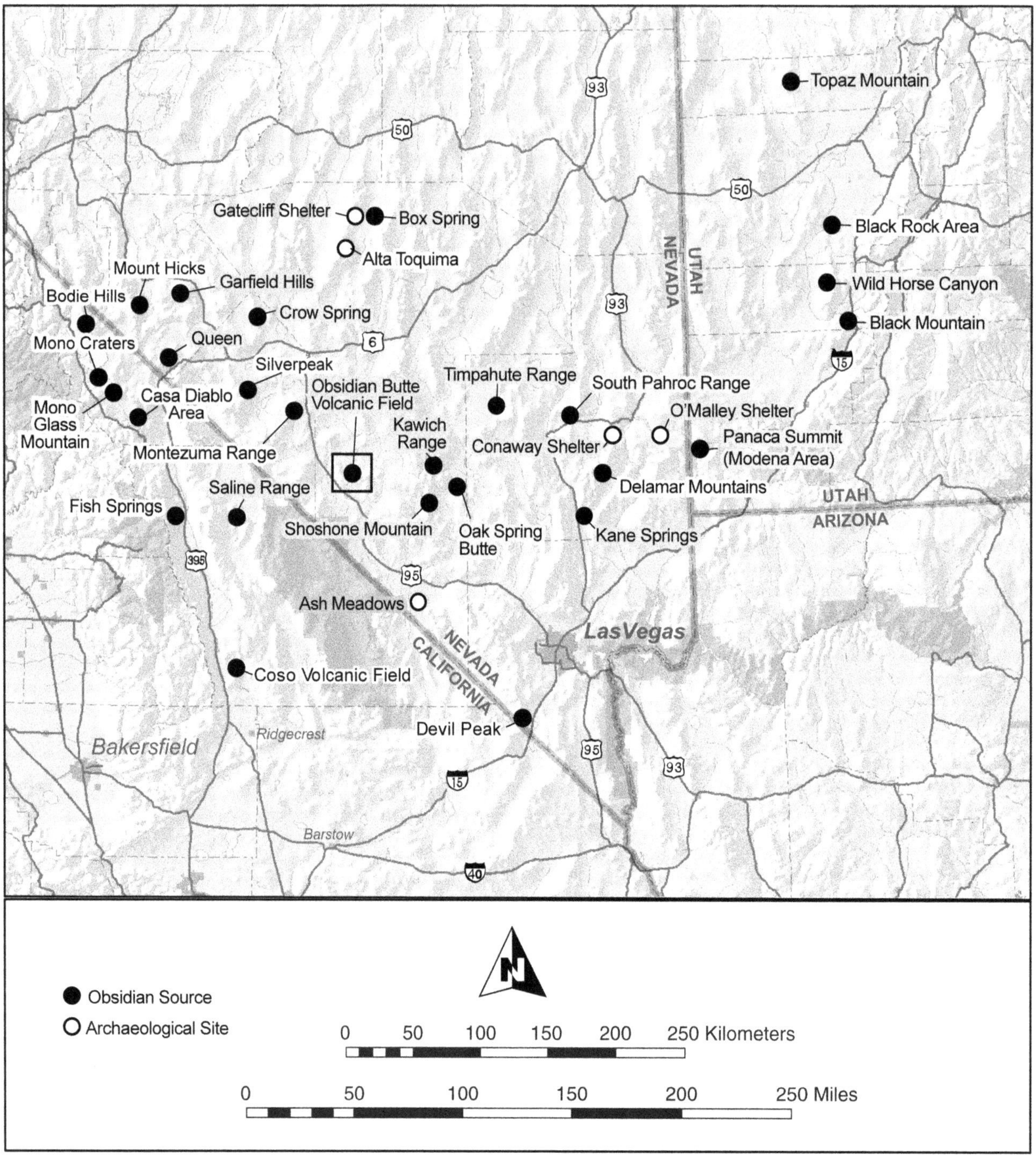

Figure 10.1. The southern Nevada study area, showing general location of regionally significant obsidian sources and selected archaeological sites (area framed around Obsidian Butte volcanic field shown in detail in Figure 10.2)

the southern and eastern flank of Obsidian Butte and subsequent analysis by EDXRF (Hughes 2001b). Analysis of 11 minor and trace elements from each of 14 sampling localities identified two (perhaps three) varieties of artefact-quality obsidian. The first, named Obsidian Butte after the nearby dominant physiographic feature, is represented at outcrops and exposures along the southern margin of Obsidian Butte and along the western margin of Tolicha Wash. This obsidian represents glass collected entirely from geologic Flow Unit 3D. The second chemical type, Airfield Canyon, occurs in the northern portion of Obsidian Butte, outcropping in the uplands and along the western rim of the northern reaches of Tolicha Wash immediately adjacent to Airfield Canyon. Obsidian of this chemical type is restricted to outcrops within geologic Flow Unit 3B. A third variety of obsidian was identified at one locality (H99-4), but a more detailed collection will be required to define it.

On the basis of the success of this study, a five-phase research design involving extensive field reconnaissance and laboratory analysis was developed and approved to expand the scope of the project beyond the immediate confines of NAFR. Descriptions and location information for 14 obsidian collection loci from the eastern and southern portions of Obsidian Butte have been presented (Hughes 2001b), and comparable information for the geologic obsidian samples collected in 2001 and 2003 appear in Hughes (2004, 4–14). Good overviews of the complex volcanic history and geology of the area appear in Ball (1907), Byers *et al.* (1976), Cornwall (1962, 1972), Eckel (1968), Ekren *et al.* (1971), and Lipman *et al.* (1966).

Sample Preparation and X-Ray Fluorescence Laboratory Conditions

Sample Preparation

Specimens from each collection location were cleaved and analysed as whole-rocks (not pressed powders). In most instances, ten samples were analysed from each locality, except in cases where the material was either marginal or clearly unsuitable for toolstone manufacture.

X-ray Fluorescence Laboratory Analysis

EDXRF analyses were conducted on geologic samples from localities H01-15 through H01-27 using a Spectrace™ 5000 (Tracor X-ray) spectrometer under conditions identical to those specified in Hughes (2001b, 35–36). Samples from localities H03-28 though H03-44 were analysed on a QuanX EC™ (Thermo Electron Corporation) EDXRF spectrometer equipped with a rhodium (Rh) X-ray tube, a 50kV X-ray generator, digital pulse processor with automated energy calibration, and a Peltier cooled solid state detector with 145eV resolution (FWHM) at 5.9keV. The X-ray tube was operated at 40.0kV using a 127mm palladium (Pd) primary beam filter in an air path to generate X-ray intensity data for elements rubidium (Rb Kα), strontium (Sr Kα), yttrium (Y Kα), zirconium (Zr Kα), and niobium (Nb Kα). Barium (Ba Kα) intensities were generated by operating the X-ray tube at 50.0kV with a 0.38mm copper (Cu) filter, while those for titanium (Ti Kα), manganese (Mn Kα), and total iron ($Fe_2O_3^T$) were generated by operating the X-ray tube at 30.0kV using a 0.025 mm Pd filter. Iron versus manganese (Fe Kα/Mn Kα) ratios were computed from data generated by operating the X-ray tube at 30.0kV with a 0.025mm Pd filter. Each subroutine was run at 200–300 deadtime-corrected seconds, with tube current (mA) scaled to the physical size of each specimen.

X-ray spectra were acquired and elemental intensities extracted for each peak region of interest. Matrix correction algorithms were then applied to specific regions of the X-ray energy spectrum to compensate for inter-element absorption and enhancement effects. Following these corrections, intensities were converted to concentration estimates by employing a least-squares calibration line established for each element from analysis of up to 30 international rock standards certified by the US Geological Survey and National Institute of Standards and Technology; the Geological Survey of Japan; the Centre de Recherches Pétrographiques et Géochimiques (France); and the South African Bureau of Standards. Further details pertaining to X-ray tube operating conditions and calibration appear in Hughes (1988, 1994a).

The trace elemental composition measurements presented herein are reported to the nearest part-per-million (ppm) to reflect the resolution capabilities of non-destructive EDXRF spectrometry (Table 10.1). The resolution limits of the EDXRF instrument for the determination of Ti is about 15ppm; Mn about 11ppm, $Fe_2O_3^T$ about 0.09%; Zn about 4ppm; Ga about 2ppm; for Rb about 4ppm; for Sr about 3ppm; Y about 2ppm; Zr about 4ppm; for Nb about 3ppm; and Ba about 11ppm (cf. Hughes 1988, 257; Hughes 1994a, 265).

Computed summary statistics can sometimes convey an inflated measurement precision, and the present data are no exception. For example, in Table 10.2 the computed sample standard deviation for Ti for Locality H01-27 is 6, yet this value is lower than the calibration-imposed limits of resolution (ca. 15ppm) for this element. Because measurement precision for any element cannot be less than calibration-imposed resolution limits, when sample standard deviations are greater than the calibration limits for an element (e.g., the value of 6 for Mn at H01-27 [Table 10.2] which is less than 11), the larger number is preferred as a more robust reflection of geochemical composition variation and measurement (analytical) error due to variations in sample size, surface and X-ray reflection geometry. Likewise discussed are the use of the coefficient of variation (CV, expressed in %), to measure element variability independent of sample means (Hughes 1984, 7–8; Hughes 1993a, 203; Hughes 1994a, 267). In the present study, CV% values also helped to isolate and evaluate such variability.

X-ray Fluorescence Analysis Results

Tables 10.1–10.4 present sample means, sample standard deviations, and coefficients of variation for the obsidian collection localities reported here. Data for other locations appear elsewhere (Hughes 2001b, 30–34, Tables 6–9; Hughes 2004, 4–14). The major chemical types (sensu Hughes 1998a, 104) identified are discussed in narrative form below and located in Figures 10.1–10.2.

The Obsidian Butte Area

The **Obsidian Butte** chemical type identified in Hughes (2001b), is represented principally at outcrops and exposures along the southern margin of Obsidian Butte and along the western margin of Tolicha Wash. The chemical signature of this obsidian is represented by glass collected from Flow Unit 3D (Hughes 2001b, Figure 4). Obsidian of this type was identified at sampling locations H99-1 through H99-3, H99-5 through H99-9, H99-14, and at locality H01-25.

Airfield Canyon obsidian occurs mainly in the northern portion of Obsidian Butte, outcropping in the uplands and along the western rim of the northern reaches of Tolicha Wash immediately adjacent to Airfield Canyon. Obsidian of this chemical type were recognised at outcrops within flow units 3B and 4 (Hughes 2001b, Figure 4). Sampling locations H99-10, H99-11, H99-12, and H99-13 were originally employed to geochemically define this glass type (Hughes 2001b); additional representatives of this variety

Table 10.1. Quantitative composition estimates for geologic obsidian samples of the Obsidian Butte chemical type, Obsidian Butte area, Nevada. Note for Tables 10.2–10.5: all trace element values (except Fe/Mn ratios) in ppm or weight percent composition ($Fe_2O_3^T$); X – sample mean, S.D. – standard deviation, CV% - coefficient of variation. Ten samples analysed from each locality

Element		H99-1	H99-2	H99-3	H99-5	H99-6	H99-7	H99-8	H99-9	H99-14	H01-25
Ti	X	763	750	745	751	762	770	735	757	754	750
	S.D.	37	48	47	33	34	23	40	38	39	53
	CV%	5	6	6	4	5	3	5	5	5	7
Mn	X	379	387	364	380	383	381	380	383	382	384
	S.D.	18	10	13	14	11	11	12	11	11	8
	CV%	5	3	4	4	3	3	3	3	3	2
$Fe_2O_3^T$	X	1.20	1.15	1.15	1.18	1.22	1.22	1.19	1.19	1.22	1.25
	S.D.	0.05	0.04	0.05	0.04	0.03	0.04	0.04	0.02	0.04	0.02
	CV%	4	4	4	3	3	3	3	2	3	2
Rb	X	166	168	166	168	168	170	166	172	168	167
	S.D.	9	5	6	7	5	7	7	5	6	6
	CV%	5	3	4	4	3	4	4	3	4	3
Sr	X	83	82	81	82	82	83	81	83	82	84
	S.D.	5	4	2	4	2	4	5	3	2	3
	CV%	6	5	3	5	2	5	6	4	3	3
Y	X	23	22	23	23	23	23	22	23	22	20
	S.D.	2	2	2	1	1	2	3	1	1	2
	CV%	9	9	9	6	4	9	14	5	4	8
Zr	X	142	140	140	141	139	142	140	141	141	141
	S.D.	6	5	3	4	3	5	5	4	4	3
	CV%	4	4	2	3	2	4	4	3	3	2
Nb	X	21	21	21	21	21	20	21	20	21	19
	S.D.	2	2	2	2	2	2	2	2	2	1
	CV%	10	10	10	10	10	10	10	10	8	6
Ba	X	505	502	491	493	470	478	495	481	476	460
	S.D.	20	16	14	21	17	18	26	18	22	19
	CV%	4	3	3	4	4	4	5	4	5	4
Fe/Mn		30–32	29–31	30–31	29–32	31–32	30–32	30–31	30–32	30–31	27–28

of obsidian were identified at H01-17, H01-18, H01-23, H01-26, and H01-27 localities.

The third major chemical variety of obsidian erupted from the Obsidian Butte volcanic field, **North Obsidian Butte**, were identified within the boundaries of flow units 3D and 4, but has a concentrated occurrence in the northern portion of Flow Unit 3B. Volcanic glasses of the North Obsidian Butte variety are distinguished from Obsidian Butte and Airfield Canyon chemical types on the basis of contrasts in Sr, Zr, and Ba composition (Figures 10.3–10.5). North Obsidian Butte glasses contain Ba in concentrations between ca. 600–800ppm, and Sr in concentrations between ca. 100–120ppm. Comparison of geochemical and artefact data show that this obsidian is the geologic and geographic counterpart of volcanic glass previously referred to as "Unknown C" (Hughes 2001b, 39, Note 4) identified at sites throughout southern Nevada (e.g., Hughes 1993b, 1994b, 1996, 1998b, 1998c). Examples of North Obsidian Butte volcanic glass were identified at localities H01-15, H01-19, H01-20, H01-21, H01-22, and H01-24.

There is trace element evidence for a fourth chemical variant, **Northern Domes Cluster**, on the basis of data from a group of small eroded domes (localities H03-36, H03-37, H03-38, and Geo-1) about 5.6km northwest of Obsidian Butte. Although their Sr composition overlaps with obsidian from the Timpahute Range it is higher than North Obsidian Butte; and it produced obsidian with higher Ba composition than either of the latter sources (Figure 10.5).

Discussion of Obsidian Butte Volcanic Field Obsidians

The four principal chemical varieties of obsidian identified

Table 10.2. Quantitative composition estimates for geologic obsidian samples of the Airfield Canyon chemical type, Obsidian Butte area, Nevada. Ten samples analysed from each locality except H01–23, H01–26, and H01–27 (n = 5 each)

Element		*H99-10*	*H99-11*	*H99-12*	*H99-13*	*H01-17*	*H01-18*	*H01-23*	*H01-26*	*H01-27*
Ti	X	640	600	610	622	750	607	752	811	812
	S.D.	20	24	29	22	63	39	57	57	6
	CV%	3	4	5	4	8	6	8	5	1
Mn	X	379	381	385	380	401	384	385	401	407
	S.D.	9	15	9	11	11	9	16	17	6
	CV%	2	4	2	3	3	2	4	4	2
$Fe_2O_3^T$	X	1.11	1.11	1.12	1.12	1.08	1.12	1.06	1.02	1.07
	S.D.	0.03	0.05	0.02	0.04	0.04	0.03	0 .04	0 .03	0.02
	CV%	3	5	2	4	3	2	4	3	2
Rb	X	173	179	178	176	155	173	156	150	147
	S.D.	6	7	2	6	4	6	7	3	4
	CV%	4	4	1	3	2	3	5	2	3
Sr	X	55	57	56	57	54	57	54	54	51
	S.D.	2	2	2	2	2	1	2	1	1
	CV%	4	4	4	4	3	3	4	2	3
Y	X	24	28	24	24	16	22	16	16	16
	S.D.	2	2	3	1	1	2	2	1	2
	CV%	9	8	13	5	5	8	10	7	10
Zr	X	121	123	122	122	123	123	123	141	134
	S.D.	3	5	1	5	2	2	3	3	1
	CV%	3	4	1	4	2	2	2	2	1
Nb	X	21	24	21	22	17	20	16	19	16
	S.D.	2	2	2	2	2	2	1	1	1
	CV%	10	8	10	9	11	10	8	4	8
Ba	X	286	310	291	314	300	277	301	327	331
	S.D.	16	15	19	17	17	15	22	14	7
	CV%	6	5	7	5	6	5	7	4	2
Fe/Mn		28–29	28–29	28–29	27–30	23–25	25–27	22–24	21–22	21–22

at, and around, Obsidian Butte, require some additional comment. While the details of the eruptive history of the obsidians here are not clear, the flows have been dated to 7.5–9.15 million years ago (Minor *et al.* 1993, 9). On the basis of the first phase of research (Hughes 2001b, 36–38), there appeared a compelling coincidence between chemistry and geography at Obsidian Butte; that is, obsidian of the Obsidian Butte variety was restricted to Flow Unit 3D, while Airfield Canyon glass was confined to Flow Unit 3B. However, as often happens when the scope of project sampling and analysis is extended, these clear-cut distinctions were not upheld by subsequent research. The results of this final study phase showed that, although obsidian of the Obsidian Butte and Airfield Canyon chemical types were *mainly* limited to flow units 3D and 3B, respectively, they both were documented in adjacent flows. This may not prove to be particularly important, since the geologists responsible for field mapping (Brian Hausback and Virgil Frizzell) were quite clear that flow unit subdivisions and boundaries were approximate (Hughes 2001b, 27).

The results of the current investigation show no one-to-one correspondence between mapped flow units; and trace element geochemistry suggests more complex geologic processes at work, or that more detailed geologic study is needed. Without delving too far afield, the observed geochemical results obtained from research in the Obsidian Butte volcanic field could be consistent with at least two scenarios, which are not mutually exclusive. First, an eruptive record in which obsidian emanated through vents which tapped different portions of the magma chamber. Depending on the time scale, the eruptive temperature(s) involved, and the degree to which the magma reservoir was drained/tapped/mixed, obsidians of slightly different chemical compositions could have been created (see Hughes

Table 10.3. Quantitative composition estimates for geologic obsidian samples of the North Obsidian Butte and North Domes cluster chemical types, Obsidian Butte area, Nevada. Ten samples analysed from each locality except H01–21 and H01–24 (n = 5 each), and Geo–1 (n = 15). North Domes cluster represented by samples from H0–36, H03–37, H03–38, and Geo–1; all others represent North Obsidian Butte chemical type

Element		***H01-15***	***H01-19***	***H01-20***	***H01-21***	***H01-22***	***H01-24***	***H03-36***	***H03-37***	***H03-38***	***Geo-1***
Ti	X	833	850	818	845	843	838	997	1030	971	972
	S.D.	39	43	51	51	57	51	35	75	58	70
	CV%	5	5	6	6	7	6	4	7	6	7
Mn	X	393	393	390	384	377	393	354	355	336	403
	S.D.	9	11	13	14	11	14	19	29	20	13
	CV%	2	3	3	4	3	4	5	8	3	3
$Fe_2O_3{}^T$	X	1.28	1.31	1.27	1.26	1.27	1.32	1.28	1.28	1.20	1.28
	S.D.	0.04	0.03	0.04	0.04	0.04	0.05	0.04	0.08	0.06	0.04
	CV%	3	2	3	3	2	4	3	6	5	3
Rb	X	160	161	162	160	160	157	146	143	135	150
	S.D.	5	6	4	4	5	4	4	7	5	6
	CV%	3	4	2	3	3	2	3	5	4	4
Sr	X	101	104	103	105	104	104	119	118	114	121
	S.D.	4	3	5	4	2	2	4	5	4	4
	CV%	4	3	5	4	2	2	3	4	3	3
Y	X	20	19	20	20	19	20	20	20	19	16
	S.D.	2	1	1	1	1	2	2	2	2	1
	CV%	8	4	5	7	6	9	10	9	10	8
Zr	X	158	159	158	161	161	159	157	157	152	163
	S.D.	5	4	3	4	5	5	4	4	4	5
	CV%	3	2	2	3	3	3	3	3	3	3
Nb	X	19	18	19	18	18	18	16	16	15	16
	S.D.	2	1	1	2	1	1	3	2	3	2
	CV%	9	6	5	8	7	6	17	11	18	12
Ba	X	602	639	625	632	659	641	760	765	782	779
	S.D.	35	26	34	37	43	38	35	35	72	42
	CV%	6	4	6	6	7	6	5	4	9	5
Fe/Mn		28–30	29–31	28–29	28–29	28–29	28–30	28–34	28–35	29–34	27–29

and Smith 1993). Alternatively, the observed differences in trace element chemistry among obsidians here could be attributable to crystal–liquid fractionation processes. The slight trending differences in trace element chemistry would reflect progressive fractionation through time. There is some evidence to support this latter alternative in the Obsidian Butte case. Macdonald *et al.* (1992, Figure 41) shows a fairly strong correlation between the elements Ba and Sr across a wide variety of tectonic settings. It can be seen from this figure that in high silica systems (i.e., SiO_2 > 70% by weight), Sr increases in concert with Ba. This is precisely what has been documented in the Obsidian Butte volcanic field, with Sr and Ba increasing from Airfield Canyon to Obsidian Butte, to North Obsidian Butte, to the Northern Domes Cluster north of Obsidian Butte (Figure 10.5). While this correspondence hardly closes the case, it does suggest that the chemical distinctions with the volcanic field may relate to a series of processes involving the evolution of the magma system 7.5–9.15 million years ago (see Hughes and Smith 1993; Macdonald *et al.* 1992, 41–42).

Other Chemical Varieties of Obsidian

Obsidian of the **Timpahute Range** chemical type occurs in an alluvial fan on the eastern side of Sand Spring Valley north of Tempiute Mountain (locality H03-28), and geochemically identical material occurs nearby at archaeological site 26LN706B. Because the primary outcrop(s) for this obsidian were not discovered during field reconnaissance, the most likely geographic landmark was used for chemical type nomenclature, although subsequent search for the primary eruptive unit may be rewarded to the east, or at the south end of the Worthington Mountains. Timpahute Range obsidian is superficially similar in trace element composition to North Obsidian Butte; but there

Table 10.4. Quantitative composition estimates for geologic obsidian samples of the Timpahute Range, Delamar Mountains, Oak Spring Butte, Kawich Range, Shoshone Mountain, and Devil Peak (east) chemical types, Nevada

Element		**H03-28**	**26LN706B**	**H03-30**	**H03-30A**	**H03-31**	**H03-32**	**H03-33**	**H03-35**	**H03-44**
Ti	X	716	592	635	661	1194	1202	768	1298	618
	S.D.	29	61	35	72	66	72	34	78	48
	CV%	4	10	6	11	6	6	4	6	8
Mn	X	436	528	204	217	1189	1181	628	303	509
	S.D.	22	24	10	14	69	73	47	14	22
	CV%	5	5	5	6	6	3	8	5	4
$Fe_2O_3^T$	X	1.41	1.39	1.30	1.33	3.97	3.91	2.30	1.40	0.93
	S.D.	0.05	0.07	0.05	0.09	0.18	0.19	0 .14	0.05	0.03
	CV%	3	5	4	7	5	5	6	3	4
Rb	X	181	187	172	178	177	173	171	191	181
	S.D.	5	9	9	10	7	8	7	4	7
	CV%	3	5	5	6	4	4	4	2	4
Sr	X	120	118	18	19	20	19	18	79	96
	S.D.	4	7	2	1	3	2	2	3	1
	CV%	3	6	9	6	13	11	12	4	1
Y	X	31	30	49	49	81	84	62	26	27
	S.D.	1	1	3	3	4	4	2	2	1
	CV%	4	2	6	6	5	4	3	7	4
Zr	X	153	155	168	171	955	960	686	212	100
	S.D.	5	5	4	5	16	33	41	5	3
	CV%	3	3	3	3	2	4	6	2	3
Nb	X	23	24	32	35	64	68	49	23	24
	S.D.	2	1	3	2	4	4	2	2	2
	CV%	10	5	8	6	6	6	4	10	10
Ba	X	578	592	97	108	6	9	15	667	293
	S.D.	18	23	8	8	7	8	14	45	15
	CV%	3	4	8	7	107	91	95	7	5
Fe/Mn		23–30	23–25	59–72	57–70	26–29	26–30	28–35	35–45	12–15

are important differences. In particular, Timpahute Range glass contains more Rb (ca. 30ppm), Y (ca. 10ppm), Mn (ca. 100ppm), Ti (ca. 200ppm), and $Fe_2O_3^T$ (ca. 0.15%), and less Ba than does North Obsidian Butte. Trace element data indicate that Timpahute Range is the geologic counterpart for obsidian previously referred to as "Unknown B", identified at numerous archaeological sites on Pahute and Rainier mesas in southern Nevada (Hughes 1993b, 1994b) [within NAFR (e.g., Hughes 1998b, sample nos. 20 and 26.1); 26CK4946 in the Lake Mead area (Hughes 1994c); 26CK5423 (Hughes 1997); and 26CK1139 (Hughes 2003)].

Delamar Mountains obsidian was defined on the basis of analysis of obsidian nodules from H03-30 and H03-30A, the former locality within Delamar Mountains proper, the latter from Coyote Spring Valley at the south end of the range. Delamar Mountains obsidian is quite distinctive geochemically, particularly in Sr depletion, modest (ca. 100ppm) Ba composition, and unusually high Fe/Mn ratio (due to relative depletion of Mn). Trace element measurements for this obsidian correspond with those of late moat rhyolites from the Kane Springs Wash Caldera (Novak 1984; Novak and Mahood 1986, 370–371). Trace element data indicate that Delamar Mountains is the geologic counterpart for obsidian previously referred to as "Unknown A", which has been identified at numerous archaeological sites in southern Nevada [e.g., 26CK4867 (Hughes 1995a, sample nos. 20 and 26.1); 26CK4946 in the Lake Mead area (Hughes 1994c); Pintwater Cave (Hughes 1996, referred to as "Unknown Variety 1"); 26CK5423 (Hughes 1997); Flaherty Rockshelter (Hughes 1999, referred to as "Unknown Variety 1"); and 26CK1139 (Hughes 2003)].

Ash-flow tuff obsidian from two localities (H03-31 and H03-32) immediately west of Oak Spring Butte was used to

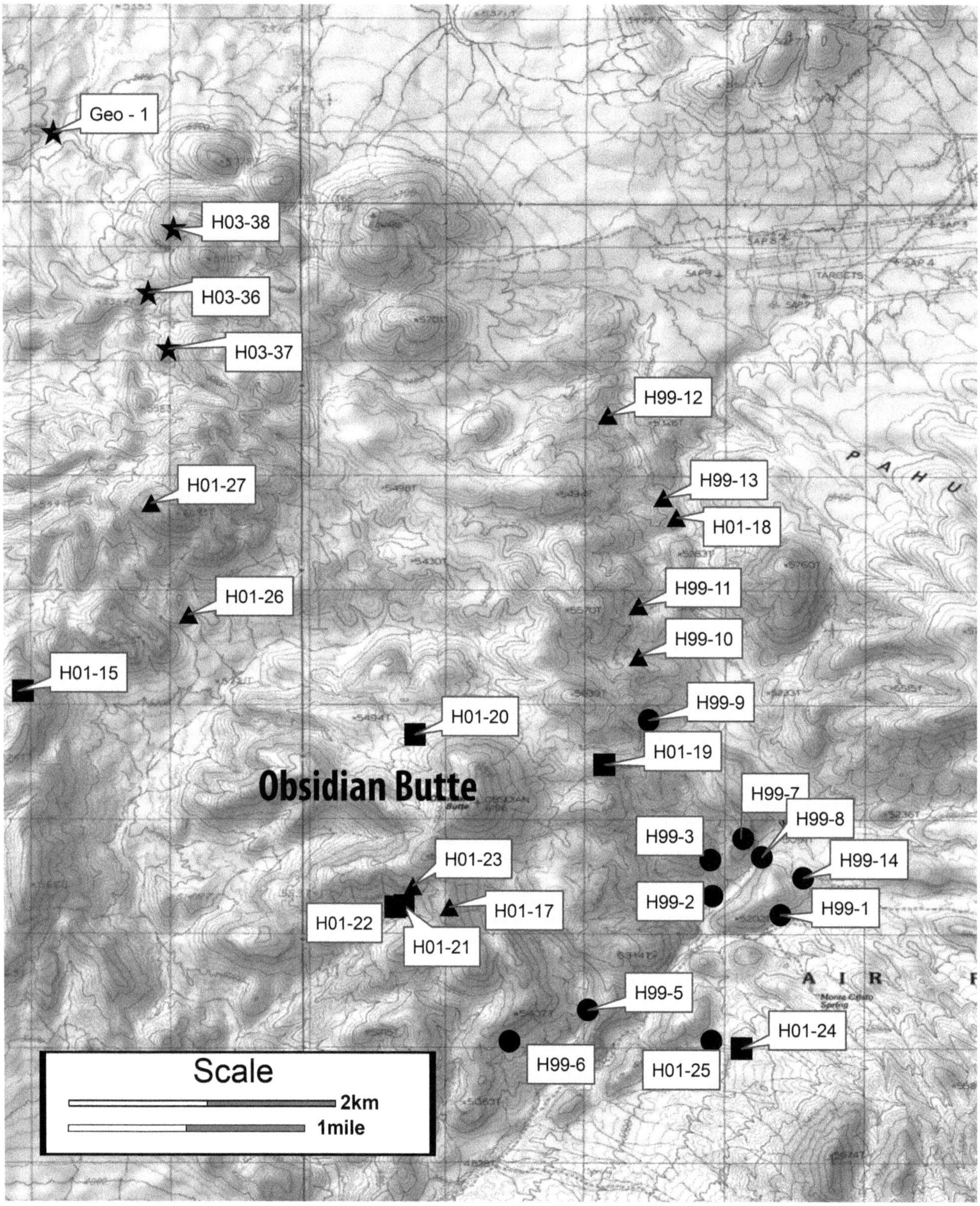

Figure 10.2. The Obsidian Butte volcanic field (locations with black dots denote collection localities for obsidian of the Obsidian Butte chemical type; filled triangles denote examples of Airfield Canyon glass, filled squares specify North Obsidian Butte localities, and black stars locate obsidian from the North Domes cluster)

identify the **Oak Spring Butte** chemical type (cf., Hughes 1995b). Here bedded tuffs with distinctly different visual characteristics were encountered; a lower, very white finely-bedded ash, capped by a thick, darker tuff. Care was taken in the field to separate obsidian from the base of the unit (H03-31) from material collected higher up (H03-32) in the formation, which may be as much as 107m thick (Rogers and Noble 1969). However, trace element data in Table 10.4 show that both collection localities (H03-31 and H03-32) represent the same chemical type. Based on mapped tuff distributions, both localities appear to occur within the Grouse Canyon member of the Belted Range tuff (Rogers and Noble 1969); but the broad range of values published for elements in common with those measured here do not

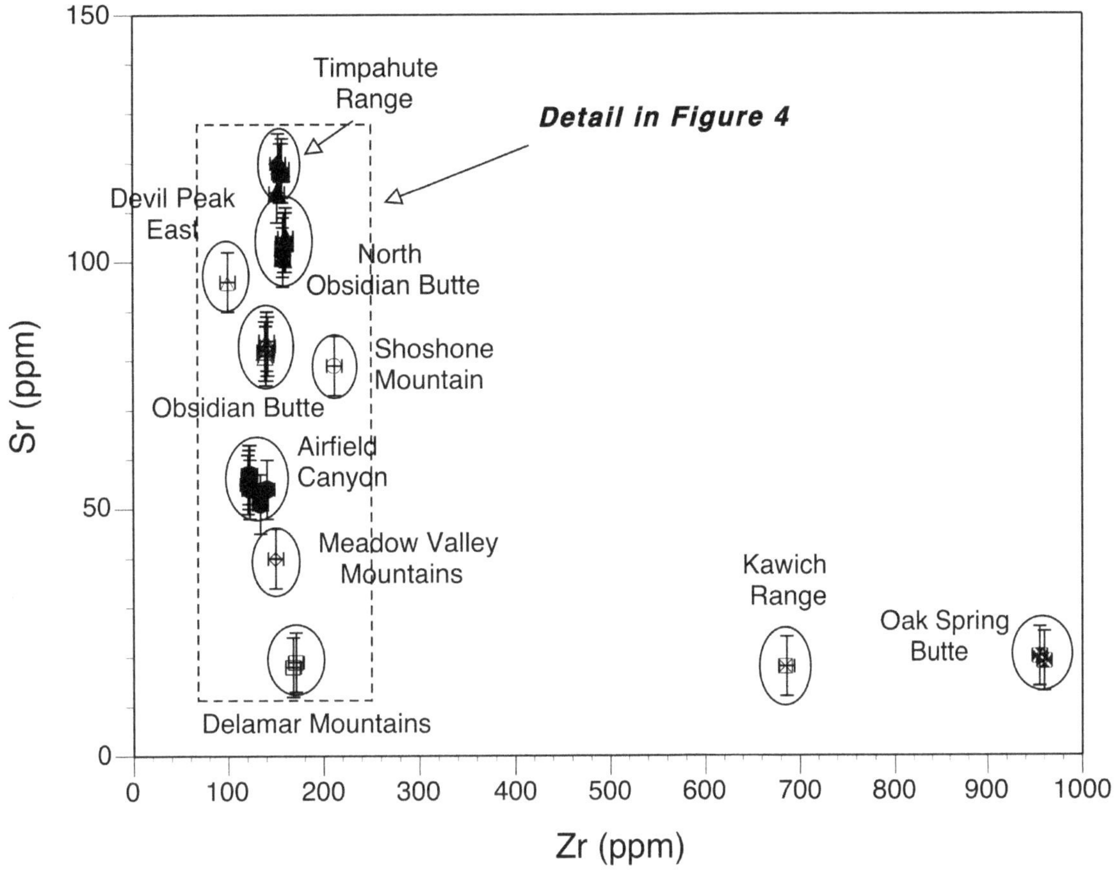

Figure 10.3. Zr vs. Sr composition of some archaeologically significant obsidian sources (chemical types) in southern Nevada (ellipses are 95% confidence interval estimates for concentration values from Tables 10.1–10.4; error bars are ±2 sigma measurement estimates for each collection locality)

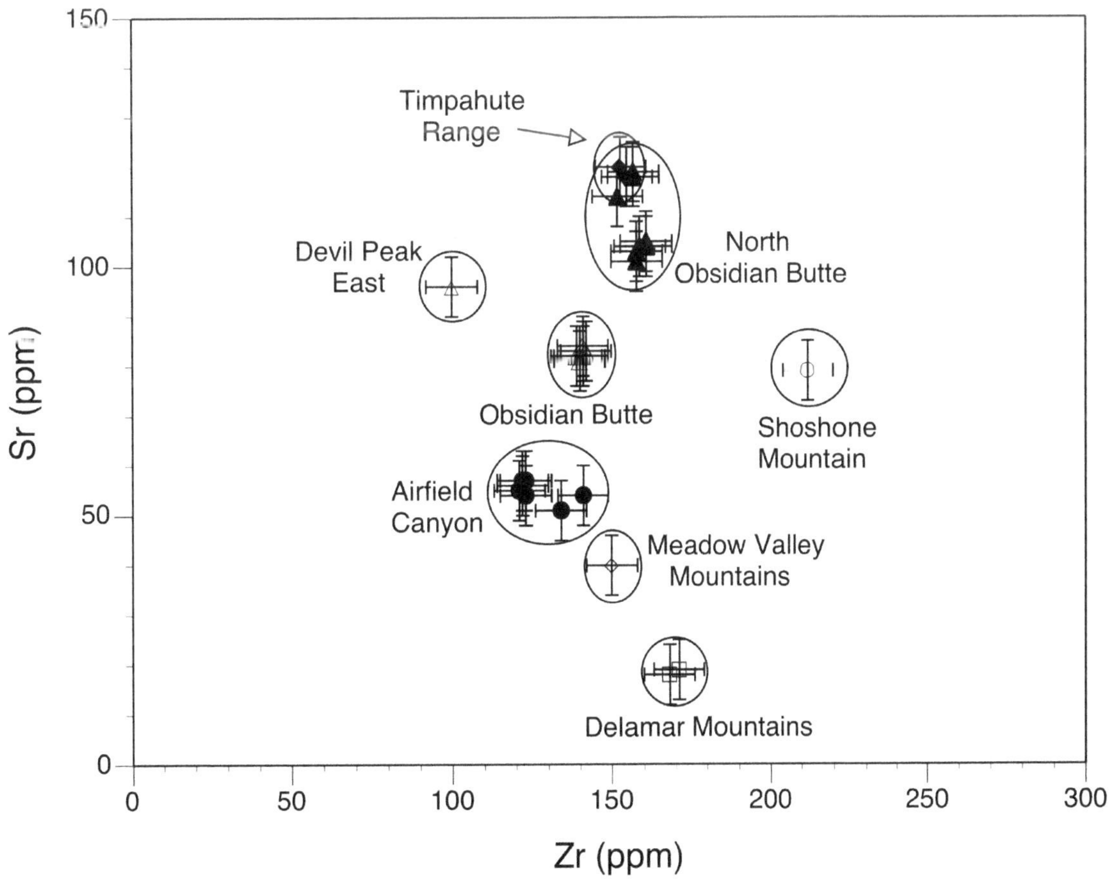

Figure 10.4. Zr vs. Sr composition of some archaeologically significant obsidian sources (chemical types) in southern Nevada containing < 300ppm Zr (ellipses are 95% confidence interval estimates for concentration values from Tables 10.1–10.4; error bars are ±2 sigma measurement estimates for each collection locality)

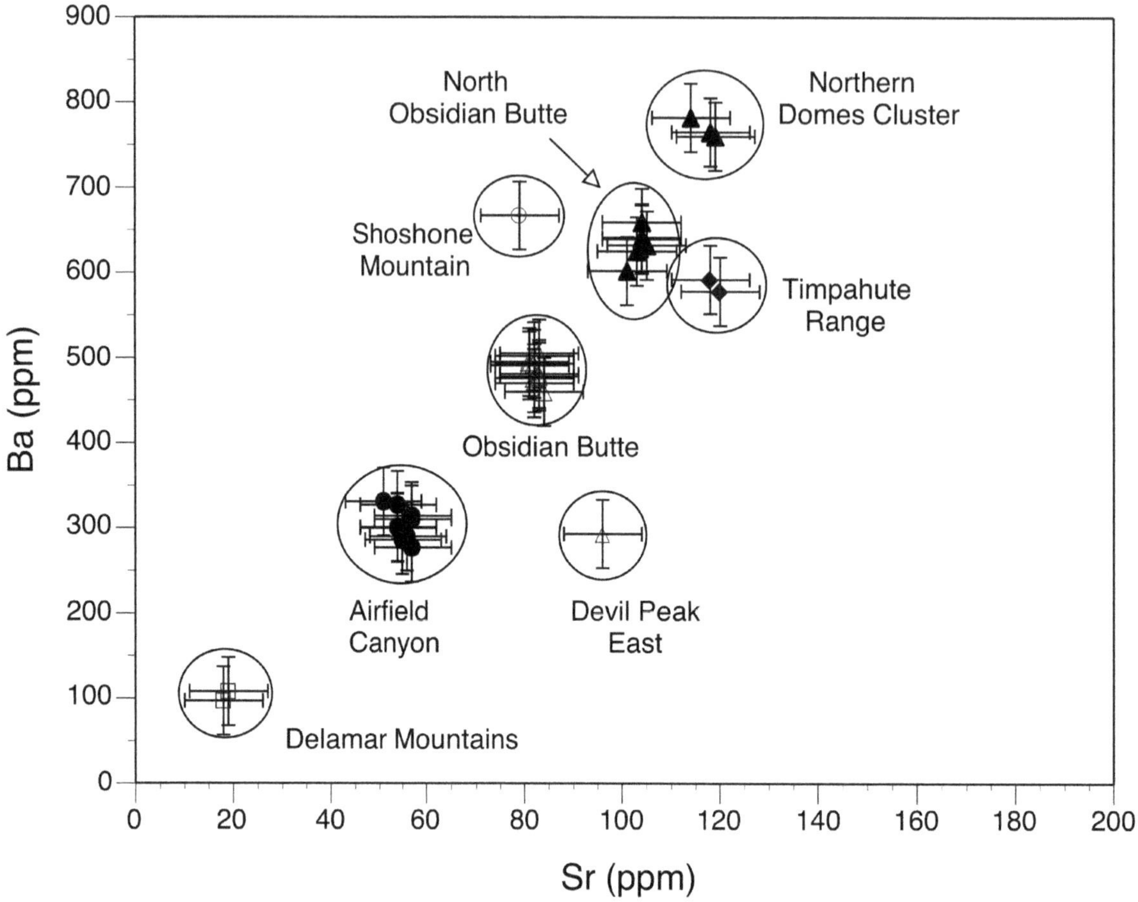

Figure 10.5. Sr vs. Ba composition of some archaeologically significant obsidian sources (chemical types) in southern Nevada containing < 300ppm Zr (ellipses are 95% confidence interval estimates for concentration values from Tables 10.3–10.4; error bars are ±2 sigma measurement estimates for each collection locality)

allow a clear chemical separation between Grouse Canyon and obsidian from Saucer Mesa (cf., Macdonald and Bailey 1973, Table 5; Noble *et al.* 1968, Tables 1 and 2; Noble *et. al.* 1977, Table 1), which could have been anticipated given the comparatively broad range in chemical composition of ash-flow tuff glasses (Hughes and Smith 1993).

Ash-flow tuff obsidian from the mouth of Apache Tear Canyon (H03-33) represents obsidian of the **Kawich Range** geochemical type. Although the parent source for the small obsidian nodules collected at H03-33 was not located, drainage patterns indicate that it can scarcely be anywhere other than to the east in the Kawich Range. Kawich Range obsidian, like other ash-flow tuff glasses, is more heterogeneous than dome-and-flow obsidians. Nonetheless, this obsidian is quite distinct from its ash-flow "brethren" at Oak Spring Butte. Specifically, Kawich Range obsidian contains considerably less Ti, Mn, $Fe_2O_3^T$, Y, and Zr than Oak Spring Butte glass (see Table 10.4). These two ash-flow tuff obsidians are clearly distinguished from all other artefact-quality obsidians in this sampling universe on the basis of high Zr composition (Figure 10.3).

Shoshone Mountain obsidian was defined on the basis of volcanic glass eroding from formation at a single, primary locality (H03-35). Trace element correspondences demonstrate that this is the primary source for obsidian found in secondary contexts in Fortymile Wash, Topopah Wash, and Yucca Wash, which was used extensively at those localities to manufacture artefacts (see Buck *et al.* 1998). Geochemical data presented here (Table 10.5) for this obsidian accord well with previously published trace element analyses (e.g., Macdonald *et al.* 1992, Appendix I, p. 142). This obsidian, previously referred to as "Fortymile, Topopah and Yucca Wash", has been identified at archaeological sites throughout southern Nevada (e.g., Hughes 1993b, 1994b, 1996, 1998b, 1998c).

Obsidian nodules from, and adjacent to, a perlite mine (H03-44) were used to chemically characterise **Devil Peak** obsidian. Shackley (1994) has previously identified two chemical variants; one on the east side of Devil Peak (Devil Peak East), the other on the west side (Devil Peak West). Since our collection was made on the east side of Devil Peak, the chemical signature pertains only to Devil Peak East. Despite the small size of the nodules here, obsidian from Devil Peak East has been identified archaeologically in southern Nevada, for example, at Keno Cave (Hughes 2002).

Trace Element Characteristics of Geologic Obsidian Samples

Narrative and tabular data have been presented on the geologic obsidian samples collected and analysed for this project, but some additional comments are in order

Table 10.5. Source-specific distribution of time-sensitive obsidian projectile points from Ash Meadows National Wildlife Refuge, Nevada. Typological attributions for projectile points based on Thomas (1981); dating based on Lyneis (1982), Sutton et al. (2007), Warren (1984), and Warren and Crabtree (1986). Gatecliff series dating from Thomas (1981)

Time Period	***AD 1200 – Historic***		***AD 500 – 1200***	***AD 500 – 2000 BC***	***1500 – 3000 BC***	***?***	
Point Type	***Cottonwood Triangular***	***Desert Side-Notched***	***Rosegate Series***	***Elko Series***	***Gatecliff Series***	***Large Side-Notched***	
Obsidian Source							***Totals***
Shoshone Mountain	7	7	9	11	3	1	38
Obsidian Butte				1		1	2
North Obsidian Butte				1			1
Airfield Canyon				1			1
Oak Spring Butte			2	2	1		5
Saline Range			1	1		1	3
Delamar Mountains		1					1
Kane Spring				1			1
West Sugarloaf	1		1	2			4
West Cactus Peak	1						1
Queen				1			1
Panaca Summit						1	1
Montezuma Range			1				1
Totals	**9**	**8**	**14**	**21**	**4**	**4**	**60**

concerning the chemical relationships and contrasts between identified chemical types. Figure 10.3 shows a plot of Zr vs. Sr composition; and Figure 10.5 presents Sr vs. Ba composition for all obsidian samples analysed (ash-flow tuff obsidians are absent from Figure 10.5 because they contain no measurable amount of Ba). Even at this gross scale, chemical contrasts are evident. Perhaps the most dramatic in Figure 10.3 is that of the high Zr composition ash-flow tuff obsidians of the Oak Spring Butte and Kawich Range geochemical types. These glasses are segregated from all others on the basis of high Zr (> 600ppm) and very low Sr concentration.

When these high Zr obsidians are removed from consideration, it is possible to highlight and illustrate more subtle distinctions among glass types. Figures 10.4–10.5 present such data for all obsidians containing < 300ppm Zr. The Zr vs. Sr plots for these obsidians show that the majority of them are quite distinct – even on the basis of only two elements. The strength of these distinctions can be evaluated by consulting Figures 10.4–10.5, which present 95% confidence interval data for these same specimens. These latter figures show, on the one hand, just how distinctive some sources are (e.g., Shoshone Mountain, Delamar Mountains, Devil Peak East, Resting Spring Range, Airfield Canyon, and Obsidian Butte), and, on the other hand, how similar others (i.e., Timpahute Range and North Obsidian Butte) are on the basis of Zr and Sr composition. Figure 10.6 show that Rb vs. Y composition effects a clear separation between Timpahute Range and North Obsidian Butte glasses. Timpahute Range obsidians also tend to have higher concentrations of Mn and $Fe_2O_3^T$ and lower Ti and Ba composition than North Obsidian Butte (see Tables 10.1 and 10.4). Figure 10.4 hints at a slight variation within the Airfield Canyon chemical type. This is a result of two localities, H01-26 and H01-27, situated near domes at the extreme northwestern distribution of the Airfield Canyon chemical type in Flow Unit 3B, which contain slightly higher concentration of Zr than other examples of the type. However, when evaluated at 2-sigma Figure 10.4 shows that this distinction may well be an analytical artefact.

Summary Comments

The exhaustive field reconnaissance and laboratory analyses conducted during this project have yielded important new knowledge. With respect to the Obsidian Butte volcanic field, it is now clear that there are at least four chemical variants of obsidian, which have been named Obsidian Butte, Airfield Canyon, North Obsidian Butte, and Northern Domes Cluster. While trace and minor element data show that the chemical signatures of Obsidian Butte and Airfield Canyon are quite delimited, these same elements show somewhat more variability within the North Obsidian Butte chemical type. Whether this variability is the by-product of crystal–liquid fractionation processes, a chemically variable magma at one point in time, or due to slight chemical differences resulting from tapping of the underlying magma chamber at different points in time, cannot be resolved on the basis of present data.

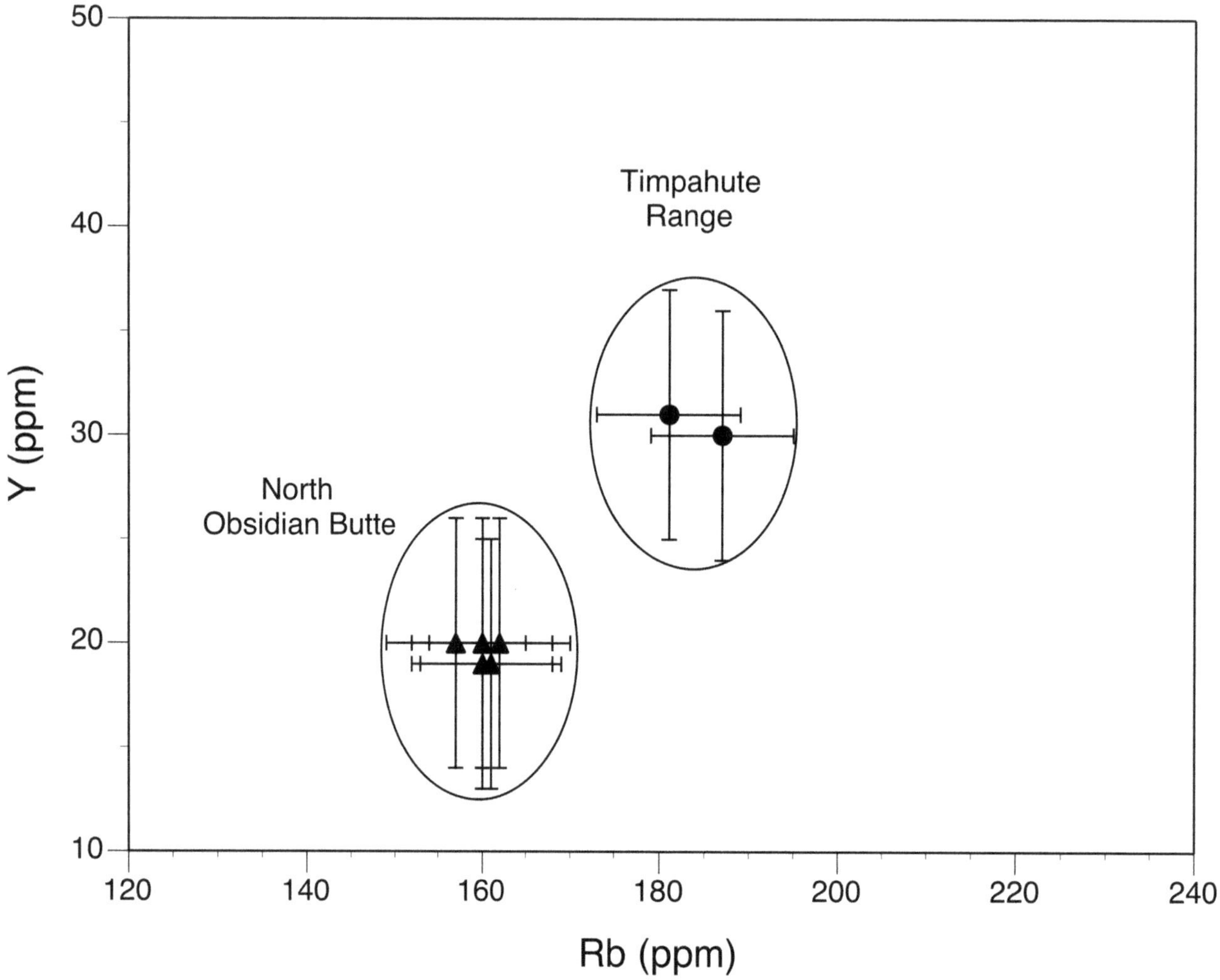

Figure 10.6. Rb vs. Y composition of Timpahute Range and North Obsidian Butte geologic samples (ellipses are 95% confidence interval estimates for concentration values from Tables 10.3–10.4; error bars are ±2 sigma measurement estimates for each collection locality)

Data from this project were critical to resolving a long-standing problem in regional sourcing studies – the issue of series of "unknowns" with no known geographic counterpart. Based on research completed here, it is now clear that obsidian previously referred to as "Unknown C" can be linked with confidence to the North Obsidian Butte chemical type. Field research and laboratory analysis of obsidian from the Delamar Mountains shows that this glass is the geologic counterpart of obsidian previously referred to as "Unknown A" and "Unknown Variety 1", while collection and analysis of obsidian from the Timpahute Range area indicates that it is the parent chemical counterpart for obsidian previously identified as "Unknown B" (see Jones *et al.* 2003, 38). In addition, chemical data generated from this project show that Shoshone Mountain is the parent geochemical type for obsidian previously termed "Fortymile, Topopah, and Yucca Wash" (Buck *et al.* 1998, 94, 220).

Obsidian from another major "source" in southern Nevada – Kane Springs – was not mentioned in this chapter, despite its obvious importance at archaeological sites in the area. Fieldwork in the Kane Springs area, done on foot and with vehicles (not via helicopter), focused on Delamar Range geologic mapping and field information provided by Robert Scott (US Geological Survey). While obsidian was found in this area (at localities H03-30 and H03-30A), the results were surprising in that they did not represent the well-known Kane Springs trace element profile (cf., Nelson and Holmes 1979, Table III; Hughes 2005, Table II.2). Because the search was for primary obsidian outcrops, obsidian from Kane Springs Wash itself was not collected because obsidian of this chemical type has already been documented therein. Obsidian matching the Kane Springs source profile has been identified in primary context in the Meadow Valley Mountains, to the east of Kane Springs Wash (Harding *et al.* 1995; Scott *et al.* 1995: 10,389), accounting for some of the trace element heterogeneity attributed to the Kane Springs Wash "source". Just as obsidian from Shoshone Mountain has been redeposited into local washes (see above) and exploited prehistorically wherever it occurred, obsidians of both the Delamar Mountains and Meadow Valley Mountains chemical types, redeposited in to Kane Springs Wash, also were exploited in primary and secondary contexts as sources for archaeological obsidian.

Important research still remains to be conducted. Analysis of archaeological samples from Conaway and O'Malley shelters (Figure 10.1), as well as at sites in southwestern Utah State, has identified artefacts made from five geochemically distinctive varieties of obsidian not represented in the current geologic sampling universe indicating that more reconnaissance, collection, and chemical analysis needs to be done in this area. Nonetheless, these trace and minor element "fingerprints" should provide a solid geochemical foundation for "sourcing" studies conducted on archaeological artefacts from throughout southern Nevada.

Archaeological Implications

The reconnaissance, collection, and laboratory analysis of these obsidians were done to generate chemical data to serve archaeological ends. Site-specific studies documenting time/space continuities and contrasts in obsidian conveyance for certain portions of the Great Basin have appeared (Hughes 1983, 1985, 1986c, 1994d; Hughes and Bennyhoff 1986; Nelson 1984; Nelson and Holmes 1979) or are in preparation (Hughes 2010a, 2010b). The details from the latter studies will be presented elsewhere (Hughes 2010a, 2010b), but it can be noted that obsidians from Obsidian Butte have been identified in the central Great Basin at Alta Toquima and other sites atop Mount Jefferson (over 160km to the north), at Gatecliff Shelter (over 180km to the north), along with volcanic glass from the Timpahute Range, over 170km to the southeast (Figure 10.1). A few additional words are in order, though, about the prehistoric use of some of the obsidians in the present study and their potential for helping us arrive at more nuanced understanding of obsidian conveyance in the area.

At Conaway and O'Malley shelters in southern Lincoln County, Nevada (see Figure 10.1), large numbers of obsidian projectile points were recovered. Fowler *et al.* (1973, 55) wrote that obsidian 'nodules occur widely over the study area, apparently derived from the volcanism which formed some of the mountains to the south and east.' Source analysis revealed that Fowler *et al.* (1973) were essentially correct; the vast majority of points from all temporal periods represented at these sites were fashioned from obsidian of the nearby Panaca Summit chemical type.

But a very different situation was discovered at archaeological sites in the Ash Meadows National Wildlife Refuge of southern Nye County, Nevada (Figure 10.1). Here archaeological survey documented 250 archaeological sites and evidence of human use of the area for perhaps 10,000 years (Lyon *et al.* 2008). Source analysis of time-sensitive projectile points from these sites (data in Hughes 2008) revealed a very different and complex set of relationships (see Table 10.5) than encountered at Conaway and O'Malley shelters. Although more than half of the projectile points from all time periods at Ash Meadows sites were made from obsidian from the closest available source (Shoshone Mountain, located about 60km northeast of Ash Meadows), there are significant differences in source use by time period. Applying the chronological framework of Warren and Crabtree (1986) (see also Lyneis 1982; Sutton *et al.* 2007; Warren 1984), prior to AD 500, when Elko Series, Gatecliff Series, and large side-notched dart points were in use, just over half (54%) of the specimens were fashioned from Shoshone Mountain volcanic glass. During the following time period (ca. AD 500–1200), marked by dominant use of Rose Spring (Rosegate Series) points, 64% of the specimens were manufactured from Shoshone Mountain obsidian. Finally, by the time Desert Side-notched and Cottonwood Triangular points were in use (from ca. AD 1200 to historic times), nearly three of every four obsidian projectile points (74%) was made from Shoshone Mountain glass.

The data in Table 10.5 also suggest some other intriguing patterns. Though, to date, Gatecliff Series and large side-notched points occur too infrequently in Ash Meadows sites to yield meaningful comparisons, note that during the time Elko Series points were in use nearly half of them (48%) were made from non-Shoshone Mountain obsidian – in other words, *not* on obsidian from the closest source. It is during Elko Series times that obsidian from the Coso Volcanic Field (located ca. 140km to the southwest) first appears at Ash Meadows sites, as do single specimens from Queen (located ca. 250km to the northwest); the Saline Range (ca. 150km northwest); and Kane Springs, Meadow Valley Mountains (ca. 150km to the east). Obsidian Butte area glasses (located > 100km to the north) make their first and only appearance in Ash Meadows sites during Elko Series times; they were absent in subsequent periods.

Though one would not want to push these results too far, given the small sample size involved, they help illustrate a more general point. Depending on a host of factors, we find that some sites (like Conaway and O'Malley shelters) show an essentially unchanging use of local obsidian, but others, like those in Ash Meadows, reveal patterning that would not be evident unless the artefacts were partitioned into appropriate temporal categories. If these Ash Meadows projectile points had not been segregated on the basis of their relatively well-known time spans (Thomas 1981; Warren and Crabtree 1986) and had been conflated into, say, a post-2000 BC category, the gradual increase in use of Shoshone Mountain obsidian through time, the sudden and brief appearance/use of Obsidian Butte glasses, the relative diversity of source use during Elko times, and the relative decrease in source-use diversity through time would have gone unrecognised. This underscores that the great research potential of the trace element compositional signatures identified here will be maximised if archaeological assemblages subjected to trace element analysis are first stratified by artefact type/class and time period, so that fine and coarse-grained change and continuity in prehistoric obsidian acquisition, use, and conveyance patterns can be identified in the southern Great Basin and elsewhere in far western North America.

Acknowledgements

This paper was adapted and revised from a longer report (Hughes 2004), which discusses other minor obsidian and ash-flow tuff sources. The research reported here would not have been possible if not for the assistance of numerous colleagues and associates. Keith Myhrer (Nellis Air

Force Base) was the project sponsor, and it is a pleasure to acknowledge his support and permission to publish the research results. Jay Newman (US Army Corps of Engineers) supported the project throughout, as did Richard Arnold (Chair, Southern Paiute tribe). Karen Gardner (Prewitt and Associates, Inc.) assisted in administrative and logistics matters. D. J. Haarklau was especially helpful as field escort and obsidian-collecting partner, and Lalovi Miller (Moapa Paiute band) and Joe Kennedy (Timbisha Shoshone tribe) provided important insights on obsidian use. Jack Stewart and Robert Scott (US Geological Survey), Donald Noble (University of Nevada at Reno), and Myron Best (Brigham Young University) furnished location information for some southern Nevada obsidians. Robert L. Smith (Scientist Emeritus, US Geological Survey), Fred W. Nelson (Brigham Young University), and George T. Jones (Hamilton College) provided constructive comments on the manuscript. Over two decades ago Lonnie Pippin first introduced me to NAFR and NTS during a bone-jarring jeep ride up a wash that I will never forget. I also appreciate the long-term interest and assistance of David Rhode and Anne DuBarton (Desert Research Institute). Robert Hafey generously provided location information for the Timpahute Range area source, and Heidi Roberts (HRA, Inc.) provided contextual information and financial support for the Ash Meadows study. Thanks also to Lynn Haarklau (Nellis Air Force Base) for all manner of assistance during the years this project took to come to fruition. Her organisation and field skills were critical to overall success of the project. Special thanks to Tammara Norton, Jay King, and Ben Hughes for preparing Figures 10.1–10.2.

References

Ambroz, J. A., M. D. Glascock, and C. E. Skinner. 2001. Chemical Differentiation of Obsidian Within the Glass Buttes Complex, Oregon. *Journal of Archaeological Science* 28, 741–746.

Ball, S. H. 1907. *A Geologic Reconnaissance in Southwestern Nevada and Eastern California* (US Geological Survey Bulletin 308). Washington, D.C., Government Printing Office.

Bowman, H. R., F. Asaro, and I. Perlman. 1973. On the Uniformity of Composition in Obsidians and Evidence for Magmatic Mixing. *Journal of Geology* 81, 312–327.

Buck, P. E., W. T. Hartwell, G. Haynes, and D. Rhode. 1998. *Archaeological Investigations at Two Early Holocene Sites near Yucca Mountain, Nye County, Nevada* (Topics in Yucca Mountain Archaeology 2). Las Vegas, NV, Desert Research Institute.

Byers, F. M., Jr., W. J. Carr, P. P. Orkild, W. D. Quinlivan, and K. A. Sargent. 1976. *Volcanic Suites and Related Cauldrons of Timber Mountain-Oasis Valley Caldera Complex, Southern Nevada* (US Geological Survey Professional Papers 919). Washington, D.C., Government Printing Office.

Cornwall, H. R. 1962. Caldera and Associated Volcanic Rocks near Beatty, Nye County, Nevada. In *Petrologic Studies: A Volume in Honor of A. F. Buddington*, edited by A. E. J. Engel, H. L. James and B. F. Leonard, 357–371. Boulder, CO, Geological Society of America.

Cornwall, H. R. 1972. *Geology and Mineral Deposits of Southern Nye County, Nevada* (Nevada Bureau of Mines and Geology Bulletin 77). Reno, NV, Nevada Bureau of Mines and Geology.

Eckel, E. B. (ed.). 1968. *Nevada Test Site* (Geological Society of America Memoirs 110). Boulder, CO, Geological Society of America.

Ekren, E. B., R. E. Anderson, C. L. Rogers, and D. C. Noble. 1971. *Geology of Northern Nellis Air Force Base Bombing and Gunnery Range, Nye County, Nevada* (US Geological Survey Professional Papers 651). Washington, D.C., Government Printing Office.

Ericson, J. E. 1981. *Exchange and Production Systems in Californian Prehistory: The Results of Hydration Dating and Chemical Characterization of Obsidian Sources* (B.A.R. International Series 110). Oxford, British Archaeological Reports.

Fowler, D. D., D. B. Madsen, and E. M. Hattori. 1973. *Prehistory of Southern Nevada* (Desert Research Institute Publications in the Social Sciences 6). Reno, NV, Desert Research Institute.

Harding, A. E., R. B. Scott, H. E. Mehnert, and L. W. Snee. 1995. Evidence of the Kane Springs Wash Caldera in the Meadow Valley Mountains, Southeast Nevada. In *Geologic Studies in the Basin and Range-Colorado Plateau Transition in Southeastern Nevada, Southwestern Utah, and Northwestern Arizona, 1992* (US Geological Survey Bulletin 2056), edited by R. B. Scott and W. C. Swadley, 135–179. Washington, D.C., Government Printing Office.

Hughes, R. E. 1983. X-ray Fluorescence Characterization of Obsidian. In *The Archaeology of Monitor Valley: 2. Gatecliff Shelter* (Anthropological Papers of the American Museum of Natural History. Vol. 59, No. 1), edited by D. H. Thomas, 401–408. New York, American Museum of Natural History.

Hughes, R. E. 1984. Obsidian Sourcing Studies in the Great Basin: Problems and Prospects. In *Obsidian Studies in the Great Basin* (Contributions of the University of California Archaeological Research Facility 45), edited by R. E. Hughes, 1–19. Berkeley, University of California.

Hughes, R. E. 1985. Obsidian Source Use at Hidden Cave. In *The Archaeology of Hidden Cave, Nevada* (Anthropological Papers of the American Museum of Natural History. Vol. 61, No. 1), edited by D. H. Thomas, 332–353. New York, American Museum of Natural History.

Hughes, R. E. 1986a. Trace Element Composition of Obsidian Butte, Imperial County, California. *Bulletin of the Southern California Academy of Sciences* 85, 35–45.

Hughes, R. E. 1986b. Energy Dispersive X-ray Fluorescence Analysis of Obsidian from Dog Hill and Burns Butte, Oregon. *Northwest Science* 60, 73–80.

Hughes, R. E. 1986c. *Diachronic Variability in Obsidian Procurement Patterns in Northeastern California and South-Central Oregon* (University of California Publications in Anthropology 17). Berkeley: University of California.

Hughes, R. E. 1988. The Coso Volcanic Field Reexamined: Implications for Obsidian Sourcing and Hydration

Dating Research. *Geoarchaeology* 3, 253–265.

Hughes, R. E. 1989. A New Look at Mono Basin Obsidians. In *Current Directions in California Obsidian Studies* (Contributions of the University of California Archaeological Research Facility 48), edited by R. E. Hughes, 1–12. Berkeley, University of California.

Hughes, R. E. 1990. Obsidian Sources at James Creek Shelter, and Trace Element Geochemistry of Some Northeastern Nevada Volcanic Glasses. In *The Archaeology of James Creek Shelter* (University of Utah Anthropological Papers 115), edited by R. G. Elston and E. E. Budy, 297–305. Salt Lake City, University of Utah Press.

Hughes, R. E. 1993a. Trace Element Geochemistry of Volcanic Glass from the Obsidian Cliffs Flow, Three Sisters Wilderness, Oregon. *Northwest Science* 67, 199–207.

Hughes, R. E. 1993b. X-ray Fluorescence Analysis of Obsidian Projectile Points from Three Archaeological Sites in the Vicinity of Pahute Mesa, Nye County, Nevada. Unpublished Geochemical Research Laboratory Letter Report 93-110. Rancho Cordova, CA.

Hughes, R. E. 1994a. Intrasource Chemical Variability of Artefact-Quality Obsidians from the Casa Diablo Area, California. *Journal of Archaeological Science* 21, 263–271.

Hughes, R. E. 1994b. X-ray Fluorescence Analysis of Projectile Points from Archaeological Sites on Pahute and Rainier Mesas, Nevada Test Site, Nye County, Nevada. Unpublished Geochemical Research Laboratory Letter Report 94-25. Rancho Cordova, CA.

Hughes, R. E. 1994c. X-ray Fluorescence Analysis of Obsidian from the Lake Mead National Recreation Area, Clark County, Nevada. Unpublished Geochemical Research Laboratory Letter Report 94-74. Rancho Cordova, CA.

Hughes, R. E. 1994d. Mosaic Patterning in Prehistoric California–Great Basin Exchange. In *Prehistoric Exchange Systems in North America*, edited by T. G. Baugh and J. E. Ericson, 363–383. New York, Plenum Press.

Hughes, R. E. 1995a. X-ray Fluorescence Analysis of Obsidian from 26CK4856 and 26CK4867 on Nellis Air Force Base, Clark County, Nevada. Unpublished Geochemical Research Laboratory Letter Report 95-13. Portola Valley, CA.

Hughes, R. E. 1995b. X-ray Fluorescence Analysis of Geologic Obsidian Samples from the Vicinity of Tubb Spring, Nevada Test Site, Nevada. Unpublished Geochemical Research Laboratory Letter Report 95-15. Portola Valley, CA.

Hughes, R. E. 1996. X-ray Fluorescence Analysis of Obsidian Artifacts from Pintwater Cave (26CK253) and 26CK2056 Located near Indian Springs, Clark County, Nevada. Unpublished Geochemical Research Laboratory Letter Report 96-52. Portola Valley, CA.

Hughes, R. E. 1997. X-ray Fluorescence Analysis of Obsidian Artifacts from 26CK5423, Clark County, Nevada. Unpublished Geochemical Research Laboratory Letter Report 97-5. Portola Valley, CA.

Hughes, R. E. 1998a. On Reliability, Validity, and Scale in Obsidian Sourcing Research. In *Unit Issues in Archaeology: Measuring Time, Space and Material*, edited by A. F. Ramenofsky and A. Steffen, 103–114. Salt Lake City, University of Utah Press.

Hughes, R. E. 1998b. X-ray Fluorescence Analysis of Artifacts and Geologic Samples from Archaeological Sites Within Nellis Air Force Base, Nevada. Unpublished Geochemical Research Laboratory Letter Report 98-94. Portola Valley, CA.

Hughes, R. E. 1998c. X-ray Fluorescence Analysis of Obsidian Artifacts in the Yucca Mountain Area, Nye County, Nevada. Unpublished Geochemical Research Laboratory Letter Report 98-15. Portola Valley, CA.

Hughes, R. E. 1999. X-ray Fluorescence Analysis of Obsidian Artifacts from Flaherty Rockshelter (26CK415), Clark County, Nevada. Unpublished Geochemical Research Laboratory Letter Report 99-59. Portola Valley, CA.

Hughes, R. E. 2001a. Energy Dispersive X-ray Fluorescence Analysis of Obsidian Artifacts from Archaeological Sites in the Carson Desert and Stillwater Mountains. In *Prehistory of the Carson Desert and Stillwater Mountains* (University of Utah Anthropological Papers 123), edited by R. L. Kelly, 241–250. Salt Lake City, University of Utah Press.

Hughes, R. E. 2001b. Geochemical Characterization of Obsidian from Outcrops on the Eastern and Southern Flanks of Obsidian Butte, Nellis Air Force Range, Nevada. In *Investigation of Geochemical Variability in Obsidian Raw Material and Artifact Sources on the North Nellis Air Force Range, Nevada* (Nellis Air Force Base Cultural Resource Reports 00-04), edited by L. Haarklau, 27–54. Nellis, NV, Nellis Air Force Base.

Hughes, R. E. 2002. Energy Dispersive X-ray Fluorescence (EDXRF) Analysis of Artifacts from Keno Cave (26CK4446), Clark County, Nevada. Unpublished Geochemical Research Laboratory Letter Report 2002-90. Portola Valley, CA.

Hughes, R. E. 2003. X-ray Fluorescence Analysis of Obsidian Artifacts from 26CK1139, Clark County, Nevada. Unpublished Geochemical Research Laboratory Letter Report 2003-6. Portola Valley, CA.

Hughes, R. E. 2004. Trace Element Characterization of Volcanic Glasses from Obsidian Butte, Nellis Air Force Range, Nevada, and Surrounding Areas. Unpublished Report Submitted to Prewitt and Associates, Inc., Austin, TX, 1 March 2004.

Hughes, R. E. 2005. Determination of the Geologic Sources for Obsidian Artifacts from Camels Back Cave, and Trace Element Analysis of Some Western Utah and Southeastern Nevada Volcanic Glasses. In *Camels Back Cave* (University of Utah Anthropological Papers 125), edited by D. N. Schmitt and D. B. Madsen, 249–256. Salt Lake City, University of Utah Press.

Hughes, R. E. 2008. Energy Dispersive X-ray Fluorescence Analysis of Artifacts Recovered from Various Archaeological Sites and Localities in the Ash Meadows National Wildlife Refuge, Nye County, Nevada. Unpublished Geochemical Research Laboratory Letter Report 2008-31. Portola Valley, CA.

Hughes, R. E. 2010a. Geochemical Identification of the

Sources for Alta Toquima Obsidian Artifacts. In *The Archaeology of Monitor Valley: 4. Alta Toquima and the Mt. Jefferson Complex* (Anthropological Papers of the American Museum of Natural History), edited by D. H. Thomas. New York, American Museum of Natural History (in press).

Hughes, R. E. 2010b. Trace Element Analysis of Obsidian from the Mt. Jefferson Tablelands. In *The Archaeology of Monitor Valley: 4. Alta Toquima and the Mt. Jefferson Complex* (Anthropological Papers of the American Museum of Natural History), edited by D. H. Thomas. New York, American Museum of Natural History (in press).

Hughes, R. E., and J. A. Bennyhoff. 1986. Early Trade. In *Great Basin* (Handbook of North American Indians. Vol. 11. W. C. Sturtevant, general editor), edited by W. L. d'Azevedo, 238–255. Washington, D.C., Smithsonian Institution.

Hughes, R. E., and R. L. Smith. 1993. Archaeology, Geology, and Geochemistry in Obsidian Provenance Studies. In *Effects of Scale on Archaeological and Geoscientific Perspectives* (Geological Society of America Special Papers 283), edited by J. K. Stein and A. R. Linse, 70–91. Boulder, CO, Geological Society of America.

Jack, R. N. 1976. Prehistoric Obsidian in California I: Geochemical Aspects. In *Advances in Obsidian Glass Studies: Archaeological and Geochemical Perspectives*, edited by R. E. Taylor, 183–217. Park Ridge, NJ, Noyes Press.

Jackson, T. L. 1989. Late Prehistoric Obsidian Production and Exchange in the North Coast Ranges, California. In *Current Directions in California Obsidian Studies* (Contributions of the University of California Archaeological Research Facility 48), edited by R. E. Hughes, 79–94. Berkeley, University of California.

Jones, G. T., C. Beck, E. E. Jones, and R. E. Hughes. 2003. Lithic Source Use and Paleoarchaic Foraging Territories in the Great Basin. *American Antiquity* 68, 5–38.

Lipman, P. W., R. L. Christiansen, and J. T. O'Connor. 1966. *A Compositionally Zoned Ash-Flow Sheet in Southern Nevada* (US Geological Survey Professional Papers 524-F). Boulder, CO, Geological Society of America.

Lyneis, M. M. 1982. Prehistory in the Southern Great Basin. In *Man and Environment in the Great Basin* (Society for American Archaeology Papers 2), edited by D. B. Madsen and J. F. O'Connell, 172–185. Washington, D.C., Society for American Archaeology.

Lyon, J. D., H. Roberts, R. Ahlstrom, C. Harper, S. Eskenazi, R. Davide, C. Fowler, and E. von Till Warren. 2008. *Shared Place: An Archaeological Survey of the Ash Meadows National Wildlife Refuge, Nye County, Nevada* (HRA Archaeological Reports 07-24). Las Vegas, NV, HRA.

Lyons, W. H., S. P. Thomas, and C. E. Skinner. 2001. Changing Obsidian Sources at the Lost Dune and McCoy Creek Sites, Blitzen Valley, Southeast Oregon. *Journal of California and Great Basin Anthropology* 23, 273–296.

Macdonald, R., and D. K. Bailey. 1973. *The Chemistry of the Peralkaline Oversaturated Obsidians* (US Geological Survey Professional Papers 440-N-1). Washington, D.C., Government Printing Office.

Macdonald, R., R. L. Smith, and J. E. Thomas. 1992. *Chemistry of the Subalkalic Silicic Obsidians* (US Geological Survey Professional Papers 1523). Washington, D.C., Government Printing Office.

Minor, S. A., D. A. Sawyer, R. R. Wahl, V. A. Frizzell, Jr., S. P. Schilling, R. G. Warren, P. P. Orkild, J. A. Coe, M. R. Hudson, R. J. Fleck, M. A. Lanphere, W. C. Swadley, and J. C. Cole. 1993. *Preliminary Geologic Map of the Pahute Mesa 30′ × 60′ Quadrangle, Nevada* (US Geological Survey Open-File Reports 93-299). Washington, D.C., Government Printing Office.

Nelson, F. W., Jr. 1984. X-ray Fluorescence Analysis of Some Western North American Obsidians. In *Obsidian Studies in the Great Basin* (Contributions of the University of California Archaeological Research Facility 45), edited by R. E. Hughes, 27–62. Berkeley, University of California.

Nelson, F. W., and R. D. Holmes. 1979. Trace Element Analysis of Obsidian Sources and Artifacts from Western Utah. In *Antiquities Section Selected Papers 6 (15)*, edited by D. B. Madsen, 65–80. Salt Lake City: Division of State History, Utah State Historical Society.

Noble, D. C., W. L. Rigot, and H. R. Bowman. 1977. Rare-Earth-Element Content of Some Highly Differentiated Ash-Flow Tuffs and Lavas. In *Ash-Flow Tuffs* (Geological Society of America Special Papers 180), edited by C. E. Chapin and W. E. Elston, 77–85. Boulder, CO, Geological Society of America.

Noble, D. C., K. A. Sargent, H. H. Mehnert, E. B. Ekren, and F. M. Byers, Jr. 1968. Silent Canyon Volcanic Center, Nye County, Nevada. In *Nevada Test Site* (Geological Society of America Memoirs 110), edited by E. E. Eckel, 65–75. Boulder, CO, Geological Society of America.

Novak, S. W. 1984. Eruptive History of the Rhyolitic Kane Springs Wash Volcanic Center, Nevada. *Journal of Geophysical Research* 89, 8603–8615.

Novak, S. W. and G. A. Mahood. 1988. Rise and Fall of a Basalt-Trachyte-Rhyolite Magma System at the Kane Springs Wash Caldera, Nevada. *Contributions to Mineralogy and Petrology* 94, 352–373.

Rogers, C.L., and D. C. Noble. 1969. *Geologic Map of the Oak Spring Butte Quadrangle, Nye County, Nevada* (US Geological Survey Maps GQ-822). Washington, D.C., Government Printing Office.

Scott, R. B., D. M. Unruh, L. W. Snee, A. E. Harding, L. D. Nealy, H. R. Blank, Jr., J. R. Budhan, and H. H. Mehnert. 1995. Relation of Peralkaline Magmatism to Heterogeneous Extension during the Middle Miocene, Southeastern Nevada. *Journal of Geophysical Research* 100(B6), 10,381–10,401.

Shackley, M. S. 1994. Intersource and Intrasource Geochemical Variability in Two Newly Discovered Archaeological Obsidian Sources in the Southern Great Basin. *Journal of California and Great Basin Anthropology* 16, 118–129.

Stross, F. H., T. R. Hester, R. F. Heizer, and R. N. Jack. 1976. Chemical and Archaeological Studies of Mesoamerican Obsidians. In *Advances in Obsidian Glass Studies: Archaeological and Geochemical Perspectives*, edited by R. E. Taylor, 240–258. Park Ridge, NJ, Noyes Press.

Sutton, M. Q., M. E. Basgall, J. K. Gardner, and M. W. Allen. 2007. Advances in Understanding Mojave Desert Prehistory. In *California Prehistory: Colonization, Culture, and Complexity*, edited by T. L. Jones and K. A. Klar, 229–245. Lanham, MD, Altamira Press.

Thomas, D. H. 1981. How to Classify the Projectile Points from Monitor Valley, Nevada. *Journal of California and Great Basin Anthropology* 3, 7–43.

Warren, C. N. 1984. The Desert Region. In *California Archaeology*, edited by M. J. Moratto, 339–430. Orlando, FL, Academic Press.

Warren, C. N., and R. H. Crabtree. 1986. Prehistory of the Southwestern Area. In *Great Basin* (Handbook of North American Indians. Vol. 11. W. C. Sturtevant, general editor), edited by W. L. d'Azevedo, 183–193. Washington, D.C., Smithsonian Institution.

Chapter 11
Procurement and Consumption of Obsidian in the Early Formative Mixteca Alta: A View from the Nochixtlán Valley, Oaxaca, Mexico

Jeffrey P. Blomster and Michael D. Glascock

Abstract: *Obsidian, through compositional sourcing analysis, provides a robust documentation of interregional interaction and political economy throughout Mesoamerica. Recently sourced obsidian data from the Early Formative (1200/1150–850 BC) site of Etlatongo, in the Mixteca Alta of Oaxaca, Mexico, are used to examine the types of material that entered this large village. The data show that Etlatongo villagers acquired obsidian from a surprising range of sources, while focusing primarily on one Central Mexican source. To place these data in a larger context, samples sourced from an earlier site in the region (Yucuita) were used, as well as samples from an earlier site in an adjacent region (Rancho Dolores Ortíz). Rancho Dolores Ortíz may have been an important node in the exchange network through which obsidian entered the Nochixtlán Valley. These data challenge previous interpretations about the involvement of the Mixteca Alta in Early Formative exchange in Oaxaca.*

Keywords: *Obsidian, Sources, Early Formative Period, Mixteca Alta, Oaxaca, Mexico*

Introduction

Throughout the vast panorama of time and space encompassed by Mesoamerica, interregional exchange plays a constant role. One material that moved through exchange networks and interaction spheres throughout pre-Hispanic Mesoamerica is obsidian. Due to its utilisation in a variety of quotidian and ritual activities, obsidian is a valuable tool for researchers throughout Mesoamerica. In addition to site-level studies that look at tool production, there has also been a focus on identifying quarries to further understand extraction, production and exchange (Cobean 2002). Most researchers continue to use sourcing studies of obsidian as a salient way to explore interaction and exchange, both within and between regions. Indeed, the larger region in which the current project is situated, Oaxaca, was the setting during the 1970s for some of the earliest sourcing research on this material in Mesoamerica, through the pioneering work of Jane Pires-Ferreira (1975; Winter and Pires-Ferreira 1976) in the Valley of Oaxaca and Robert Zeitlin (1978, 1979) in the Isthmus of Tehuantepec. This paper examines from which sources Early Formative villagers in the Nochixtlán Valley procured obsidian.

Mesoamerica refers to a region crossing modern national boundaries in which disparate groups shared core traditions, including calendrical practices, cosmologies, and sacrifice (Figure 11.1). Archaeologists divide the pre-Hispanic past of highland Mesoamerica, beginning with sedentary villages, into three major periods: Formative or Pre-Classic (1600 BC – AD 300), Classic (AD 300–900), and Post-Classic (AD 900–1521). Throughout the vast span of pre-Hispanic occupation of Mesoamerica, the extraction and consumption of obsidian, the production of tools, and its exchange within and between regions reflect changing political economies and shifting strategies, alliances, and control. The procurement and use of obsidian occurred within a larger social and economic context. Understanding the movement of obsidian provides crucial data on the exchange networks in which different regions of Mesoamerica engaged from the Early Formative to Late Post-Classic, throughout dynamic cycles of political change and development.

The sourcing of obsidian from sites in the modern Mexican state of Oaxaca (Figure 11.2) is particularly useful for several reasons. Although the earliest sourcing studies remain problematic due to the technology employed in the 1970s, there has been a rich history of obsidian sourcing in various parts of Oaxaca, encompassing different chronological phases (see Elam 1993; Joyce *et al.* 1995; Pires-Ferreira 1975; Winter 1989a). Of especial importance, no obsidian sources have been documented anywhere in Oaxaca, despite much earlier talk of an "unknown Oaxacan source" (Cobean *et al.* 1991; Pires-Ferreira 1975). Because of the lack of an obsidian source in Oaxaca, this material was invariably imported into villages and represents interaction and exchange. Chemical characterisation of obsidian from archaeological sites informs on consumption and participation in interregional exchange networks. The great distances involved between sites in Oaxaca and obsidian sources in Mexico and Guatemala (see Figures 11.1 and 11.3) renders unlikely the prospect of regular direct journeys and extraction of obsidian by Oaxacan villagers.

In this paper compositional data are presented on the obsidian sources utilised in one part of Oaxaca, the Nochixtlán Valley, during the Early Formative period, divided here into two phases: Cruz A (1500–1200 BC) and Cruz B (1200/1150–850 BC). Both phases are times when interaction among villages intensified throughout Mesoamerica and socio-political complexity can first be observed in several regions of Oaxaca with the presence of emerging chiefdoms (Blomster 2004a; Marcus and Flannery 1996; Winter 1989b). For now, largely ignored are the exchange mechanisms by which the obsidian arrived in the Nochixtlán Valley and how it was distributed within sites, concentrating instead on the obsidian sources

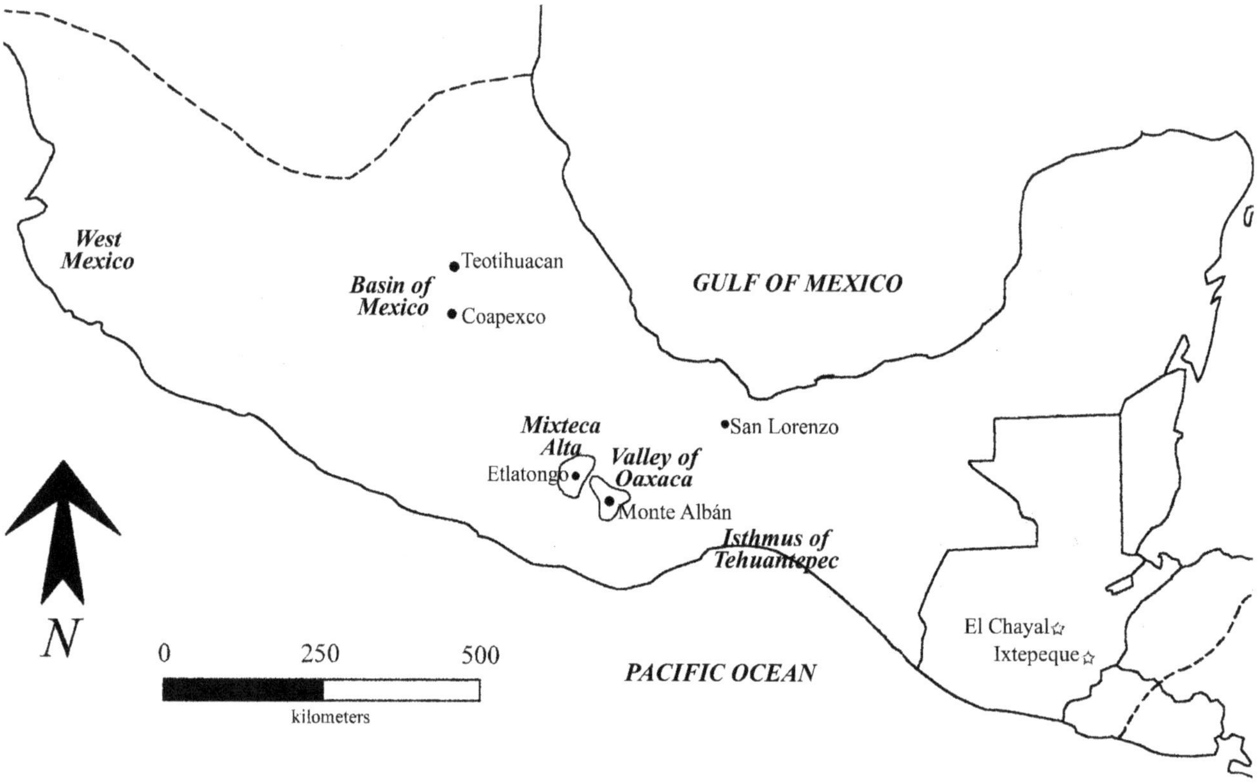

Figure 11.1. Map of Mesoamerica, showing major sites mentioned in the text (solid circles), and Guatemalan obsidian sources (clear stars). Regions are shown in bold and italic letters. Solid lines mark modern national boundaries, while the two dashed lines represent the northern and southern boundaries of Mesoamerica

themselves and the different focus on them between Cruz A and Cruz B phases.

Presented here are compositional data for 275 Early Formative obsidian samples (Appendices 1 and 2), part of a larger project of obsidian analysis that now includes 415 samples (Blomster 2004b). The vast majority of obsidian samples (n = 210) come from excavations at the site of Etlatongo, probably the centre of a small chiefdom in the Nochixtlán Valley during the Cruz B phase (Blomster 1998, 2004a; see below). Etlatongo, however, has not provided samples from the earlier Cruz A phase. While the current project focuses on Etlatongo, samples were analysed from two earlier sites to better understand in which networks these villagers participated, and to explore differences between the Cruz A and Cruz B phases. Forty-five obsidian samples were analysed from Cruz A occupations at Yucuita, a site approximately 8km to the north of Etlatongo. An additional 20 samples came from Rancho Dolores Ortíz, a Cruz A site located approximately 200km northeast of Yucuita in the Cuicatlán Cañada, a region that served to connect disparate regions throughout the long history of exchange networks in Oaxaca (see Figure 11.2). From an initial project started in 1994 (Blomster and Glascock 2002) to samples analysed in 2009, these data were acquired during three periods of analysis at the University of Missouri Research Reactor (MURR), primarily with Neutron Activation Analysis (hereafter – NAA; see Appendix 1). For two Early Formative samples from Etlatongo, ET92413 and ET92417, X-ray Florescence (XRF) was utilised (see Appendix 2). Both methods are described below.

Etlatongo and the Early Formative in the Nochixtlán Valley

The Mixteca Alta encompasses a series of small, irregular valleys surrounded by mountains; these valleys account for only 20% of the land mass in this region (Smith 1976, 24). While rich information about Post-Classic Mixtec religion and politics comes from the pre-Hispanic Mixtec themselves, recorded in a series of painted books called "codices," the archaeology of this region has not been at the same frequency and scale as in the Valley of Oaxaca, particularly well documented from the Late Formative through the Post-Classic at Monte Albán, due to numerous field seasons at that Zapotec urban center first initiated by Alfonso Caso in 1931 (see Caso *et al.* 1967). Within the Mixteca Alta, the focus here is on the largest of its valleys, the Nochixtlán Valley, which Ronald Spores systematically surveyed in the 1960s (Spores 1972). Located at 2000m above sea level, the Nochixtlán Valley contains at least three documented Cruz A and three Cruz B sites. The exact number of Cruz A and Cruz B sites in the Nochixtlán Valley remains unclear as the 1960s survey did not distinguish between three currently recognised phases, Cruz A, Cruz B, and Cruz C, ranging from 1500–700 BC (Winter and Blomster 2008).

While there has been substantial investigations on the origins of socio-political complexity in the contemporaneous Valley of Oaxaca (Flannery and Marcus 1994; Marcus and Flannery 1996; Winter 1989b), much less research has been devoted to this topic in the Nochixtlán Valley (but see Blomster 2004a; Spores 1984, 2001;

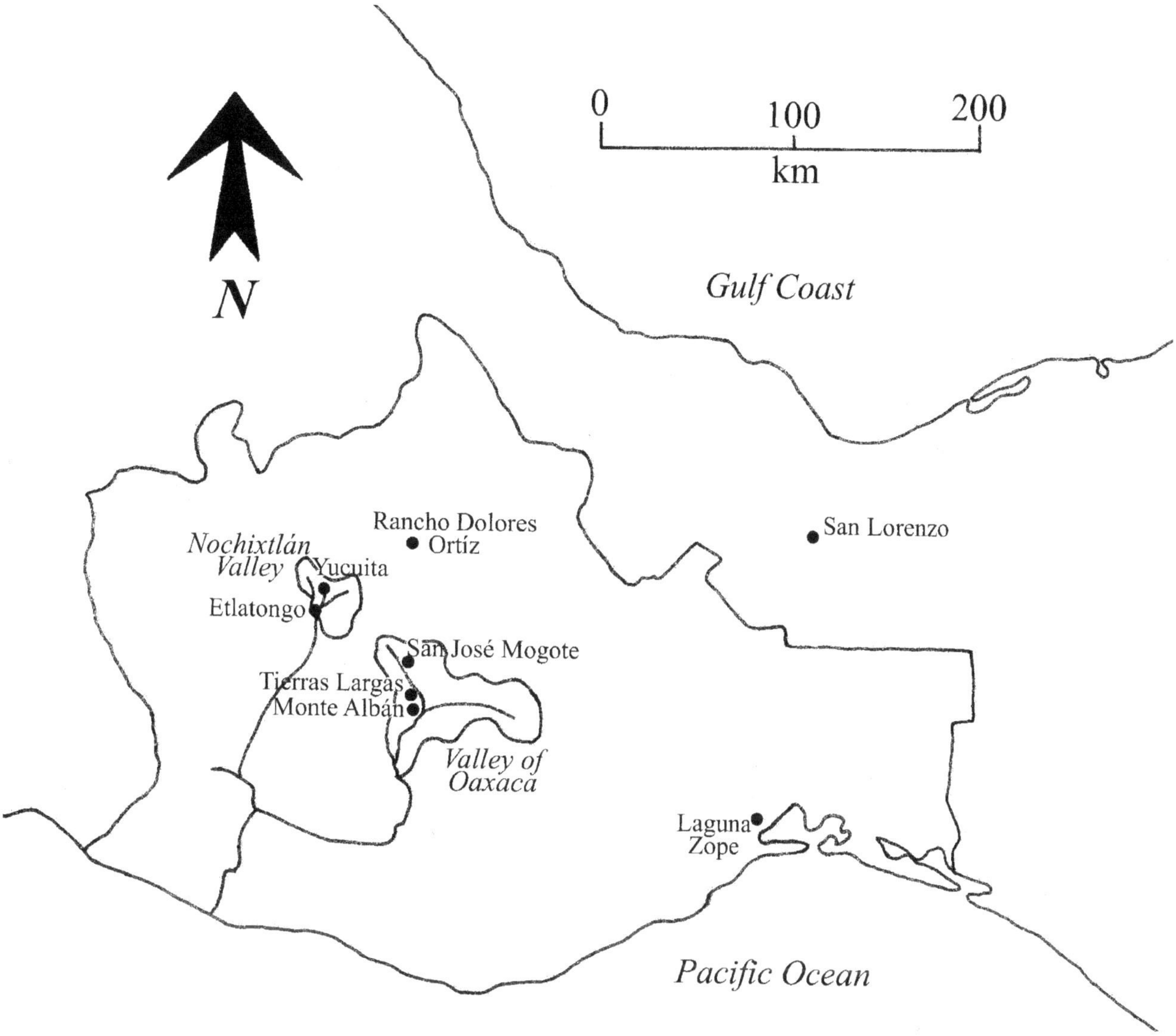

Figure 11.2. Map of Oaxaca State, showing sites from which obsidian samples were sourced and other imporant sites for reference

Winter 1982, 1984; Zárate Morán 1987), providing fewer comparative data for the current Etlatongo focus. Indeed, some scholars who have focused on the contemporaneous Valley of Oaxaca (Drennan 1983; Marcus 1989), perhaps due to comparatively fewer Mixtec investigations, have essentially characterised the Mixteca Alta as a periphery, apparently uninvolved in the kinds of socio-political complexity and interregional interaction evinced at San José Mogote, the site of a small chiefdom in the northwest sector of the Valley of Oaxaca (Figure 11.2). In this paper, the Etlatongo obsidian data are deployed to evaluate this hypothesis.

The largest Cruz A site in the Nochixtlán Valley – Yucuita – was first identified by Spores (1972) and its Early Formative component further investigated by Plunket (1990) and Winter (1982, 1984). Spores (1984) excavated at Yucuita (Site N203K), which lies on the south slope of the Yucuita Hill located about 350m east of the Yucuita River (Figure 11.4), exposing sections of small structures at the site. In a subsequent intensive survey, Plunket (1990, 358) found Cruz A sherds on the surface over an area of 65ha at Yucuita. What seems like an extremely large site may be an agglomeration of a few ranchos; only additional excavation can determine the actual size of Cruz A Yucuita. Winter (1982) excavated portions of two Cruz A households represented by three bell-shaped pits, features L3, C90, and C2. All 45 obsidian samples from Yucuita analysed in this project come from these three features, with three samples from Feature C2, 20 from Feature C90, and 22 from Feature L3.

The presence of obsidian at Cruz A Yucuita shows how involved early villagers were with interregional interaction. Throughout Oaxaca, obsidian appears in the archaeological record by 1300 BC (Winter 1984, 1989a). Unlike some regions without local raw materials useful for chipped-stone production, such as portions of the Gulf Coast of Mexico, the early presence of obsidian at Yucuita seems unnecessary given the presence of dependable chert sources. From Yucuita, there is a high quality chert outcrop just 2km to the northwest, at what would become an important resource and quarry incorporated into the Classic period hilltop urban centre of Yucuñudahui. Obsidian has been estimated as 18% of the total chipped stone inventory, by piece, at Cruz A Yucuita (Winter 1984, Table 9.1), evincing

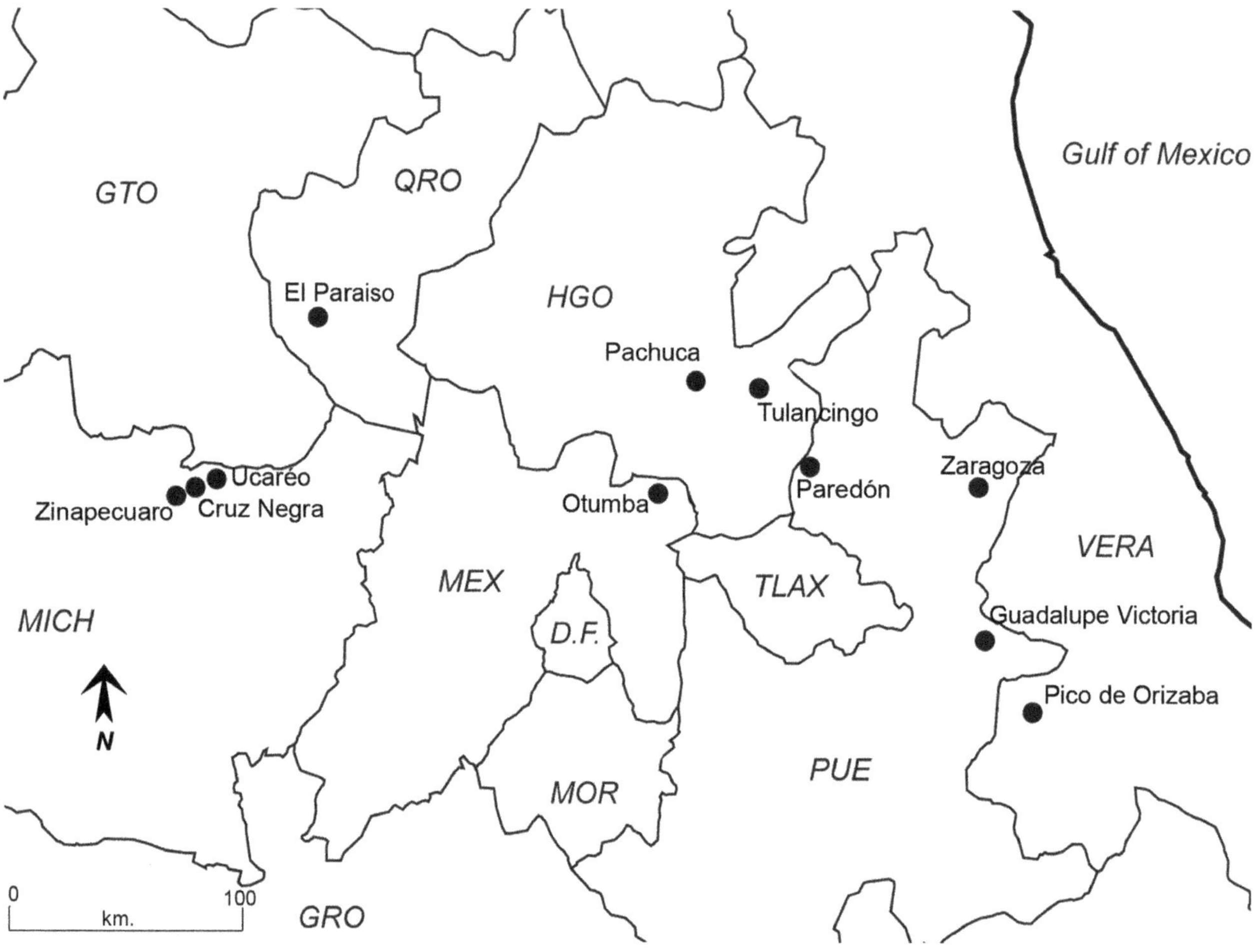

Figure 11.3. Map of Mexican obsidian sources discussed in text; modified from Smith et al. (2007, Figure 1)

substantial utilisation of this necessarily imported resource. This early extensive use of obsidian challenges "least effort models" and other primarily formalist economic ways in which early villagers are often characterised. Acquisition of obsidian formed an important social act through which early villagers at Yucuita connected with their contemporaries within and beyond the Nochixtlán Valley (Mauss 1954; Sahlins 1972).

No primary contexts dating to the Cruz A phase have been found at Etlatongo by two different projects (Blomster 1998, 2004a; Zárate Morán 1987). The presence, however, of scattered Cruz A sherds in several early Cruz B contexts can be used to infer the presence of a Cruz A rancho or small village at Etlatongo.

Yucuita's population precipitously declined by the start of the Cruz B phase, at which time Etlatongo became the largest site in the Nochixtlán Valley (Blomster 2004a). Located north of the confluence of two rivers (the Yanhuitlán and Yucuita; see Figure 11.4), Etlatongo grew during the Cruz B phase to approximately 26.2ha, a figure determined by both excavation (sondage and robotage) and intensive surface survey (Figure 11.5). The surface survey of Etlatongo revealed evidence of outlying barrios that could further extend the size of the Cruz B village. Socio-political complexity at Etlatongo was comparable to that documented for San José Mogote (Marcus and

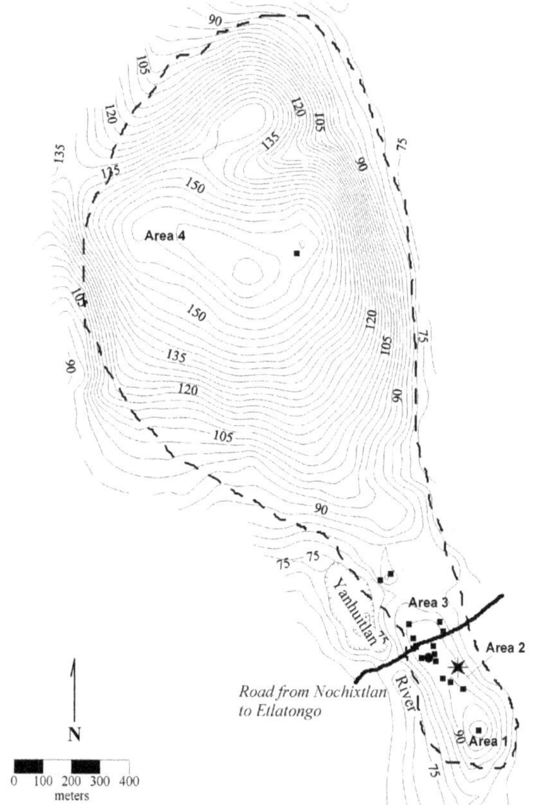

Figure 11.5. Map of Etlatongo; squares represent excavated units (not to scale). The four research areas/areas are also indicated. The dashed line represents the maximum site boundary for all occupational phases

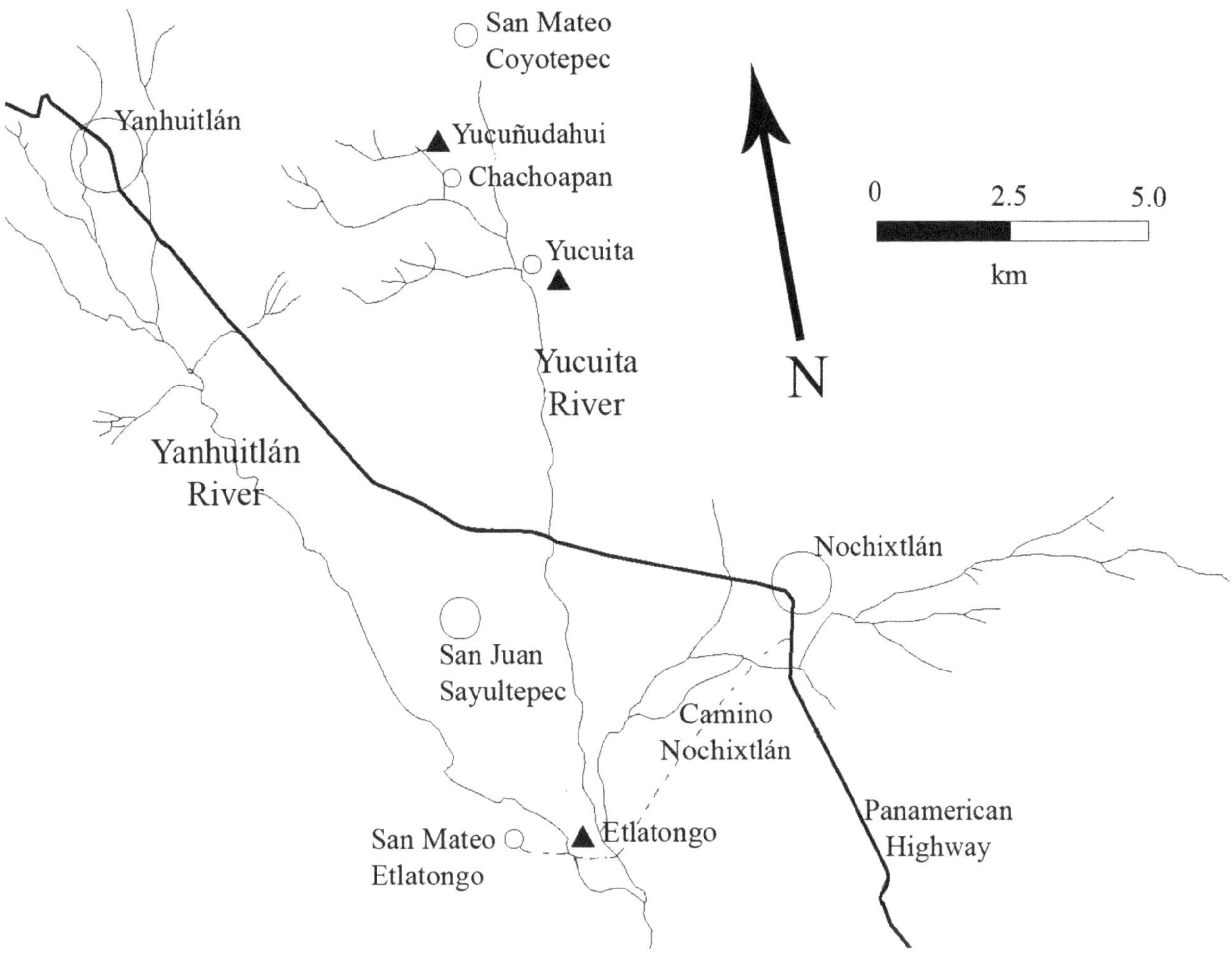

Figure 11.4. Setting of the sites of Etlatongo and Yucuita within the Nochixtlán Valley. Black triangles are archaeological sites, while clear circles are modern towns, roughly adjusted for size

Flannery 1996). While the amount of integration with the surrounding hamlets and villages in the Nochixtlán Valley remains unclear, the size of Etlatongo and the clear evidence of increasing socio-political complexity suggest that the site became the head of a small chiefdom after 1150 BC (Blomster 2004a).

The Cruz B occupation is generally buried by three metres of subsequent inhabitation, making reconstructions of Etlatongo's Early Formative occupation problematic at best. There does, however, appear to be variety in the types of architecture and use of space at the site. An area in the southern portion of Etlatongo (Area 1; see Fig. 11.5), dominated by a large mound with a substantial Cruz B component, has been interpreted as public space. Several portions of the Cruz B site have what is interpreted as "higher" status occupations (the term higher is used rather than simply "high" or "elite" to reinforce that social differentiation lay along a continuum – see Blomster 2004a). A concentration of higher status structures was found in the northern portion of what was designated as Research Area 2, probably representing successive structures built by the same higher status family (Figure 11.5). The earliest primary occupation excavated at Etlatongo is represented by one structure with five floors/resurfacing episodes and three features (Blomster 1998, 2004a).

As a sample of the kinds of contexts from which the obsidian analysed in this paper come, some details are provided here of two associated contexts. The first floor of the earliest documented occupation was placed on a nearly 50cm high deposit of platform fill (referred to as B. 714), elevating it substantially above the surrounding land. Calibrated using two sigma statistics, a sample for radiocarbon dating from Feature 29 (also referred to as B. 715), a bell-shaped pit, produced a range of ca. 1300–900 cal BC (Blomster 2004a, 105). Based on its position – the deepest of several metres of Cruz B deposits – and the ceramics, the materials in Feature 29 probably date to just prior to 1150 BC. Because this feature extended beyond the confines of the excavation unit, presenting the possibility of collapse, only approximately a quarter of the feature was excavated to its base (Blomster 1998, 478–479).

Not including figurines and other non-vessel ceramic objects, 368 sherds were recovered from B. 715, and 428 from B. 714 (see Table 11.1). For the analysis of ceramics from Etlatongo, generally the emphasis is on minimum number of vessels (MNV) for each context. These data are also presented in Table 11.1. These two contexts contained relatively few examples of chipped-stone fragments; including obsidian, B. 715 yielded 21 and B. 714 yielded a total of 23. As a frequency of total chipped stone, obsidian constitutes only 4.7 % of the B. 715 assemblage, but

Table 11.1. Contents of two Cruz B contexts from Etlatongo

Contents	B. 715	B. 714
Total ceramic sherds*	368	428
MNV	31	32
Total sherds, weight (g)	5439.2	4834.6
Chipped stone	21	23
Chipped stone, weight (g)	218.0	160.5
Obsidian	1	5
Obsidian, weight (g)	10.1	8.3
Obsidian to chipped stone	4.7%	21.7%
Obsidian to chipped stone, weight	4.6%	5.2%
Ratio, obsidian to sherds	0.002	0.012
Ration, obsidian to sherds, by weight	0.002	0.002

*Does not include figurines or non-vessel fragments

substantially more in B. 714 where obsidian is 21.7% of the chipped-stone assemblage. This compares with a frequency, by piece, of 13% from contemporaneous House 1 at San José Mogote, and 22.8% for House 4 (Parry 1987, Table 8). Comparisons by weight are roughly equal between the two Etlatongo contexts (Table 11.1), as the one piece of obsidian in B. 715 is large as shown in Figure 11.6 (on the right). Ratios of obsidian-to-sherd frequencies range from 0.002 to 0.012, below contemporaneous examples from San José Mogote, with 0.066 for House 1 and 0.156 for House 4 (Flannery and Marcus 2005; Parry 1987).

NAA, XRF, and Obsidian Samples

Obsidian sources have been identified only in two main areas of volcanism in Mesoamerica: from east to west, one extends from Veracruz State to Nayarit State (Figure 11.3), while the other traverses Western Honduras to the Pacific coast of Guatemala and El Salvador. Many of the obsidian sources in modern Mexico and Guatemala were used for thousands of years. The complex of associated mines, workshops and habitation areas linked with them generally remain poorly understood (Cobean 2002). As much of the geology between the loci of obsidian on the Pacific Coast remains poorly known, it remains possible that there are obsidian sources that remain undocumented, including within Oaxaca (Cobean 2002, 31). The data presented here are part of a larger obsidian sourcing project, which actually includes 415 sourced obsidian fragments (Blomster 2004b). None of these samples come from an unknown or undocumented source. It appears unlikely that at this late date there is still a significant but undocumented obsidian source in Oaxaca.

The Archaeometry Laboratory at MURR has been conducting obsidian provenance research for almost 30 years. During this time, more than 12,000 source samples from around the world have been collected and analysed. Through NAA, chemical "fingerprints" have been developed for approximately 40 obsidian sources throughout Mesoamerica (Cobean *et al.* 1991; Cobean 2002; Glascock *et al.* 1998, 29). NAA has proven to be the most accurate technique for associating archaeological obsidian artefacts with their source (Smith *et al.* 2007, 430).

Figure 11.6. Sample of obsidian recovered from the earliest Cruz B contexts excavated in 1992 at Etlatongo: B. 715 (on the right) and B. 714 (the three on the left)

Because the chemical variability is greater between obsidian sources than within a single source (Elam 1993, 9), obsidian fragments retrieved from archaeological excavations can be matched to a specific source.

All but two of the obsidian samples presented here were analysed and characterised by NAA at MURR. Samples were prepared and analysed according to procedures that have been previously described (Ambroz 1997; Cobean *et al.* 1991; Glascock *et al.* 1998). After the original samples were crushed to create a number of interior fragments (approximately 25–50mg in size), the fragments were inspected under a magnifier to eliminate those with crush fractures, metallic streaks, etc. For the abbreviated procedure (Glascock *et al.* 1994) used with the short irradiation employed for the majority of the samples submitted from Etlatongo, a sample weighing about 100mg was placed into a clean polyethylene vial. The samples were subjected to a five-second irradiation, a 25 minute period of decay, and counted for 12 minutes using a high-purity germanium (HPGe) detector. This procedure measures six short-lived elements: Al, Ba, Cl, Dy, Mn, K, and Na.

A second and longer irradiation procedure was utilised for three samples from Etlatongo. The specimens, weighing 250–300mg, were placed into high-purity quartz vials and irradiated in bundles of about 30–35 obsidian samples for 70 hours. After decaying for about eight days, the long

Table 11.2. Comparison of Cruz B obsidian sources from Etlatongo with Cruz A sites in the Nochixtlán Valley and Cuicatlán Cañada. Obsidian sources organised by region. Number in parentheses represents the number of sourced samples. Except for values < 1, all percentages have been rounded off to whole numbers

	Cruz A and Cruz B Villages		
Obsidian Source, State	***Rancho Dolores Ortíz, Cruz A (n = 20)***	***Yucuita, Cruz A (n = 45)***	***Etlatongo, Cruz B (n = 210)***
Paredón, Puebla	0	2% (1)	65% (137)
Otumba, Mexico	0	0	19% (39)
Tulancingo, Hidalgo	0	0	0.5% (1)
Guadalupe Victoria, Puebla	90% (18)	82% (37)	8% (16)
Pico de Orizaba, Veracruz	5% (1)	16% (7)	1% (2)
Ucaréo, Michoacán	0	0	5% (11)
Cruz Negra, Michoacán	0	0	0.5% (1)
El Chayal, Guatemala	5% (1)	0	1% (2)
Ixtepeque, Guatemala	0	0	0.5% (1)
Total Sources	**3**	**3**	**9**

irradiation samples were loaded on a sample changer where they were counted for 1800 seconds each to measure the medium-lived elements: Ba, La, Lu, Nd, Sm, U, and Yb. Three weeks later the long irradiation samples were counted again for 10,000 seconds to measure the long-lived elements: Ce, Co, Cs, Eu, Fe, Hf, Rb, Sb, Sc, Sr, Ta, Tb, Th, Zn and Zr. Standards made from SRM-278 Obsidian Rock and SRM-1633a Flyash were similarly prepared and irradiated for calibration and quality control of the analytical data, respectively.

At the conclusion of each irradiation and counting procedure, bivariate scatterplots were used and the results compared against 95% confidence intervals generated using data from obsidian source samples analysed at MURR. All obsidian samples submitted as part of this project were sourced by NAA with two exceptions also reported here. Due to its non-destructive nature, lower costs, and the vastly improved capabilities of modern instrumentation, it was possible for two recently submitted Cruz B Etlatongo samples to use XRF, which is usually successful unless the samples are small, the possible sources are chemically similar to one another, or the artefacts come from as yet unknown or totally unexpected sources. In the case of small samples, the physics of XRF must be well-understood to properly interpret the data and make corrections. If any of these particular difficulties or limitations occurs, NAA is utilised.

The two Etlatongo samples were analysed using an XRF spectrometer (see Appendix 2). The spectrometer employed in this study is an Elva-X XRF made by Elvatech Corporation located in the Ukraine. It is equipped with an air-cooled tungsten anode tube with a 140μm Be window and a thermoelectrically cooled Si-PIN diode detector. The detector has a resolution of 180eV for the 5.9keV peak. The beam dimensions are approximately 4–5 mm. The X-ray tube was operated at 40kV using a tube current of about 20–25μA, yielding a count rate of about 6000 counts per second for most samples. Measurement times were 180 seconds. Peak deconvolution and element concentrations were accomplished using the Elva-X spectral analysis package which enabled measurement of 11 elements (K, Ti, Mn, Fe, Zn, Ga, Rb, Sr, Y, Zr, and Nb). The instrument was calibrated using compositional data from a series of well-characterised source samples in the MURR obsidian reference collection, including 11 Mesoamerican sources (El Chayal, Ixtepeque, San Martin Jilotepeque, Guadalupe Victoria, Pico de Orizaba, Otumba, Paredón, Sierra de Pachuca, Ucaréo, Zaragoza, and Zacualtipan). Consensus values for these obsidian sources were previously determined by NAA and XRF at MURR and in other laboratories. Artefacts larger than 1cm across are suitable for routine analysis without correction.

Selection of Etlatongo Obsidian Samples

To obtain a sample not formed by judgmental selection by the researchers, obsidian fragments were selected based on context rather than any inherent property of the obsidian piece itself. With the 2009 analysis of three additional Cruz B samples from Etlatongo, all obsidian fragments from the selected Cruz B contexts have been analysed and sourced, for a total of 210 samples. These discrete contexts include bell-shaped pits and other pit features, middens, platform fill, and deposits associated with floors and other house features. The contents of two of these contexts (B. 715 and B. 714) are detailed in Table 11.1. As all obsidian fragments from these contexts were analysed, the percentages in Table 11.2 accurately reflect the relative frequencies of obsidian sources represented in these contexts. The majority of samples represent debitage (63%), with the remainder made up of blades (29%) and flake tools (8%).

Cruz B Obsidian Procurement and Consumption at Etlatongo

Cruz B villagers at Etlatongo consumed obsidian from nine different sources, representing three major regions (see Table 11.2). While brief references are made with obsidian data from the contemporaneous chiefly centre of San José Mogote in the Valley of Oaxaca, as noted above, methodological problems with those sourced data preclude a more detailed comparison at this time.

Central Mexican Obsidian at Etlatongo

Etlatongo villagers relied primarily on obsidian from Paredón, in the northeast corner of the Valley of Mexico (Figure 11.3). While usually cited as in the State of Puebla, Paredón actually represents an obsidian source system extending into the states of Puebla and Hidalgo (Cobean 2002, Figure 2.2). Paredón is located southeast of two even larger source systems in Hidalgo: Pachuca and Tulancingo. Obsidian from Paredón, a transparent gray colour with occasional darker bands or streaks, constitutes 65% (n = 137) of the total Cruz B sample from Etlatongo. Obsidian from Paredón was an important source for early prismatic blade production, with blades from this source appearing frequently at Early Formative San Lorenzo in the Gulf of Mexico coast, and Coapexco in the Basin of Mexico (Cobean 2002, 53). Blades made from Paredón obsidian also constitute part of the Cruz B sample reported here from Etlatongo. Paredón obsidian, which occurs near ground surface and thus is relatively easily accessible, was underestimated in earlier trace element analyses, due to the difficulty in chemically distinguishing it from Otumba obsidian (Charlton *et al.* 1978). Perhaps due to these technological limitations, obsidian from Paredón has not been documented from Early Formative Valley of Oaxaca sites (Pires-Ferreira 1975; Winter and Pires-Ferreira 1976).

The focus on Central Mexican obsidian at Etlatongo continues with the second most frequent source, represented by 39 fragments from Otumba, comprising 19% of the Cruz B sample (Table 11.2). Otumba obsidian, from a source system southwest of Paredón, is generally a dark semitranslucent gray, but there is much variety, including an opaque gray (Cobean 2002, 59). Only obsidian from Paredón and Otumba register frequencies in the double digits at Etlatongo. The reliance on these Central Mexican sources is striking, as combined they constitute 84% of the total Cruz B obsidian sample from Etlatongo. Otumba obsidian is the dominant source reported for contemporaneous San José Mogote in the Valley of Oaxaca (where it was called Barranca de los Estetes); it represents just under half of the obsidian sourced from that chiefly centre (Pires-Ferreira 1975, Table 6).

One obsidian fragment in the Etlatongo sample comes from Tulancingo, a source system approximately 20km north to northwest of Paredón with at least six different obsidian outcrops. Tulancingo obsidian, opaque black or gray, is probably underrepresented in earlier chemical analyses of obsidian due to some overlap with the nearby Pachuca source system (Cobean 2002, 49). Perhaps as a result, obsidian from this source has not been reported from the Valley of Oaxaca.

Obsidian from the Gulf Coast at Etlatongo

Villagers at Etlatongo consumed obsidian from two sources associated with the Gulf Coast, Guadalupe Victoria (State of Puebla) and Pico de Orizaba (State of Veracruz). Both obsidian sources are located near the borders of the states of Veracruz and Puebla, with Guadalupe Victoria approximately 25km northwest of Pico de Orizaba, also the name of the volcano that is the third highest peak in Mexico. Obsidian from these two sources appear visually similar: Pico de Orizaba obsidian is a relatively transparent gray with numerous fine dark gray bands visible in the matrix, while Guadalupe Victoria obsidian is cloudier (Cobean and Stocker 2002, 139). At least six obsidian outcrops have been reported on the slopes of Pico de Orizaba, representing two or three chemically and geologically distinct source systems. Well-preserved obsidian mines, including mine shafts with wooden ladders, have been documented on the north slope of the volcano, known as the Ixtetal Valley mines (Cobean and Stocker 2002, 131). The study of the Ixtetal Valley mines revealed that the kind of high quality obsidian necessary for tool production only occurs in relatively narrow horizontal bands, requiring substantial quarrying in order to access these bands (Cobean and Stocker 2002, 137). In contrast, no primary obsidian flows have been found in the Guadalupe Victoria area; as currently understood, the source consists of vast quantities of obsidian cobbles exposed in barrancas (i.e., deep gullies with steep sides) and stream beds (Cobean and Stocker 2002, 167).

Combined, obsidian from these two Gulf Coast sources contributes 18 fragments, or 9%, of the analysed Cruz B obsidian at Etlatongo. Most of these samples (n = 16) are from Guadalupe Victoria; considering the importance of this source throughout Early Formative Oaxaca and Mesoamerica in general, its 8% frequency at Etlatongo seems low. At San José Mogote, Guadalupe Victoria comprises 11% of the small sample (n = 44) of sourced obsidian. In the previous phase in the Valley of Oaxaca, Guadalupe Victoria obsidian forms a majority of the analysed assemblage from the hamlet of Tierras Largas (Pires-Ferreira 1975, Table 6; Winter 1972). At the Isthmus of Tehuantepec centre of Laguna Zope, Guadalupe Victoria obsidian remains a majority of the sourced assemblage (n = 47) contemporaneous with Cruz B Etlatongo (Zeitlin 1978; Zeitlin 1979, Table IV-2). While the lower quality Guadalupe Victoria obsidian fragments in the sample are restricted to debitage and two flake tools, both Pico de Orizaba obsidian specimens from Etlatongo are blade fragments.

One important Gulf Coast obsidian source not represented in the large Cruz B Etlatongo sample is Zaragoza in the State of Puebla. Zaragoza obsidian, initially misidentified in the 1970s as from Altotonga, Veracruz (Pires-Ferreira 1975, Table 6), provides a minor component of the obsidian assemblage analysed from contemporaneous San José Mogote, and is the only sourced obsidian at that Valley of Oaxaca chiefly centre that has not been found at Cruz B Etlatongo. This nearly opaque black obsidian does occur nearly 500 years later at Etlatongo (Blomster 2004b).

West Mexican Obsidian at Etlatongo

In terms of frequency, two sources from West Mexico contribute a combined 5.5% of the Cruz B obsidian at Etlatongo. All West Mexico samples, except for one, are from Ucaréo in Michoacán State, part of a source system that became important during this time. Ucaréo obsidian is a dark translucent gray, often with fine parallel gray bands (Cobean 2002, 67). Artefacts from the Ucaréo–Zinapécuaro source system have been found throughout most of Mesoamerica. Ucaréo has been estimated as second in importance only to the Pachuca quarries for later Mesoamericans. Ucaréo obsidian, along with Paredón obsidian, played a central role in the spread of prismatic blade technology throughout Early Formative Mesoamerica (Cobean 2002, 63–65). Survey of the Ucaréo–Zinapécuaro source system, where three obsidian flows (one of which is the Ucaréo flow, from which the Etlatongo obsidian comes) and cobble deposits cover at least 80km^2, has documented over a thousand quarries and different types of quarries and habitation sites (Healan 1997). Previously misidentified in earlier studies as from the Zinapécuaro flow (Pires-Ferreira 1975), Ucaréo obsidian comprises nearly a third of the sample sourced from San José Mogote, suggesting very different foci in exchanges networks between contemporaneous Mixtecs in the Nochixtlán Valley and Zapotecs in the Valley of Oaxaca.

In addition, one fragment in the Etlatongo sample comes from Cruz Negra, Michoacán, which is one of the three obsidian flows in the Ucaréo–Zinapécuaro source system southwest of Ucaréo (see above). Originally material from this flow was labelled "Ucaréo 2," but it has since been possible to identify Cruz Negra obsidian as compositionally different from the Ucaréo flow (Cobean 2002; Healan 1997). The appearance of a fragment of Cruz Negra obsidian at Etlatongo is significant. To this date, ET92122 (Appendix 1) is the only Cruz Negra sample sourced at MURR that comes from an archaeological context.

Finally, more distant obsidian sources are present at Cruz B Etlatongo, but in very small quantities. Guatemalan sources represent only a combined 1.5% (n = 3) of the Cruz B obsidian sample from Etlatongo. These Guatemalan fragments – from El Chayal and Ixtepeque – were excavated at a possible public area at Etlatongo (see above; Blomster 2004a). Guatemalan obsidian, from El Chayal, formed a minor component – also 2% – of the sample sourced from San José Mogote (Pires-Ferreira 1975, Table 6).

The NAA data show that while ancient villagers at Etlatongo relied primarily on the Central Mexico source of Paredón, and Otumba to a lesser extent, they had access to a variety of exchange networks that connected them to sources from the Gulf Coast (Guadalupe Victoria and Pico de Orizaba), West Mexico (Ucaréo and Cruz Negra), and rarely Guatemala (El Chayal and Ixtepeque). With one exception, all Cruz B contexts analysed from Etlatongo showed an overwhelming emphasis on Paredón, with vast differences between households in presence and frequencies of the other sources.

Cruz A Obsidian Procurement: Yucuita and Rancho Dolores Ortíz

To determine the nature and extent of interaction networks and the antiquity of the Cruz B focus on Paredón obsidian at Etlatongo, analysis of earlier – Cruz A – obsidian is crucial. Because no primary Cruz A occupations and contexts have been documented at Etlatongo (see above), obsidian from the Nochixtlán Valley site with a well-documented Cruz A occupation – Yucuita – was used. The analysis consisted of 45 obsidian samples from three Cruz A storage pits excavated by Winter (1982) in the 1970s at Yucuita (see above), including all obsidian fragments from two features that yielded the most samples (n = 42). In addition, to explore potential routes of obsidian into the Cruz A Nochixtlán Valley, 20 samples were analysed from Rancho Dolores Ortíz, a site whose important role in Early Formative exchange networks is suggested by the high amount (60%) of obsidian relative to other kinds of chipped stone (Winter 1984, 1989a). All obsidian samples sourced from Rancho Dolores Ortíz are from House 1, excavated by M. Winter in 1975. As with Yucuita, Rancho Dolores Ortíz experienced a substantial population decline after the Cañada equivalent of the Cruz A phase, with no occupational remains identified between 1300 and 700 BC (Winter 1984, 188).

The NAA results, compared with those from Cruz B Etlatongo, appear in Table 11.2. These Cruz A compositional data demonstrate a clear pattern – the great importance of Gulf Coast sources, especially Guadalupe Victoria, at both Yucuita and Rancho Dolores Ortíz. These Cruz A data, although based on smaller sample sizes than from Etlatongo, suggest villagers accessed fewer obsidian sources, with only three sources present in both Cruz A assemblages compared to nine obsidian sources identified at Cruz B Etlatongo.

Two Gulf Coast sources – Guadalupe Victoria and Pico de Orizaba – dominate the Yucuita assemblage, where obsidian from these two sources constitutes 98% of the sample, mostly comprised of Guadalupe Victoria obsidian. These data support Winter's (1984, 185) visual sourcing of obsidian from Yucuita as originating primarily from Guadalupe Victoria. The only non-Gulf Coast source represented by the Yucuita sample is one fragment from Paredón, which later became so dominant at Cruz B Etlatongo. Obsidian from more distant locales – West Mexico and Guatemala – is not present in the sample from Cruz A Yucuita. The Yucuita obsidian sample is comprised primarily of flakes (including several bipolar examples) and shatter.

The general pattern exhibited at Cruz A Yucuita is repeated with the assemblage of obsidian sourced from Rancho Dolores Ortíz. Once again, Gulf Coast sources overwhelmingly dominate this small sample, with 95% originating from that region, all but one from Guadalupe Victoria. These data support earlier compositional analysis of three samples from Rancho Dolores Ortíz, all of which were sourced to Guadalupe Victoria (Winter 1989a, Table 1). Interestingly, the non-Gulf Coast obsidian does not come from Central Mexico – Paredón is completely absent – but from a much more distant locale: El Chayal, Guatemala. Given the scarcity of Guatemalan obsidian in the much larger Etlatongo sample, the single fragment of El Chayal obsidian in the small Cruz A sample appears significant. As noted above, obsidian occurs in high frequency at Rancho Dolores Ortíz; the site probably played a crucial role in Cruz A exchange networks. The site's strategic location allowed occupants, who appear to have been more involved in obsidian exchange than their contemporaries in the Nochixtlán Valley, to funnel obsidian from the Gulf Coast region into Oaxaca, as well as tapping into networks that included Guatemalan obsidian. Some of the Guatemalan obsidian may have come via a trade route from the Isthmus of Tehuantepec, where El Chayal obsidian became so important in the subsequent Golfo phase (roughly equivalent to Cruz B) at Laguna Zope (Zeitlin 1978, 1979).

It is important to note that while only obsidian is considered in this paper, one should not assume Early Formative villagers at this time were as focused as the authors are on this one resource. Each household engaged in numerous subsistence and crafting activities in a larger domestic economic context; indeed, some of the obsidian-heavy contexts from Rancho Dolores Ortíz also yielded abundant freshwater clam shell ornament fragments (Winter 1984, 187). Some Early Formative households at Etlatongo were engaged in the production of pottery, ceramic figurines, shell ornaments, and a repertoire of non-obsidian chipped stone tools, what some archaeologists have called "multicrafting" (Hirth 2006).

Conclusion

The Cruz B data from Etlatongo reveal a transformation in obsidian procurement from earlier patterns in the Nochixtlán Valley. The previously dominant source from the Gulf Coast, the low quality Guadalupe Victoria obsidian, represents only 8% of the total Cruz B Etlatongo obsidian assemblage. Instead, Etlatongo villagers focused on Central Mexican sources, with 65% of the obsidian deriving from Paredón. The sample reported here documents an explosion in the number of sources procured and consumed, tripling from three at Cruz A Yucuita to nine at Cruz B Etlatongo. The increase in number of sources acquired and the shift in focus to Central Mexican obsidian occurred during a transformation in obsidian tools, with the appearance of prismatic blades during this time (see Clark 1987; Clark and Lee 1984, 255; Cobean *et al.* 1971, 666). These changes in sources and the nature of material exchange reflect the dramatic transformations in interregional interaction and social complexity emblematic of the Cruz B phase, a time of increasing socio-political complexity throughout Mesoamerica (Blomster 2004a). Etlatongo became an important chiefly centre during the Cruz B phase and pursued exchange networks different from those in the Cruz A phase.

Previous researchers have consigned the Nochixtlán Valley and the Mixteca Alta into a more peripheral role in interregional interaction and socio-political complexity compared with the contemporaneous Valley of Oaxaca (Drennan 1983; Marcus 1989). The data presented here for obsidian from Cruz B Etlatongo directly contradict this hypothesis in terms of interregional interaction. Rather than encountering fewer obsidian sources in the Nochixtlán Valley than in the Valley of Oaxaca, the sample of 210 Cruz B obsidian fragments reveals that nine discrete sources were accessed, while five were identified in the smaller (n = 44) San José Mogote sample, with two probable unidentified sources increasing the potential number of sources at San José Mogote to seven. At Cruz B Etlatongo, five sources (Paredón, Tulancingo, Pico de Orizaba, Cruz Negra, and Ixtepeque) either were not used or unidentified in the Valley of Oaxaca (perhaps Ixtepeque is the "Unknown Guatemalan" source at San José Mogote, and Tulancingo may be the "Unknown Oaxaca" source). The larger sample impacts data richness, a phenomenon observed by Elam (1993, 104–109), with minor sources more likely to be identified in larger data sets. Minor obsidian sources, however, also appear in the Valley of Oaxaca data, where only three obsidian sources contribute over 10% each of the obsidian analysed from San José Mogote (Pires-Ferreira 1975, Table 6).

Early Formative villagers in the Nochixtlán Valley pursued a myriad of exchange strategies, with access to some sources not yet documented in the Valley of Oaxaca. Indeed, the only obsidian source used at the contemporaneous chiefly centre of San José Mogote not encountered in the larger Etlatongo sample is from Zaragoza, which further emphasizes the different interaction spheres in which villagers from Etlatongo and San José Mogote participated. While Etlatongo villagers focused on Central Mexican sources, their contemporaneous "neighbours" at San José Mogote, although also heavily invested in Central Mexican obsidian (albeit with a different focus – Otumba), obtained nearly a third of their obsidian from West Mexico, a region which played only a minor role (5.5%) at Etlatongo.

The exact movement of obsidian to Etlatongo and the underlying exchange mechanisms remain subjects beyond the scope of the current paper. The presence of obsidian blades at Etlatongo occur in the context of a lack of obsidian core fragments; while this may suggest some tools arrived as finished pressure blades (Clark 1987), the amount of obsidian debitage evinces significant obsidian tool production at Etlatongo.

In terms of exchange routes, important nodes in the exchange network lay beyond the Nochixtlán Valley to the south, at the important Isthmus village of Laguna Zope, and at least during the Cruz A phase, to the northeast, where the Cuicatlán Cañada site of Rancho Dolores Ortíz played an important role in the movement of Gulf Coast and Guatemalan obsidian, a role which appears to have been abandoned during the Cruz B phase. Radical transformations in interaction spheres and exchange routes occurred during the Cruz B phase, with the shift from Gulf Coast to Central Mexican obsidian sources in the Nochixtlán Valley, a time when imported pottery from the Gulf Coast Olmec site of San Lorenzo appears at Etlatongo (Blomster *et al.* 2005). Interaction spheres through which obsidian moved may have overlapped before entering the Nochixtlán Valley. Etlatongo villagers may not have dealt specifically with discrete networks which moved Gulf Coast, Central Mexican, and West Mexican obsidian. Other non-local materials also arrived at Etlatongo, including a magnetite mirror from the Valley of Oaxaca and marine shell (Blomster 1998, 2004a; Pires-Ferreira 1975; Winter 1984). Etlatongo villagers engaged in a complex and dynamic web of interactions, with different households negotiating access to resources with various levels of networks linking them to contemporaries throughout Mesoamerica.

Acknowledgements

This research has been supported by the Foundation for the Advancement of Mesoamerican Studies, Inc.; the University Facilitating Fund, George Washington University; and a US National Science Foundation grant (BCS-0102325) to the Archaeometry Lab at MURR. This research could not have been performed without the cooperation and permission of the Consejo de Arqueología of the Instituto Nacional de Antropología e Historia (INAH). We extend our thanks to everyone involved on both the Oaxaca and Mexico City levels of INAH in helping us export these samples to MURR, especially Nelly Robles García, Eduardo López Calzada and Joaquín García-Bárcena. Special thanks is due to Marcus Winter in Oaxaca for providing obsidian samples from his excavations at Yucuita and Rancho Dolores Ortíz. Philip Perazio was a patient teacher in classifying the formal attributes of the obsidian assemblage. This essay has benefited greatly from constructive critiques of an earlier draft by Mike Ohnersorgen and Geoffrey Braswell. We are responsible for any remaining errors. Finally, we wish to thank the many people at San Mateo Etlatongo who have assisted in all stages of the research, plus the many colleagues and friends who have participated in the research or offered advice throughout the project.

References

Ambroz, J. A. 1997. Characterization of Archaeologically Significant Obsidian Sources in Oregon by Neutron Activation Analysis. Unpublished MS Thesis. Department of Chemistry, University of Missouri, Columbia, MO.

Blomster, J. P. 1998. At the Bean Hill in the Land of the Mixtec: Early Formative Social Complexity and Interregional Interaction at Etlatongo, Oaxaca, Mexico. Unpublished PhD Dissertation. Department of Anthropology, Yale University, New Haven, CT.

Blomster, J. P. 2004a. *Etlatongo: Social Complexity, Interaction, and Village Life in the Mixteca Alta of Oaxaca, Mexico.* Belmont, CA, Wadsworth.

Blomster, J. P. 2004b. Diachronic and Synchronic Analyses of Obsidian Procurement in the Mixteca Alta, Oaxaca. Unpublished Final Report submitted to the Foundation for the Advancement of Mesoamerican Studies, Inc., Crystal River, FL.

Blomster, J. P., and M. D. Glascock. 2002. Obsidian Exchange in Formative Period Oaxaca: A View from the Mixteca Alta. Paper presented at the 67th Annual Meeting of the Society for American Archaeology, Denver, CO.

Blomster, J. P., H. Neff, and M. D. Glascock. 2005. Olmec Pottery Production and Export in Ancient Mexico Determined through Elemental Analysis. *Science* 307, 1068–1072.

Caso, A., I. Bernal, and J. R. Acosta. 1967. *La Cerámica de Monte Albán* (Memorias del Instituto Nacional de Antropología e Historia 13). Mexico City, Instituto Nacional de Antropología e Historia.

Charlton, T. H., D. C. Grove, and P. K. Hopke. 1978. The Paredón, Mexico, Obsidian Source and Early Formative Exchange. *Science* 201, 807–809.

Clark, J. E. 1987. Politics, Prismatic Blades, and Mesoamerican Civilization. In *The Organization of Core Technology*, edited by J. Johnson and C. Morrow, 259–284. Boulder, CO, Westview Press.

Clark, J. E., and T. A. Lee, Jr. 1984. Formative Obsidian Exchange and the Emergence of Public Economies in Chiapas, Mexico. In *Trade and Exchange in Early Mesoamerica*, edited by K. G. Hirth, 235–274. Albuquerque, University of New Mexico Press.

Cobean, R. H. (ed.). 2002. *A World of Obsidian: The Mining and Trade of Volcanic Glass in Ancient Mexico* (Serie Arqueologia de Mexico). Mexico City and Pittsburgh, PA, Instituto Nacional de Antropologia e Historia/University of Pittsburgh.

Cobean, R. H., M. D. Coe, E. A. Perry, K. K. Turekian, and D. P. Kharkar. 1971. Obsidian Trade at San Lorenzo Tenochtitlan, Mexico. *Science* 174, 666–671.

Cobean, R. H., and T. L. Stocker. 2002. Obsidian Sources on or Near the Slopes of Pico de Orizaba Volcano. In *A World of Obsidian: The Mining and Trade of a Volcanic Glass in Ancient Mexico*, edited by R. H. Cobean, 131–182. Mexico City and Pittsburgh, PA, Instituto Nacional de Antropologia e Historia/University of Pittsburgh.

Cobean, R. H., J. R. Vogt, M. D. Glascock, and T. L. Stocker. 1991. High-Precision Trace-Element Characterization of Major Mesoamerican Obsidian Sources and Further Analyses of Artifacts from San Lorenzo Tenochtitlan, Mexico. *Latin American Antiquity* 2, 69–91.

Drennan, R. D. 1983. Ritual and Ceremonial Development at the Early Village Level. In *The Cloud People: Divergent Evolution of the Zapotec and Mixtec Civilizations*, edited by K. V. Flannery and J. Marcus, 46–50. New York, Academic Press.

Elam, J. M. 1993. Obsidian Exchange in the Valley of Oaxaca, Mexico, 2500–500 BP. Unpublished PhD Dissertation. Department of Anthropology, University of Missouri, Columbia, MO.

Flannery, K. V., and J. Marcus. 1994. *Early Formative Pottery of the Valley of Oaxaca, Mexico* (Memoirs of the University of Michigan Museum of Anthropology 27). Ann Arbor, University of Michigan.

Flannery, K. V., and J. Marcus. 2005. *Excavations at San José Mogote 1: The Household Archaeology* (Memoirs of the University of Michigan Museum of Anthropology 40). Ann Arbor, University of Michigan.

Glascock, M. D., G. E. Braswell, and R. H. Cobean. 1998. A Systematic Approach to Obsidian Source Characterization. In *Archaeological Obsidian Studies: Method and Theory*, edited by M. S. Shackley, 15–65. New York and London, Plenum Press.

Glascock, M. D., H. Neff, K. Stryker, and T. N. Johnson. 1994. Sourcing Archaeological Obsidian by an Abbreviated NAA Procedure. *Journal of Radioanalytical and Nuclear Chemistry* 180, 29–35.

Healan, D. M. 1997. Pre-Hispanic Quarrying in the Ucaréo–Zinapécuaro Obsidian Source Area. *Ancient Mesoamerica* 8, 77–100.

Hirth, K. 2006. Modeling Domestic Craft Production at Xochicalco. In *Obsidian Craft Production in Ancient Central Mexico*, edited by K. Hirth, 275–286. Salt Lake City, University of Utah Press.

Joyce, A. A., J. M. Elam, M. D. Glascock, H. Neff, and M. Winter. 1995. Exchange Implications of Obsidian Source Analysis from the Lower Río Verde Valley, Oaxaca, Mexico. *Latin American Antiquity* 6, 3–15.

Marcus, J. 1989. Zapotec Chiefdoms and the Nature of Formative Religions. In *Regional Perspectives on the Olmec*, edited by R. J. Sharer and D. C. Grove, 148–197. New York, Cambridge University Press.

Marcus, J., and K. V. Flannery. 1996. *Zapotec Civilization: How Urban Society Evolved in Mexico's Oaxaca Valley*. New York, Thames and Hudson.

Mauss, M. 1954. *The Gift*. London, Cohen.

Parry, W. J. 1987. *Chipped Stone Tools in Formative Oaxaca, Mexico: Their Procurement, Production and Use* (Memoirs of the University of Michigan Museum of Anthropology 20). Ann Arbor, University of Michigan.

Pires-Ferreira, J. W. 1975. *Formative Mesoamerican Exchange Networks with Special Reference to the Valley of Oaxaca* (Memoirs of the University of Michigan Museum of Anthropology 7). Ann Arbor, University of Michigan.

Plunket, P. 1990. Patrones de Asentamiento en el Valle de Nochixtlán y Su Aportación a la Evolución Cultural en la Mixteca Alta. In *Lecturas Históricas del Estado de Oaxaca. Vol. 1. Época Prehispánica*, edited by M. Winter, 349–378. Mexico City, Instituto Nacional de Antropología e Historia.

Sahlins, M. 1972. *Stone Age Economics*. Chicago, Aldine.

Smith, C. E., Jr. 1976. *Modern Vegetation and Ancient Plant Remains of the Nochixtlán Valley, Oaxaca* (Vanderbilt University Publications in Anthropology 16). Nashville, TN, Vanderbilt University.

Smith, M. E., A. L. Burke, T. S. Hare, and M. D. Glascock. 2007. Sources of Imported Obsidian at Postclassic Sites in the Yautepec Valley, Morelos: A Characterization Study Using XRF and INAA. *Latin American Antiquity* 18, 429–450.

Spores, R. A. 1972. *An Archaeological Settlement Survey of the Nochixtlán Valley, Oaxaca* (Vanderbilt University Publications in Anthropology 1). Nashville, TN, Vanderbilt University.

Spores, R. A. 1984. *The Mixtecs in Ancient and Colonial Times*. Norman, University of Oklahoma Press.

Spores, R. A. 2001. Estudios Mixtecos, Ayer, Hoy y Mañana: Dónde Stábamos, Dónde Estamos, Hacia Dónde Vamos? In *Procesos de Cambio y Conceptualización del Tiempo: Memoria de la Primera Mesa Redonda de Monte Albán*, edited by N. M. Robles García, 167–181. Mexico City, Instituto Nacional de Antropología e Historia.

Winter, M. 1972. Tierras Largas: A Formative Community in the Valley of Oaxaca. Unpublished PhD Dissertation. Department of Anthropology, University of Arizona, Tucson, AZ.

Winter, M. 1982. *Guía Zona Arqueológica de Yucuita*. Oaxaca City, Mexico, Centro INAH Oaxaca.

Winter, M. 1984. Exchange in Formative Highland Oaxaca. In *Trade and Exchange in Early Mesoamerica*, edited by K. G. Hirth, 179–214. Albuquerque, University of New Mexico Press.

Winter, M. 1989a. La Obsidiana en Oaxaca Prehispánica. In *La Obsidiana en Mesoamérica* (Colección Científica 176), edited by M. Gaxiola and J. E. Clark, 345–361. Mexico City, Instituto Nacional de Antropología e Historia.

Winter, M. 1989b. *Oaxaca: The Archaeological Record*. Mexico City, Minutiae Mexicana.

Winter, M., and J. P. Blomster. 2008. Religión e Interacción: Oaxaca y los Olmecas. In *Olmeca: Balance y Perspectivas. Memoria de la Primera Mesa Redonda*, edited by M. T. Uriarte and R. González Lauck, 205–226. Mexico City, Universidad Nacional Autónoma de México.

Winter, M., and J. W. Pires-Ferreira. 1976. Distribution of Obsidian among Households in Two Oaxacan Villages. In *The Early Mesoamerica Village*, edited by K. V. Flannery, 306–311. New York, Academic Press.

Zárate Morán, R. 1987. *Excavaciones de un Sitio Preclásico en San Mateo Etlatongo, Nochixtlán, Oaxaca, México* (B.A.R. International Series 322). Oxford, British Archaeological Reports.

Zeitlin, R. N. 1978. Long-Distance Exchange and the Growth of a Regional Center: An Example from the Southern Isthmus of Tehuantepec, Mexico. In *Prehistoric Coastal Adaptations*, edited by B. L. Stark and B. Voorhies, 183–210. New York, Academic Press.

Zeitlin, R. N. 1979. Prehistoric Long-Distance Exchange on the Southern Isthmus of Tehuantepec, Mexico. Unpublished PhD Dissertation. Department of Anthropology, Yale University, New Haven, CT.

Appendix 1.

Elemental concentrations and source names for Oaxacan obsidian artefacts through NAA (values are in ppm unless otherwise indicated)

Sample #	*Site Name*	*Al (%)*	*Ba*	*Cl*	*Dy*	*K (%)*	*Mn*	*Na (%)*	*Source Name, State*
ET92001	Etlatongo	6.56	0	1003	7.56	4.37	372	2.99	Paredón, Puebla
ET92002	Etlatongo	6.23	0	1129	7.27	4.33	365	2.93	Paredón, Puebla
ET92003	Etlatongo	6.30	0	1137	7.52	4.15	366	2.93	Paredón, Puebla
ET92004	Etlatongo	6.26	0	1072	7.50	4.24	365	2.92	Paredón, Puebla
ET92005	Etlatongo	6.40	0	1074	7.59	4.14	364	2.93	Paredón, Puebla
ET92006	Etlatongo	6.48	0	1108	7.83	4.21	371	2.97	Paredón, Puebla
ET92007	Etlatongo	6.12	0	1058	7.78	4.21	375	3.02	Paredón, Puebla
ET92008	Etlatongo	6.55	0	1114	7.86	4.32	374	3.00	Paredón, Puebla
ET92009	Etlatongo	6.12	0	1093	7.51	3.91	354	2.89	Paredón, Puebla
ET92010	Etlatongo	6.38	0	1074	7.52	3.89	365	2.95	Paredón, Puebla
ET92011	Etlatongo	6.35	0	1044	7.46	4.15	361	2.88	Paredón, Puebla
ET92012	Etlatongo	6.21	0	1080	8.12	3.90	370	2.99	Paredón, Puebla
ET92013	Etlatongo	6.91	672	543	2.16	3.29	563	3.18	Pico de Orizaba, Veracruz
ET92014	Etlatongo	6.80	818	405	2.12	3.70	582	3.25	Pico de Orizaba, Veracruz
ET92015	Etlatongo	6.30	915	463	1.99	3.21	507	3.17	Guadalupe Victoria, Puebla
ET92016	Etlatongo	6.88	0	1137	7.54	4.13	366	2.95	Paredón, Puebla
ET92017	Etlatongo	6.66	972	734	1.54	3.46	532	3.33	Guadalupe Victoria, Puebla
ET92018	Etlatongo	6.62	0	1127	8.04	4.18	374	3.02	Paredón, Puebla
ET92019	Etlatongo	7.52	823	753	3.39	3.42	404	3.09	Otumba, State of Mexico
ET92020	Etlatongo	6.28	0	1140	7.71	4.21	367	2.96	Paredón, Puebla
ET92021	Etlatongo	7.04	745	757	2.76	3.27	398	3.05	Otumba, State of Mexico
ET92022	Etlatongo	6.45	142	563	3.89	4.16	177	2.82	Ucaréo, Michoacan
ET92023	Etlatongo	6.65	141	501	3.79	4.25	175	2.82	Ucaréo, Michoacan
ET92024	Etlatongo	6.82	907	546	2.40	3.62	654	3.07	El Chayal, Guatemala
ET92025	Etlatongo	6.26	48	458	3.69	3.99	172	2.73	Ucaréo, Michoacan
ET92026	Etlatongo	6.29	84	509	4.05	4.37	176	2.71	Ucaréo, Michoacan
ET92027	Etlatongo	6.63	139	437	3.81	4.27	175	2.79	Ucaréo, Michoacan
ET92028	Etlatongo	7.13	994	710	1.92	3.40	453	2.90	Ixtepeque, Guatemala
ET92029	Etlatongo	6.30	93	1194	7.94	4.87	378	2.69	Paredón, Puebla
ET92030	Etlatongo	6.76	125	537	4.00	3.87	172	2.87	Ucaréo, Michoacan
ET92031	Etlatongo	6.86	0	956	7.64	3.94	367	2.97	Paredón, Puebla
ET92032	Etlatongo	6.46	0	1011	8.23	4.36	365	2.88	Paredón, Puebla
ET92033	Etlatongo	6.57	0	834	7.99	4.16	355	2.86	Paredón, Puebla
ET92034	Etlatongo	7.11	0	1013	7.90	4.35	366	2.96	Paredón, Puebla
ET92035	Etlatongo	6.61	0	1039	7.70	3.83	359	2.91	Paredón, Puebla
ET92036	Etlatongo	6.73	87	950	8.62	4.03	373	3.03	Paredón, Puebla
ET92037	Etlatongo	6.84	0	927	8.18	4.27	367	2.97	Paredón, Puebla
ET92038	Etlatongo	6.77	0	1216	7.86	3.91	366	2.92	Paredón, Puebla
ET92039	Etlatongo	6.49	0	972	7.94	4.08	361	2.92	Paredón, Puebla
ET92040	Etlatongo	6.63	0	976	7.26	4.11	363	2.94	Paredón, Puebla
ET92041	Etlatongo	6.55	0	930	7.61	4.02	361	2.93	Paredón, Puebla
ET92042	Etlatongo	7.40	914	628	3.10	3.08	644	3.01	El Chayal, Guatemala
ET92043	Etlatongo	7.06	717	510	2.79	3.17	390	3.02	Otumba, State of Mexico
ET92044	Etlatongo	6.59	0	1004	7.63	3.90	358	2.93	Paredón, Puebla
ET92045	Etlatongo	6.46	0	976	7.39	4.14	359	2.92	Paredón, Puebla
ET92046	Etlatongo	6.49	0	739	7.84	4.14	370	2.97	Paredón, Puebla
ET92047	Etlatongo	6.46	55	785	7.98	4.06	369	2.94	Paredón, Puebla
ET92048	Etlatongo	6.72	0	800	8.07	4.21	372	2.99	Paredón, Puebla
ET92049	Etlatongo	6.33	0	783	7.84	4.16	369	2.97	Paredón, Puebla
ET92050	Etlatongo	6.72	0	812	8.04	4.44	371	2.99	Paredón, Puebla
ET92051	Etlatongo	6.99	0	828	8.78	4.26	372	3.00	Paredón, Puebla
ET92052	Etlatongo	6.56	0	751	8.06	4.15	368	2.95	Paredón, Puebla

Sample #	Site Name	Al (%)	Ba	Cl	Dy	K (%)	Mn	Na (%)	Source Name, State
ET92053	Etlatongo	6.59	47	797	8.18	4.25	371	2.99	Paredón, Puebla
ET92054	Etlatongo	6.92	815	396	1.57	3.32	528	3.29	Guadalupe Victoria, Puebla
ET92055	Etlatongo	6.64	53	765	7.63	4.22	370	3.00	Paredón, Puebla
ET92056	Etlatongo	6.50	0	786	8.17	4.09	367	2.96	Paredón, Puebla
ET92057	Etlatongo	6.63	0	741	7.66	4.10	370	2.99	Paredón, Puebla
ET92058	Etlatongo	6.47	0	771	7.66	4.23	369	2.96	Paredón, Puebla
ET92059	Etlatongo	7.86	795	281	2.98	3.56	404	3.09	Otumba, State of Mexico
ET92060	Etlatongo	6.72	52	735	8.32	4.18	374	2.99	Paredón, Puebla
ET92061	Etlatongo	6.50	37	793	8.17	4.40	369	2.97	Paredón, Puebla
ET92062	Etlatongo	7.41	0	764	7.76	4.42	371	2.99	Paredón, Puebla
ET92063	Etlatongo	6.66	0	834	8.26	4.09	374	3.01	Paredón, Puebla
ET92064	Etlatongo	7.76	772	334	3.35	3.67	406	3.11	Otumba, State of Mexico
ET92065	Etlatongo	6.47	114	741	7.92	4.32	370	2.98	Paredón, Puebla
ET92066	Etlatongo	6.83	0	751	8.15	4.15	369	2.99	Paredón, Puebla
ET92067	Etlatongo	7.32	802	324	3.45	3.82	404	3.12	Otumba, State of Mexico
ET92068	Etlatongo	6.37	114	610	7.80	4.60	370	2.82	Paredón, Puebla
ET92069	Etlatongo	6.74	0	731	7.65	4.39	369	2.98	Paredón, Puebla
ET92070	Etlatongo	6.59	44	640	7.53	3.82	365	2.94	Paredón, Puebla
ET92071	Etlatongo	6.74	0	646	7.49	3.89	372	2.99	Paredón, Puebla
ET92072	Etlatongo	6.87	663	270	3.26	3.49	405	3.12	Otumba, State of Mexico
ET92073	Etlatongo	6.54	96	646	8.67	4.16	371	2.94	Paredón, Puebla
ET92074	Etlatongo	6.57	0	616	8.11	4.03	368	2.96	Paredón, Puebla
ET92075	Etlatongo	6.51	76	571	8.64	4.07	374	3.02	Paredón, Puebla
ET92076	Etlatongo	6.08	833	252	3.03	3.62	356	3.03	Otumba, State of Mexico
ET92077	Etlatongo	6.49	0	658	8.86	4.49	378	3.01	Paredón, Puebla
ET92078	Etlatongo	6.83	896	333	1.99	3.15	530	3.28	Guadalupe Victoria, Puebla
ET92079	Etlatongo	6.53	43	652	8.31	4.20	377	3.00	Paredón, Puebla
ET92080	Etlatongo	6.15	0	600	8.75	4.14	375	2.99	Paredón, Puebla
ET92081	Etlatongo	6.42	71	628	9.00	3.91	380	3.04	Paredón, Puebla
ET92082	Etlatongo	6.92	790	248	3.32	3.70	408	3.13	Otumba, State of Mexico
ET92083	Etlatongo	6.59	0	553	8.22	4.28	371	2.96	Paredón, Puebla
ET92084	Etlatongo	6.65	51	611	8.85	4.27	371	2.98	Paredón, Puebla
ET92085	Etlatongo	6.45	88	625	8.23	4.17	370	2.90	Paredón, Puebla
ET92086	Etlatongo	6.89	79	628	7.88	4.12	376	3.01	Paredón, Puebla
ET92087	Etlatongo	6.45	82	534	8.44	3.97	375	2.99	Paredón, Puebla
ET92088	Etlatongo	6.75	0	633	8.62	3.91	370	2.95	Paredón, Puebla
ET92089	Etlatongo	6.43	0	637	8.00	4.11	369	2.96	Paredón, Puebla
ET92090	Etlatongo	6.65	0	563	8.53	4.35	376	3.02	Paredón, Puebla
ET92091	Etlatongo	6.92	51	608	1.14	4.22	378	3.02	Paredón, Puebla
ET92092	Etlatongo	6.64	198	228	4.18	3.81	171	2.76	Ucaréo, Michoacan
ET92093	Etlatongo	7.05	0	605	8.89	4.06	379	3.03	Paredón, Puebla
ET92094	Etlatongo	6.54	90	651	8.66	4.19	376	3.00	Paredón, Puebla
ET92095	Etlatongo	6.20	0	603	8.58	4.23	372	2.97	Paredón, Puebla
ET92096	Etlatongo	6.84	0	509	8.17	4.28	369	2.96	Paredón, Puebla
ET92097	Etlatongo	6.56	0	659	9.03	4.18	374	2.99	Paredón, Puebla
ET92098	Etlatongo	6.55	0	602	8.32	3.96	372	2.99	Paredón, Puebla
ET92099	Etlatongo	7.02	0	655	8.00	4.13	379	3.08	Paredón, Puebla
ET92100	Etlatongo	7.31	805	255	2.60	3.43	391	3.03	Otumba, State of Mexico
ET92101	Etlatongo	6.96	0	707	8.44	4.18	378	3.04	Paredón, Puebla
ET92102	Etlatongo	7.58	817	284	3.28	3.41	405	3.12	Otumba, State of Mexico
ET92103	Etlatongo	6.42	0	654	7.66	4.32	366	2.95	Paredón, Puebla
ET92104	Etlatongo	6.53	37	601	7.21	4.69	366	2.60	Paredón, Puebla
ET92105	Etlatongo	7.04	698	269	3.48	3.55	400	3.02	Otumba, State of Mexico
ET92106	Etlatongo	7.33	836	328	2.87	3.37	402	3.08	Otumba, State of Mexico
ET92107	Etlatongo	6.16	0	674	8.51	4.17	371	2.96	Paredón, Puebla
ET92108	Etlatongo	6.46	0	705	7.64	4.22	365	2.93	Paredón, Puebla
ET92109	Etlatongo	6.42	101	683	7.62	4.13	373	2.98	Paredón, Puebla

Sample #	***Site Name***	***Al (%)***	***Ba***	***Cl***	***Dy***	***K (%)***	***Mn***	***Na (%)***	***Source Name, State***
ET92110	Etlatongo	6.22	0	710	7.18	4.06	364	2.92	Paredón, Puebla
ET92111	Etlatongo	6.17	133	263	3.42	4.12	171	2.77	Ucaréo, Michoacan
ET92112	Etlatongo	6.60	62	707	7.63	4.18	370	2.99	Paredón, Puebla
ET92113	Etlatongo	6.33	35	671	7.51	4.27	366	2.93	Paredón, Puebla
ET92114	Etlatongo	7.59	736	259	3.09	3.34	403	3.06	Otumba, State of Mexico
ET92115	Etlatongo	7.55	818	279	3.29	3.34	407	3.12	Otumba, State of Mexico
ET92116	Etlatongo	7.17	783	281	3.85	3.41	400	3.05	Otumba, State of Mexico
ET92117	Etlatongo	6.82	52	688	7.91	4.22	372	2.99	Paredón, Puebla
ET92118	Etlatongo	6.31	73	644	7.65	4.11	370	2.98	Paredón, Puebla
ET92119	Etlatongo	6.78	729	294	3.79	3.42	401	3.08	Otumba, State of Mexico
ET92120	Etlatongo	6.71	0	708	8.38	4.27	374	2.99	Paredón, Puebla
ET92121	Etlatongo	6.89	791	367	2.01	3.29	517	3.14	Guadalupe Victoria, Puebla
ET92122	Etlatongo	6.94	0	230	7.82	3.76	243	3.21	Cruz Negra, Michoacan
ET92123	Etlatongo	6.54	0	697	8.00	4.33	374	3.03	Paredón, Puebla
ET92124	Etlatongo	6.26	0	661	7.40	4.18	366	2.93	Paredón, Puebla
ET92125	Etlatongo	6.32	0	623	7.67	4.37	366	2.95	Paredón, Puebla
ET92126	Etlatongo	6.38	37	764	7.87	4.67	377	3.05	Paredón, Puebla
ET92127	Etlatongo	6.60	52	752	8.34	4.67	374	3.04	Paredón, Puebla
ET92128	Etlatongo	6.60	955	354	1.95	3.90	537	3.27	Guadalupe Victoria, Puebla
ET92129	Etlatongo	6.74	117	765	7.89	4.74	379	3.06	Paredón, Puebla
ET92130	Etlatongo	7.26	822	333	3.13	3.68	404	3.11	Otumba, State of Mexico
ET92131	Etlatongo	7.14	957	396	1.82	3.93	551	3.43	Guadalupe Victoria, Puebla
ET92132	Etlatongo	6.22	0	786	8.16	4.51	364	2.96	Paredón, Puebla
ET92133	Etlatongo	6.32	930	356	1.83	3.32	539	3.37	Guadalupe Victoria, Puebla
ET92134	Etlatongo	7.24	739	312	3.37	3.49	408	3.15	Otumba, State of Mexico
ET92135	Etlatongo	8.11	810	289	3.10	3.53	406	3.13	Otumba, State of Mexico
ET92136	Etlatongo	7.19	863	382	1.88	3.58	529	3.32	Guadalupe Victoria, Puebla
ET92137	Etlatongo	7.63	810	303	3.72	3.73	403	3.10	Otumba, State of Mexico
ET92138	Etlatongo	6.77	0	770	7.88	4.45	367	2.98	Paredón, Puebla
ET92139	Etlatongo	7.06	0	728	8.00	4.61	373	3.02	Paredón, Puebla
ET92140	Etlatongo	6.53	0	746	7.83	4.54	373	3.03	Paredón, Puebla
ET92141	Etlatongo	6.35	0	833	8.74	4.62	377	3.05	Paredón, Puebla
ET92142	Etlatongo	6.38	0	691	7.62	4.60	371	2.99	Paredón, Puebla
ET92143	Etlatongo	7.10	797	315	3.42	3.76	406	3.11	Otumba, State of Mexico
ET92144	Etlatongo	7.38	778	276	2.82	3.86	405	3.13	Otumba, State of Mexico
ET92145	Etlatongo	7.95	785	316	3.37	4.03	405	3.12	Otumba, State of Mexico
ET92146	Etlatongo	7.08	929	305	1.90	3.57	518	3.27	Guadalupe Victoria, Puebla
ET92147	Etlatongo	7.51	979	392	1.65	3.70	562	3.53	Guadalupe Victoria, Puebla
ET92148	Etlatongo	7.22	795	332	3.07	3.84	403	3.13	Otumba, State of Mexico
ET92149	Etlatongo	7.41	732	324	3.34	3.77	401	3.11	Otumba, State of Mexico
ET92150	Etlatongo	7.41	824	292	3.25	3.68	405	3.11	Otumba, State of Mexico
ET92151	Etlatongo	7.50	878	306	3.25	3.85	412	3.17	Otumba, State of Mexico
ET92152	Etlatongo	7.65	766	309	3.47	3.53	405	3.13	Otumba, State of Mexico
ET92153	Etlatongo	7.37	698	315	3.55	3.69	405	3.12	Otumba, State of Mexico
ET92154	Etlatongo	7.28	731	317	3.39	3.53	408	3.15	Otumba, State of Mexico
ET92155	Etlatongo	7.65	695	330	3.32	3.65	409	3.16	Otumba, State of Mexico
ET92156	Etlatongo	6.97	0	705	7.70	3.75	368	2.97	Paredón, Puebla
ET92157	Etlatongo	6.86	0	750	8.31	4.06	365	2.97	Paredón, Puebla
ET92158	Etlatongo	6.41	76	715	7.89	4.05	364	2.96	Paredón, Puebla
ET92159	Etlatongo	6.52	48	716	7.46	4.12	368	2.98	Paredón, Puebla
ET92160	Etlatongo	6.74	0	759	7.95	4.18	372	2.99	Paredón, Puebla
ET92161	Etlatongo	6.60	61	761	7.88	4.12	376	3.02	Paredón, Puebla
ET92162	Etlatongo	6.80	0	695	8.48	3.90	372	2.98	Paredón, Puebla
ET92163	Etlatongo	6.83	0	739	8.45	4.33	381	3.08	Paredón, Puebla
ET92164	Etlatongo	6.45	0	696	7.73	4.27	369	2.94	Paredón, Puebla
ET92165	Etlatongo	6.57	0	743	8.17	3.85	368	2.98	Paredón, Puebla
ET92166	Etlatongo	6.67	0	728	7.33	4.07	366	2.94	Paredón, Puebla

Sample #	*Site Name*	*Al (%)*	*Ba*	*Cl*	*Dy*	*K (%)*	*Mn*	*Na (%)*	*Source Name, State*
ET92167	Etlatongo	6.81	0	707	7.89	3.97	368	2.98	Paredón, Puebla
ET92168	Etlatongo	6.60	0	714	8.16	4.10	375	3.02	Paredón, Puebla
ET92169	Etlatongo	6.39	0	745	7.81	4.08	366	2.98	Paredón, Puebla
ET92170	Etlatongo	7.87	790	313	3.03	3.19	397	3.05	Otumba, State of Mexico
ET92171	Etlatongo	6.85	0	802	7.85	3.82	370	2.97	Paredón, Puebla
ET92172	Etlatongo	7.33	918	311	3.19	3.35	402	3.12	Otumba, State of Mexico
ET92173	Etlatongo	7.16	956	306	1.67	3.19	517	3.24	Guadalupe Victoria, Puebla
ET92174	Etlatongo	6.82	0	768	8.30	4.18	376	3.04	Paredón, Puebla
ET92175	Etlatongo	6.81	0	778	8.84	3.98	365	2.96	Paredón, Puebla
ET92176	Etlatongo	6.64	0	762	7.58	3.91	366	2.94	Paredón, Puebla
ET92177	Etlatongo	6.69	742	685	15.53	3.81	404	3.57	Tulancingo, Hidalgo
ET92178	Etlatongo	6.77	83	682	7.93	4.01	370	2.99	Paredón, Puebla
ET92179	Etlatongo	6.75	39	718	8.06	3.96	370	2.98	Paredón, Puebla
ET92180	Etlatongo	6.60	58	649	8.45	3.98	363	2.94	Paredón, Puebla
ET92181	Etlatongo	6.95	844	346	3.70	3.09	390	3.00	Otumba, State of Mexico
ET92182	Etlatongo	6.32	0	709	7.56	4.73	368	2.76	Paredón, Puebla
ET92183	Etlatongo	6.72	91	713	8.23	4.07	370	2.97	Paredón, Puebla
ET92184	Etlatongo	6.83	118	746	7.80	4.07	369	2.96	Paredón, Puebla
ET92185	Etlatongo	7.05	1015	390	1.75	3.19	516	3.23	Guadalupe Victoria, Puebla
ET92186	Etlatongo	6.55	0	733	8.10	3.85	365	2.97	Paredón, Puebla
ET92187	Etlatongo	6.30	72	604	8.00	4.28	366	2.96	Paredón, Puebla
ET92188	Etlatongo	7.68	669	229	2.98	3.31	400	3.11	Otumba, State of Mexico
ET92189	Etlatongo	6.54	148	285	3.95	3.97	174	2.80	Ucaréo, Michoacan
ET92190	Etlatongo	6.40	0	697	7.75	4.17	370	2.99	Paredón, Puebla
ET92191	Etlatongo	7.17	141	244	3.84	4.04	175	2.87	Ucaréo, Michoacan
ET92192	Etlatongo	7.16	825	270	1.73	3.46	525	3.31	Guadalupe Victoria, Puebla
ET92193	Etlatongo	7.05	777	243	1.94	3.60	533	3.32	Guadalupe Victoria, Puebla
ET92194	Etlatongo	6.37	50	665	8.07	4.31	376	3.04	Paredón, Puebla
ET92196	Etlatongo	6.78	0	656	8.69	4.13	374	2.99	Paredón, Puebla
ET92197	Etlatongo	6.64	0	651	8.01	4.20	370	2.97	Paredón, Puebla
ET92198	Etlatongo	6.47	79	656	7.97	3.91	376	3.03	Paredón, Puebla
ET92199	Etlatongo	6.22	0	693	8.47	4.22	373	2.98	Paredón, Puebla
ET92200	Etlatongo	6.77	727	282	3.30	3.41	408	3.05	Otumba, State of Mexico
ET92201	Etlatongo	6.49	50	628	7.13	4.10	364	2.94	Paredón, Puebla
ET92202	Etlatongo	7.21	951	358	1.57	3.41	540	3.39	Guadalupe Victoria, Puebla
ET92203	Etlatongo	6.40	0	744	7.82	4.21	368	2.97	Paredón, Puebla
ET92204	Etlatongo	7.15	655	249	3.26	3.37	401	3.02	Otumba, State of Mexico
ET92205	Etlatongo	6.59	0	682	8.09	4.22	378	3.04	Paredón, Puebla
ET92206	Etlatongo	6.30	79	675	8.66	4.14	376	2.96	Paredón, Puebla
ET92207	Etlatongo	6.49	111	245	3.57	4.05	170	2.77	Ucaréo, Michoacan
ET92208	Etlatongo	6.46	0	694	8.28	3.80	377	3.00	Paredón, Puebla
ET92316	Yucuita	6.73	951	353	1.62	3.75	531	3.26	Guadalupe Victoria, Puebla
ET92317	Yucuita	7.35	879	325	1.54	3.43	523	3.27	Guadalupe Victoria, Puebla
ET92318	Yucuita	6.90	1086	316	1.52	3.37	521	3.25	Guadalupe Victoria, Puebla
ET92319	Yucuita	6.65	886	288	0.96	3.17	513	3.20	Guadalupe Victoria, Puebla
ET92320	Yucuita	6.68	988	377	2.18	3.18	520	3.26	Guadalupe Victoria, Puebla
ET92321	Yucuita	6.53	1064	318	1.67	3.55	510	3.18	Guadalupe Victoria, Puebla
ET92322	Yucuita	6.75	915	373	1.41	3.34	523	3.24	Guadalupe Victoria, Puebla
ET92323	Yucuita	6.73	906	270	1.25	3.44	510	3.13	Guadalupe Victoria, Puebla
ET92324	Yucuita	6.66	715	235	1.42	3.65	547	3.08	Pico de Orizaba, Veracruz
ET92325	Yucuita	7.00	887	297	1.73	3.53	519	3.27	Guadalupe Victoria, Puebla
ET92326	Yucuita	6.45	670	227	1.94	3.63	566	3.16	Pico de Orizaba, Veracruz
ET92327	Yucuita	7.19	932	311	1.68	3.51	530	3.32	Guadalupe Victoria, Puebla
ET92328	Yucuita	6.63	943	286	1.66	3.44	531	3.30	Guadalupe Victoria, Puebla
ET92329	Yucuita	6.66	791	233	1.80	3.77	552	3.03	Pico de Orizaba, Veracruz
ET92330	Yucuita	7.00	911	348	2.08	3.49	529	3.28	Guadalupe Victoria, Puebla
ET92331	Yucuita	7.08	859	379	1.92	3.64	522	3.25	Guadalupe Victoria, Puebla

Sample #	*Site Name*	*Al (%)*	*Ba*	*Cl*	*Dy*	*K (%)*	*Mn*	*Na (%)*	*Source Name, State*
ET92332	Yucuita	7.09	774	216	1.43	3.53	564	3.18	Pico de Orizaba, Veracruz
ET92333	Yucuita	7.22	871	304	1.83	3.31	525	3.34	Guadalupe Victoria, Puebla
ET92334	Yucuita	7.00	890	361	1.93	3.68	518	3.25	Guadalupe Victoria, Puebla
ET92335	Yucuita	7.28	972	318	2.00	3.70	525	3.12	Guadalupe Victoria, Puebla
ET92336	Yucuita	7.32	990	343	1.92	3.44	533	3.34	Guadalupe Victoria, Puebla
ET92337	Yucuita	7.23	679	219	1.53	3.67	568	3.20	Pico de Orizaba, Veracruz
ET92338	Yucuita	6.98	866	294	2.34	3.40	520	3.24	Guadalupe Victoria, Puebla
ET92339	Yucuita	7.17	867	360	2.35	3.41	528	3.31	Guadalupe Victoria, Puebla
ET92340	Yucuita	6.59	974	364	1.96	3.48	524	3.27	Guadalupe Victoria, Puebla
ET92341	Yucuita	6.21	0	617	8.55	4.44	367	2.96	Paredón, Puebla
ET92342	Yucuita	6.82	822	272	1.26	3.23	520	3.28	Guadalupe Victoria, Puebla
ET92343	Yucuita	7.36	739	265	1.53	3.59	568	3.19	Pico de Orizaba, Veracruz
ET92344	Yucuita	7.29	737	254	2.50	3.55	573	3.25	Pico de Orizaba, Verzcruz
ET92345	Yucuita	6.74	927	273	1.59	3.46	524	3.28	Guadalupe Victoria, Puebla
ET92346	Yucuita	6.68	866	264	1.48	3.48	516	3.21	Guadalupe Victoria, Puebla
ET92347	Yucuita	6.89	846	281	1.47	3.39	529	3.33	Guadalupe Victoria, Puebla
ET92348	Yucuita	7.11	795	266	1.44	3.33	526	3.31	Guadalupe Victoria, Puebla
ET92349	Yucuita	6.84	839	298	2.03	3.31	518	3.20	Guadalupe Victoria, Puebla
ET92350	Yucuita	7.34	870	322	1.39	3.33	519	3.31	Guadalupe Victoria, Puebla
ET92351	Yucuita	6.84	902	350	1.97	3.45	518	3.21	Guadalupe Victoria, Puebla
ET92352	Yucuita	6.99	913	292	1.84	3.25	512	3.22	Guadalupe Victoria, Puebla
ET92353	Yucuita	6.90	1001	331	1.97	3.52	533	3.32	Guadalupe Victoria, Puebla
ET92354	Yucuita	6.67	1118	334	1.31	3.69	538	3.36	Guadalupe Victoria, Puebla
ET92355	Yucuita	7.46	985	350	2.12	3.51	525	3.27	Guadalupe Victoria, Puebla
ET92356	Yucuita	7.07	1041	309	1.38	3.35	534	3.34	Guadalupe Victoria, Puebla
ET92357	Yucuita	7.33	1050	312	2.31	3.34	530	3.32	Guadalupe Victoria, Puebla
ET92358	Yucuita	7.40	1003	327	2.16	3.25	537	3.37	Guadalupe Victoria, Puebla
ET92359	Yucuita	7.00	972	307	1.58	3.37	512	3.18	Guadalupe Victoria, Puebla
ET92360	Yucuita	7.01	1067	300	2.14	3.45	518	3.25	Guadalupe Victoria, Puebla
ET92392	R. Dolores Ortíz*	7.27	891	376	2.46	3.20	638	2.99	El Chayal, Guatemala
ET92393	R. Dolores Ortíz	7.11	801	374	1.82	3.32	516	3.24	Guadalupe Victoria, Puebla
ET92394	R. Dolores Ortíz	7.15	938	329	2.08	3.12	522	3.23	Guadalupe Victoria, Puebla
ET92395	R. Dolores Ortíz	6.86	946	328	1.63	3.48	523	3.26	Guadalupe Victoria, Puebla
ET92396	R. Dolores Ortíz	6.54	859	349	1.63	3.14	512	3.19	Guadalupe Victoria, Puebla
ET92397	R. Dolores Ortíz	7.09	824	314	2.34	3.11	527	3.29	Guadalupe Victoria, Puebla
ET92398	R. Dolores Ortíz	6.91	934	338	2.14	3.20	516	3.22	Guadalupe Victoria, Puebla
ET92399	R. Dolores Ortíz	6.66	611	216	1.89	3.31	549	3.08	Pico de Orizaba, Veracruz
ET92400	R. Dolores Ortíz	7.37	935	394	1.50	3.07	519	3.26	Guadalupe Victoria, Puebla
ET92401	R. Dolores Ortíz	6.71	836	341	1.74	3.03	516	3.20	Guadalupe Victoria, Puebla
ET92402	R. Dolores Ortíz	7.40	987	342	2.34	3.31	521	3.26	Guadalupe Victoria, Puebla
ET92403	R. Dolores Ortíz	6.71	894	333	1.73	3.12	516	3.19	Guadalupe Victoria, Puebla
ET92404	R. Dolores Ortíz	7.20	924	332	1.54	3.51	521	3.25	Guadalupe Victoria, Puebla
ET92405	R. Dolores Ortíz	7.04	977	377	2.03	3.36	527	3.30	Guadalupe Victoria, Puebla
ET92406	R. Dolores Ortíz	7.70	874	365	1.81	3.45	529	3.29	Guadalupe Victoria, Puebla
ET92407	R. Dolores Ortíz	7.13	964	336	1.69	3.55	520	3.26	Guadalupe Victoria, Puebla
ET92408	R. Dolores Ortíz	6.91	963	352	1.72	3.24	517	3.23	Guadalupe Victoria, Puebla
ET92409	R. Dolores Ortíz	6.70	952	319	1.84	3.27	513	3.23	Guadalupe Victoria, Puebla
ET92410	R. Dolores Ortíz	6.74	937	335	1.66	3.21	515	3.22	Guadalupe Victoria, Puebla
ET92411	R. Dolores Ortíz	7.05	882	377	1.65	3.51	518	3.22	Guadalupe Victoria, Puebla
ET92412	Etlatongo	6.73	0	1061	8.47	4.27	372	2.96	Paredón, Puebla

*Rancho Dolores Ortíz.

Appendix 2.

Elemental concentrations and source names for Early Formative obsidian artefacts through XRF (values are in ppm unless otherwise indicated)

Sample #	*K (%)*	*Ti*	*Mn*	*Fe*	*Zn*	*Ga*	*Rb*	*Sr*	*Y*	*Zr*	*Nb*	*Source Name, State*
ET92413	4.39	963	290	9947	64	18	177	3	47	232	42	Paredón, Puebla
ET92417	3.76	1184	481	9170	47	19	129	151	18	152	15	Otumba, State of Mexico

Chapter 12
Geochemical Characterisation of Obsidian in Western Mexico: the Sources in Jalisco, Nayarit, and Zacatecas

Michael D. Glascock, Phil C. Weigand, Rodrigo Esparza López, Michael A. Ohnersorgen, Mauricio Garduño Ambriz, Joseph B. Mountjoy, and J. Andrew Darling

Abstract: *A comprehensive geochemical study of obsidian sources from the states of Jalisco, Nayarit, and Zacatecas located in western Mexico was undertaken. This paper, reports on the results of chemical analysis for more than 1000 obsidian samples taken from source localities in these states using X-ray Fluorescence and Neutron Activation Analysis. The chemical results are compared to the geographic data to establish a geochemical database useful for archaeological interpretation. Recent artefact studies from sites in Jalisco and Nayarit have demonstrated the potential of the database for studying procurement and exchange in a long under-appreciated region.*

Keywords: *Obsidian, Source Analysis, Jalisco, Nayarit, Zacatecas, Western Mexico*

Introduction

Obsidian acquisition, exchange, and utilisation in the core regions of Mesoamerica (i.e., central Mexico, the Valley of Oaxaca, the Yucatan Peninsula, and the highlands of Guatemala) have been the focus of archaeological investigation for several decades and command a large and expanding literature (Braswell and Glascock 2002; Charlton 1969; Cobean *et al.* 1971, 1991; Gaxiola and Clark 1989; Hirth *et al.* 2006; Joyce *et al.* 1995; Moholy-Nagy 2003; Pires-Ferriera 1976; Santley *et al.* 2001; Sidrys 1976; Stark *et al.* 1992). Knowledge of the locations of obsidian sources and their geochemistry (Aoyama *et al.* 1999; Braswell and Glascock 1992; Cobean 2002; Sheets *et al.* 1990) have matured to the point that it is now possible to source nearly 100% of the artefacts from sources located in central Mexico, Guatemala and Honduras found on archaeological sites in Mesoamerica and elsewhere (Barker *et al.* 2002).

On the other hand, the lack of published information concerning obsidian sources in the states of Jalisco, Nayarit, and Zacatecas, has continued to be a problem for archaeologists working there. For many years, a publication by Ericson and Kimberlin (1977) and dissertation by Mahood (1980) represented the only published chemical data on obsidian from western Mexico. Ignorance of the sources of obsidian and their compositional profiles limited efforts to develop inferential models for western Mexico aimed at recognising procurement strategies, understanding the dynamics of past land use, and identifying exchange networks.

For decades, most Mesoamerican scholars have considered western Mexico to be a peripheral region of Mesoamerica; and awareness of the regional archaeology was generally limited to the unique tomb burial chambers and associated ceramic offerings (Townsend 1998; Weigand 1985, 1993). Although other intellectual interests prevailed, such as the origins and evolution of Mesoamerican and Mayan civilisation, to some extent disinterest in West Mexican prehistory, up until the 1970s, can also be attributed to historical, economic, and political factors, not the least of which was tourism, that tended to draw attention to the central highlands and the Maya lowlands (Bell 1974; Bernal 1980; Foster and Gorenstein 2000; Schondube 1980). As a result, much of western Mexico's prehistory was largely overlooked (however, see review: Pollard 1997).

During the fall of 2002, a conference was organised by Drs David Grove, Phil Weigand, and Eduardo Williams, with sponsorship from FAMSI (Foundation for the Advancement of Mesoamerican Studies, Inc.), at which a number of archaeologists from Mexico and the USA met in the city of Guadalajara to evaluate what was known about western Mexico prehistory in its own terms. The region was recognised as having had early precocious cultural developments similar to those that happened elsewhere in Mesoamerica. The prehistoric occupants of western Mexico developed their own forms of socio-political complexity, and there was evidence for interactions with other regions, including central Mexico, the US Southwest, and the Pacific coasts of Central America and South America.

Western Mexico was recognised as a mineral-rich area by the Spanish soon after the Conquest. In addition to silver and lead, the region was found to be rich with minerals such as azurite, malachite, cuprite and quartz (Bakewell 1971; Weigand 1994; Weigand and García de Weigand 1994). The lake valleys contained high-quality clays for the production of kaolin and brick (Aronson 1996). These resources were complemented by an abundance of water, pine and oak forests, migratory birds, fish, reptiles, and salt. Aside from obsidian, the area held a plentiful profile of resources capable of supporting a sizeable population.

As western Mexico's importance is becoming more widely recognised, the tempo of archaeological research being conducted there is increasing and important new data are beginning to emerge. The purpose of this report is to summarise the research conducted in recent years towards locating and characterising the obsidian sources in western Mexico.

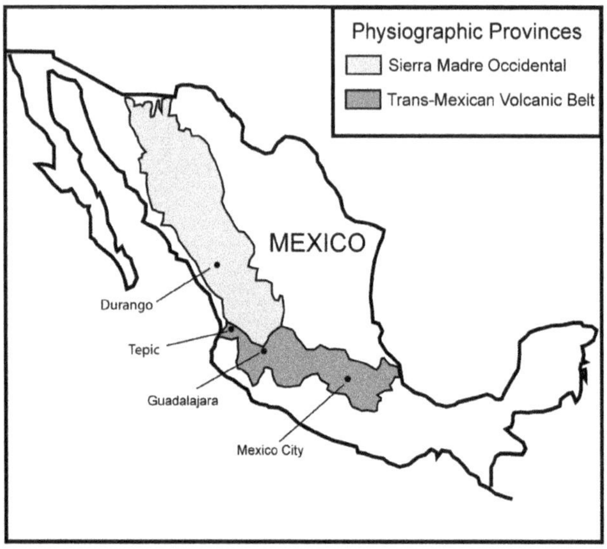

Figure 12.1. Map of Mexico showing the physiographic provinces, adapted from Darling and Hayashida (1995)

Geography

Western Mexico can be divided into two broad physiographic provinces that intersect northwest of the city of Guadalajara (Figure 12.1). The Sierra Madre Occidental (hereafter – SMO) dominates northwestern Mexico and is comprised predominantly of rhyolitic ash-flow tuffs and silicic lava dating from 34 to 23 million years ago. The Trans-Mexican Volcanic Belt (hereafter – TMVB) trends from NW to SE across central Mexico and consists of a chain of volcanic centres dating from the Pleistocene through recent times. Geologists who study plate tectonics (Righter *et al.* 1995; Rosas-Elguera *et al.* 1996) describe the intersection as a superposition of subduction and continental rifting of tectonic regimes that have been taking place since the latter Cenozoic.

Subduction of the Farallon Plate beneath the North American Plate resulted in formation of Rivera and Cocos oceanic plates, as well as the SMO and TMVB. Continued motion of the Rivera Plate induced an uplifted zone in western Mexico known as the "Jalisco Block" as shown in Figure 12.2. The boundaries of the Jalisco Block are defined by the Tepic-Zacoalco (hereafter – TZR), the Colima (hereafter – CR), and the Chapala (hereafter – CHR) rifts. The TZR and CHR separate the SMO from the Jalisco and Michoacan blocks, and CR is the boundary between the Jalisco and Michoacan blocks. As a consequence of the complex plate tectonics, the Jalisco Block and its boundaries represent one of the most active volcanic zones in the world since the Pliocene. In particular, the states of Colima, Jalisco, and Nayarit are populated by large stratovolcanic centres at Colima, La Primavera, Tequila, Ceboruco, and Sanganguey (Figure 12.2). Colima and Ceboruco are the only volcanoes still active today, although there are indications of low level activity around the Volcan de Tequila and within the Caldera de Coli (i.e., the geological name of La Primavera), among other localities (such as geysers, sulfuric gas vents, etc.). With so much volcanic activity occurring in such a small region, there is no surprise that a large number of obsidian sources were created. Our evidence suggests that the number and density of obsidian deposits in the vicinity of the Jalisco Block and its boundaries rivals that of better known volcanic regions such as Oregon, New Zealand, the Kamchatka Peninsula and the Rift Valley of East Africa. The vast majority of western Mexico's obsidian outcrops are along the Tequila/Coli axis.

Methodology

The number of obsidian outcrops, deposits, and pits in the region reaches into the thousands; and most of these show evidence of prehistoric mining activity. Since the 1980s, the authors and various colleagues have systematically collected and analysed geological samples of obsidian from hundreds of widely dispersed outcrops, quarries, and stream beds. Special attention was paid to those locations containing artefact-quality obsidian and showing evidence of tool manufacture. To date, more than 1000 source samples and 500 artefacts from Jalisco and the neighbouring states of Nayarit and Zacatecas have been collected and analysed by Neutron Activation Analysis (hereafter – NAA). More recently, a large proportion of the original source samples were re-analysed by X-ray Fluorescence (hereafter – XRF).

Both analytical techniques are ideally suited to study obsidian because they are capable of measuring the critical elements present in obsidian that uniquely differentiate between obsidian sources. Both methods are multi-elemental and require minimal sample preparation. XRF measures ten to 12 elements and can be performed quickly and non-destructively on samples larger than $1cm^2$. NAA measures more than 25 elements, most with greater precision and accuracy than XRF; and it can be used to analyse obsidian fragments weighing only a few milligrams. Both analytical techniques are currently available at the Archaeometry Laboratory of the University of Missouri Research Reactor (MURR).

NAA Procedures

Descriptions of the analytical procedures for NAA have been reported in greater detail (Cobean *et al.* 1991; Glascock *et al.* 1998) and will be described only briefly here. Samples for NAA were prepared by removing a number of obsidian fragments from the interiors of the source samples. Typically, this amounted to about 100mg of sample material for short irradiations and about 250mg for long irradiations. The short and long NAA samples were weighed into clean, high-purity vials made of polyethylene and quartz, respectively.

Samples in polyethylene vials are irradiated for five seconds in a thermal neutron flux of 8×10^{13} n cm^{-2} s^{-1}. The samples are allowed to decay for 25 minutes before starting a 12-minute measurement. The short irradiation measurement permits determination of seven short-lived elements: Al, Ba, Cl, Dy, K, Mn, and Na. The long irradiation samples in quartz vials are measured twice after a 70-hour irradiation by a neutron flux of 5×10^{13} n cm^{-2} s^{-1}. The first measurement for 30 minutes per sample occurs

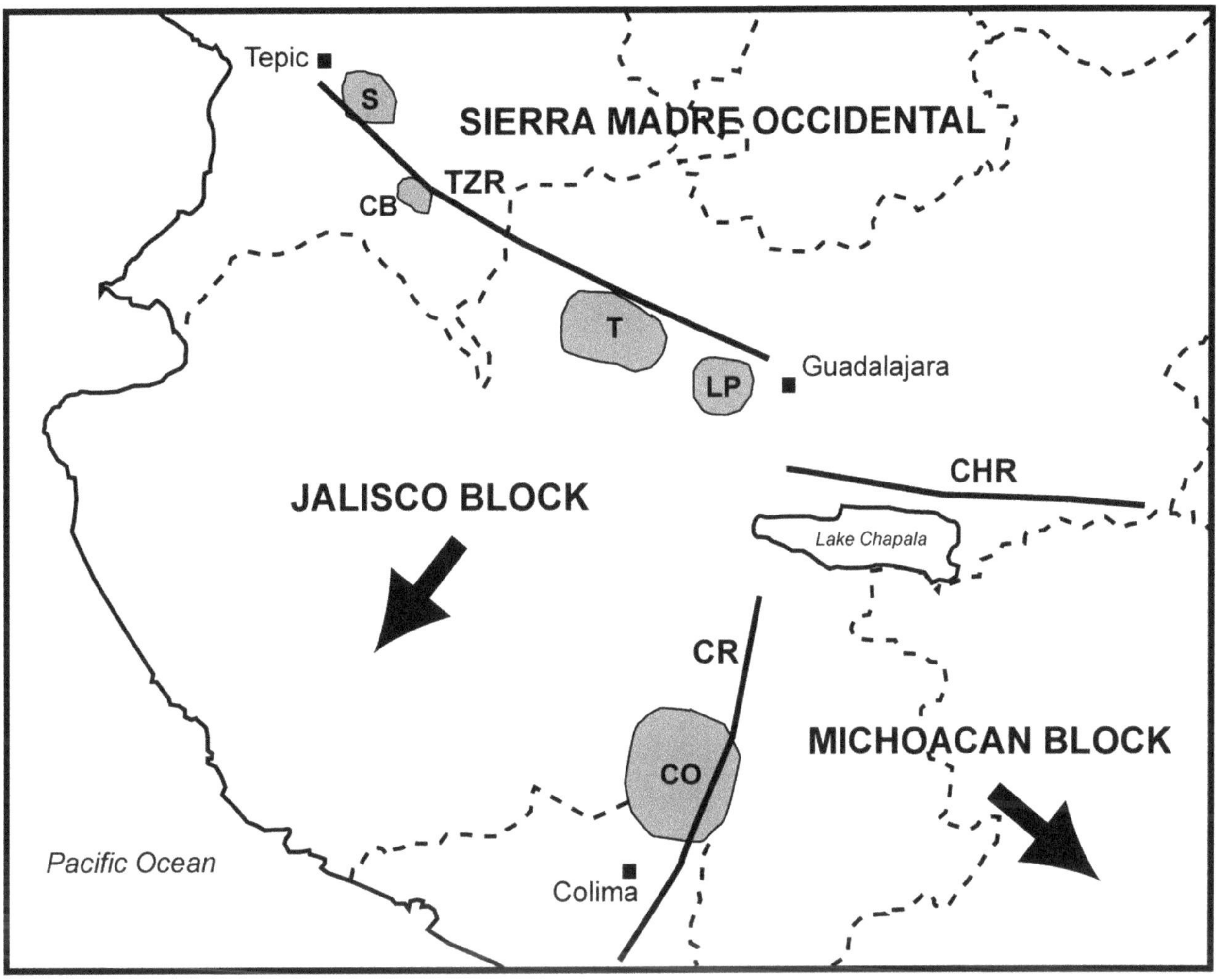

Figure 12.2. Map shows the geodynamic setting of the Jalisco Block and its related boundaries and major volcanoes. The rift boundaries and volcanoes shown are: CR – Colima Rift; TZR – Tepic-Zacoalco Rift; CHR – Chapala Rift; S – Sanganguey Volcano; CB – Ceboruco Volcano; T – Tequila Volcano; LP– La Primavera Volcano; and CO – Colima Volcano. The directions of motion for the Jalisco and Michoacán blocks are indicated

between seven and eight days after the end of irradiation to measure seven medium-lived elements: Ba, La, Lu, Nd, Sm, U, and Yb; and the second measurement for 2.5 hours per sample takes place between four and five weeks after the end of irradiation to measure 15 long lived elements: Ce, Co, Cs, Eu, Fe, Hf, Rb, Sb, Sc, Sr, Ta, Tb, Th, Zn, and Zr. Barium can be measured following either a short- or long-irradiation; however, the data measured after long-irradiation is usually, although not always, superior.

XRF Procedures

X-ray Fluorescence is performed using an Elva-X Energy Dispersive X-ray Fluorescence (EDXFR) spectrometer. The Elva-X is a lightweight EDXRF that fits on a table top. The spectrometer is equipped with an air-cooled tungsten anode X-ray tube with a 140 micron Be window and thermoelectrically cooled Si-PIN diode detector. The detector has a resolution of 180eV for the 5.9keV X-ray from iron. The beam dimensions are 3 × 4 mm. The X-ray tube is operated at 40kV using a tube current of about 25μA which allows the measurement of 11 elements in most samples: K, Ti, Mn, Fe, Zn, Ga, Rb, Sr, Y, Zr, and Nb. Measurement times are 180 seconds. Peak deconvolution and element concentrations are accomplished using the ElvaX spectral analysis package. The instrument was calibrated using a series of well-characterised obsidian source samples from the MURR reference collections, including 11 Mesoamerican and three Peruvian sources. Samples suitable for analysis by XRF should be larger than 0.8cm across for reliable results.

Descriptions of Obsidian Sources

Previous obsidian source characterisation studies (Ambroz *et al.* 2001; Cobean *et al.* 1991; Glascock *et al.* 1998) have shown that sources may contain material from multiple eruptive sequences and have more than one elemental signature due to the evolution of the magma between eruptions. Ideally, a sufficient number of samples should be collected from each geologically distinct deposit (primary or secondary) such that chemical analysis can assess the overall variability within each source. A number of the sources in this study are chemically similar, especially those in the State of Jalisco where the sources are geographically close to one another. Although ground conditions did not always allow optimal sampling at all locations, it was conducted as rigorously as possible with the number of

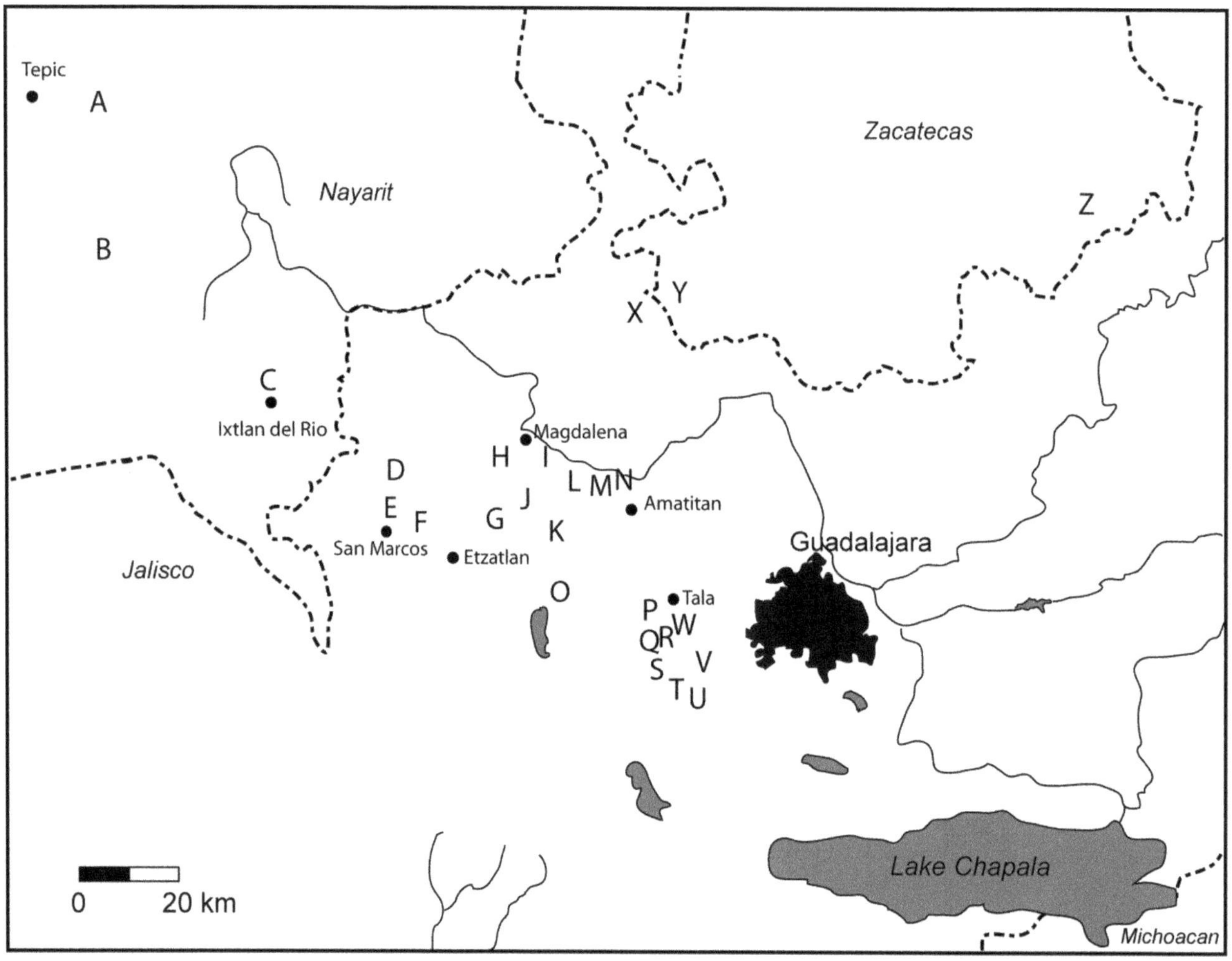

Figure 12.3. Map showing the locations of obsidian sources mentioned in this work. The sources are as follows: A – Volcan Las Navajas (2 subsources); B – San Leonel; C – Ixtlan del Rio; D – Llano Grande; E – Osotero; F – Hacienda de Guadalupe; G – San Juanito de Escobedo; H – La Quemada; I – Cinco Minas/Magdalena; J – La Joya; K – La Providencia; L – Santa Teresa; M – Tequila; N – La Pila; O – La Mora/Teuchitlan; P – Huaxtla; Q – Boquillas; R – La Primavera; S – San Juan De Los Arcos; T – Ahuisculco; U – Navajas; V – Ixtepete; W – San Isidro; X – La Lobera; Y – Huitzila; and Z – Nochistlan

samples collected from each source area ranging from six to 200 specimens. A map showing the locations of the 26 source areas that were characterised in this study is shown in Figure 12.3.

Table 12.1 lists the source names, source abbreviation, number of compositional groups, name(s) of person(s) who sampled the source (see also text below), and an assessment of the quality of the obsidian from each source. The assessment of quality is based on the fracturing properties of the glass, including the frequency of inclusions and friability of the glass. Excellent quality glass has few inclusions and will produce fine, sharp-edged tools. Poor quality glass has numerous inclusions and bubbles which make the glass very fragile to the point of crumbling easily. More complete details regarding the sampling of materials from individual source areas are explained below.

Volcan Las Navajas, Nayarit [VNN]

The Volcan Las Navajas obsidian source is located about 18km east of the city of Tepic in the State of Nayarit. The source zone begins immediately north of the village of San Luis de Lozada. The primary outcrop extends north from the village for a distance of about 5km, along the eastern edge of the collapsed crater, and eventually turns to the northeast and continues downslope from the main crater for a distance of 5km. Large outcrops of obsidian are found along this path with nodules ranging in size from 2cm pebbles to large blocks of obsidian up to 30cm in length. The glassy obsidian has an excellent fracturing quality which produces fine tools that are an opaque (or murky) green colour, with a slightly translucent appearance along the edges of thin pieces. Evidence for mining of obsidian from Volcan Las Navajas during pre-Hispanic times is extensive, with large accumulations of waste material from tool production found throughout. Fourteen source samples were collected by Mauricio Garduño Ambriz at intervals downslope from the primary outcrop. Chemical analysis of the source samples by NAA and XRF identified two distinct compositions.

San Leonel, Nayarit [SLN]

Samples of obsidian from San Leonel, Nayarit were collected by geologist John Pierson. Scattered source

Table 12.1. Obsidian sources located in the states of Jalisco, Nayarit, and Zacatecas (western Mexico)

Map ID	Source Name and Abbreviation	Description and Comments	Quality
A	Volcan Las Navajas, Nayarit [VNN]	Glass has a murky green colour with minimal inclusions	Excellent
B	San Leonel, Nayarit [SLN]	Glass is dark gray in colour. J. Pierson gave the samples to J. Mountjoy who submitted them for analysis	Medium
C	Ixtlán del Rio, Nayarit [IRN]	Glass has a dull gray colour with streaky appearance	Excellent
D	Llano Grande, Jalisco [LGJ]	Glass comes in a wide range of colours: blue, yellow, red, dark green, gray, black, and brown. Most nodules are a mixture of three colours	Excellent
E	Osotero, Jalisco [OSJ]	Glass comes in multiple colours with a high proportion of red and brown. Samples were collected from Osotero, San Marcos, Las Fuentes, and San Sebastian	Medium
F	Hacienda de Guadalupe, Jalisco [GEJ]	Glass is black with a high proportion of crystalline inclusions which make it very brittle making and not very useful for tool production	Medium
G	San Juanito de Escobedo, Jalisco [JEJ]	Glass is black in colour, but the material is very brittle	Poor
H	La Quemada, Jalisco [LQJ]	Glass is highly opaque but with a dark green colour	Medium
I	Cinco Minas/Magdalena, Jalisco [CMJ]	Glass is extremely brittle	Medium
J	La Joya, Jalisco [JJ]	Glass is usually dark green in colour and has excellent fracturing characteristics	Excellent
K	La Providencia, Jalisco [PRJ]	Glass is very brittle and not useful for tool manufacture	Poor
L	Santa Teresa, Jalisco [STJ]	Glass is gray in colour	Mixture of excellent and medium
M	Tequila, Jalisco [QJ]	Glass is very brittle	Poor
N	La Pila, Jalisco [PJ]	Glass is black	Excellent
O	La Mora/Teuchitlán, Jalisco [TJ]	Glass is black	Excellent
P	Huaxtla, Jalisco [HXJ]	Glass is black	Medium
Q	Boquillas, Jalisco [BQJ]	Glass is black	Medium
R	La Primavera, Jalisco [LPJ]	Glass is black. Samples from La Primavera were collected by C. Beekman and the remaining samples from Estancia and Zapopan were collected by P. Weigand and R. Esparza	Excellent
S	San Juan de los Arcos, Jalisco [SJLA]	Glass is black	Excellent
T	Ahuisculco, Jalisco [AJ]	Glass is dark gray	Excellent
U	Navajas, Jalisco [NJ]	Glass is dark gray	Excellent
V	Ixtepete, Jalisco [IXJ]	Glass is dark gray	Medium
W	San Isidro, Jalisco [SIJ]	Glass is dark green	Excellent
X	La Lobera, Jalisco [LLJ]	Glass comes in a variety of colours	Excellent
Y	Huitzila, Zacatecas [HZZ]	Glass comes in a variety of colours	Excellent
Z	Nochistlán, Zacatecas [NCZ]	Glass is dark gray	Medium

samples were gathered from ridges and benches of a farming community located approximately 5.5km west of the village of San Leonel. San Leonel is situated approximately 24km south of the city of Tepic along Mexico Highway 15. The sample collection area is also located 9km east of the village of La Curva. The source samples were given to Joseph Mountjoy who submitted them to MURR for characterisation. The glass is generally a dark gray colour. Information about possible usage and mining activity are unknown.

Ixtlán del Rio, Nayarit [IRN]

The main outcrop for Ixtlán del Rio obsidian source is located on the northern edge of the town of Ixtlán del Rio situated in southeastern Nayarit. The source was studied by Ericson and Kimberlin (1977) who collected obsidian from a deposit located along the main highway passing through the town. A number of pre-Hispanic workshops are located adjacent to the source zone. The obsidian from Ixtlán del Rio is a dull gray colour with some streaks, but with little or no evidence of crystalline inclusions. The source nodules are generally smaller than 10cm diameter, but the Ixtlán del Rio obsidian has excellent flaking qualities. The source samples analysed in this study were collected by Joseph Mountjoy, and chemical analysis indicates that the source is homogeneous with one composition.

Osotero, Jalisco [OSJ]

The Osotero obsidian source is an extensive outcrop located in the hills northwest of the city of Etzatlán. Samples were collected by Phil Weigand from deposits located near the village of San Marcos, Las Fuentes, San Sebastián Oeste and San Sebastián Casco, all of which support existence of a single homogeneous composition for this source. Most of the Osotero outcrop consisted of heavily weathered cobbles ranging up to 35cm in diameter. The glass has a medium fracturing quality. The Osotero source has the highest proportion of red and brown obsidian in the region. These colours of obsidian are found in similar proportions at the ceremonial centres of El Arenal and Peñol de Santa Rosalía suggesting use of Osotero obsidian was most prevalent during the Formative and Post-Classic periods.

Hacienda de Guadalupe, Jalisco [GEJ]

The Hacienda de Guadalupe source is located on low hills immediately south of Osotero not far from Palo Verde. Samples from the Hacienda de Guadalupe source were collected by Phil Weigand. The raw material contains a high proportion of crystalline inclusions, is very brittle, and would not be useful for tool manufacture. There is no evidence that obsidian from this source was mined.

Llano Grande, Jalisco [LGJ]

The Llano Grande source area is located far into the mountains northwest of the Etzatlán Lake valley, in an area close to the Nayarit border. A number of depressions with basalt hammerstones nearby suggest the presence of ancient quarries. Due to a great deal of erosion, the actual number of depressions is difficult to assess. The obsidian from this source comes in a multitude of colours: blue, yellow, red, dark green, gray, black, brown gray, etc. Many of the nodules are a mixture of up to three colours. One sample has five colours within the same nodule. The variety of colours and the excellent fracturing quality made the obsidian from this source very popular during the Formative and Classic periods. Source specimens were collected from the Llano Grande and nearby Arroyo Chacuaco by P. Weigand and R. Esparza. Chemical analysis of the source samples indicates the source has a single homogeneous composition.

San Juanito de Escobedo, Jalisco [JEJ]

Samples of obsidian were collected from source outcrops located near the pueblo of San Juanito de Escobedo by Phil Weigand. The obsidian was generally of very poor quality and it is unlikely to have been used for pre-Hispanic tool manufacture. Analysis of samples from this source determined the glass has a single compositional profile.

La Quemada, Jalisco [LQJ]

The La Quemada source area is located approximately 10km northwest of the city of Magdalena near the pueblo of La Quemada. Obsidian of medium quality is found on a low ridge. The glass is completely opaque with a dark green colour. A number of surface depressions and the presence of a small number of hammerstones suggest that limited mining activity probably occurred here. The La Quemada and Ojo Caliente source samples were collected by Phil Weigand. Chemical analysis indicates the source has a single homogeneous compositional profile.

Cinco-Minas/Magdalena, Jalisco [CMJ]

Samples from the Cinco-Minas source area were collected from near the village by this same name by Phil Weigand. Cinco-Minas is located in the Municipio de Magdalena; and during historic times it was an active area for mining silver and gold. Mining operations have now ceased. The area is marked by a few minor depressions that suggest prehistoric mining for obsidian. The source samples were of poor to medium quality glass; and compositional analysis indicates a single homogeneous source.

La Joya, Jalisco [JJ]

The La Joya mining complex is one of the largest sources of obsidian in Mesoamerica and has been described in detail by Weigand and Spence (1982). The deposit covers an area of approximately 5km^2 with evidence for more than 1000 mining pits and deposits. Obsidian from La Joya is usually a dark green colour with excellent fracturing qualities. The multi-coloured or 'rainbow' variety of obsidian is abundant here; and it is this type of obsidian that is being heavily mined today for use in making jewelry. Based on the number of mining pits and their sizes, Weigand and Spence (1982) calculated that approximately 13,000 tons of obsidian was extracted from La Joya over a period of 1200 years with extensive quarrying during the Classic and Post-Classic periods. The average amount of material excavated was estimated to be 11 tons per year. Samples for chemical analysis were collected from more than 100 locations around the source by P. Weigand and R. Esparza. The analytical results find that La Joya obsidian has a single homogeneous compositional profile.

La Providencia, Jalisco [PRJ]

The La Providencia source is located about 18km east of Etzatlán and 10km northeast of Ahulaulco de Mercado. The glass is very brittle and it would be difficult to produce useful tools. There is no evidence of pre-Hispanic mining activity using this source material. A number of samples were collected by P. Weigand and R. Esparza. Chemical analysis indicates that the obsidian has a single compositional group.

Santa Teresa, Jalisco [STJ]

Samples from the massive Santa Teresa obsidian source were collected by P. Weigand and R. Esparza from a number of locations near the town of Santa Teresa, including the villages of Huitzilapa, Lupita, and La Minita. The glass is a gray colour and fractures to produce high quality tools. Evidence of mining this source area is extensive during the Formative, Classic and Post-Classic periods. More than 200 samples were collected and analysed to determine that the source has a single homogeneous composition.

Tequila, Jalisco [QJ]

The Tequila obsidian source is located about 5km west of the town of Tequila. The obsidian is of poor quality with many white flecks indicating a very brittle glass. There is no evidence of prehistoric mining activity at this source. Samples from this source were collected by Phil Weigand. Analysis indicates a single chemical composition. Although Cobean *et al.* (1991) reportedly visited this area for his samples from the Tequila source, the compositional profiles are completely different.

La Pila, Jalisco [PJ]

The La Pila source area is also located in the Municipio de Tequila. The outcrop is located in rugged terrain and quite complex. There is evidence of mining activity. The glass is of excellent quality for manufacturing tools. Source samples were collected by Phil Weigand. Compositional analysis yields a single chemical profile.

La Mora/Teuchitlán, Jalisco [TJ]

A majority of the mines and associated workshops for the La Mora/Teuchitlán source are located on a hill about 2km behind the village of La Mora, and 3km west of the monumental late Formative period precinct of the Guachimontones de Teuchitlán, where nearby extensive workshops using this source are located. P. Weigand and R. Esparza collected samples from a number of locations including from a contemporary quarry known as Pedregál. Large numbers of mining pits are found along the hillside, but many have been damaged by agricultural activity. Hammerstones, although not frequent, are found adjacent to the mines and pits. The black-coloured glass is of excellent quality for tool manufacture. Analysis of the source samples indicates a single compositional fingerprint for the La Mora/Teuchitlán source.

Huaxtla, Jalisco [HXJ]

The Huaxtla obsidian source area is located in the Municipio de Tala and is a part of the La Primavera massif. Obsidian is abundant with many large boulders. The outcrop has been disturbed by modern agriculture but there is limited evidence of hammerstones and waste flakes. Because the quality of the material was medium, it is likely that the obsidian was not exchanged but instead its use was local. Samples for analysis were collected by Phil Weigand. The glass has a black colour and by chemical analysis it has been determined to have a single composition.

Boquillas, Jalisco [BQJ]

The Boquillas source area is located in the Municipio de Tala and is very modest in size, about 100m^2. It was used during the Santa Crúz de Bárcenas (AD 900 to 1250) and Etzatlán (AD 1250 to Conquest) phases. Obsidian from Boquillas was used for every type of tool production except fine prismatic blades, which at this site appear to have been imported from La Joya. There is workshop activity all over the area. Samples for analysis were collected by Phil Weigand. The analytical results found that the Boquillas source is represented by a single compositional group.

La Primavera, Jalisco [LPJ]

The La Primavera source was surveyed and samples were collected by Beekman (1996). Because the La Primavera volcanic complex is quite extensive, C. Beekman's samples were collected only from the Cañon de las Flores mines. The glass is of medium quality and compositional analysis yields a single fingerprint for this subsource of the La Primavera volcano.

San Juan de los Arcos [SJLA]

The source area known as San Juan de los Arcos was heavily mined during pre-Hispanic times. More than 250 mines were located with evidence of many more existing as subterranean. Massive exploitation of the high quality glass from this source is apparent. Evidence for workshops on the site is widespread with numerous basalt hammerstones and piles of waste from tool production scattered over a wide area. The source is also unique in that it was apparently used to produce large quantities of microblades, as well as extremely thin flakes used for the manufacture, by a unique technique, of the obsidian jewelry so characteristic of the late Formative western Mexico (Clark and Weigand n.d.). Samples for compositional analysis were collected by P. Weigand and R. Esparza. Compositional analysis of the source samples determined a single homogeneous composition.

Ahuisculco, Jalisco [AJ]

The Ahuisculco obsidian source is located in the Municipio de Tala. The source area gets its name from the local village that overlays many of the best areas of the outcrop. The earliest materials represent burials from the late Formative El Arenal phase (ca. 300 BC to AD 200); however, some Classic and Post-Classic materials are also evident. Phil Weigand collected samples from deposits at opposite ends of the village where pre-Hispanic mining activity is evident. The obsidian is gray is colour and has excellent fracturing properties. Analysis of the source samples indicates a single homogeneous compositional group.

Navajas, Jalisco [NJ]

The Navajas source area is quite extensive. There are numerous depressions and pits that clearly indicate ancient mining activity. The largest quarry located to date (30m in diameter and 6m deep) is here. An estimated 2000m^3 of artefact-quality obsidian could have been extracted from this single pit. Several mining tunnels were also identified with depths ranging up to 8m. Hammerstones are found with high frequency. On the highest hill, overlooking the valley floor and lower hills where the obsidian was mined, is the location of the Navajas ceremonial precinct consisting of a series of structures dating to the late Formative and early Classic periods (El Arenal and Ahualulco phases;

350 BC to AD 450). A number of small Post-Classic complexes dot the entire region. Obsidian from Navajas is of high quality and it was employed for a wide range of artefacts, including some use in the late Formative for jewelry making. Samples were collected by P. Weigand and R. Esparza from along the edge of the outcrop where most of the mines and quarries are located. Chemical analysis determined a single homogeneous composition for the obsidian from Navajas.

Ixtepete, Jalisco [IXJ]

Samples for the Ixtepete source were collected from an alluvial deposit located on the outskirts of Guadalajara by Andrew Darling. Darling (1998) described the source as consisting of water-rolled obsidian nodules, generally not larger than 5cm in diameter. A single compositional profile was identified.

San Isidro, Jalisco [SIJ]

The San Isidro source outcrops are located on the southwestern edge of the Caldera de Coli, within the Primavera Forest Preserve. There is considerable evidence that San Isidro was an important source of high-quality obsidian. The San Isidrio zone is very large with numerous trenches and tunnel mines extending into the volcanic dome. One of the tunnel mines is 35m long and 5m wide at the bottom. It is estimated that 550m^3 of material could have been removed from this mine alone. Huge piles of talus exist on the downhill slope around this mine and others in the San Isidro zone. The glass is mostly black and opaque green, but obsidian in reddish-brown and black colours with some white inclusions are also found. Samples for analysis at MURR were collected by Phil Weigand. A single compositional profile for San Isidro obsidian was identified; however, the compositional profile is similar to that for Hacienda de Guadalupe.

La Lobera, Jalisco [LLJ] and Huitzila, Zacatecas [HZZ]

The La Lobera and Huitzila obsidian sources are located on opposite sides of the border between the states of Jalisco and Zacatecas. Samples were collected by Darling (1993, 1998). The excellent quality glass is found in a variety of colours, including rainbow-coloured. Evidence for exploitation of the sources during pre-Hispanic times is extensive (Darling and Glascock 1998; Darling and Hayashida 1995; Jimenez Betts and Darling 2000).

Nochistlán, Zacatecas [NCZ]

The Nochistlán obsidian source area is located about 85km to the east-northeast of the Huitzila–La Lobera deposits. The source is located on the eastern flanks of the Sierra of the same name and appears to be related to the Cerro San Miguel volcano. Evidence for prehistoric use of Nochistlán obsidian is not apparent among the outcrops or mixed gravel deposits, but artefacts from the local archaeological site of Cerro el Tuiche have the same composition as the Nochistlán source (Darling 1993). The samples Noshistlán were collected by Andrew Darling.

Results

More than 1000 obsidian source samples from sources in Jalisco, Nayarit, and Zacatecas were analysed by XRF and NAA during this study. The compositional data were tabulated with a database programme to enable simultaneous use of the chemical data and geographic information for group identification and refinement. The compositional groups were identified by examination of multiple bivariate plots of the elements from which groups of chemically-related samples were recognised that corresponded to specific geographic locations.

Figure 12.4 shows a bivariate plot of the compositional groups based on the XRF data for Rb and Zr. Figure 12.5 shows a bivariate plot of the compositional groups based on the NAA data for Cs and Hf. The individual compositional groups are displayed as 95% confidence ellipses calculated for each source after removal of the few ambiguous "outlier" specimens that failed to agree statistically with the remaining specimens. The vast majority of the compositional groups in Figures 12.4–12.5 are well-separated from one another, especially when using the data from NAA. Two clusters of compositional groups (i.e., QJ-STJ and GEJ-SIJ-LPJ-LGJ) are overlapping on Figure 12.5. However, by examination of additional bivariate plots using the NAA data for Ba, Eu, Fe, La, Mn, Rb, Sc, Sm, Th, and Zn, successful differentiation is demonstrated between the overlapping compositional groups as shown in Figures 12.6–12.7.

In summary, 27 unique compositional groups have been identified using a combination of XRF and NAA. The group means and standard deviations from NAA and XRF are listed in Table 12.2. The reliability of these compositional groups for sourcing was examined by using a series of multivariate methods to calculate the probability of an incorrect assignment for source specimens (Glascock *et al.* 1998). These multivariate methods are useful for identifying the critical elements for discriminating between individual sources and the most efficient and economical analytical methods necessary to reliably match artefacts to the correct sources. To examine the differences between compositional groups, the group means listed in Table 12.2 for all 28 elements measured by NAA were used to perform a hierarchical aggregate cluster analysis. Squared mean Euclidean distances for all 27 of the sources were calculated and clustering was achieved by nearest neighbour distance calculations on the distance matrix. The resulting dendrogram, plotted as a function of relative dissimilarity, is shown in Figure 12.8. From the dendrogram it is clear that when using NAA, it is relatively easy to differentiate between all of the sources.

XRF and NAA have shown that compositionally similar obsidian sources in western Mexico located mainly in the State of Jalisco and also in the states of Nayarit and Zacatecas can be differentiated on the basis of source

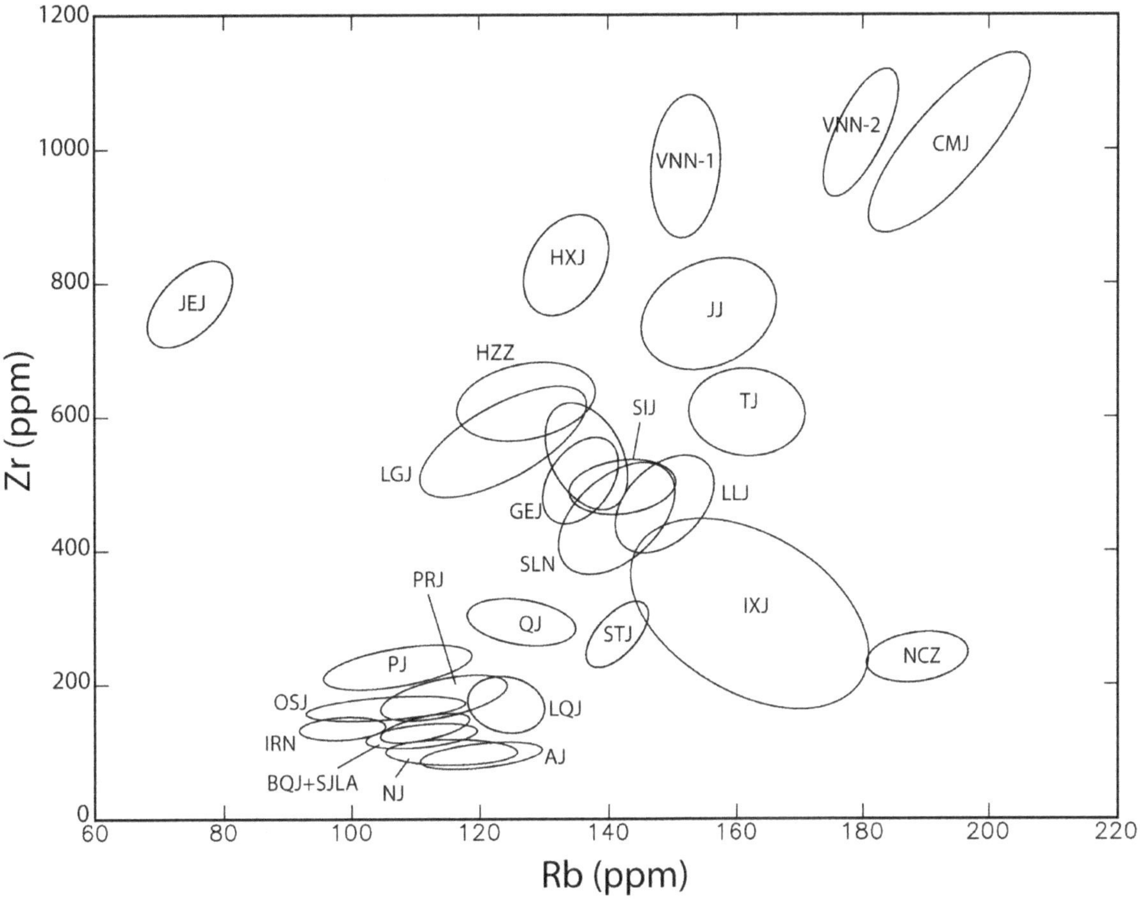

Figure 12.4. Bivariate plot of Rb vs. Zr from XRF showing 95% confidence ellipses calculated for individual obsidian sources in Jalisco, Nayarit, and Zacatecas

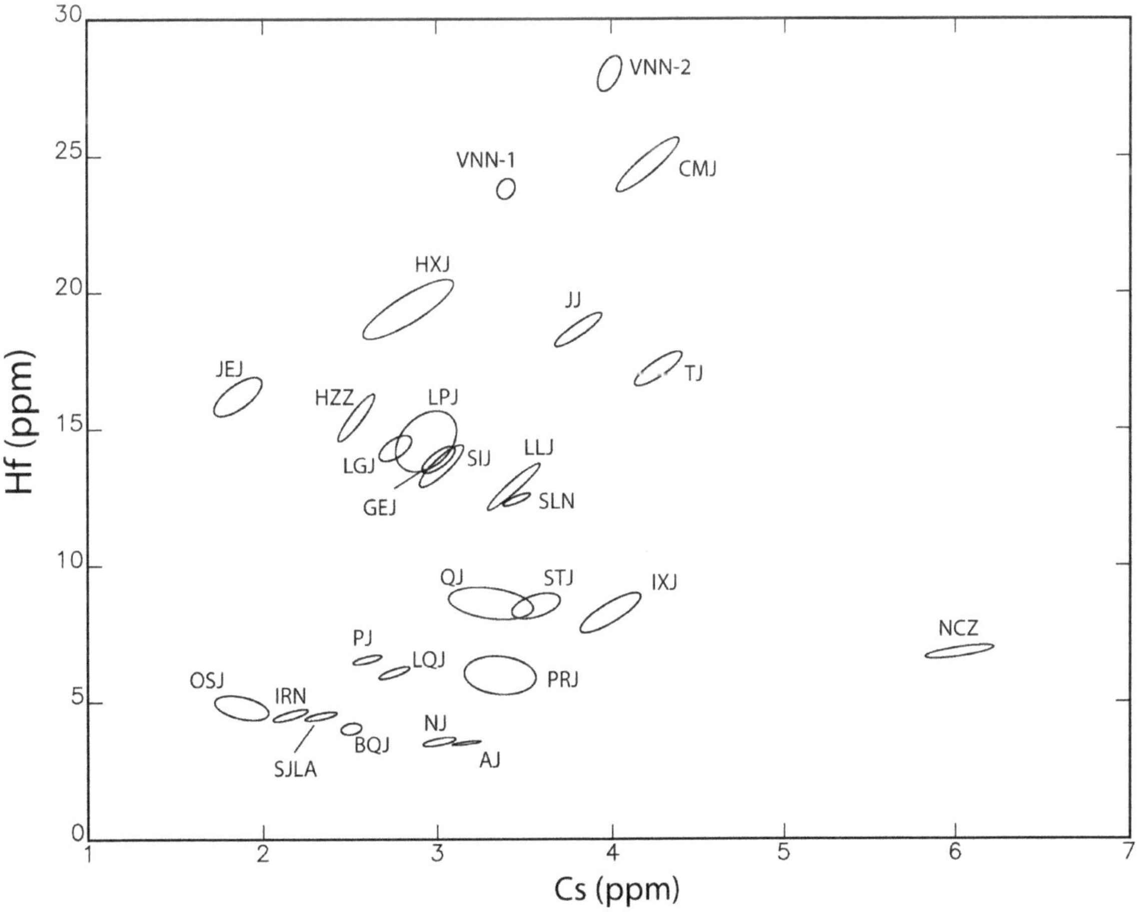

Figure 12.5. Bivariate plot of Cs vs. Hf from NAA showing 95% confidence ellipses calculated for individual obsidian sources in Jalisco, Nayarit, and Zacatecas

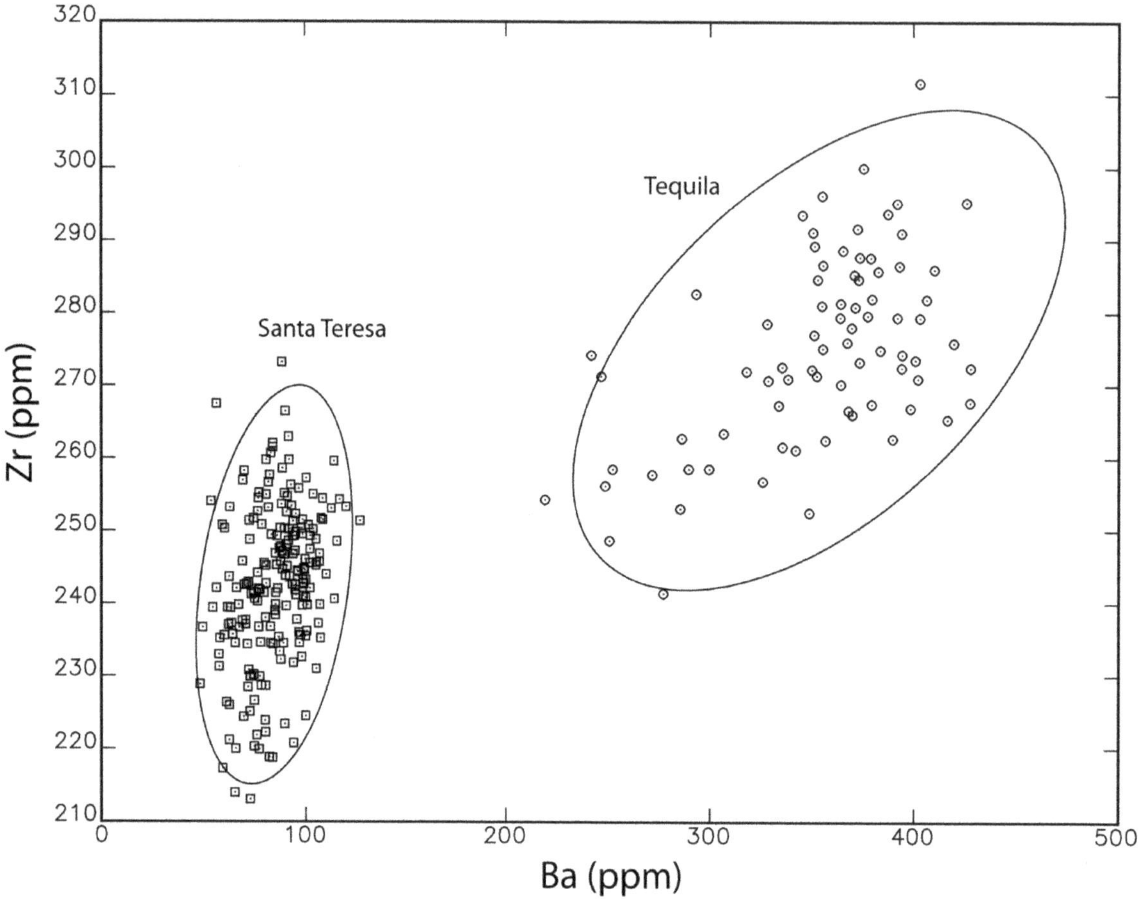

Figure 12.6. Bivariate plot of Ba vs. Zr from NAA showing data for obsidian samples from the Tequila and Santa Teresa sources surrounded by 95% confidence ellipses

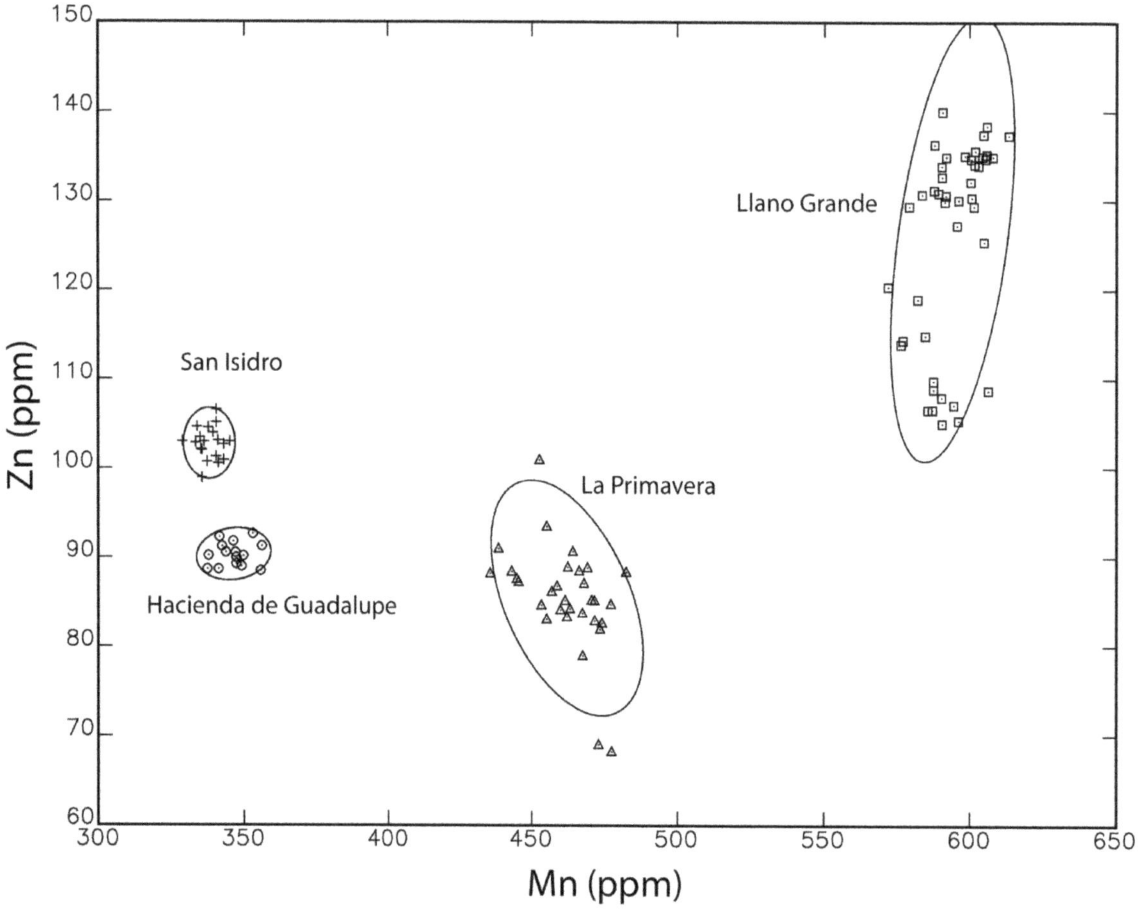

Figure 12.7. Bivariate plot of Mn vs. Zn from NAA showing data for obsidian samples from the Hacienda de Guadalupe, Llano Grande, La Primavera, and San Isidro sources surrounded by 95% confidence ellipses

Table 12.2. Element concentration means and standard deviations for chemical groups of obsidian from sources in Jalisco, Nayarit, and Zacatecas measured by NAA and XRF. The concentrations are listed in parts-per-million (ppm), unless % is noted

	VNN-1	***VNN-2***	***SLN***	***IRN***	***LGJ***
Element	***V. Las Navajas-1, Nayarit***	***V. Las Navajas-2, Nayarit***	***San Leonel, Nayarit***	***Ixtlán del Rio, Nayarit***	***Llano Grande, Jalisco***
NAA Results	(n = 9)	(n = 5)	(n = 9)	(n = 17)	(n = 43)
Na (%)	4.45 ± 0.08	4.62 ± 0.10	3.28 ± 0.04	2.86 ± 0.02	3.64 ± 0.05
Cl	961 ± 37	1189 ± 33	477 ± 43	430 ± 38	863 ± 69
K (%)	3.19 ± 0.25	3.06 ± 0.15	3.79 ± 0.16	3.99 ± 0.18	3.78 ± 0.21
Sc	0.22 ± 0.04	0.13 ± 0.03	0.54 ± 0.01	1.49 ± 0.04	2.07 ± 0.03
Mn	1260 ± 23	1361 ± 19	257 ± 3	248 ± 11	594 ± 10
Fe (%)	4.28 ± 0.03	4.42 ± 0.04	1.23 ± 0.01	0.86 ± 0.05	1.81 ± 0.02
Co	0.128 ± 0.006	0.141 ± 0.011	0.016 ± 0.010	1.12 ± 0.13	0.04 ± 0.03
Zn	242 ± 3	270 ± 4	100 ± 2	41 ± 2	126 ± 11
Rb	157 ± 11	185 ± 2	146 ± 1	102 ± 2	125 ± 2
Sr	< 15	< 15	< 15	98 ± 13	17 ± 9
Zr	906 ± 15	1078 ± 9	422 ± 5	145 ± 8	503 ± 14
Sb	0.55 ± 0.01	0.62 ± 0.02	0.38 ± 0.01	0.60 ± 0.04	0.41 ± 0.03
Cs	3.40 ± 0.01	3.99 ± 0.01	3.46 ± 0.03	2.16 ± 0.04	2.76 ± 0.04
Ba	48 ± 7	85 ± 19	59 ± 32	537 ± 9	33 ± 6
La	92.6 ± 0.8	107 ± 1	53.5 ± 0.3	27.3 ± 0.3	54.5 ± 0.7
Ce	194 ± 2	225 ± 2	111 ± 1	48.5 ± 1.0	114 ± 2
Nd	80.6 ± 2.5	88.1 ± 2.5	47.4 ± 1.3	16.1 ± 0.5	49.8 ± 6.1
Sm	17.5 ± 0.2	19.4 ± 0.2	9.36 ± 0.14	3.00 ± 0.06	10.7 ± 0.3
Eu	2.64 ± 0.02	3.33 ± 0.03	0.239 ± 0.003	0.309 ± 0.006	0.430 ± 0.008
Tb	3.19 ± 0.02	3.69 ± 0.04	1.52 ± 0.01	0.41 ± 0.01	1.64 ± 0.11
Dy	20.2 ± 0.8	23.2 ± 1.1	9.55 ± 0.30	2.36 ± 0.30	10.6 ± 0.7
Yb	10.6 ± 0.2	12.1 ± 0.1	5.13 ± 0.10	1.72 ± 0.05	6.11 ± 0.35
Lu	1.49 ± 0.01	1.69 ± 0.01	0.76 ± 0.02	0.33 ± 0.01	0.89 ± 0.02
Hf	23.8 ± 0.2	28.0 ± 0.3	12.4 ± 0.1	4.52 ± 0.09	14.3 ± 0.2
Ta	6.21 ± 0.04	7.56 ± 0.05	2.21 ± 0.02	1.59 ± 0.03	2.39 ± 0.03
Th	17.1 ± 0.8	21.5 ± 0.2	14.2 ± 0.1	11.3 ± 0.2	13.0 ± 0.1
U	8.02 ± 0.16	9.73 ± 0.45	4.69 ± 0.28	4.04 ± 0.12	4.37 ± 0.89
XRF Results	(n = 9)	(n = 5)	(n = 9)	(n = 21)	(n = 12)
K (%)	2.73 ± 0.42	3.66 ± 0.28	3.63 ± 0.20	3.54 ± 0.12	3.62 ± 0.17
Ti	2330 ± 54	2265 ± 104	570 ± 128	1047 ± 73	1066 ± 84
Mn	948 ± 108	1034 ± 51	173 ± 6	223 ± 50	347 ± 64
Fe (%)	4.89 ± 0.46	3.94 ± 0.36	1.21 ± 0.09	0.80 ± 0.03	1.85 ± 0.16
Zn	262 ± 18	249 ± 18	105 ± 10	34 ± 4	137 ± 12
Ga	24 ± 1	26 ± 1	18 ± 1	14 ± 1	19 ± 1
Rb	152 ± 2	180 ± 3	141 ± 4	99 ± 3	124 ± 6
Sr	9 ± 2	10 ± 1	<2	10 ± 4	3 ± 1
Y	134 ± 19	104 ± 19	53 ± 8	15 ± 3	63 ± 8
Zr	974 ± 49	1024 ± 44	448 ± 38	135 ± 8	563 ± 38
Nb	112 ± 8	108 ± 7	38 ± 7	22 ± 4	47 ± 6

specimens. One would anticipate a similar outcome for obsidian artefacts from the region. This has been demonstrated using a collection of 103 artefacts from sites in Jalisco submitted for analysis by Phil Weigand (Glascock *et al.* 2005). The artefact analyses found that a total of 12 different chemical groups of obsidian were represented. Nine of the 12 chemical groups are sources identified in this study, but three groups from unknown sources were found and these were represented by 12 of the 103 artefacts (about 12%).

More recent analyses on 105 obsidian artefacts (Glascock *et al.* 2007) along the coast of Nayarit, from the site of Chacalilla, by M. Ohnersorgen and M. Garduño Ambriz show heavy use of the green obsidian from the local Volcan las Navajas source and the more distant Ixtlan del Rio and La Joya sources during the Post-Classic, indicating strong economic ties to these particular areas. Furthermore, there is evidence of blade production at Chacalilla, in particular using source material from Volcan las Navajas and La Joya, suggesting that these materials may have been further distributed locally.

Table 12.2 (continued)

	OSJ	*GEJ*	*JEJ*	*LQJ*	*CMJ*
Element	***Osotero, Jalisco***	***Hacienda de Guadalupe, Jalisco***	***San Juanito de Escobedo, Jalisco***	***La Quemada, Jalisco***	***Cinco Minas, Jalisco***
NAA Results	(n = 101)	(n = 15)	(n = 6)	(n = 21)	(n = 9)
Na (%)	3.04 ± 0.07	3.31 ± 0.04	3.89 ± 0.12	3.20 ± 0.05	3.68 ± 0.02
Cl	464 ± 179	797 ± 66	439 ± 43	594 ± 45	1492 ± 58
K (%)	3.81 ± 0.27	3.73 ± 0.16	3.65 ± 0.24	3.91 ± 0.18	3.52 ± 3
Sc	1.79 ± 0.04	0.61 ± 0.03	7.76 ± 0.09	2.11 ± 0.03	0.51 ± 0.02
Mn	299 ± 18	346 ± 6	803 ± 38	371 ± 5	372 ± 3
Fe (%)	0.83 ± 0.05	1.31 ± 0.02	2.50 ± 0.05	0.85 ± 0.01	1.88 ± 0.03
Co	0.62 ± 0.07	0.035 ± 0.007	1.58 ± 0.10	0.20 ± 0.03	0.04 ± 0.02
Zn	32 ± 6	90 ± 1	75 ± 1	43 ± 2	152 ± 3
Rb	104 ± 3	142 ± 2	80 ± 1	130 ± 2	196 ± 4
Sr	128 ± 30	< 15	186 ± 30	< 15	< 15
Zr	177 ± 12	457 ± 8	735 ± 13	172 ± 6	918 ± 28
Sb	0.30 ± 0.02	0.38 ± 0.01	0.63 ± 0.03	0.49 ± 0.02	0.54 ± 0.05
Ba	1.88 ± 0.06	25 ± 5	1159 ± 40	64 ± 8	32 ± 10
Cs	916 ± 21	3.01 ± 0.04	1.86 ± 0.06	2.76 ± 0.04	4.21 ± 0.07
La	25.9 ± 0.7	54.6 ± 1.1	35.5 ± 0.3	39.7 ± 0.5	55.4 ± 0.8
Ce	45.4 ± 1.1	117 ± 3	70 ± 1	76 ± 1	12 ± 2
Nd	18.4 ± 6.6	45.9 ± 1.1	29.7 ± 1.2	28.1 ± 1.1	56.2 ± 2.3
Sm	2.72 ± 0.10	10.2 ± 0.1	6.1 ± 0.1	5.34 ± 0.07	14.3 ± 0.2
Eu	0.381 ± 0.028	0.096 ± 0.003	2.18 ± 0.08	0.079 ± 0.003	0.311 ± 0.008
Tb	0.34 ± 0.02	1.49 ± 0.01	0.94 ± 0.11	0.77 ± 0.01	2.60 ± 0.05
Dy	2.23 ± 0.30	10.2 ± 0.5	5.41 ± 0.45	4.83 ± 0.32	17.4 ± 0.3
Yb	1.60 ± 0.08	6.30 ± 0.22	3.46 ± 0.05	3.11 ± 0.11	10.0 ± 0.2
Lu	0.29 ± 0.01	0.88 ± 0.01	0.54 ± 0.01	0.51 ± 0.01	1.54 ± 0.06
Hf	4.80 ± 0.08	13.9 ± 0.2	16.2 ± 0.3	6.07 ± 0.09	24.7 ± 0.4
Ta	1.32 ± 0.03	3.57 ± 0.03	1.81 ± 0.02	2.61 ± 0.03	5.04 ± 0.08
Th	10.5 ± 0.4	16.3 ± 0.2	7.36 ± 0.14	14.3 ± 0.2	22.2 ± 0.3
U	3.32 ± 0.55	6.55 ± 0.26	3.24 ± 0.33	5.4 ± 0.2	9.7 ± 0.7
XRF Results	(n = 15)	(n = 7)	(n = 5)	(n = 10)	(n = 6)
K (%)	3.57 ± 0.14	3.48 ± 0.09	3.67 ± 0.22	3.72 ± 0.12	3.45 ± 0.09
Ti	1028 ± 103	729 ± 154	3875 ± 428	543 ± 67	810 ± 170
Mn	248 ± 64	184 ± 7	802 ± 181	146 ± 56	296 ± 15
Fe (%)	0.82 ± 0.06	1.35 ± 0.10	2.65 ± 0.21	0.83 ± 0.04	1.98 ± 0.17
Zn	38 ± 3	111 ± 9	158 ± 13	43 ± 6	199 ± 21
Ga	15 ± 1	17 ± 1	31 ± 1	15 ± 1	24 ± 1
Rb	105 ± 6	136 ± 3	75 ± 3	124 ± 3	194 ± 6
Sr	127 ± 17	< 2	229 ± 7	< 2	3 ± 1
Y	13 ± 3	61 ±12	30 ± 5	27 ± 8	116 ± 16
Zr	165 ± 8	505 ± 29	770 ± 29	170 ± 19	1009 ± 61
Nb	17 ± 5	62 ± 6	41 ± 7	35 ± 7	79 ± 8

Conclusions

In western Mexico, especially along the northern flanks of the Jalisco Block, there existed a series of major obsidian mining complexes that were active from Formative to late Post-Classic times. The Tequila/Coli axis has the evidence for the most intensive extraction activities during this 2500–3000 year time period. Some of the complexes, such as La Joya, Teuchitlán-La Mora, and San Juan de los Arcos, had specialised workshops nearby, which themselves were immense. The prismatic blade workshop at Las Cuevas is the largest workshop located to date within all of western Mexico, covering approximately 15 hectares and with millions of pieces of obsidian debris on the surface. These mining areas have, in the composite, thousands of quarries. Several have chambered underground mines of significant proportions, mostly for exotic colours such as green and red. The range of colours is impressive: green, blue, grey, black, yellow, white, red, orange, and mixtures thereof, solid colours or translucent, including gold and silver tinged material as well as the nowadays highly popular 'rainbow' variety. Overall, the quality of these obsidians is good to excellent. Since the Tequila/Coli axis is also the nuclear area for western Mexico's first experiment in civilisation, called the Teuchitlán Tradition, it seems obvious that this high

Table 12.2 (continued)

	JJ	*PRJ*	*STJ*	*QJ*	*PJ*
Element	***La Joya, Jalisco***	***La Providencia, Jalisco***	***Santa Teresa, Jalisco***	***Tequila, Jalisco***	***La Pila, Jalisco***
NAA Results	(n = 152)	(n = 39)	(n = 195)	(n = 77)	(n = 10)
Na (%)	3.47 ± 0.15	3.05 ± 0.05	3.32 ± 0.08	3.41 ± 0.08	3.34 ± 0.03
Cl	1152 ± 133	516 ± 55	823 ± 71	700 ± 76	549 ± 43
K (%)	3.35 ± 0.30	4.07 ± 0.20	4.16 ± 0.21	4.14 ± 0.18	3.98 ± 0.17
Sc	0.74 ± 0.02	2.44 ± 0.07	2.33 ± 0.11	3.48 ± 0.06	3.99 ± 0.05
Mn	572 ± 23	293 ± 9	439 ± 7	452 ± 13	336 ± 2
Fe (%)	1.90 ± 0.02	0.99 ± 0.03	1.01 ± 0.02	1.10 ± 0.04	1.00 ± 0.01
Co	0.04 ± 0.01	1.07 ± 0.05	0.19 ± 0.02	0.37 ± 0.04	0.35 ± 0.01
Zn	131 ± 11	33 ± 7	63 ± 6	63 ± 7	46 ± 2
Rb	157 ± 2	117 ± 2	145 ± 2	129 ± 4	105 ± 1
Sr	< 15	65 ± 6	< 15	23 ± 7	48 ± 6
Zr	733 ± 28	210 ± 13	243 ± 11	275 ± 13	238 ± 8
Sb	0.98 ± 0.09	0.70 ± 0.02	0.88 ± 0.06	0.82 ± 0.03	0.37 ± 0.01
Cs	38 ± 8	3.36 ± 0.08	3.56 ± 0.05	3.31 ± 0.10	2.60 ± 0.03
Ba	3.81 ± 0.05	480 ± 25	85 ± 15	354 ± 48	944 ± 16
La	66.9 ± 1.2	32.4 ± 0.9	50.4 ± 0.8	44.2 ± 1.2	33.6 ± 0.3
Ce	138 ± 2	58.7 ± 1.3	102 ± 2	88.5 ± 2.5	64.1 ± 0.9
Nd	60.6 ± 2.6	19.5 ± 0.8	38.0 ± 5.5	32.0 ± 4.6	24.1 ± 1.4
Sm	12.5 ± 0.2	3.80 ± 0.10	7.99 ± 0.16	6.85 ± 0.21	4.70 ± 0.06
Eu	0.150 ± 0.006	0.359 ± 0.018	0.154 ± 0.014	0.364 ± 0.045	0.674 ± 0.008
Tb	1.88 ± 0.05	0.51 ± 0.02	1.16 ± 0.05	0.97 ± 0.06	0.64 ± 0.01
Dy	11.8 ± 0.7	3.14 ± 0.33	7.53 ± 0.45	6.29 ± 0.49	3.95 ± 0.24
Yb	7.16 ± 0.16	2.26 ± 0.07	4.75 ± 0.28	3.95 ± 0.23	2.40 ± 0.11
Lu	1.08 ± 0.03	0.39 ± 0.01	0.73 ± 0.02	0.62 ± 0.02	0.40 ± 0.02
Hf	18.6 ± 0.2	5.98 ± 0.28	8.52 ± 0.19	8.62 ± 0.24	6.56 ± 0.07
Ta	3.74 ± 0.05	1.73 ± 0.03	2.81 ± 0.05	2.34 ± 0.07	1.51 ± 0.02
Th	15.7 ± 0.3	13.3 ± 0.3	14.2 ± 0.2	12.5 ± 0.4	11.2 ± 0.1
U	6.2 ± 0.8	4.0 ± 0.5	4.8 ± 1.2	3.8 ± 0.9	3.8 ± 0.3
XRF Results	(n = 23)	(n = 11)	(n = 15)	(n = 12)	(n = 13)
K (%)	3.48 ± 0.22	3.81 ± 0.13	3.84 ± 0.08	3.75 ± 0.11	3.81 ± 0.20
Ti	1070 ± 99	1144 ± 72	793 ± 122	1003 ± 164	1109 ± 107
Mn	333 ± 76	201 ± 40	184 ± 62	201 ± 56	190 ± 39
Fe (%)	1.93 ± 0.18	0.94 ± 0.06	1.04 ± 0.05	1.09 ± 0.05	0.99 ± 0.08
Zn	153 ± 15	42 ± 4	67 ± 6	68 ± 4	50 ± 6
Ga	20 ± 1	16 ± 1	17 ± 1	17 ± 1	17 ± 2
Rb	156 ± 5	114 ± 5	141 ± 2	126 ± 4	107 ± 5
Sr	5 ± 1	67 ± 4	4 ± 1	23 ± 11	64 ± 3
Y	71 ± 13	19 ± 3	46 ± 9	40 ± 4	22 ± 4
Zr	754 ± 38	180 ± 15	275 ± 23	293 ± 16	226 ± 15
Nb	62 ± 7	26 ± 6	47 ± 6	43 ± 3	24 ± 6

profile of obsidian procurement is one of the explanations for the rise of this early complex social system.

Acknowledgements

We thank Chris Beekman, John Millhauser and other colleagues for providing some of the source samples analysed during this study. We also appreciate funding support from the US National Science Foundation (grants BCS-0504015 and BCS-0418823) which made this work possible.

References

Ambroz, J. A., M. D. Glascock, and C. E. Skinner. 2001. Chemical Differentiation of Obsidian within the Glass Buttes Complex, Oregon. *Journal of Archaeological Science* 28, 741–746.

Aoyama, K., T. Tashiro, and M. D. Glascock. 1999. A pre-Columbian Obsidian Source in San Luis, Honduras: Implications for the Relationship between Late Classic Maya Political Boundaries and the Boundaries of Obsidian Exchange Networks. *Ancient Mesoamerica* 19, 237–249.

Aronson, M. 1996. Technological Change: Ceramic

Table 12.2 (continued)

	TJ	*HXJ*	*BQJ*	*LPJ*	*SJLA*	*AJ*
Element	***Teuchitlán/La Mora, Jalisco***	***Huaxtla, Jalisco***	***Boquillas, Jalisco***	***La Primavera, Jalisco***	***San Juan los Arcos, Jalisco***	***Ahuisculco, Jalisco***
NAA Results	(n = 52)	(n = 16)	(n = 20)	(n = 15)	(n = 26)	(n = 39)
Na (%)	3.51 ± 0.12	3.42 ± 0.04	2.98 ± 0.04	3.37 ± 0.06	3.16 ± 0.04	2.77 ± 0.10
Cl	880 ± 134	719 ± 57	528 ± 37	902 ± 149	585 ± 96	491 ± 70
K (%)	3.65 ± 0.30	4.01 ± 0.19	3.82 ± 0.20	3.97 ± 0.16	3.89 ± 0.23	4.04 ± 0.21
Sc	0.09 ± 0.01	1.33 ± 0.23	1.90 ± 0.03	1.53 ± 0.14	2.19 ± 0.03	2.07 ± 0.03
Mn	298 ± 4	538 ± 22	298 ± 4	462 ± 12	387 ± 5	258 ± 3
Fe (%)	1.33 ± 0.02	1.77 ± 0.07	0.73 ± 0.01	1.48 ± 0.03	0.74 ± 0.01	0.57 ± 0.01
Co	0.04 ± 0.05	0.28 ± 0.25	0.33 ± 0.01	0.05 ± 0.01	0.29 ± 0.26	0.19 ± 0.01
Zn	133 ± 8	112 ± 10	34 ± 5	85 ± 6	33 ± 2	28 ± 4
Rb	168 ± 3	138 ± 4	107 ± 1	141 ± 3	109 ± 1	121 ± 1
Sr	< 15	< 15	99 ± 14	< 15	52 ± 10	21 ± 9
Zr	553 ± 15	709 ± 18	137 ± 7	501 ± 26	146 ± 9	114 ± 7
Sb	0.82 ± 0.06	0.32 ± 0.02	0.42 ± 0.01	0.35 ± 0.02	0.36 ± 0.01	0.58 ± 0.04
Cs	4.26 ± 0.05	2.84 ± 0.10	2.51 ± 0.02	2.94 ± 0.07	2.33 ± 0.04	3.17 ± 0.03
Ba	30 ± 8	44 ± 17	689 ± 8	33 ± 9	723 ± 14	215 ± 7
La	48.1 ± 0.6	98.5 ± 4.9	29.0 ± 0.4	74.4 ± 1.5	30.4 ± 0.7	29.6 ± 0.4
Ce	105 ± 2	199 ± 10	50.4 ± 0.7	151 ± 3	54.9 ± 0.98	51.0 ± 0.7
Nd	48.5 ± 2.8	80.7 ± 5.6	16.7 ± 3.7	64.6 ± 4.8	18.8 ±1.8	15.3 ± 2.2
Sm	11.5 ± 0.3	15.4 ± 0.2	2.77 ± 0.03	11.7 ± 0.3	3.35 ± 0.05	2.95 ± 0.07
Eu	0.115 ± 0.004	0.267 ± 0.008	0.355 ± 0.005	0.119 ± 0.006	0.380 ± 0.006	0.187 ± 0.003
Tb	1.98 ± 0.04	2.11 ± 0.06	0.31 ± 0.03	1.65 ± 0.05	0.42 ± 0.01	0.35 ± 0.02
Dy	12.4 ± 0.6	14.0 ± 0.7	2.17 ± 0.29	10.2 ± 0.5	2.60 ± 0.26	2.22 ± 0.27
Yb	6.98 ± 0.16	8.07 ± 0.38	1.48 ± 0.04	6.22 ± 0.25	1.73 ± 0.07	1.48 ± 0.05
Lu	1.01 ± 0.04	1.18 ± 0.05	0.26 ± 0.01	0.94 ± 0.03	0.32 ± 0.01	0.29 ± 0.01
Hf	17.2 ± 0.2	19.4 ± 0.4	4.03 ± 0.09	14.6 ± 0.4	4.49 ± 0.06	3.51 ± 0.03
Ta	3.65 ± 0.04	4.43 ± 0.21	1.53 ± 0.02	3.64 ± 0.09	1.77 ± 0.02	1.77 ± 0.02
Th	14.7 ± 0.2	15.5 ± 0.3	10.9 ± 0.1	15.9 ± 0.3	11.0 ± 0.1	12.8 ± 0.1
U	8.0 ± 1.0	6.1 ± 1.2	3.6 ± 0.3	5.4 ± 0.9	3.4 ± 0.5	4.2 ± 0.4
XRF Results	(n = 16)	(n = 5)	(n = 13)	(n = 7)	(n = 12)	(n =10)
K (%)	3.57 ± 0.14	3.41 ± 0.13	3.62 ± 0.13	3.63 ± 0.29	3.76 ± 0.14	3.77 ± 0.16
Ti	581 ±109	1276 ± 146	906 ± 52	906 ± 149	940 ± 74	618 ± 108
Mn	211 ± 7	272 ± 15	201 ± 71	227 ± 52	184 ± 71	89 ± 14
Fe (%)	1.32 ± 0.07	1.89 ± 0.05	0.74 ± 0.04	1.47 ± 0.08	0.77 ± 0.03	0.61 ± 0.03
Zn	143 ± 10	168 ± 8	37 ± 6	115 ± 8	40 ± 4	31 ± 4
Ga	19 ± 1	21 ± 1	14 ± 1	18 ± 2	15 ± 1	14 ± 1
Rb	162 ± 4	134 ± 3	111 ± 4	137 ± 3	111 ± 3	120 ± 4
Sr	< 2	12 ± 7	92 ± 2	< 2	55 ± 2	21 ± 2
Y	76 ± 10	85 ± 6	12 ± 3	59 ± 11	13 ± 4	13 ± 4
Zr	607 ± 30	827 ± 34	124 ± 9	541 ± 36	136 ± 10	94 ± 9
Nb	60 ± 5	82 ± 4	19 ± 5	63 ±6	22 ± 5	20 ± 5

Mortuary Technology in the Valley of Atemajac from the Late Formative to the Classic Periods. *Ancient Mesoamerica* 7, 163–169.

Bakewell, P. J. 1971. *Silver Mining and Society in Colonial Mexico, Zacatecas 1546-1700*. Cambridge, Cambridge University Press.

Barker, A. W., C. E. Skinner, M. S. Shackley, M. D. Glascock, and J. D. Rogers. 2002. Mesoamerican Origin for an Obsidian Scraper from the pre-Columbian Southeastern United States. *American Antiquity* 67, 103–108.

Beekman, C. S. 1996. The Long-Term Evolution of a Political Boundary: Archaeological Research in Jalisco, Mexico. Unpublished PhD Dissertation, Vanderbilt University, Nashville, TN.

Bell, B. (ed.). 1974. *The Archaeology of West Mexico*. Ajijic, Jalisco, Sociedad Estudios Avanzados de Occidente de Mexico A.D.

Bernal, I. 1980. *A History of Mexican Archaeology. The Vanished Civilizations of Middle America*. London, Thames and Hudson.

Braswell, G. E., and M. D. Glascock. 1992. A New Obsidian Source in the Highlands of Guatemala. *Ancient Mesoamerica* 3, 47–49.

Braswell, G. E., and M. D. Glascock. 2002. The Emergence of Market Economies in the Ancient Maya World:

Table 12.2 (end)

	NJ	*IXJ*	*SIJ*	*LLJ*	*HZZ*	*NCZ*
Element	***Las Navajas, Jalisco***	***Ixtepete, Jalisco***	***San Isidro, Jalisco***	***La Lobera, Jalisco***	***Huitzila, Zacatecas***	***Nochistlán, Zacatecas***
NAA Results	(n = 18)	(n = 3)	(n=19)	(n = 22)	(n = 50)	(n = 15)
Na (%)	2.76 ± 0.04	3.29 ± 0.13	3.28 ± 0.04	3.47 ± 0.04	3.42 ± 0.03	2.69 ± 0.44
Cl	491 ± 51	1114 ± 36	632 ± 27	860 ± 92	704 ± 79	840 ± 140
K (%)	3.97 ± 0.16	3.76 ± 0.38	3.63 ± 0.18	3.52 ± 0.20	3.73 ± 0.17	4.35 ± 0.44
Sc	1.99 ± 0.02	0.79 ± 0.24	0.54 ± 0.01	1.29 ± 0.02	0.80 ± 0.02	2.18 ± 0.07
Mn	250 ± 3	305 ± 5	338 ± 4	404 ± 19	429 ± 27	336 ± 19
Fe (%)	0.59 ± 0.01	0.96 ± 0.11	1.27 ± 0.02	1.40 ± 0.01	1.69 ± 0.02	1.32 ± 0.01
Co	0.25 ± 0.01	0.05 ± 0.02	0.021 ± 0.003	0.07 ± 0.02	0.04 ± 0.01	0.97 ± 0.05
Zn	22 ± 4	59 ± 8	103 ± 2	100 ± 6	112 ± 5	60 ± 3
Rb	118 ± 2	163 ± 6	141 ± 2	151 ± 2	128 ± 2	186 ± 2
Sr	26 ± 5	< 15	< 15	< 15	< 15	101 ± 11
Zr	113 ± 7	254 ± 58	444 ± 12	445 ± 12	563 ± 8	249 ± 5
Sb	0.60 ± 0.03	0.47 ± 0.06	0.34 ± 0.01	0.49 ± 0.02	0.40 ± 0.03	0.70 ± 0.03
Cs	3.01 ± 0.04	3.86 ± 0.30	3.03 ± 0.05	3.48 ± 0.04	2.52 ± 0.03	6.02 ± 0.05
Ba	308 ± 10	42 ± 8	31 ± 11	36 ± 9	23 ± 7	689 ± 13
La	28.4 ± 0.3	35.2 ± 0.2	52.1 ± 0.6	52.2 ± 0.5	78.4 ± 1.9	52.5 ± 1.5
Ce	49.1 ± 0.7	79 ± 1	111 ± 2	105 ± 2	156 ± 3	105 ± 2
Nd	20 ± 6	30 ± 2	45 ± 2	42 ± 2	63 ± 4	38 ± 5
Sm	2.74 ± 0.041	8.1 ± 0.3	9.9 ± 0.3	9.45 ± 0.23	12.9 ± 0.5	8.70 ± 0.18
Eu	0.213 ± 0.004	0.050 ± 0.017	0.089 ± 0.003	0.284 ± 0.006	0.183 ± 0.004	0.736 ± 0.007
Tb	0.31 ± 0.01	1.25 ± 0.09	1.52 ± 0.03	1.45 ± 0.03	1.84 ± 0.05	1.04 ± 0.04
Dy	2.16 ± 0.19	8.8 ± 1.1	9.8 ± 0.4	9.4 ± 0.4	11.6 ± 0.5	6.7 ± 0.5
Yb	1.36 ± 0.03	5.14 ± 0.05	6.14 ± 0.19	6.03 ± 0.16	7.23 ± 0.26	3.76 ± 0.12
Lu	0.27 ± 0.01	0.84 ± 0.03	0.88 ± 0.02	0.90 ± 0.01	1.05 ± 0.01	0.62 ± 0.01
Hf	3.55 ± 0.07	9.5 ± 1.9	13.7 ± 0.3	13.1 ± 0.1	15.3 ± 0.2	6.78 ± 0.07
Ta	1.72 ± 0.03	3.52 ± 0.18	3.60 ± 0.06	2.81 ± 0.03	3.17 ± 0.04	1.32 ± 0.02
Th	12.3 ± 0.2	17.4 ± 1.1	16.0 ± 0.3	16.4 ± 0.2	14.1 ± 0.2	21.0 ± 0.2
U	4.6 ± 1.0	3.6 ± 0.6	6.5 ± 0.3	4.0 ± 0.8	3.4 ± 0.2	4.7 ± 0.5
XRF Results	(n = 12)	(n = 4)	(n = 14)	(n = 16)	(n = 17)	(n = 10)
K (%)	3.74 ±0.16	3.74 ± 0.16	3.57 ± 0.17	3.70 ± 0.14	3.71 ± 0.19	4.21 ± 0.40
Ti	611 ± 103	564 ± 160	751 ± 133	767 ± 132	1047 ± 116	1220 ± 109
Mn	114 ± 31	155 ± 24	182 ± 7	197 ± 35	227 ± 13	229 ± 48
Fe (%)	0.62 ±0.03	1.06 ± 0.14	1.30 ± 0.06	1.35 ± 0.11	1.63 ± 0.14	1.24 ± 0.05
Zn	29 ± 3	83 ± 18	108 ± 8	99 ± 11	129 ± 14	61 ± 3
Ga	15 ± 1	16 ± 2	17 ± 1	18 ± 1	20 ± 1	22 ± 3
Rb	116 ± 5	162 ± 9	142 ± 4	149 ± 4	127 ± 5	189 ± 4
Sr	30 ± 1	< 2	< 2	< 2	3 ± 1	107 ± 4
Y	14 ± 3	47 ± 7	56 ± 9	55 ± 8	64 ± 9	40 ± 6
Zr	99 ± 9	305 ± 65	495 ± 19	469 ± 33	623 ± 27	240 ± 17
Nb	22 ± 5	52 ± 5	60 ± 5	42 ± 7	55 ± 6	22 ± 4

Obsidian Exchange in the Terminal Classic, Yucatan, Mexico. In *Geochemical Evidence for Long-Distance Exchange*, edited by M. D. Glascock, 33–52. Westport, CT, Bergin and Garvey.

Charlton, T. H. 1969. On the Identification of pre-Hispanic Obsidian Mines in Southern Hidalgo. *American Antiquity* 34, 176–177.

Clark, J., and P. C. Weigand. n.d. Obsidian Lapidary without Polishing. In *Obsidiana Mesoamericana*, edited by L. Gonzalez. Mexico City, I.N.A.H. (in press).

Cobean, R. H. (ed.). 2002. *A World of Obsidian: The Mining and Trade of Volcanic Glass in Ancient Mexico* (Serie Arqueologia de Mexico). Mexico City and Pittsburgh, PA, Instituto Nacional de Antropologia e Historia/University of Pittsburgh.

Cobean, R. H., M. D. Coe, E. A. Perry, Jr., K. K. Turekian, and D. P. Kharkar. 1971. Obsidian Trade at San Lorenzo Tenochtitlan, Mexico. *Science* 174, 141–146.

Cobean, R. H., J. R. Vogt, M. D. Glascock, and T. L. Stocker. 1991. High-Precision Trace-Element Characterization of Major Mesoamerican Obsidian Sources and Further Analyses of Artifacts from San Lorenzo Tenochtitlan, Mexico. *Latin American Antiquity* 2, 69–91.

Darling, J. A. 1993. Notes on Obsidian Sources of the Southern Sierra Madre Occidental. *Ancient Mesoamerica* 4, 245–253.

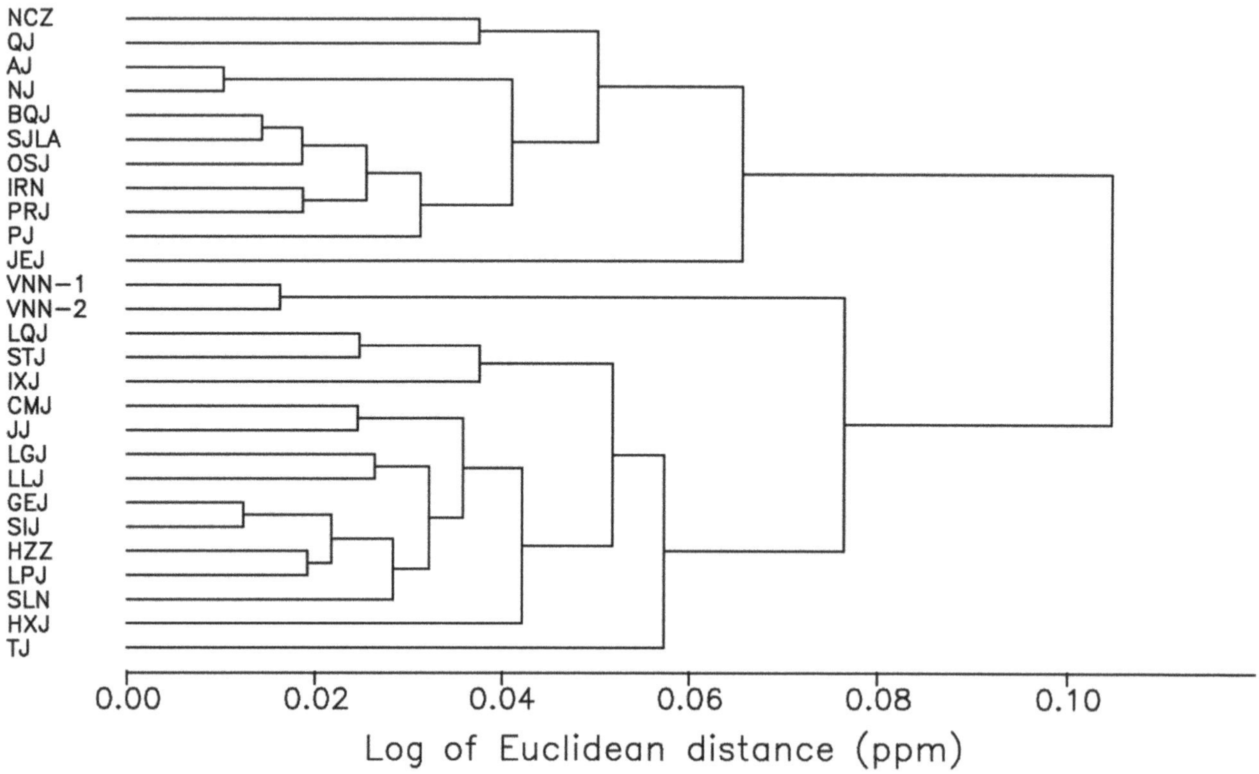

Figure 12.8. Dendrogram generated from hierarchical aggregative cluster analysis of the compositional group means used to illustrate the relative similarities and differences between obsidian sources in Jalisco, Nayarit, and Zacatecas

Darling, J. A. 1998. Obsidian Distribution and Exchange in the North-Central Frontier of Mesoamerica, A.D. 0–1500. Unpublished PhD Dissertation, University of Michigan, Ann Arbor, MI.

Darling, J. A., and M. D. Glascock. 1998. Acquisition and Distribution of Obsidian in the North-Central Frontier of Mesoamerica. In *Rutas de Intercambio en Mesoamérica*, edited by E. C. Rattray, 345–364. Méxcio, D.F., Instituto de Investigaciones Antropológicas, Universidad Nacional Autónoma de México.

Darling, J. A., and F. M. Hayashida. 1995. Compositional Analysis of the Huitzila and La Lobera Obsidian Sources in the Southern Sierra Madre Occidental. *Journal of Radioanalytical and Nuclear Chemistry* 196, 245–254.

Ericson, J. E., and J. Kimberlin. 1977. Obsidian Sources, Chemical Characterization and Hydration Rates in West Mexico. *Archaeometry* 19, 157–166.

Foster, M. S., and S. Gorenstein (eds). 2000. *Greater Mesoamerica. The Archaeology of West and Northwest Mexico*. Salt Lake City, University of Utah Press.

Gaxiola, G., and J. E. Clark (eds). 1989. *La Obsidiana en Mesoamerica* (Serie Arqueología 176). México, D.F., Instituto Nacional de Antropología e Historia.

Glascock, M. D., G. E. Braswell, and R. H. Cobean. 1998. A Systematic Approach to Obsidian Source Characterization. In *Archaeological Obsidian Studies: Method and Theory*, edited by M. S. Shackley, 15–65. New York and London, Plenum Press.

Glascock, M. D., R. J. Speakman, P. C. Weigand, M. A. Ohnersorgen, and M. Garduno-Ambriz. 2007. A Comprehensive Database for Obsidian Sources in Western Mexico. Poster presented at the 72nd Annual Meeting of the Society for American Archaeology, Austin, TX.

Glascock, M. D., P. C. Weigand, and R. Esparza. 2005. Obsidian Exploitation and Exchange in Western Mesoamerica. Paper presented at the 70th Annual Meeting of the Society for American Archaeology, Salt Lake City, UT.

Hirth, K. G., G. Bondar, M. D. Glascock, A. J. Vonarx, and T. Daubenspeck. 2006. Supply-Side Economics: An Analysis of Obsidian Procurement and the Organization of Workshop Provisioning. In *Obsidian Craft Production in Ancient Central Mexico*, edited by K. G. Hirth, 115–136. Salt Lake City, University of Utah Press.

Jimenez Betts, P. F., and J. A. Darling. 2000. Archaeology of Southern Zacatecas: The Malpaso, Juchipila, and Valparaiso-Bolaños Valleys. In *Greater Mesoamerica. The Archaeology of West and Northwest Mexico*, edited by M. S. Foster and S. Gorenstein, 155–180. Salt Lake City, University of Utah Press.

Joyce, A. A., J. M. Elam, M. D. Glascock, H. Neff, and M. Winter. 1995. Exchange Implications of Obsidian Source Analysis from the Lower Rio Verde Valley, Oaxaca, Mexico. *Latin American Antiquity* 6, 3–15.

Mahood, G. A. 1980. The Geological and Chemical Evolution of a Late Pleistocene Rhyolitic Center: The Sierra La Primavera, Jalisco, Mexico. Unpublished PhD Dissertation. University of California–Berkeley, CA.

Moholy-Nagy, H. 2003. Source Attribution and the Utilization of Obsidian in the Maya Area. *Latin American Antiquity* 14, 301–310.

Pires-Ferriera, J. W. 1976. Obsidian Exchange in Formative Mesoamerica. In *The Early Mesoamerican Village*, edited by K. W. Flannery, 292–311. New York, Academic Press.

Pollard, H. P. 1997. Recent Research in West Mexican Archaeology. *Journal of Archaeological Research* 5, 345–384.

Righter, K., I. S .E. Carmichael, T. A. Becker, and P. R. Renne. 1995. Pliocene–Quaternary Volcanism and Faulting at the Intersection of the Gulf of California and the Mexican Volcanic Belt. *Geological Sosiety of America Bulletin* 107, 612–626.

Rosas-Elguera, J., L. Ferrari, V. H. Garduno-Monroy, and J. Urrutia-Fucugauchi. 1996. Continental Boundaries of the Jalisco Block and their Influence in the Pliocene–Quaternary Kinematics of Western Mexico. *Geology* 24, 921–924.

Santley, R. S., T. P. Barrett, M. D. Glascock, and H. Neff. 2001. Pre-Hispanic Obsidian Procurement in the Tuxtla Mountains, Southern Veracruz, Mexico. *Ancient Mesoamerica* 12, 49–63.

Schondube, O. 1980. Epoca Prehispanica. In *Historia de Jalisco*, edited by J. Ma. Muriá, 111–130. Guadalajara, Goviernor del Estado de Jalisco.

Sheets, P., K. G. Hirth, F. W. Lange, F. H. Stross, F. Asaro, and H. V. Michel. 1990. Obsidian Sources and Elemental Analyses of Artifacts in Southern Mesoamerica and the Northern Intermediate Area. *American Antiquity* 55, 144–158.

Sidrys, R. V. 1976. Classic Maya Obsidian Trade. *American Antiquity* 41, 449–464.

Stark, B. L., L. Heller, M. D. Glascock, J. M. Elam, and H. Neff. 1992. Obsidian-Artifact Source Analysis for the Mixtequilla Region, South-Central Veracruz, Mexico. *Latin American Antiquity* 3, 221–239.

Townsend, R. F. (ed.). 1998. *Ancient West Mexico. Art and Archaeology of the Unknown Past*. New York, Art Institute of Chicago and Thames and Hudson.

Weigand, P. C. 1985. Evidence for Complex Societies during the Western Mesoamerican Classic period. In *The Archaeology of West and Northwest Mesoamerica*, edited by M. Foster and P. C. Weigand, 47–91. Denver, CO, Westview Press.

Weigand, P. C. 1993. *Evolución de una Civilización Prehispánica: Arqueología de Jalisco, Nayarit, y Zacatecas*. Zamora, Mexico, El Colegio de Michoacán.

Weigand, P. C. 1994. Observations of Ancient Mining within the Northwestern Regions of the Mesoamerican Civilization, with Emphasis on Turquoise. In *In Quest of Mineral Wealth, Aboriginal and Colonial Mining, and Metallurgy in Spanish America* (Geoscience and Man. Volume 33), edited by A. K. Craig and R. C. West, 21–35. Baton Rouge, LA, Department of Geography and Anthropology, Louisiana State University.

Weigand, P. C., and A. García de Weigand. 1994. Minería Prehispánica en Jalisco. *Estudios Jaliscienses* 17, 5–21.

Weigand, P. C., and M. W. Spence. 1982. The obsidian Mining Complex at La Joya, Jalisco. *Anthropology* 6(1–2), 175–188.

Chapter 13
Thoughts and Inference on Prehistoric Obsidian Source Exploitation in the Pacific Rim and Beyond

M. Steven Shackley

Abstract: *'In these last twenty years, the use of obsidian archaeometry has witnessed boundless expansion, partly due to the concomitant advances in computer technology and instrumental chemistry, and partly due to a recognition by archaeologists that archaeometrists provide much more than mere measurement.' (Shackley 1998a, 1).*

Keywords: *Obsidian, Source Analysis, Methodology, Geoarchaeology, Pacific Rim, North America, Mesoamerica*

Introduction

The commentary in this Chapter is a result of reading the 11 chapters that comprise this volume on twenty-first century applications of obsidian source provenance to archaeological problems. This is a personal view and in a minor way compares the current research to that reported in the 1998 volume *Archaeological Obsidian Studies: Method and Theory* (Shackley 1998b). Have we made significant and reliable progress in obsidian provenance research in the decade plus since the 1998 volume? Is there a direction signalled by the current research? This forms the perspective of my comments.

It is rather axiomatic that "this is the only book of its kind" which all publishers and authors note in publishing today, but this volume on the range of obsidian studies in a twenty-first century archaeology mainly in the Pacific Rim region is certainly unique. In 1998, when some of the authors here collaborated on the volume *Archaeological Obsidian Studies: Method and Theory*, few could imagine the hold that obsidian studies would have on the archaeological world (Shackley 1998b). I am not here to be only prescient, but to outline what I think obsidian studies, and indeed archaeological science has done and will do for archaeological method and theory. Of course, I am not the only one who has tackled this issue; and if one looks at the few attempts to observe the relationship between archaeology and the natural and physical sciences that wholly underpin it, the ideology is rather eclectic within the discipline (Goldberg 1988; Green 1998; Joyce 2010; Killick 2008; Killick and Goldberg 2009; Pollard and Bray 2007). In an effort to support and defend the application of obsidian studies to an ever expanding theoretical archaeology, I will take a rather larger view, returning to more specifics at the end.

While the title of this volume suggests only a Pacific Rim focus, it reaches well beyond those shores, and is better for it. The chapter by C. Dillian takes an intensive look at archaeological obsidian on the US Atlantic coast, and R. Hughes' work in the southern Great Basin of North America, while not Pacific Rim studies, significantly enhance the perspective.

A Personal Background View

Taking the lead from a mentor and friend, the late New Mexico native Roger C. Green (1998), I will interject a bit about my own experience in the volcanic glass field and 20 years of teaching at a major research university in an Anthropology Department that has not always been that supportive of science in archaeology.

In my own region of interest, the North American Southwest, the use of obsidian source provenance to solve methodological and theoretical issues in prehistory was rejected 25 years ago, or at least ignored by most archaeologists. Ceramics were seen as the most important data set to solve these problems, of course overlooking the first 12,000+ years of prehistory in this part of the world including one of the earliest dated sites in North America, Murray Springs (cf., Boyer and Robinson 1956; Jack 1971; Schreiber and Breed 1971). Slowly, as obsidian studies began to illuminate issues of social identity, gender relations, and more post-processual areas of procurement and exchange that could not be addressed with only ceramic data, obsidian research became more "normative" in the Southwest (Shackley 2005). Now, few large research or Cultural Resource Management projects are designed without an obsidian provenance component. Indeed, a recent multimillion dollar US NSF-funded project focusing on migration and social networks during the Classic period, awarded to both the University of Arizona and the Center for Desert Archaeology, is primarily based on obsidian source provenance in concert with ceramic studies to elucidate migration and social relations during this probable socially volatile period. This very evolution of a recognition by archaeologists of the exemplary utility of obsidian studies *with* ceramic analyses has transformed our concept of prehistory in the Southwest. This project might be considered "post-processual" by many Southwestern archaeologists; but like Rosemary Joyce, I prefer the label social archaeology to define this kind of work (Joyce 2010). This philosophical change in the attitude toward glass studies in Southwestern archaeology can be mirrored in many of the regions focused on by the authors in this volume, most notably R. Hughes for the Great Basin (Chapter 10).

Archaeological Science and Social Archaeology

Many in archaeology would characterise me, and many of the authors here as "scientific" or "processual" archaeologists, labels that they are happy to apply, but mean little to me. True, I have degrees in geology and anthropology, so both could be thrown my way. In a recent volume I published in a university press when applying obsidian provenance studies to social and gender identity in the pre-Classic Hohokam Southwest, one of the reviewers took exception: 'Come on Shackley, you do not do that stuff'. Was I being untrue to my physical science heritage by rallying scientific evidence to understand and explain the social differences among men and the technological and social differences between men and women in that society?

Recently, Rosemary Joyce and Julie Hendon have used the XRF analysis and source provenance of obsidian in Late and Terminal Classic Honduras:

> *'to argue that rather than pursuing a site-wide strategy of minimizing cost, or creating a single monopolistic relationship with a single site providing all the obsidian used at the site, the people of Cerro Palenque obtained obsidian in small quantities from a wide range of sources, many of them quite distant. She [Hendon] argues that for the people of Cerro Palenque, this cosmopolitanism of social relations, mediated through exchange of obsidian, was what was of greater social value.... Based on a variety of analyses, we found that people in the Ulua Valley oriented their ritual actions toward geographic points on the horizon, mountains that were also sources of valued stone resources. In a situation where stones used for items of daily use indexically signified their places of origin, the gathering together of obsidian from many sources transformed Cerro Palenque into a place containing many places... This is the kind of richness that XRF [and many other instrumental techniques] promises archaeology of the twenty-first century, and is already delivering. It is a way of interrogating materiality that can open windows into past social experiences that are limited only by our ability to imagine how materials that XRF can be applied to mediate the lives of people across space and over time.' (Joyce 2010; bracketed comments are mine).*

In Hendon and Joyce's case, these avowed social archaeologists recognise that to address the understanding of prehistory with a twenty-first century perspective, it could very well be that *only* techniques of archaeological science can truly illuminate them.

In the early 1990s when I arrived at Berkeley, the boundaries between processual and post-processual archaeology were being sharply drawn, particularly by the typically energised graduate students. Emblematic during shifts in perspective, the old is seen as less useful even useless, and the new the only way to go. My archaeological science courses were seen as the old stuff, while visiting lecturers Alyson Wylie and Ian Hodder, the new, exciting stuff, even though those colleagues did not see it that way.

Archaeology really has never generated its own body of theory. During my career we have borrowed theoretical views from the natural sciences (cultural ecology, optimal foraging, historical ecology, behavioural ecology), sociology and social psychology (cognitive approaches and much of what is called post-processualism), and mathematics and statistics. This is not the place to argue whether borrowing theory from other disciplines has been good or bad for archaeology. However, as we move into the twenty-first century, methods and theory in the archaeological sciences will form the discipline, not only because the techniques will focus the problems, but because they are becoming normative.

Some of the chapters in this volume have elucidated the archaeological relevance of source provenance data. Most provide a look at obsidian sources in regions that have not received much attention in the English literature. This is also good. Fully, one-half of the chapters deal with obsidian source provenance in Northeast Asia, particularly Japan and Korea. This is one of the greatest contributions to this literature. Few of us in the English-speaking world have the necessary acumen in Japanese or Korean to interpret or follow research in this region. My thanks to these authors for bringing their work to the English-speaking linguistic world.

The Regionally Focused Studies in this Volume

The volume editors have introduced the authors and their chapters and I will not repeat that here. I do want to examine the regions covered in the volume within the context of obsidian studies generally and what could be the future of these methods and technologies.

The Instrumental Technology of Obsidian Source Provenance

The technology, methodology, and theoretical perspective on obsidian provenance expanded exponentially in the latter half of the twentieth century (Shackley 1998b; 2005, 2008). Gone are the days when NAA and laboratory (desktop) XRF (including PIXE) were the dominant instrumental techniques used to determine elemental composition of obsidian. Now LA-ICP-MS and portable XRF (PXRF) are allowing scholars to analyse much smaller sizes non-destructively, and in the latter case, move the laboratory to the field. This will irrevocably change obsidian studies for the foreseeable future, and mainly for the better.

For over three decades, XRF has remained the dominant non-destructive technology for obsidian studies. The newer digital detectors, smaller targets, small aperture X-ray tube collimators and Windows-driven software have immeasurably improved the efficiency of this technique (Shackley 2010a). There is also a fortuity in the elements in the mid-Z X-ray region that are analysed with great precision by XRF and those elements in that region that are sensitive incompatible elements in discriminating volcanic rocks (Cann 1983; Hughes and Smith 1993; Shackley 2005, 2010a). However, XRF while sensitive down to about 8mm in minimum sample diameter has not been effective at measuring very small debitage (< 5mm). Why is this important? Eerkens *et al.* (2007) noted that when analysing debitage smaller than 10mm in various Great Basin archaeological contexts that both the distance to source and the diversity of sources detected was greater (see also Shackley 1990 for a discussion in the US Southwest). This is probably because these hunter-gatherers while travelling over the landscape are rejuvenating cutting and scraping tools and in-haft projectile points; and the resulting debitage is quite small. However, recent experiments at Berkeley with the ThermoScientific *Quant'X* EDXRF instrument using a 3.5mm tube collimator, suggests that obsidian down to at least 5mm and likely to 3mm can be reliably analysed. With the advances in digital technology and software, as well as decreasing costs, XRF will remain one of the best non-destructive options for the analysis of obsidian and glass (Shackley 2010b). In this volume, eight of the 11 studies used XRF as at least one of the instrumental techniques in the research (Izuho and Hirose in Chapter 2 review other studies that also used NAA and EPMA – electron microprobe). Neutron activation analysis was used in five of the studies and PIXE, EPMA, and LA-ICP-MS one each.

In the 1998 volume, I asked the question: 'Which instrument is the best?' Obviously, the question is still with us (Shackley 1998a, 7). I noted that NAA, XRF, and PIXE were dominant, followed by ICP-MS (Shackley 1998a, 8). I would say that today, 12 years later, the relative mix is XRF, NAA, PIXE, with LA-ICP-MS becoming more common partly because it is becoming much more available. It will be interesting in some future volume to see what the mix of instrumental technology will be dominant. And keep in mind that often the choice of instrumentation is primarily based on availability. If you are a student at the University of Missouri you have a choice of NAA, XRF, and LA-ICP-MS; at the University of California at Berkeley you have a choice of EDXRF, WXRF, EPMA, environmental electron microscopy (ESEM), and LA-ICP-MS; but at many institutions a scholar has only one option. This is often due to historical methodological trends; XRF was dominant in geology programmes in the 1970s through the 1990s when ICP-MS became more useful for geological and then archaeological problems.

Perhaps coming from applications in ceramic studies, LA-ICP-MS promises to provide another analytical technique that is in essence non-destructive, although some museum staff will have to be convinced. The range of elements detected is great, and the most advantageous attribute is the ability to analyse very small areas down to the micron (Speakman and Neff 2005). While the instrument cost is high for ICP-MS, it is becoming more readily available at universities worldwide. This is ably demonstrated by C. Phillips study in the Kurile Islands (see Chapter 7).

Regionally Focused Discussion

Rather than outline each chapter individually, I will take a regional approach, looking at the advances in each region: Northeast Asia, Mesoamerica, and North America.

Y. Kuzmin has been working on Northeast Asian obsidian provenance studies for many years. Not too long ago, the database for archaeological obsidian in the Russian Far East was essentially non-existent. Now, as Y. Kuzmin demonstrates, from the Palaeolithic to Neolithic, people were moving at least 1000km based on source provenance studies. This is a region of extreme climatic irregularities; and obsidian studies, in part, have shown that humans have been overcoming these difficulties for tens of thousands of years. Y. Kuzmin provides an excellent review of Northeast Asian research and I will not repeat it here; but suffice it to say it is remarkable as to what has been accomplished mainly since the 1990s.

A prime example of this pioneering research is the Kamchatkan studies discussed by A. Grebennikov and coauthors. The Kamchatka Peninsula is one of the most remote places in the world. The "sources" in the study are mainly based on archaeological obsidian, partly because moving around in this region is so difficult. Still transport of obsidian over 500km is evident in the research.

C. Phillips, in examining nearly 2000 artefacts in the Kurile Islands, found that obsidian was imported to the islands from Kamchatka and Hokkaido, a considerable distance by boat in often rough seas. As the author notes, this kind of database is crucial in understanding the colonisation of the islands through time. Again, obsidian provenance studies provide data more reliable than many other archaeological data sets.

Perhaps one of the more descriptively valuable contributions to the volume is the review by M. Izuho and W. Hirose of obsidian studies on Hokkaido Island. I visited Hokkaido when I was stationed at Okinawa Island in the 1960s; and it is a wild and beautiful part of Japan. The value here to me is the review comparing the various methods used in the analysis of archaeological obsidian and the attempt to integrate them all. This chapter should have utility for scholars for some time to come.

Analysis by T. Tsutsumi of over 5000 samples is a remarkable study of Japanese obsidian. As in Chapter 5 by N.-C. Cho and coauthors on Korea, much is offered that is missing in the obsidian provenance English literature. On the main Japanese island of Honshu, T. Tsutsumi notes that obsidian has been exploited for at least 35,000 years and exchange over 200km during the Jomon was common. Obsidian studies in Japan are very useful in understanding one of the earliest hunter-gatherer societies in the world. Along with the Hokkaido study, this work will serve for years as a basis for understanding of early hunter-gatherer exchange and group interaction.

Fortuitously, I have some knowledge of Korean Palaeolithic

archaeology. I was a graduate school colleague of now Professor Seonbok Yi of Seoul National University; and we spent many hours discussing the vagaries of being hunter-gatherer archaeologists at a US Southwestern university where hunter-gatherer studies were not rigour. Y. Kuzmin noted the relationship between what is now the Korean Peninsula and what is now eastern Russia during the Palaeolithic. Chapter by N.-C. Cho and coauthors is a welcome review of obsidian studies in this region. They note that there are few studies that reveal obsidian use during the Palaeolithic, but this could be simple sampling error based on Y. Kuzmin's results to the north. The problem with research in North Korea frustrates obsidian research certainly, but study by N.-C. Cho and coauthors is a welcome addition to our knowledge.

Last, but not least of the Asian studies, is study by H. Obata and coauthors of the Ryukyu Archipelago. Of all regions of Japan, they note that this part has received the least intensive obsidian research. But, they have rectified this in recent work. Nearly all the obsidian was procured from a single source on Kyushu Island. Outstanding is the research on lithic technology that indicates that conservation of this valuable stone was evident in these early sites. The small proportion of obsidian in the sites and the chronological evidence suggests that obsidian was used sporadically as a toolstone, but valued nonetheless.

Obsidian provenance studies in Mesoamerica are quite well-embedded in archaeological research. From the Formative into the Post-Classic periods, obsidian production was practically at the industrial level. Transport of obsidian through socially well-developed exchange networks is often over thousands of kilometres, generally as finished blades. It is relatively common to find obsidian from Sierra de Pachuca in Hidalgo State of Mexico north of Mexico City thousands of kilometres south in Honduras. And while local sources of obsidian may occur, social dynamics are in play such that it is economically or socially required to import obsidian tools from many kilometres distant from a foreign power. So, obsidian provenance studies allow archaeologists to peer into very complex social networks and a market economy that cannot necessarily be seen in other regions (see Joyce's 2010 comments above).

This is why studies by M. Glascock and coauthors are so crucial in understanding the social and political dynamics in Mesoamerica. While many areas of Mesoamerica have been a focus of study, the authors have looked at more than 1000 samples in Jalisco, Nayarit, and Zacatecas states of West Mexico, not generally a preferred region of study. The database alone is worth its weight of an edited volume, but the authors go beyond just providing data. This is a long-term study by many of the authors here, and some of the obsidian "mines" in this region, as with some of the metal exploitation is the largest in Mexico. As I discussed above, this is industrial production of state societies that control the means of production. Without the research here, inferences about the quantity, quality, and level of state control would be impossible.

The other Mesoamerican project in the volume by J. Blomster and M. Glascock is another look at Early Formative obsidian procurement and production on Oaxaca. This is an extension of J. Blomster's excellent doctoral work on social complexity in this region (Blomster 1998). While the residents of Etlatongo "accessed" obsidian from many sources, as noted above a single source was dominant. However, while a single source may have been dominant, the great diversity of sources was ignored in early studies. J. Blomster's work turns much of what we know from that region on its head. This is what we should be doing with obsidian studies as stated most eloquently by Joyce (2010).

Finally, the two studies from North America, my region of interest, are illustrative of a number of issues. My colleague, R. Hughes, using the very large Nellis Air Force Base study in Nevada along with projectile point form and temporal style research, illustrates cultural change through the long prehistory in the Great Basin to great effect. This is one of R. Hughes's strengths and goes back to his dissertation research in the North American Northwest and Great Basin (Hughes 1984, 1986). This combination of well-grounded obsidian source provenance work and lithic technology is a hallmark of western North American social archaeology. Much of this work has been either by Richard or in concert with his research, to provide a 'nuanced understanding of obsidian conveyance in the area' (Hughes, this volume). While he notes that the sample sizes at some sites are relatively small, the evidence points to significant shifts in procurement over time, and thus changes in socially prescribed behaviour through time. Not unlike J. Blomster's research in Early Formative Mexico, obsidian provenance studies provide a confidently source assigned window into social behaviour in the past.

Finally, the last study by C. Dillian is perhaps the most illustrative. While obsidian during state formation and market economies in Mesoamerica moved great distances across the continent, C. Dillian discovered that similar long-distance transport occurred in prehistory all the way from the Pacific coast to the Atlantic coast within North America. While we may argue about this, in Mesoamerica state societies and the industrial production of obsidian was quite different from the dominance of hunter-gatherer social organisation and exchange during much of prehistory north of Mesoamerica. As C. Dillian infers, much of this material must have been transmitted through a "down-the-line" exchange and may have taken generations. While many of the studies in this volume point to the technological utility and social value of obsidian, none demonstrate that value more than a prehistoric mid-Atlantic value of a raw material that is only available thousands of kilometres west and not necessarily better than locally available toolstone.

C. Dillian concludes with a statement that could be used for this volume as well:

> *'Realistically, many people in the northeastern United States may not have ever even seen obsidian, which suggests that a fixed definition of worth and exchange was unlikely. Though the utility of such*

objects and the pragmatic use-value of such high-quality raw material would have been apparent to any flintknapper, I suggest instead that the value of obsidian in the prehistoric northeast was based on its foreign, exotic, and possibly even otherworldly aura, and the draw of that exotic appeal transcends time and space.' (Dillian, this volume).

The papers in this volume do move us forward beyond where we were in 1998. I suggest that those pioneering studies, particularly in North Asia, are generating a database that will lend itself to real social archaeological understanding; so that in another decade a volume like this will look decidedly different.

References

Blomster, J. P. 1998. At the Bean Hill in the Land of the Mixtec: Early Formative Social Complexity and Interregional Interaction at Etlatongo, Oaxaca, Mexico. Unpublished PhD Dissertation. Department of Anthropology, Yale University, New Haven, CT.

Boyer, W. W., and P. Robinson. 1956. Obsidian Artifacts of Northwestern New Mexico and Their Correlation with Source Material. *El Palacio* 63, 333–345.

Cann, J. R. 1983. Petrology of Obsidian Artefacts. In *The Petrology of Archaeological Artefacts*, edited by D. R. C. Kempe and A. P. Harvey, 227–255. Oxford, Clarendon Press.

Eerkens, J. W., J. R. Ferguson, M. D. Glascock, C. E. Skinner, and S. A. Waechter. 2007. Reduction Strategies and Geochemical Characterization of Lithic Assemblages: A Comparison of Three Case Studies from Western North America. *American Antiquity* 72, 585–597.

Goldberg, P. 1988. The Archaeologist as Viewed by the Geologist. *Biblical Archaeologist* 51, 197–202.

Green, R. C. 1998. A 1990s Perspective on Method and Theory in Archaeological Volcanic Glass Studies. In *Archaeological Obsidian Studies: Method and Theory*, edited by M. S. Shackley, 223–236. New York and London, Plenum Press.

Hughes, R. E. 1984. Obsidian Sourcing Studies in the Great Basin: Problems and Prospects. In *Obsidian Studies in the Great Basin* (Contributions of the University of California Archaeological Research Facility 45), edited by R. E. Hughes, 1–20. Berkeley, University of California.

Hughes, R. E. 1986. *Diachronic Variability in Obsidian Procurement Patterns in Northeastern California and South-Central Oregon* (University of California Publications in Anthropology 17). Berkeley: University of California.

Hughes, R. E., and R. L. Smith. 1993. Archaeology, Geology, and Geochemistry in Obsidian Provenance Studies. In *Scale on Archaeological and Geoscientific Perspectives* (Geological Society of America Special Papers 283), edited by J. K. Stein and A. R. Linse, 79–91. Boulder, CO, Geological Society of America.

Jack, R. N. 1971. The Source of Obsidian Artifacts in Northern Arizona. *Plateau* 43, 103–114.

Joyce, R. A. 2010. Is There a Future for XRF in a 21st Century Archaeology? In *X-Ray Fluorescence Spectrometry (XRF) in Geoarchaeology*, edited by M. S. Shackley. New York, Springer Publishing (in press).

Killick, D. 2008. Archaeological Science in the USA and Britain. In *Archaeological Concepts for the Study of the Cultural Past*, edited by A. Sullivan, 40–64. Salt Lake City, University of Utah Press.

Killick, D., and P. Goldberg. 2009. A Quiet Crisis in American Archaeology. *The SAA Archaeological Record* 9(1), 6–10.

Pollard, A. M., and P. Bray. 2007. A Bicycle Made for Two? The Integration of Scientific Techniques into Archaeological Interpretation. *Annual Review of Anthropology* 36, 245–259.

Schreiber, J. P., and W. J. Breed. 1971. Obsidian Localities in the San Francisco Volcanic Field, Arizona. *Plateau* 43, 115–119.

Shackley, M. S. 1990. Early Hunter-Gatherer Procurement Ranges in the Southwest: Evidence from Obsidian Geochemistry and Lithic Technology. Unpublished PhD Dissertation, Arizona State University, Tempe, AZ.

Shackley, M. S. 1998a. Current Issues and Future Directions in Archaeological Volcanic Glass Studies: An Introduction. In *Archaeological Obsidian Studies: Method and Theory*, edited by M. S. Shackley, 1–14. New York and London, Plenum Press.

Shackley, M. S. (ed.). 1998b. *Archaeological Obsidian Studies: Method and Theory*. New York and London, Plenum Press.

Shackley, M. S. 2005. *Obsidian: Geology and Archaeology in the North American Southwest*. Tucson, University of Arizona Press.

Shackley, M. S. 2008. Archaeological Petrology and the Archaeometry of Lithic Materials. *Archaeometry* 50, 194–215.

Shackley, M. S. 2010a. An Introduction to X-Ray Fluorescence Spectrometry for Archaeologists. In *X-Ray Fluorescence Spectrometry (XRF) in Geoarchaeology*, edited by M. S. Shackley. New York, Springer Publishing (in press).

Shackley, M. S. (ed.). 2010b. *X-Ray Fluorescence Spectrometry (XRF) in Geoarchaeology*. New York, Springer Publishing.

Speakman, R. J., and H. Neff (eds). 2005. *Laser Ablation-ICP-MS in Archaeological Research*. Albuquerque, University of New Mexico Press.

Index